W0193149

Future of Business and Finance

The Future of Business and Finance book series features professional works aimed at defining, describing and charting the future trends in these fields. The focus is mainly on strategic directions, technological advances, challenges and solutions which may affect the way we do business tomorrow, including the future of sustainability and governance practices. Mainly written by practitioners, consultants and academic thinkers, the books are intended to spark and inform further discussions and developments.

Florian Kaefer

Sustainability Leadership in Tourism

Interviews, Insights, and Knowledge from Practice

 Springer

Florian Kaefer
Sustainability Leaders Project
Zurich, Switzerland

ISSN 2662-2467   ISSN 2662-2475   (electronic)
Future of Business and Finance
ISBN 978-3-031-05313-9   ISBN 978-3-031-05314-6   (eBook)
https://doi.org/10.1007/978-3-031-05314-6

This Springer imprint is published by the registered company Springer Nature Switzerland AG
The registered company address is: Gewerbestrasse 11, 6330 Cham, Switzerland

To my son Lucas, and all future generations. May this book help to keep your world—and vacations—enjoyable.

Acknowledgements

My utmost gratitude to all the individuals who have generously shared their insights, ideas and learnings by contributing to this book, thereby helping professionals around the world to become sustainability champions and supporters of their business or community.

A big thank you to the many collaborators who have supported SLP over the years, especially: Dr. Natàlia Ferrer Roca for her encouragement and priceless support during the early years, Tanya Makarova for offering a helping hand whenever needed and Supriya A N for being such a great social media ninja and editorial assistant.

Prashanth Mahagaonkar at Springer has been a great editor to work with, always replying promptly and offering valuable support throughout the process of preparing and publishing the book, for which I am grateful indeed.

Disclaimer

The interviews and content in this book are largely based on the website, https://sustainability-leaders.com/. All interviews have been re-published here with the consent of the interviewees.

Contents

About the Author

Florian Kaefer, Ph.D., co-founded the Sustainability Leaders Project (SLP) in 2014 together with Natàlia Ferrer Roca, Ph.D., now the leading knowledge platform focused on sustainable tourism leadership. Over the last 8 years he has interviewed more than 260 sustainability changemakers and champions of sustainable tourism, many of whom are now part of the SLP expert panel which meets every few months to discuss timely questions, with insights published on https://sustainability-leaders.com/.

Dr. Kaefer obtained his Ph.D. (Management Communication) from the Waikato University Management School in New Zealand. He holds a master's degree in sustainable development (Exeter University, UK) and a bachelor in tourism management (Brighton University, UK, in collaboration with Angell Academy, Germany).

He also founded and leads The Place Brand Observer (PlaceBrandObserver.com), an online magazine and consultancy focused on place brand development, management and communication—and is the author of the book *An Insider's Guide to Place Branding: Shaping the Identity and Reputation of Cities, Regions, Countries.* Published by Springer in 2021, the book is part of the publisher's Management for Professionals series.

Currently based in Switzerland near Zurich, Florian was awarded 'TOP 30 Champion of Environmental Sustainability in Hospitality and Tourism 2021', by the International Hospitality Institute, USA. He is a public speaker, coach and advises tourism businesses and destinations (FlorianKaefer.info).

Introduction to Sustainability Leadership in Tourism

How can we make sure our communities are welcoming, our visitor economies resilient and our destinations sustainable?

Much of my work as founder and editor of the *Sustainability Leaders Project* (SLP) so far has centred around how destinations and tourism businesses can prepare for a future in which the sustainability—economic, environmental and social—will be more and more important for attracting and retaining visitors.

This book is my attempt to provide some answers to this crucial question, by sharing insights into the art of sustainability as practiced by leaders and changemakers in tourism. It offers you a unique glimpse behind the scenes of sustainable tourism as it is—or should be—practiced around the world. The book is a summary, or perhaps better, a snapshot, of state-of-the-art thinking on the topic, based on my observations as SLP editor and convenor of our expert panel.[1]

This book differs from others on the topic of sustainable tourism in that it is not a practitioner's reflection. Nor is it an academic textbook offering a comprehensive, presumably neutral overview of the field (if this is even possible). As a specialised journalist, advisor and observer of both academia and practice, I have over the last 8 years been focusing on connecting people and ideas—across disciplines and borders. This book is therefore my own take on sustainable tourism leadership and what it takes to master sustainability, and it reflects what I consider important and good practice, based on my conversations with many of world's leading thinkers and practitioners dedicated to developing and nurturing tourism businesses and destinations.

I find it important to point this out, as we all perceive this world differently and there is not necessarily just one 'right' way to do things, even though—in tourism and elsewhere—some approaches are more likely to succeed than others, when applied in practice.

---

[1]For more about the panel, visit https://sustainability-leaders.com/panel/

My understanding of sustainable tourism is largely based on my conversations with over 260 professionals (and counting) around the world whom we have interviewed over the years, on Sustainability-Leaders.com, as well as the insightful contributions by our panel (representing many different countries and backgrounds). You will find interview portraits of some of the most active contributors to the topic in Part II of the book, to give you a bit of an idea of who they are and how they approach tourism and sustainability.

My interest in sustainability and tourism emerged during my master's studies in sustainable development at Exeter University in England, which I completed in 2010. Having previously studied tourism management in Germany and England, I have always been fascinated by discussions and approaches which venture out of disciplinary silo-think, and which connect the dots across disciplines and industries. Both 'tourism' and 'sustainability' are complex terms and subject to various, differing interpretations by individuals, communities of practice and schools of thought, and thus comfortably fit these characteristics.

My desire to connect academics (thinkers) with practitioners (doers) and to build bridges across disciplines and departments is what led me to establish SLP in 2014, together with Natàlia Ferrer Roca. Since then, Sustainability-Leaders.com has become a leading online knowledge hub on sustainability leadership in tourism, particularly aimed at business owners and managers, and those in charge of developing and managing tourist destinations.

I hope that you find this book both useful and inspiring and that it serves your own purpose—be it to help your business or destination community to strengthen its sustainability or resilience, or to pursue your own studies and explorations of the fascinating world of travel and tourism.

This book is arranged in two parts. First Part: Following this brief introduction (Chap. 1), in the second chapter we jump right into the topic at hand by exploring what sustainable tourism is and why it matters—and how the term relates to similar concepts, namely responsible, transformative or regenerative tourism. We address the questions: what characterises a sustainable tourism leader—and what does it take to become a changemaker? What motivates professionals to focus their career on tourism and sustainability?

In Chap. 3 we then look at the one-million-dollar question: how to succeed with sustainable tourism, including views on why many sustainability initiatives fail (the pitfalls and challenges), and how to increase your chances at success. Closely related: how to measure the effectiveness and impact of sustainability initiatives in tourism, in addition to some key bits of advice from tourism makers and shapers.

Chapter 4 is dedicated to the question: what comes next in tourism sustainability? We discuss some of the trends most likely to affect tourism success in the future. And, for the researchers among you, I will point out some of the main open questions that are still in need of investigation.

In Part II, I introduce you to the individuals whose collective insights informed this book and the advice shared in it. Although, to my mind, the persons featured are among the most active contributors to sustainable tourism as research subject or practice, they are only a small group among the wider community of professionals whose work and ideas touch on the topic. You will find many more interview portraits with inspiring and trailblazing tourism managers, advisors and researchers on Sustainability-Leaders.com. I invite and encourage you to reach out to those forward-thinking professionals—and to keep a close eye on their work, especially if you aspire to become a tourism leader or sustainability changemaker.

As much as I have enjoyed putting together this book, by the time you read it, some parts might already be outdated. My advice for you is to stay up to date on the latest insights, strategies and examples of innovative sustainability practices by keeping an eye on the website and the SLP social media channels, especially LinkedIn.[2]

---

[2]https://www.linkedin.com/company/sustainabilityleadersproject/

Part I

Sustainable Tourism Essentials

Understanding Sustainable Tourism

What is sustainable tourism? This question might seem easy to answer, but it really is not, as there exist many different interpretations of the term. While there is no 'exact' definition (and academics love to come up with new concepts all the time), I know from my conversations with practitioners that they tend to find some more useful than others. Chapter 1 offers you a summary of the meaning and purpose of 'sustainability', in the context of tourism—based on what I have learned through the interviews and the panel sessions which we have conducted at SLP. We will also explore how 'sustainable' tourism differs from responsible, transformative or other forms of tourism, what the characteristics are of a sustainability leader or tourism changemaker, and what motivates some of the most successful of them to dedicate their career to tourism and sustainability.

2.1   What Is Sustainable Tourism?

How did 'sustainable tourism' come about? When were the seeds of sustainability planted for it to be different from mainstream tourism?

Tourism has come a long way from the nomadic lifestyle of early humans to explorers, followed by curious travellers trying to learn about new cultures. With the passage of time, tourism has taken many forms and in turn, has shaped many destinations.

When flying became accessible, people started travelling long distances. Enthusiastic youngsters tried their hands at voluntourism and budget travellers went backpacking. The rich were not left behind, enjoying their own small-group luxury wildlife discoveries among other resource-intense experiences.

Fast forward and we are facing the ugly face of this growing popularity of travel—overtourism. Where things have spiralled out of control, industry stakeholders are getting serious about doing tourism differently. This time with a close eye on respecting locals and the environment.

© The Author(s), under exclusive license to Springer Nature Switzerland AG 2022   7
F. Kaefer, *Sustainability Leadership in Tourism*, Future of Business and Finance,
https://doi.org/10.1007/978-3-031-05314-6_2

Glenn Jampol, Chairperson of the Global Ecotourism Network (GEN) and an ecotourism pioneer in Costa Rica, in our panel session[1] explained that sustainable tourism was a response in the 1990s to the need for tourism businesses and travellers to support authentic and small-scale activities, implement interpretative guiding, abide by ethical business practices, as well as socio-economic participation in the local community.

It also entails reducing the visitors' carbon footprint. It had to encompass personalised service to the clients, education and training for both the local staff and the guests, the support and conservation of natural resources, and proactive efforts to increase the well-being of local and indigenous people.

Although the spirit of sustainable tourism has remained the same since then, the terminology has been spun in different directions. And: today everybody is talking about it—academics, locals, businesses, travel media, influencers, visitors and a few committed governments.

So, before we delve deeper into the world of sustainable tourism, let us start by taking a look at what it actually means.

In line with the objective of the book, we spoke with a wide range of people who understand the immediate need for sustainable tourism. We asked them what they think is the definition of sustainable tourism, and we found a few common themes in their responses.[2]

Not surprising, there were no conflicting views or debatable opinions. In a way, all of them said the same, and the meaning of sustainable tourism remained fairly constant across the answers provided by the panel. The common themes that connected all the responses were:

1. Perpetuity of tourism
2. It is all about the locals
3. Resource efficiency
4. Protecting the environment
5. In line with the SDGs
6. Profitable

The foundational values of sustainable tourism include operating in a way that optimises positive benefits for people and place, argued Audrey Scott.

Furthermore, sustainable tourism minimises the negative impacts and thereby addresses the long-term (future) needs of travellers, the industry, the environment, and host communities, as Brian Mullis put it. Sustainability is about acknowledging future ambitions, needs—and intergenerational justice.

---

[1] https://sustainability-leaders.com/sustainable-tourism-redefined/

[2] All responses in full are available at https://sustainability-leaders.com/sustainable-tourism-redefined/

Perpetuity

A keen forethought for future generations is evident from most of the responses. The longevity of the industry has to be secured so that future generations too can benefit from the beauty of the planet. Lisa Choegyal and Vicky Smith aptly call sustainable tourism a style of tourism that is non-extractive, ensuring it contributes to the environment, ecology, culture or social setting in a way that enables it to continue ad infinitum.

Maintaining the perpetuity of tourism as an activity ensures retaining the virtues of a destination that make it attractive for visiting in the first place.

Darrell Wade considers a business sustainable when the financial, environmental, social and governance principles that underpin it are in harmony with each other and not competing.

For Natalia Naranjo, sustainable tourism generates long-term benefits for the environment, the territory and communities at a destination. Similarly, Xavier Font pointed out that tourism which puts the long-term well-being of the places that are visited ahead of the short-term enjoyment of the visitors can be considered sustainable.

Anne De Jong finds that everyone involved should benefit from tourism, and it should protect the world for future generations.

It Is All About the Locals

There is growing awareness about how locals have been left behind in the tourism growth story. Tourism for many years has been devoid of seeking local stakeholder opinion, with a frequent absence of locals in the tourism supply chain. It has been a one-sided affair. But now, things have started to change.

There is a push to pivot tourism to be more inclusive and to involve locals in every stage of tourism planning: Empowering locals—not only supporting them socially and economically but acknowledging and celebrating them as the torchbearers of the destination. After all, it is they who will safeguard their culture and environment on behalf of the tourism industry.

Frankie Hobro stressed that sustainable tourism must contribute positively socio-economically and culturally to the area in which it is based. It needs to develop in consultation with the community and other stakeholders. Only then can communities enjoy tangible and measurable benefits.

In straightforward terms, as Fiona Jeffery put it, sustainability in a travel and tourism context is about the ability to benefit local communities and bringing an enhanced appreciation of the destination and its culture and people to travellers.

Local value addition indeed is key to a happy (and therefore more likely welcoming) local population and tourism employees. For this to happen, a supportive governmental framework is essential, according to Christian Baumgartner.

Gianna Moscardo and Marcus Cotton shared a similar opinion during our panel discussion, that sustainable tourism respects and supports employees through thick and thin and ensures that tourism benefits the community rather than being merely extractive.

Resource Efficiency

Our Planet Earth is blessed with finite resources. For us to pass the baton of tourism to the next generation, it is imminent that we cease all activities that are detrimental to the environment and to present our world in a condition better than what we currently have. As Jorge Moller Rivas put it in the panel session, it is our actions now that will reflect on the quality of life of our grandkids.

The thing is, tourism does not function in a silo. It is an industry that competes with others for scarce resources such as land, labour, capital, water, and energy—often at the cost of the local communities.

James Crockett pointed out practical examples of resource efficiency, such as ensuring water consumption and solid waste are within the limits of what the destination can manage.

In short, sustainability in tourism means continuing the viability of a destination in the future. The need to stay within the planet's capacity when delivering tourism services is also something Rebecca Hawkins and Sonya Gottlebe emphasised.

Peter Richards summarised it well by saying that sustainable tourism should enable us to continue life on this planet within the means available to us and our future generations, hopefully leaving the world a bit better than when we started (which, as we will see in the next chapter, is now referred to as regenerative tourism).

He further added that the textbook definition of sustainability as 'development that meets the needs of the present without compromising the ability of future generations to meet their own needs' is not too bad. Although indeed, it is very human-focused and should include a stronger and perhaps more direct message not to exhaust natural resources.

Protecting the Environment

Tourism and the natural world are interdependent and for tourism experiences to be eternal, protecting the environment is paramount. In this regard, sustainable tourism is about travelling without leaving a negative impact on the environment.

Many panel respondents highlighted that tourism needs to be managed in a way that benefits the natural environment. If it is to be considered sustainable, it minimises its ecological and climatic impacts.

Joanna Van Gruisen rightly pointed out that if tourism damages the site, ecologically, socially or culturally, negatively impacting the destination's character and appeal, then it is unsustainable.

In Line with the SDGs

By now, we have a fairly good idea of what sustainability entails. A good way to structure the many expectations and ambitions are the UN Sustainable Development Goals.

Alex Pastollnigg recommends defining 'sustainable tourism' holistically as per the 17 SDGs. This comprehensive set of goals does not only refer to environmental and climate sustainability (which are, of course, important), but notably also social sustainability, financial/economic sustainability, and—most often overlooked—the impact on SDG 16 (world peace) paired with SDG 4 (quality education—for locals and, importantly, also for travellers).

Anna Spenceley seconds this opinion by asking why to reinvent the wheel when the UNWTO definition suffices? It is straightforward and all-encompassing in its meaning.

Leaning on the SDGs sustainable tourism should:

- Make optimal use of environmental resources that constitute a key element in tourism development, maintaining essential ecological processes and helping to conserve natural heritage and biodiversity.
- Respect the socio-cultural authenticity of host communities, conserve their built and living cultural heritage and traditional values, and contribute to inter-cultural understanding and tolerance.
- Ensure viable, long-term economic operations, providing socio-economic benefits to all stakeholders that are fairly distributed, including stable employment and income-earning opportunities and social services to host communities, and contributing to poverty alleviation.

She further adds that for more detail, the Global Sustainable Tourism Council's criteria[3] can be used as baseline standards for sustainable tourism.

Aivar Ruukel takes a different approach. According to him, there is no such thing as 'sustainable tourism'. The correct term would be sustainable development of tourism. The meaning of it is to apply sustainable development principles—like the SDGs—to the tourism sector. This highlights the importance of using those in sustainable tourism discourse and practice.

Profitable
While the above-mentioned themes cover the social and environmental aspects of the triple bottom line, financial viability should not be forgotten. It is important to make money, not least to be in a position to help and support—local people, climate change mitigation, protection of heritage, nature or wildlife.

Willem Niemeijer therefore rightly emphasised that while sustainable tourism contributes in measurable ways to the conservation of the environment and benefits the community, it also has to be profitable.

In other words, sustainable tourism seeks maximum benefit and minimum cost per tourist, monetary or otherwise, according to Jonathan Tourtellot. It is about linking sustainability measures with economic ones, Christof Burgbacher highlighted.

Key Takeaways Sustainable tourism: what is it all about? It is about supporting the local economy, offering community-based travel experiences where travellers can participate and learn, reducing carbon footprint and plastic use, protecting and

---

[3] https://www.gstcouncil.org

conserving nature and wildlife, while still making a profit. Sounds complicated and challenging? Well, it certainly is.

Especially because tourism, as we know, is not a single entity but a network of various establishments, each with its purpose. All stakeholders have a vested interest in the outcome yet are needed to ensure the long-term success of tourism as we know it. Peter Richards aptly calls it a partnership that makes the world a better place for all living beings.

And one thing has to be clear: tourism will never be entirely sustainable. Even the occasions where it gets close to this goal are rare. Joanna Van Gruisen in the panel pointed out that, when looked at globally, sustainable tourism is something of an oxymoron, in the sense that any tourism involving international flight travel is likely to be contributing more CO_2 than it can return, however green a holiday is.

Perhaps, to make the term more accessible, it helps to look at sustainable tourism at a more local scale and judge: if what is attracting the tourist remains as an attraction over time, then the tourism can be considered sustainable.

I particularly like how Elisa Spampinato[4] understands sustainable tourism, as a way of thinking and operating that considers all the areas on which tourism has an impact. It also considers time as a critical factor in terms of resource management. It is a dynamic balance that must be constantly maintained through the interaction of all the actors involved.

2.2   Sustainable, Responsible, Transformative: Where Is the Difference?

Growing awareness about the need for more sustainability in travel and tourism has given rise to many forms and terms, especially: sustainable tourism, responsible tourism, transformative tourism, and regenerative tourism. Each of them with the promise to be the answer to all tourism-related problems. But what do these terms actually mean?

We asked our industry experts[5] to help us break down each one of them. Answers by the panel illustrate that all four terms are interlinked and overlap, especially also when used in communications and tourism marketing discourse.

Sustainable Tourism
Sustainable tourism is a broader concept (compared to the other terms) that includes responsible and regenerative tourism—many panel respondents point this out.

---

[4]For full answers of all panel participants, see https://sustainability-leaders.com/sustainable-tourism-redefined/

[5]https://sustainability-leaders.com/sustainable-responsible-regenerative-tourism-explained/

Essentially, responsible, transformative and regenerative tourism cover aspects which are also part of sustainable tourism.

According to Audrey Scott, sustainable tourism dabbles with criteria and certification with regard to the environment more than other forms of tourism. While it is important to maintain a minimum standard of 'do no harm', she finds that there should be a greater focus on 'providing positive value' and finding ways to measure socio-economic and other benefits.

For Elizabeth Becker, author of the much-acclaimed book *Overbooked*,[6] sustainable tourism necessarily involves the whole community and hopefully, forces the community—from local to national—to agree on a plan. This need for a holistic, non-segmented approach was also highlighted by Amine Ahlafi.

I like Elisa Spampinato's association of sustainable tourism with balance, that is, how to maintain a balance between all the aspects of life affected by the sector, over time.

Sustainable tourism also ties in well with the Sustainable Development Goals,[7] argued Alex Pastollnigg, which makes it fairly straightforward as a concept, because those 17 sustainability parameters are now clearly defined.

Responsible Tourism

Judging by panel responses, responsible tourism puts an emphasis on the traveller, whereas sustainable tourism focuses on the entire travelism system and its many stakeholders and participating organisations.

Alex Pastollnigg, Vicky Smith and Marcus Cotton added that it is about taking responsibility for the decisions made in tourism and the impacts created as a result. Peter Richards feels that responsible tourism forces us to consider who needs to work together to put the good intentions of sustainable tourism into practice. That is, who is responsible for making it happen.

Elisa Spampinato elaborated that responsible tourism is a form of self-reflection and involves a lot of self-questioning. It focuses on the awareness of the impacts of our actions, habits, and the choices we make while travelling. For her, it also applies in the first instance to the tourists' behaviour, but also to the whole hospitality sector and all the stakeholders involved.

Are our choices creating more social disparities, pollution, discrimination through our attitudes? These are some of the questions that a responsible tourist would ask themselves, together with (and this especially applies to tour operators):

- How is the local community involved in your operations, how is the interaction with the local community structured?

---

[6]More about the book, how it came about and its key insights, in our interview with Elizabeth, available at https://sustainability-leaders.com/elizabeth-becker-interview/

[7]https://sdgs.un.org/goals

- Do they participate in the designing of the experience?
- What is left to the local community at the end of the process?

Responsible tourism, as Elisa Spampinato observed, is a form of education and an ethical set of behaviours, towards the place, natural environment, people and their cultural heritage.

Transformative Travel

The name already implies it: transformative travel is supposed to create a marked change in the behaviour of travellers as consumers and seeks to change the demand side of the tourism equation.

Brian Mullis put it brilliantly, that transformational travel is an approach that can be designed and implemented to support travellers who are intentionally travelling to stretch, learn and grow into new ways of being and engaging with the world. This involves inspiring travellers to be more mindful of where they go, how they consume, and how they leave and continue to engage with a destination.

Joanna Van Gruisen feels that transformative tourism is more likely to be linked with slow tourism. Tourism practitioners aim to open the eyes of the traveller to a new world perspective, through the different cultural local context.

Elisa Spampinato considers transformative tourism particularly important at this moment in time because it helps us reflect on how we approach the tourism experience, as tourists, as well as tourism professionals.

It is a matter of consciously getting out of the bubble in which many of us live, and 'using' tourism as a gate to reconnect on a higher level. What transformational tourism teaches us, ultimately, is to let go of our ego and embrace tourism as an open and unexpected encounter with the unknown. And from there we can grow as human beings and expand our consciousness.

Further, Gianna Moscardo finds that transformative travel is based on a set of values that are strongly tied to certain cultures and levels of affluence. She wonders whether it is much more like a way of apologising for unsustainable tourism than a genuine attempt to make tourism better in terms of sustainability.

For Peter Richards, transformative travel puts the onus for responsible tourism and sustainability back into the hands of the tourists, embedded in a more self-aware and spiritual approach. While certainly powerful and with the potential to become a strong movement to inspire changemakers, he asks whether it feels a bit too self-consciously 'committed', to be able to reach the mainstream.

Regenerative Tourism

And last but not the least, regenerative tourism. Regenerative tourism proposes that we can go beyond aiming for balance and actually design tourism to fix, repair and give back more than it takes. It is underpinned by the certainty that it is too late to search for zero impact, as we have overpassed the limits of consumption and need alternatives that bring back preeminent natural conditions, feels Mariana Madureira.

Hence, Marta Mills calls on industry leaders to focus on regenerative tourism that moves sustainability up a level. It is no longer only about minimising the negative impacts but going beyond and restoring that damage to make the places heal and flourish.

Given the pace and scale of damage that is being inflicted on our environment, Glenn Jampol equally thinks that sustaining is no longer an option and now, it must be about renewal. Due to the current existential danger of damaging human-made climate change, the desecration of our natural areas and resources, and the urgent need to update the concept of travelling to encompass regenerative forces and behaviour, he now prefers to use regenerative tourism to label our philosophy.

Brian Mullis explained that regenerative tourism is a sustainable, inclusive, and fluid process that is context-specific and guided by principles that involve stakeholders collectively, creating the conditions for the tourism system to improve the overall health and well-being of all living systems in a destination over time, enabling them to thrive and for tourism to create net positive impacts.

Marcus Cotton agrees that regenerative tourism looks at sustainability in a more proactive light, seeking not merely to mitigate adverse impacts but to ensure positive regeneration of the environment and communities.

Joanna Van Gruisen during the panel discussion eloquently explained how regenerative tourism looks beyond sustainability. It recognises that we have gone too far in damaging the planet and aspires to give back. It is a way of seeing tourism as a living system within a wider carpet of life. This is where our focus should presently be.

Likewise, James Crockett sees in regenerative tourism the next evolution. It requires that tourism helps to restore habitat, ecosystems and cultures to a former, healthier state. It is the latest and most comprehensive attempt to make tourism as transformative in delivering net benefits to a destination as possible.

Also, Vicky Smith finds that regenerative tourism not only creates positive change but conditions for renewal, revitalisation and onward evolution. As a result, change in attitudes, targets, measures and priorities cannot function the same way as before.

Elisa Spampinato continues that regenerative tourism gives us an innovative interpretation. It seems to have gathered some attitudes, practises and approaches already existing at the grassroots level. The regenerative tourism approach is holistic at its core, its structure is flexible, horizontal and inclusive, and strongly practice-oriented.

She is confident that regenerative tourism could answer many of the questions that sustainable tourism has not been able to answer yet. Its strength relies on focusing on the connections that exist between the isolated silos and its aim to create a network of practices that inspire collaborative work. It looks at society and organisations as living organisms, which are parts of a bigger picture.

Key Takeaways What is in a name? While different schools of thought believe in the future and success of various forms of tourism, we have others who wonder if labelling is needed at all. Quite a few panel participants criticised too much attention on concepts, instead believing in putting things into practice and making a positive difference. It is indeed high time to focus on action and to avoid distractions from the real issue at hand—the need for impact. Leaders should channel their energy on the issues they can tackle, through clear goals and measurable results. As Shannon Guihan noted, without measurable goals bringing positive impacts, what does it matter what we call it? Furthermore, without the consumer on board, where are we? The great definition debate certainly does not do the consumer any favours.

Megan Epler Wood calls for setting our priorities right. She believes we are in a period of crisis and should respond as such. Having lived through some ten different definitions of sustainable tourism in her 30-year career, she thinks it is best to look at the nuts and bolts of our situation and get very serious about how to respond to it.

2.3   What Characterises Sustainability Leaders and Changemakers?

In the previous chapters, we looked at what sustainable tourism is, and how the term relates to responsible, transformative and regenerative tourism. This chapter is all about leaders and changemakers—what makes them special, their characteristics and traits.

The Booking.com 2021 Sustainable Travel Report[8] affirms that 83% of global travellers think sustainable travel is vital, with almost two-thirds saying the pandemic has made them want to travel more sustainably in the future. This is positive news for the industry but there will be an inevitable rise in greenwashing, to cater to the growing demand.

This is why society needs true leaders and changemakers now more than ever, who are committed to the cause of sustainability and not just individuals riding the wave of this new demand, trying to improve profit margins without any actual sustainability improvements on the ground.

But how to distinguish true changemakers and game changers from those merely paying lip service to sustainability?

Let us begin with what leaders and changemakers are—and how the two differ. Anna Spenceley in our panel discussion put forward the following two definitions: a

---

[8] https://globalnews.booking.com/bookingcoms-2021-sustainable-travel-report-affirms-potential-watershed-moment-for-industry-and-consumers/

sustainability leader[9] is someone who applies mindful actions and behaviours, embracing a global worldview to recognise the connection between the planet and humanity; thereby, through personal and organisational choices, effects positive environmental and social change. A changemaker[10] is someone who is taking creative action to solve a social problem.

Paul Peeters added that leaders guide their followers towards sustainable development, whereas a changemaker is someone who challenges and changes tourism thinking or practice, even without many followers.

Antonis Petropoulos described a changemaker as someone 'who at the beginning is mocked like a mad person, who creates new, alternative paths, concepts, networks, platforms, products, experiences, campaigns, infrastructure and so on, against everyone's "wise" advice'.[11]

Because the global community is more separated than we would perhaps like to admit, in terms of worldviews and perceptions of reality, one of the things that tourism changemakers and sustainability leaders do is to show the connections and the possible ways of mutual interaction, both of which we need for scaling up sustainability in tourism and elsewhere.

So, what defines true leaders in tourism sustainability? What qualities do they possess that make them trustworthy and inspiring? Going through the panel's answers, we can distil the following key qualities:

- Committed, goal-oriented
- Inspiring
- Trustworthy: walk the talk
- Strategic thinkers
- Innovative, creative problem-solvers
- Risk-takers
- Lifelong learners
- Good listeners

Let us take a quick look at each of those qualities and leadership characteristics.

Committed, Goal-Oriented

Leaders are decisive and strongly determined to set goals and to reach them, be it to develop destinations sustainably or to deal with climate change. No matter how crazy the ideas and goals may seem, we need ambitious solutions to keep our planet and communities liveable.

---

[9] https://www.td.org/insights/what-is-sustainability-leadership

[10] https://www.evansville.edu/changemaker/downloads/more-than-simply-doing-good-defining-changemaker.pdf

[11] https://sustainability-leaders.com/tourism-sustainability-leader-characteristics/

In other words, leaders or changemakers have the clarity of mind to think and take action in the right direction, on behalf of the planet and its people. They are known to take bold steps, pioneering new ideas that inspire others, as Willem Niemeijer noted.

Richard Hammond observed that a sustainability leader is well aware of threats—like the climate and biodiversity emergencies—and has a clear idea about how to develop tourism accordingly.

It is easy to get overwhelmed by self-doubt, bureaucracy and funding shortages. A leader will continue the uphill journey despite these challenges, suggested Kelly Bricker. This commitment to the cause is certainly a trait common to most leaders and changemakers whom we have interviewed over the years.

The pandemic has once again shown that we live in a globalised and highly dynamic world. Being goal-oriented helps in achieving objectives but when the challenges evolve, leaders should be willing to go back to the drawing board and rethink decisions. Staying focused in such situations and not giving up requires a good amount of willpower, admitted Reto Fry.

What is more, when faced with criticism, leaders must be prepared to stick their necks out to put things into practice to the best of their ability, Lisa Choegyal added. It requires grit and an unflinching attitude to overcome obstacles, to implement sustainability without any compromise.

Navigating opposition is a test for anyone, and changemakers are likely to encounter opposition frequently. Marta Mills said that a leader needs to be patient and not get discouraged by likely opposition. It is natural that people are afraid of change, so a changemaker is bound to cause fear and opposition. To succeed, they need the ability to persuade and encourage others to come to join them on their sustainability journey.

Gianna Moscardo agrees that traits like resilience and grit keep them going, along with compassion and patience.

Glenn Jampol: changemakers are the last bodyguards of the planet and protect and eschew its beautiful ecology and symbiosis. They paint the landscape of how the future might look and improve it for the better for future travellers.

For this, leaders must appreciate sustainability as a holistic concept and resist perfectionism, and black-or-white thinking, urged Alex Pastollnigg.

Antonis Petropoulos finds that a leader combines a philosophical attitude to life with an entrepreneurial spirit and the political will to 'change the world', to make it a better place for all, and is ready to take all sorts of risks to achieve it.

Inspiring

Tourism is booming, at least until the pandemic hit international and domestic travel According to WTTC,[12] the travel and tourism sector contributed 10% to global GDP

---

[12] https://wttc.org/Research/Economic-Impact

in 2019. With the industry bound for recovery following the pandemic-induced travel restrictions, there is a lot of work to be done to transform the industry to become sustainable.

Society reflects the attributes of a leader, and they are bound to be imitated by others. Sonja Gottlebe finds that it is their lifestyle that resonates and inspires others to act. They are permanent ambassadors of sustainability through the way they travel and manage their businesses.

Action speaks volumes, and through their lifestyle and business decisions, leaders indeed can set an example and inspire others. For Shannon Guihan, a leader can engage team members, no matter the team type or size, to feel empowered to support the change that is afoot—with both ideas and action.

Likewise, Greg Bakunzi finds that a leader inspires the community to adopt friendly tourism. An inspiring figure is encouraging and does not shame when the team fails. Another essential element is kindness, through which leaders and changemakers inspire people beyond their own team. As Peter Richards put it in the panel discussion, kindness builds trusting relationships across gaps usually characterised by misunderstanding and mistrust, and it shows the potential to create beautiful and valuable work.

Trustworthy: Walk the Talk

In addition to a strong, goal-oriented commitment and the ability to inspire others, Anne De Jong defines a leader as someone who is *genuinely* in the game to lead the transformation towards a better tourism industry. True changemakers practice what they preach with regard to sustainability and conscious living. They lead by example.

Brian Mullis added that leaders are individuals who have consistently demonstrated their ability to foster tangible positive social, economic and conservation outcomes from tourism at scale.

Strategic Thinkers

Shannon Guihan believes that sustainability leaders are strategic thinkers that can take a long view, that prioritise the potential for impact. Likewise, Darrell Wade pointed out that a seasoned leader considers the different stakeholders and the outcomes they are each seeking—looking to maximise common interest areas and build awareness of differences.

A move towards a more sustainable destination or business requires a significant societal shift—and for this, we need buy-in and participation at all levels of an organisation. Leaders, therefore, acknowledge the importance of stakeholder engagement. Holding continuous dialogue with stakeholders to keep them in the loop regarding any tourism development is the hallmark of a good leader.

No doubt, it is a juggling act to talk with and understand everybody's needs and concerns, and implement a tourism development plan that resonates with every segment.

In short, passion balanced with pragmatism is what it takes to successfully lead in tourism and sustainability, added Marcus Cotton.

Strategic thinking also includes a good understanding of the complex requirements and challenges that are associated with sustainability in tourism, reflects Christian Baumgartner:

- Leaders know good examples of best practice, know about the technicalities and solutions, and how to implement those.
- Leaders can develop and implement goals and measures in a participatory multi-stakeholder dialogue.

A changemaker is also not afraid of (growth) boundaries and pursues tourism success criteria beyond arrivals and overnight stays.

Innovative, Creative Problem-Solvers

Solving a crisis requires a multi-pronged approach. An industry like tourism which works in tandem with other verticals like energy, transport, water, and aviation cannot employ a cookie-cutter approach to its problems.

The COVID-19 pandemic, overtourism, natural disasters and other crises which affect tourism cannot be solved by waving a magic wand. A sustainability leader or changemaker is solutions-oriented and creative in solving both grand and small problems.

Because every business and every destination is different, sustainability can never be applied with a stencil. Creativity is key: leaders have the ability to find measures that suit a specific company or destination, make sustainability tangible for travellers and help them avoid greenwashing.

Most destinations are plagued by overtourism, unregulated tourism expansion, pollution, high real estate prices, among a long list of other issues. The rate at which the tourism industry is transforming to a sustainable one is slow compared to the suffering of the local communities and their destinations.

Shivya Nath further pointed out that an apt leader is someone who can look beyond the conventional definition of tourism and explore other business models, redefine growth, draw on other industries and think out of the box to make tourism equally beneficial to the local people, the planet and those who choose to travel.

Leadership calls for innovative thinking so that the demerits of profit-driven tourism are mitigated. Creative problem solving helps in achieving sustainability sooner, and more effectively.

A leader does not sit still. They are keen to exchange ideas and best practices that can help them achieve their goals sooner and better. Erik van Dijk finds them innovative, always looking for new changes as there will always be challenges to make the world better.

Alex Pastollnigg seeks leaders who are innovative while being entrepreneurial. Elisa Spampinato expects leaders to look at the same problem from different angles, hear opinions, answer the same question with the eyes of various actors.

On the other hand, a leader can see and care beyond their immediate tourism enterprise. Joanna Van Gruisen continued that a leader is someone who thinks ahead of their time, out of the box, and is not afraid to tread a new path. Someone who has an imagination that can see positive opportunities, can make singular connections and believes in the possibility of transformations.

Risk-Takers

The trait of a sustainability leader or changemaker to be innovative pushes them to think hard and out of the box. It is not unexpected of them to come up with ideas that require them to push the boundaries. Taking risks is part of the job they do. Putting innovative ideas into action involves taking calculated risks.

Megan Epler Wood explained that leadership takes courage and so does change-making. In this field, it often requires raising issues that may not be ideally suited to the current government and industry environment. It is difficult to say the industry needs reforms. All of these ideas are about change, and it can be a challenge to even raise some of these ideas.

From a sustainable tourism point of view, Aivar Ruukel feels that leaders are those who are brave enough, not only to talk about sustainability but also to act. Practice what they preach and lead the way. He urged, 'Time is running out, so the more radical the action, the better!'

Lifelong Learners

To advance in one's career, staying up to date on the latest developments in tourism is crucial. Every day is a new day. Each day brings its own set of challenges and opportunities to learn. The past is the best teacher, and the past failures are the best way to learn and overcome new obstacles.

Among the virtues of a leader is their openness to learn. Be it a new skill or interest to learn about a sustainable tourism journey of a distant region. The next big idea could be lurking anywhere. Kelly Bricker consequently referred to sustainability leaders as lifetime learners.

Ally Dragozet explained that a thought leader in this space is always up to date with the latest trends and strategies to take sustainable tourism to the next level.

Good Listeners

How does one market the sustainable tourism journey to an interested audience? Before that, how does one capture the attention of travellers and industry players?

Online and offline media are frequent sources where we look for inspiration, and there is an exponential growth in content these days. To weed out claims of success and seek inspiration from actual changemakers, individuals must know how to translate sustainable tourism into stories, as mentioned by Elizabeth Becker on the panel.

Jonathan Tourtellot added that a changemaker has the perseverance and willingness to listen to many points of view. Like residents, tourists, industry, scientists, and destination stewards—be they representatives of government, private sector or civil society.

A true leader has the requisite skills to bring together different types of people and stakeholders to work together towards change, finds Audrey Scott.

A compelling narrative influences change. It stirs action among enthusiasts. A sustainability leader is a storyteller.

My takeaway from the panel's thoughts on the characteristics of leaders and changemakers is that those are people who can inspire and influence the larger population. And: leaders are not born. They are made. As Marta Mills put it so well, it takes time to mould a leader. Changemakers need to be bold and brave to challenge the status quo. That usually involves a lot of quiet work and small steps, many conversations, good communication skills and a strong mind to stay on path when opposition gets fierce.

2.4 What Motivates Leaders to Focus on the Sustainability of Tourism?

There are choices galore to pick a career in tourism. The last time we checked, the pre-pandemic WTTC report[13] from 2019 mentioned that travel and tourism supported 319 million jobs. These numbers speak volumes about the reach tourism has and how diverse the options to build a career with it.

When one is spoilt for choice with so many options, what makes leaders and changemakers pick sustainability to pivot the course of their career? Is it their love for nature, optimism to build a better world, personal experiences, the urge to fight climate change, protect biodiversity, promote social injustice—or something entirely different?

We asked our panellists[14] what prompted them to choose sustainable tourism as their career choice. Here are the main reasons put forward by this diverse group of leaders and changemakers:

- To create social and economic equity for the locals
- Having personally witnessed environmental degradation
- To future-proof the tourism industry

---

[13] https://wttc.org/Portals/0/Documents/Reports/2019/City Travel and Tourism Impact Extended Report Dec 2019.pdf

[14] Full answers are available at https://sustainability-leaders.com/why-career-in-tourism-sustainability/

To Create Social and Economic Equity for the Locals

So far, many destinations have brought in tourists, ran the show, and reaped the economic benefits of the industry without the consent or sign-off from the local community. This form of exploitative tourism has driven a wedge in the social and economic aspirations of the locals, increasing the divide between the industry and local stakeholders.

As Jorge Moller mentioned, the role of communities is of utmost importance to develop tourism—and to an extent which they deem requisite. This is critical to building a symbiotic relationship with each other, instead of the extractive nature of the sector currently.

Gianna Moscardo in her panel answer shared an incident from a tourism conference in Africa that she attended, supposedly focused on sustainability. Here she had found representatives of a village who waited for days outside the conference venue waiting to speak to anyone who could be of help to them. They had been forcibly displaced to make way for a protected natural area and an ecolodge. That project proudly proclaimed its environmental credentials but ironically had ignored the devastating social and human impacts that resulted from its existence.

Beatriz Barreal too had similar experiences when she tried to bring a social change in Quintana Roo, Mexico, by developing tourism to provide economic opportunities to the locals, following a steep decline in the mental health and well-being of locals in the region.

Such instances only prove the need for a greater rethink in the industry. Having offered tourism experiences in protected areas, Greg Bakunzi noted that local communities can be crucial while working in National Parks, which in turn motivated him to focus on community-based tourism. Marcus Cotton has had a similar experience at Tiger Tops and Tiger Mountain Pokhara Lodge in Nepal, where tourism facilitates conservation initiatives in a protected area as well as supporting local communities.

Having Personally Witnessed Environmental Degradation

Witnessing social injustice or economic hardship of locals clearly is a strong motivator for choosing sustainability as career focus. Environmental degradation is another one.

In the 1980s, when the first signs of mass tourism had started to appear, Jonathan Tourtellot observed how tourism was changing the world—in the negative sense. He saw it happen in Crete and northern Iceland—early warnings. Costa Rica's Glenn Jampol was shocked to observe society's unhealthy and unsettling obsession with monetary goals while sacrificing the quality of air, water and planet to achieve them. When he started his small inn and coffee farm project in 1985, the main objective for him was to minimise business impact on the environment, to participate in the local community, hire only local staff and learn how to reduce the operation's carbon footprint.

Paul Peeters too shared a similar experience. The roots of his work as a changemaker for sustainability stem from the 1980s when an acid rain crisis started to destroy the forests in his home country, the Netherlands. It inspired him to steer his career towards sustainable transport, always trying to 'invent' sustainable futures.

Childhood memories can be so vivid sometimes that they remain fresh in our thoughts. Rachel Dodds, when 13 years old, saw raw sewage flowing into the ocean on a beautiful beach in Mexico. This motivated her to build a career driven by her passion for change and to help all aspects of the industry to become more sustainable.

When Shannon Guihan began her career as a sea kayak guide in Newfoundland on the east coast of Canada, she noticed the fisheries in decline and tourism making inroads. The pace at which tourism was developing was rapid and unplanned, hardly involving local communities. This was a moment of realisation for her to focus her career on sustainable tourism.

Growing up in popular tourist destinations, Xavier Font, Christian Baumgartner and Antonis Petropoulos all had first-hand experiences with tourism that was making a negative impact, such as being disrespectful to local needs and desires, including taking good care of the natural environment. This made the sustainable development of tourism a natural choice for their careers.

To Future-Proof the Tourism Industry
Way back in 1996, Fiona Jeffery felt that if the industry did not change its trajectory, it would destroy its business model. During those days, international expansion, economic growth, and jobs took precedence over environmental issues. To bring about a change in priorities in the industry, she launched World Responsible Tourism Day with the UN World Tourism Organization (UNWTO).

The hospitality industry is not far behind: known for its wasteful overindulgences sometimes. Christof Burgbacher shared his personal experience with the panel of having worked for several years in the international hotel industry in different countries. During this time he noticed the huge wastefulness with respect to food, water, energy, cleaning agents and the low level of social responsibility (care for employees and the local population). To convert these negative effects of tourism into positive ones was his greatest motivation to dedicate himself to sustainable tourism.

Finally, Audrey Scott found motivation from what she observed in all her travels around the world: the scale at which tourism was growing, a recipe for disaster. She noticed that tourism had the potential to damage the environment and quality of life for the locals but how—when done respectfully and responsibly—it presented an incredible opportunity to do good and bring positive change for the local communities and conservation of nature and wildlife. This inspired her to work to advance tourism that 'does good', promoting the values of respect, caring and connection for local people and nature as the norm.

Clearly, there is still a long way to go until we get close to this point, but around the world there is a growing focus on developing and managing tourism businesses and destinations sustainably.

Having looked at what sustainable tourism is and the characteristics of changemakers and leaders, as well as their motivations to focus their career on tourism sustainability (the 'why'), it is high time to move on to the next big question: how to do it successfully?

Keys to Sustainable Tourism Success

Just like we can sometimes learn more from someone's failures than from their successes (and should therefore appreciate and celebrate entrepreneurial failure as an important part of learning and gaining experience), sustainable tourism initiatives which did not live up to expectations can tell us a great deal about how to do it differently. Most people and organisations do not like to talk about their failures though, so finding this kind of information can be tricky. I am truly grateful to the SLP panel of tourism practitioners and sustainability leaders for not shying away from such, more delicate, questions, allowing us in turn to identify how to make sure sustainable tourism initiatives can be successful.

3.1   Pitfalls and Challenges

To some extent, failure may seem inevitable, considering the ambitious goals which drive most sustainable tourism initiatives. But why exactly are we still so bad at achieving sustainability at scale?

In our panel session on the topic,[1] Elisa Spampinato pointed out that one of the main pitfalls that prevent tourism businesses from succeeding is forgetting that sustainability is a collective journey, whose direction is determined by the destination's very specific circumstances and priorities.

Indeed, tourism is a multi-dimensional concept that requires collaboration across industry verticals, continuous dialogue, exchange of best practice and innovation, constant modifications to fit the requirements of certifications and growing expectations, among many other factors. Jonathon Day calls sustainability a complicated activity, akin to keeping many plates spinning at the same time. It is a

---

[1] Full answers not yet published on Sustainability-Leaders.com at the time of writing but will be available on the website soon.

© The Author(s), under exclusive license to Springer Nature Switzerland AG 2022   27
F. Kaefer, *Sustainability Leadership in Tourism*, Future of Business and Finance,
https://doi.org/10.1007/978-3-031-05314-6_3

commitment to a responsible and forward-thinking way of doing business—not just a technical fix or an easy add-on.

This is when things start to get complex because large-scale transformation will obviously meet resistance.

Though everyone knows and talks about it, very few have implemented measures to make their business environmentally or economically sustainable—not to mention socially just. In other words, contributing more than what one takes from the destination, community and the natural environment.

To understand the hesitant adoption of sustainability measures in tourism, we need to know more about why adapting to the new requirements is such a challenge. Our SLP panel members have a wide range of backgrounds and experiences, allowing them to decipher what pitfalls are preventing tourism businesses or destinations from succeeding with environmental, social and economic sustainability.

Whereas it is difficult to pin down the exact challenges as many things could be conflicting, depending on the business or destination—attitude and beliefs clearly play a role, Rebecca Hawkins explained. She further pointed out that only very occasionally tourism operations are downright irresponsible. More often than not, however, there is a combination of conflicting priorities (such as between service standards and sustainability criteria), business models (e.g., properties that are managed by one organisation and owned by another), and misleading incentives (rewarding consumption rather than conservation).

Let us now take a closer look at the main difficulties in achieving tourism sustainability, as reported by the panel:

- Lack of leadership and political will
- Insufficient stakeholder engagement
- Lack of a clear vision
- Putting financial profit above everything else
- Post-COVID economic hardship
- Missing sense of urgency
- Overtourism
- Risk (and temptation) of greenwashing
- Lack of awareness

Lack of Leadership and Political Will
Destinations are run by some form of government, which tend to be the main decision-makers. To set the wheel of sustainability in motion requires honest participation from governments. Elisa Spampinato stressed that we cannot work on environmental, social, economic, and cultural dimensions of sustainable tourism if we do not include an additional one to the equation: the political. Many panel members concur that the main stumbling block in the switch to sustainability is the lack of local policies supporting it. This requires a concerted effort from all tourism participants to convince governments at all levels to enact and enforce rules for sustainable tourism, said Elizabeth Becker.

Moreover, destinations will fail to reach sustainability goals if they lack a critical mass of supporting sustainable tourism businesses. Antonis Petropoulos continued, if destinations do not have a competent DMO that can coordinate these businesses and if public tourism policy is only paying lip service to sustainability, permanently fixated on arrival numbers and expenditure per head, then reaching the goal of planet- and community-friendly travel is like shooting in the dark.

To top it all, many destinations are plagued by weak government support, corruption, undermining competitive environments, weak demand by customers, lack of access to modern technologies at a reasonable price, among other reasons, pointed out Peter Richards.

Brian Mullis feels that we need more governments working across ministries and with all of the players in the tourism value chain (e.g., private sector, NGOs, communities, etc.) to unlock systems value.

On a more practical note: political cycles mostly last 3–5 years, but real sustainability change at a destination takes 10–20 years. Therefore, political will to change things is often lacking, Rachel Dodds explained. And those who want change are often not in control of the things that need to change.

Insufficient Stakeholder Engagement

Another setback that we constantly hear about is the lack of equal representation of stakeholders in tourism. As Rachell Dodds suggested, all stakeholders are not equal in terms of power.

The thing is, changing the balance of power and opening up spaces to new stakeholders could greatly contribute to sustainability, if they are given more space in the decision-making process.

Jorge Moller Rivas finds that public policy cannot go ahead without the involvement of the community. It is an equitable process only when local communities play an active role.

Elisa Spampinato explained that there are big economic interests involved in the tourism business and hence, a huge disparity of power in its management. In fact, most of the people that directly feel the impact of tourism have no voice in shaping the industry.

Not all is lost as there are encouraging examples of innovative governance, like the municipality of Barcelona, among others, which show that new solutions to the democratisation of the process can be found.

Elisa continues that urban communities have easy access to digital platforms which unlocks a lot of opportunities whereas traditional, indigenous, and ancestral rural communities lack this technological advantage. They are left out of the loop and mostly left alone to face the consequence of deregulated tourism activities and the effects of climate change. Enabling easy digital access goes a long way for the latter, giving them a voice.

However, it is not just the local population who deserve to be at the table, but also employees and other partners as they ultimately have to accept and implement the measures, according to Christof Burgbacher. If a participation process is designed correctly, it can also generate a large number of ideas and creative approaches.

Lack of a Clear Vision

Even when all the factors are in place for sustainability to take off, businesses without a clear roadmap for execution are bound to fail.

If hotels, for instance, only pick the low-hanging fruits like reducing linen and towel washing to save water, then this will not lead to the kind of change needed. The industry as a whole needs to rise above superficiality and take a tougher stance to create value for host communities. It needs to rethink business models and relationships with stakeholders.

Xavier Font seconds this opinion that the urgent often gets in the way of the important. The industry is currently aiming at reaping short-term benefits without being aware of the long-term consequences of their actions.

Self-interest is another primary reason for this gap in action. Darrell Wade mentioned that most businesses are inherently creating a short-term business model that will not have sustainability as a priority. People consider their own needs but do not recognise those of others or the impacts of their own actions.

Christian Baumgartner added that convincing the decision-makers involved to think regionally instead of operationally is critical. He suggests that businesses think and plan long-term and complex instead of one-dimensional.

If we take a bird's-eye view of the industry, offerings from private businesses often depend on customer demand. It so happens that the industry waits and observes for trends before it implements changes based on market demand.

Christian Baumgartner reminded the panel that the industry must not wait for consumers to express the desire for more sustainability. Shivya Nath reflected his views that placing the burden of choosing sustainable travel on the consumer is not the right approach towards achieving sustainability.

Apart from the lack of vision and growth strategy, the question of ownership or accountability arises when destinations are mismanaged. Just, who is responsible?

It is in the nature of how tourism works that a shared sustainability vision is such a challenge. It occurs across so many different sectors and spaces that a lot of tourism is conducted without any overarching organisation in charge of it.

This leads us to the next sustainability challenge—a large chunk of tourism is organised by businesses which have no connection to, or interest (other than financial gain) in the destinations that they send tourists to and make money from. They have no incentives for travellers to behave well and bear very little in the way of negative consequences if those behave badly.

Putting Financial Profit Above Everything Else

What do a multinational company and a small, family enterprise have in common? In most cases, they are driven by the need, or desire, to make a financial profit, ensuring shareholder returns and sufficient cash flow, which drives decisions.

Vicky Smith, in her panel contribution lamented that this pursuit of profit at the expense of others has been a long-standing problem. Greed and self-interest take precedence over societal needs. She further explained that the human condition to take the path of least resistance stops businesses from treading the sustainability path. An unwillingness to change because it is harder work.

Shivya Nath underlined that business models that prioritise volume over all else, ignoring planetary boundaries, are risking the future of the very industry.

While profit margins are important, businesses need to find a way where sustainability fits into their strategies and operations and actually makes them better, said Anne De Jong. So, creating a situation where sustainability is fully integrated into the business is a must.

The 'more tourists, more turnover, more profit' mantra needs to go. The challenge is to abandon the idea of endless growth on a resource-limited planet. All tourism professionals should understand that tourism is not an industry but a living system. When changing how we see ourselves and our sector, we can change everything else too.

This brings us to the next problem which requires a change in business attitudes. The concept of the 'customer is king' should have been ditched long ago. It encourages the mentality of overconsumption. Rebecca Hawkins questions the firmly and long-held belief that if the customer wants it, tourism as a service industry has to provide it. Or, in economic terms, that where there is demand, that the tourism industry should supply it.

On the other hand, financial sustainability is a must for achieving sustainable tourism, Glenn Jampol stressed. Without a profit, businesses cannot survive and therefore the possibility to do good is erased. So, all tourism businesses—whether regenerative or conventional—must first and foremost create viable business platforms and seek to understand who their clients are and who they will be.

Post-COVID Economic Hardship

The COVID-19 pandemic pushed the sustainability agenda back considerably. It shook the world, and the resulting economic crisis is now raising its ugly head.

The loss of income due to travel restrictions has pushed many tourism businesses to the brink of bankruptcy. To stay afloat, many are now purely concerned about profits, with little capacity or will to bother about other aspects of sustainability.

Shannon Stowell: 'Right now the economic realities of a recovering world will be a real setback for many. Some headway was being made with single-use plastics for instance, and this area seems to be regressing because of Covid'.

Lisa Choegyal explained that in this extreme post-covid economic and social suffering when tourism returns, it will be tempting to cut corners in desperation to survive and succumb to market forces. We are already seeing this in unsustainable under-cutting and price slashing, for example. Many operations have been forced to lay off staff without pay, causing enormous hardship and threatening the quality of the product once visitors return. The challenge will be to stick to sustainable tourism principles.

Clearly, economic sustainability is essential to be able to stay afloat in times like these. Sonja Gottlebe, who—like many others on the panel—has witnessed loss of revenue due to the pandemic, feels that this has shown the limits and fragility of tourism all over the world.

A Missing Sense of Urgency

Humanity is racing against time. As the clock ticks, biodiversity is lost, and sea levels continue to rise. It is difficult to gauge the immediacy of something as important as climate change or loss of biodiversity, as it does not have the palpable shock factor that the COVID-19 pandemic has had—or at least not everywhere. The pandemic was a jolt to everyone almost simultaneously, whereas the effects of climate change or environmental degradation are varied and scattered across time and places. And a few places even benefit from the wrath of the climate emergency. As long as there is no sense of real urgency for the environment or climate with the general public, we will likely hit walls going forward—in tourism and elsewhere.

Overtourism

Overtourism seems to be synonymous now with tourism in many destinations. Unintended—or unforeseen—the presence of an excessive number of tourists is creating its own set of problems for destinations to deal with, especially with regard to sustainable development.

According to Kelly Bricker, the growing population and consequently, the number of travellers has overwhelmed the tourism system.

Joanna Van Gruisen explained that, however sustainable the operation of a tourist company is, its very success can invite others who may not entirely share the same sustainable philosophy. Nothing can kill a destination's identity and reputation, its brand, faster than overtourism.

Reto Fry suggested that digital marketing and social media have a big potential to cause overtourism, which cannot be sustainable anymore. For instance, when destinations use their unique mountain lake for a campaign—such as Lake Cauma in Flims, Eastern Switzerland[2]—'Instagram travellers' are likely to flood the spot. Nature and locals have to pay the price for this mass invasion and may feel overwhelmed, despite the economic gains.

The Risk (and Temptation) of Greenwashing

To avoid problems like the one Reto mentioned above, do we need responsible digital marketing?

Not only are too ambitious marketing communications sparking overtourism, they have also given rise to greenwashing, compounding problems for sustainable tourism. Since disinformation spreads like wildfire, Richard Hammond calls for separating the green from the greenwash. Beatriz Barreal shared that in Latin America, one of the main pitfalls is greenwashing. Willem Niemeijer pointed out that even if it is done unwittingly, it needs to be rooted out. Third-party certification can help businesses and destinations to avoid this trap.

James Crockett proposed a more balanced approach when marketing the sustainability measures undertaken by an organisation. Virtue signalling and a desire to be seen as doing good are needed. The most important stuff happens behind the

---

[2]Which I used to swim in during summers and which is very close to my current home.

scenes and there are some great inclusive components that need a song and dance to promote and spread the word to generate buy-in, but it must not exaggerate the actual achievements made—or work being done.

Lack of Awareness

Lack of awareness about the need for sustainability seems to be another issue plaguing the tourism industry. Marta Mills, Anna Spenceley and Beatriz Barreal in our panel session all pointed out that a great deal of the challenges relates to a lack of awareness of what needs to be done to become more sustainable, and gets compounded when skills, resources and effort come into play.

António Abreu observes a weak understanding of the role that sustainability should play in the business. For him, too often we see action without a solid background. We often listen to people saying that they know, do, and they are champions, but, in reality, they have no clue about it.

António also pointed out the failure to acknowledge the need for qualified workers. The tourism sector is very resistant to accepting the need to include other professionals and other skills. This is especially the case when it comes to environmental issues. Hotel managers, for instance, tend to consider that anyone in the organisation is able to assume professional and technical roles, instead of recruiting qualified people. For the restaurant, they want the best chef, but for handling environmental issues, anyone can do it. It is a basic mistake that we see every day everywhere, António laments.

Antonis Petropoulos too observes lack of real commitment to sustainable principles (such as the SDGs) on the part of management and employees, along with a lack of training. Amine Ahlafi, therefore, calls for a change first and foremost in the mentality of managers and human resources in charge of the management of tourism activities. He recommends that an updated training module in sustainable development and its impact on business and on ecosystems should be in place.

Secondly, businesses are often under the assumption that sustainability costs dearly. Erik van Dijk and Audrey Scott feel otherwise, stating that sustainable tourism is not as expensive as people think.

But Frankie Hobro sympathises: many businesses are concerned about viability, as a sustainable operation can require a lot of short-term investment with little immediate return. And some businesses cannot survive long enough to benefit from the long-term gains when faced with non-sustainable competition. A lack of support for 'green development' and funding contributes to this problem, as the sustainable option often costs more than the quickest and easiest option.

Audrey added that sustainability should be thought of as a long journey that will likely accompany the management of a business or destination for as long as it exists. New approaches, technologies and ecological realities are ever-changing. However, many tourism businesses or destinations will not know where to actually start and they can get overwhelmed by the complexity of certifications, feeling that sustainability is 'all or nothing'.

The concept of 'sustainability' clearly seems daunting to many. Shannon Guihan shared that many businesses in the industry, mostly SMEs, are uncertain where to begin. She suggests those already engaged in sustainable tourism come forward and offer resources to help peers to chart their own sustainability journey, rather than intimidate them from joining the effort.

Key Takeaways How can we overcome the many difficulties in achieving tourism sustainability? Frankie Hobro—reflecting on her own experience in the UK—shared that in the past, there has not been much encouragement for tourism to be sustainable but fortunately, that is changing now with consumer pressure and expectations in an evolving market. It is good news that the younger generations show genuine concern over their future on our planet and how our everyday actions contribute to it.

Clearly, already successful sustainability trailblazers should encourage and support those who want to follow suit, lead by example and show that it is worth taking the risks, that success is possible.

3.2 Keys to Success

Judging by the many potential pitfalls and hurdles that destinations and tourism businesses will likely encounter on their path towards more sustainability, the question is: how on earth can we achieve a more sustainable tourism development and economically viable, sustainable businesses and destinations?

Here are some constructive ways, according to our panellists,[3] how the tourism industry can achieve its sustainability objectives:

- Have a clear vision, with sustainability at the core of business models
- Collaborate and involve the locals
- Keep an open mind and never stop learning
- Good governance

Clear Vision and Sustainability at the Core of Business Models
A good business plan is the bedrock of a successful sustainable business. To operate like a well-oiled machine, businesses need to get the basics right: in tourism as everywhere else. Seasoned business owner of Finca Rosa Blanca Coffee Farm and Inn in Costa Rica, Glenn Jampol has found that success depends on three basic areas (in addition to some luck):

1. A clear idea of what your goals are and why you want to fulfil them
2. A solid business plan that is based on research, comparative information and learning about successful case studies
3. Understanding your competitors, your potential client base and the financial realities that could burden your development

---

[3]Full answers not yet published on Sustainability-Leaders.com at the time of writing but will be available on the website soon.

Glenn further suggests never to base the business model on being the best or most responsible tourism company. Rather, first create an efficient business that satisfies or exceeds client expectations. Then bind it all together, weave in the educational, regenerative and ethical cloth into the business to enhance the client's experience and add value to your product.

And when it comes to marketing your product, Darell Wade of award-winning Intrepid Travel suggests to 'create an amazing product that gets the customers to do the sales and marketing for you—it's both cheaper and more sustainable'.

Brian Mullis finds that a strong business case or proven pilot that speaks to the head, heart and gut can lead a business to success, followed by a great deal of patience and perseverance to go along with it.

Clearly, all tourism businesses are unique but the one objective that should be central to all of them is their commitment to sustainability. Willem Niemeijer rightly urges to make sustainability part of the way you operate—a part of the mission.

Sustainability is not an accessory to add to the core business idea. It should be the decisive factor. It is the synergy that binds the different elements of running a business together. Sustainability plays a role in talent attraction, supply chain management, operations and branding.

According to Jonathan Tourtellot, business owners should have the cultural inclination towards thinking of the common good, especially in terms of natural and cultural heritage. Paul Peeters suggests taking a global and long-term view and being inclusive of all tourism.

Also, we need to start prioritising quality like types of interaction, total spend etc., over quantity, added Shivya Nath.

Last but not least, financial stability. Lisa Choegyal stressed that one of the first rules of sustainable tourism is that a business operation must be financially viable. Without that, one cannot be in a position to achieve environmental, conservation or social goals.

Collaborate and Involve the Locals

Tourism has grown considerably in the past two decades and generates employment for millions of people. By involving locals in every step, sustainable tourism can be used as a force to ensure the social, economic and environmental well-being of destinations. Supporting locals is an idea that resonates with all our panellists since generating jobs for locals has a positive ripple effect within society. Not just offering jobs but also involving them in stakeholder meetings.

There are immense benefits to collaborating. Sharing ideas, exchanging industry best practices, and learning from each other helps everybody. There is certainly truth in Amine Ahlafi's statement, that the mobilisation of all the destination stakeholders is crucial for a common positive impact for all.

Xavier Font, Ally Dragozet and Shivya Nath all highlighted the importance of collaboration within the industry and with other industries.

Mariana Madureira suggested that tourism businesses and destination planners should not try to do everything alone. Being part of sustainability and tourism networks is also very important to exchange experiences and create partnerships for potential new and innovative projects.

Resilience

A lot of groundwork is needed to achieve holistic, sustainable development. It is an uphill task and a continuous one to work towards the overall development of a destination. Any change is met with resistance and disapproval.

Though the insecurity that arises from sweeping changes is understandable, convincing your stakeholders of the larger positive outcomes of a proposed change in operations, or investment, is important. Being resilient and unwavering about one's beliefs in a business idea and ethics is fundamental. Elizabeth Becker finds that the problem with championing sustainability is not lack of ideas—it is the courage to do what is required despite push-backs from powerful industry groups.

Keep an Open Mind and Never Stop Learning

Learning is fundamental to succeeding and should be seen as an investment for the future. Christian Baumgartner, therefore, feels that one of the ways to succeed is to be willing to learn and accept external expertise.

Anna Spenceley suggests making the most of freely available resources from institutions including the Global Sustainable Tourism Council, the IUCN's Tourism and Protected Areas Specialist Group and the Travalyst coalition partners.

Learning for (1) one's personal growth and (2) educating others goes hand in hand. Convincing newcomers to join the sustainability journey is easier and achievable if the benefits of sustainable development are broadly taught, to everybody.

Sonja Gottlebe in our panel conversation stressed the importance of education and the need to raise awareness about sustainable behaviour. She talked about her involvement in several bush schools around the lodges she has set up and how she has begun to see small but positive changes. Kids are starting to educate their parents and help them better understand what tourism can do.

Joanna Van Gruisen finds that the key is to implement appropriate behaviour as the best way to run tourism. This may sound obvious but, as mentioned earlier, too many are influenced by the 'customer is always right' principle and react from the outside in. Businesses that are under the assumption that to maximise profit they need to provide what the tourist wants, whether sustainable or not, are well advised to change their perspective.

Joanna on the panel shared her experience of how curbing the use of bottled water at her boutique hotel was perceived by guests: 'from inception we refused to provide plastic water bottles at all; this was at a time when this was untried. The industry told us this was impossible as the tourist would not accept it. We provide freshly filtered RO water instead. In our 11 years of operation, there has only been one guest who found this an issue'.

Educating customers by gently forcing them to step out of their comfort zone is a challenge, but goes a long way in terms of behaviour change: it sets the right trends at the beginning and establishes a narrative within the industry, albeit slowly.

Good Governance

Moving from individual to systemic, the fourth key to success is to ensure good governance. Deficiency in good local leadership is a common problem across sectors and geographies. Tourism organisations can benefit if local leadership understands the urgency to achieve sustainable tourism, and national leadership to ensure it is achieved, said Elizabeth Becker.

There is indeed a dire need for leadership at a local level at destinations, noted Natalia Naranjo.

Jonathan Tourtellot too stressed the need for a determined local leadership. He further added that a charismatic local leader works great for a while but there is a need for subordinates with similar responsibilities to keep the momentum going.

Key Takeaways A clear vision, commitment and determination lead to victory. We can learn as much from victories as we can from failures. Develop the passion to contribute to the greater good, combined with patience and long-term vision. Narrating her own journey, Frankie Hobro reflected, 'I also find it helps that I am extremely curious, and I like challenges. I like to be the person who finally succeeds at something when others have said it can't be done'.

The sustainable tourism journey is not going to be an overnight success story. It requires passion, unflinching determination and a long-term commitment to toil until the objective is achieved.

No matter how hard the uphill battle is, the fruits of hard work are for the destination, businesses and locals to relish.

3.3   How to Measure Success

How to measure the effectiveness and impact of sustainable tourism initiatives? This is the 1-million-dollar question—certainly one I have heard very often as editor of SLP. Those asking it are mostly professionals in charge of businesses or destinations. In partnership with The Place Brand Observer,[4] we asked our global panel of sustainable tourism experts:[5] *what does it take for a destination to be considered successful, in terms of sustainability? And how to determine and measure sustainable tourism success?*

The panel's responses[6] highlight the many ways in how success can be measured. On a cautionary note, 'success' means different things to different people and organisations, so will ever only be established or achieved through the unique boundaries of the person seeking to reach it. Subjective in nature, success

---

[4] https://placebrandobserver.com

[5] https://sustainability-leaders.com/panel/

[6] https://sustainability-leaders.com/destination-sustainability-how-to-measure-sustainable-tourism-success/

nevertheless encourages and motivates, and therefore can stir change, be it considered positive or negative.

Although there are many ways to conceptualise success, it seems that within the tourism industry some themes stand out, in order to help perform and plan for sustainability. For our panel, the following are the key points to keep an eye on in terms of interpreting and measuring success:

- Understanding sustainability as a process, a journey
- Determining community well-being
- Collaborations and partnerships
- Holistic approach to sustainability measurement
- Choosing the right metrics and standards

Understanding Sustainability as a Process, a Journey
The time factor is key to sustainability and often ignored when trying to determine success. Jonathon Day: 'I think it is fair to say that sustainability is a journey—not a destination. It is a process, rather than an outcome. I expect that the best destinations would never feel that they have ticked off the box for sustainability and can now move on to the next thing. There are always ways to improve'.[7]

Only through time and continuous effort, as well as adaption, is sustainability effective. Raj Gyawali, like others on the panel, also believes that sustainability is a process and 'not an end result of something you do'. This speaks to a key aspect in sustainability work: it is not just about what you are doing, but rather the 'movement is what needs to be watched to know if a destination is successful or not'.[8]

As tourism organisations (of any size) evolve and become increasingly aware of how they can contribute towards regenerative tourism practices, there is a sense of feeling overwhelmed on how to 'be sustainable'. This notion of being successful at sustainability over time encourages practical implications for the efforts being done or being planned.

Steve Noakes, who has worked extensively with destinations and businesses in Australasia and South-East Asia, shared the following:

> If a destination has committed to (say) undertake a sustainable destination approach as offered by one the certification bodies accredited with the Global Sustainable Tourism Council (GSTC),[9] that can be considered a successful start. If they continue the process of becoming a certified destination, that's another degree of success. If they can maintain that certification over time, that's another degree of success.

---

[7] https://sustainability-leaders.com/destination-sustainability-how-to-measure-sustainable-tourism-success/

[8] Ibid.

[9] https://www.gstcouncil.org/

If sustainability is a journey, then markers set through intervals of time help clarify measures of sustainability success. For example, reducing carbon emissions and becoming carbon neutral is a task that cannot be completed immediately. It requires a multi-level government and stakeholder approach to ensure that social behaviour, policies, regulations, and infrastructure are built into the plan. Further, it requires informing and facilitating community change and buy-in. Without community buy-in and awareness, we risk not effecting the change necessary to reap the rewards we want to see for the future.

Determining Community Well-Being

The key to a thriving tourism industry is making sure that it benefits host communities by elevating economic opportunities, supporting socio-cultural systems, and maintaining healthy ecosystems on which communities depend. Tourism does not 'happen' in an isolated social vacuum, it contributes and influences the social and ecological fabric of cities, regions or countries.

What is more, authentic tourism experiences (which are now so sought after) are the product of community pride, culture and sharing stories that reflect the place and people. And from the experts' perspective, the key to destination sustainability is resident well-being and quality of life.

In the words of Gianna Moscardo,

> Ultimately destination residents invite or accept tourists to their communities because they hope that this economic activity will improve their quality of life or their well-being. If tourism is not making a net contribution to destination community well-being, then it's not a sustainable activity.[10]

Gianna's point is important in that tourism, like any economic activity, should account for how it impacts people within the community, and further the natural environment. A great way to calibrate sustainability efforts and initiatives is to build trust with community members; this includes all levels of the organisation and residents. It is important to hear what people are saying in terms of their own economic and social well-being but to also gauge the overall health of the environment and community.

> Regardless of the size of the destination, a deep understanding of how tourism is contributing positively to help improve the welfare of the residents must be indicated and publicly shared.—Masaru Takayama[11]

Some of the most influential voices are those who work at the coalface of tourism. Those voices need to be incorporated into the feedback loop, rather than always hearing from the management or executive levels:

---

[10] https://sustainability-leaders.com/destination-sustainability-how-to-measure-sustainable-tourism-success/

[11] Ibid.

I think also that destinations should avoid stereotypical definitions and measurements and try to get a more 'bottom up' sense of what local people think and feel and experience. Often these questions are funneled through an academic prism, or measured in an unrealistic way, or answered only by certain people at a certain 'level'. I believe that, in order to get a 'buy in' from the whole community, you need to trust the views of the whole community—even if they may know little academically about concepts like sustainability.—Gavin Bate[12]

Less generic and more specific: while sustainability should be incorporated into multiple areas of tourism operations and planning, it also needs to be specific to the community's needs, and issues.

Gavin's point really shows the breadth and span that should be included when measuring and implementing sustainability initiatives. Further to his point, applying global standards and practices is great, but they should be adapted to meet the unique challenges that communities are facing.

Tourism is a system embedded in a larger socio-cultural system. The likelihood of success of sustainable tourism implementation is significantly increased when tourism is taking place in a destination community that is committed to sustainability; a community that is concerned about the environment—has recycling programs, supports renewable energy etc.—and is conscious of social and heritage issues, thus empowering tourism to adopt sustainable tourism principles.—Jonathon Day[13]

Others added further ideas on how community buy-in can assist the tourism industry in realising its sustainability goals. Brian Mullis, for instance, suggests incentivising businesses to improve safety, quality and sustainability practices to meet visitor expectations and achieve ready-to-export standards.

Incentives can be practical solutions, but also come with challenges, therefore should be thoughtfully considered to ensure meaningful change, and have self-sufficiency as ultimate objective.

Collaboration and Partnerships

Because the tourism industry impacts and is impacted by so many sectors (transportation, food and beverage, agriculture, place development, accommodation, travel trade, etc.) it has long involved collaborations among different actors. When these partnerships work well, they can be an effective way to enhance interactions among tourism players and work towards achieving sustainable development.

A destination that is addressing its main issues related to maintaining and improving environmental and cultural resources, and is working towards sustainability with stakeholders and community, is a successful destination. Probably the best way to measure its success is to evaluate and continuously improve a participative action plan of the destination management organization with the stakeholders.—Natalia Naranjo[14]

---

[12] Ibid.

[13] https://sustainability-leaders.com/destination-sustainability-how-to-measure-sustainable-tour ism-success/

[14] Ibid.

Anna Alaman spoke about the multi-sectoral nature of the tourism industry and how 'we can be dispersed in so many touristic issues (economic, environmental, socio-cultural) that sometimes we lack creating and measuring impact on any'.[15] Anna's point emphasises exactly why partnerships and collaborations are an important approach when thinking about and planning for successful sustainable tourism development.

One way to frame effective collaborations and partnerships is through the *United Nations Development Goals*. Goal 17, for example, aims to 'revitalize partnerships for sustainable development'. The targets within this goal speak of the importance of multi-sector partnerships that mobilise and share knowledge, expertise, technology and financial resources, to support the achievement of the sustainable development goals in all countries.

Brian Mullis' perspective aligns with SDG 17 in that he feels that 'foster[ing] inter-ministerial and multi-stakeholder collaborations and establish[ing] and implement[ing] structured, mutually beneficial partnerships across the tourism value chain' can lead to measured successful sustainability in tourism.[16]

Peter Richards in his contribution to the panel emphasised these same values, while commenting on the importance of cross-sector consultation/collaboration/concrete actions which show that people from different corners can work together.

However, inclusive partnerships can be difficult to achieve. When multiple organisations and groups allay together, there are at times tensions around a unified vision and ways of operating, thus leading to potential clashes between long-term goals for sustainable development projects.

Holistic Approach to Sustainability Measurement
Planet, people, profit. A phrase that most of us have probably heard. But how often do all three of those words get included at the stage of planning and implementing destination sustainability programs and initiatives? Another way to think about the 3Ps—as Philippe Moreau shared in his answer—is the framework adopted by 'The Long Run',[17] based on what they call the 4Cs (conservation, community, culture and commerce).

Whether it is the 3Ps, the 4Cs or a variation thereof, we can all agree that there is no 'quick fix' to the sustainability challenges our societies face, and that is because of the interconnectedness and complexity of each issue.

Sustainable tourism is a 'wicked problem'—each response, though guided by general principles, will be unique. And each action will generate unique responses, and often unearth new issues. The best destinations learn from the best examples and apply the 'learnings' to their own situation.—Jonathon Day

---

[15] Ibid.

[16] https://sustainability-leaders.com/destination-sustainability-how-to-measure-sustainable-tourism-success/

[17] https://www.thelongrun.org/

Jonathon's point further explains the complex and systemic challenges sustainable tourism and the industry as a whole face. Wicked problems are issues so complex and dependent on so many factors that it is hard to grasp what exactly the problem is, or how to tackle it. Wicked problems are like a tangled mess of thread; it is difficult to know which to pull first. Gender equality, security of food and energy supply, climate change and poverty can all be classed as wicked problems. No need to mention, this makes success measurement challenging.

Interdisciplinary and multidisciplinary research is an essential aspect for innovation when thinking about wicked problems. Gavin Bate suggests:

> ...rather than just the simplistic view of economic benefits bringing more money in.... intangible factors are often best seen through the medium of stories and anecdotes and public comment. So I would say the measurement of sustainability should not be so empirical but more holistic. Less generic and more specific to an area or region.[18]

Using a mixed method of both qualitative and quantitative inquiry offers the most robust information and helps to show the larger picture of a destination sustainability challenge or issue.

Amine Ahlafi also commented on the importance of the 'symbiosis between [human] and nature', and how that relationship should be considered the 'capital of a sustainable destination'. Amine further said that this relationship should not be limited to 'one-off actions'. Rather it should be 'seen in all the links of the value chain of the tourist product and in a strategic way'.

Echoing Amine's perspective, Gianna Moscardo referred to the capitals approach to community well-being:

> ...community well-being depends upon a healthy ecosystem, a vital and diverse economy and a sense of social well-being and these are in turn based upon good stock of natural capital, social capital, human capital, cultural capital, financial capital, built and political capital. So, destinations need to measure some aspect of all these capitals as well as resident support for tourism to demonstrate if they are sustainable or not.[19]

Choosing the Right Success Metrics and Measurement Standards

It is clear to most destination leaders now that tourism can no longer simply focus on 'how many' tourists visit a destination, or even 'how much' money was spent. In fact, those measures are part of what led to some of the industry's strongest headaches, especially overcrowding.

There are many governing and certifying bodies in the world that support and encourage sustainable tourism best practices. While their approaches and requirements may not necessarily be equally good, it is a consensus among our panelists that using some sort of metric or global standard to measure against is needed for determining success.

---

[18] https://sustainability-leaders.com/destination-sustainability-how-to-measure-sustainable-tourism-success/
[19] Ibid.

While adopting global standards is a good start, it is best to choose indicators and sustainability initiatives that are relevant to the destination. For instance, if overtourism is not an issue affecting a specific location, then it might not be necessary to allocate resources for it. However, overtourism—overcrowding or tourism fatigue among residents—should still be accounted for in long-term strategic plans, so that it does not become an issue in the future.

> Destinations need to set their own relevant sustainable KPIs. At the Slovenian Tourist Board, we have established the Green Scheme of Slovenian Tourism (GSST),[20] which serves as a tool for destinations and service providers to evaluate and improve their sustainability endeavours, based on global criteria.—Maja Pak[21]

Putting together strategic plans for a destination seems to be the 'ground zero' for realising and implementing sustainable tourism development and determining success over time. These strategic plans are formulated to address local issues and reflect the concerns of residents—while still taking global sustainability best practices and guidelines into account.

As Rachel Dodds pointed out, 'without metrics or indicators establishing a baseline, nothing can be measured'. To her mind, 'this should be in the form of a policy or plan that has specific, measurable indicators outlining acceptable limits of change (LAC). A multi-stakeholder approach should be taken to establish these'.[22]

Key Takeaways Sustainable tourism is not just a trendy word to include in your quarterly destination performance reports. It is about a genuine concern and care for the social, ecological, and cultural systems that the tourism industry is embedded in and depends on. Tourism has the power to contribute towards sustainable development and can support mitigation plans for the complex challenges our societies face.

As we heard from our expert panellists, 'having' success as a sustainable tourism destination takes several factors for consideration. First, in order to adopt sustainability into destination development, one must understand that it is not about 'checking a box' and 'reaching' a sustainability designation. Rather, sustainability happens over time. It takes time to achieve meaningful change, at a macro and micro level.

Once perception of sustainability has shifted from being considered a destination to it being more of a journey, it is important to utilise partnerships and collaborations to achieve common goals and objectives. Those can assist in leveraging resources in order to build awareness and community buy-in. Without community infrastructure and support it can be hard to implement effective sustainability measures.

Community well-being is at the heart of sustainable tourism. Without a healthy community and ecosystem, tourism cannot be successful in the long term.

---

[20] https://www.slovenia.info/en/business/green-scheme-of-slovenian-tourism

[21] https://sustainability-leaders.com/destination-sustainability-how-to-measure-sustainable-tourism-success/

[22] Ibid.

Lastly, ensure you build strategies and plans to help guide the efforts being made. Without a standard to follow, baseline measures, and continuous recording, there is no evidence of sustainable tourism successes.

In sum, measuring the effectiveness of sustainable tourism can be tricky due to its complexity, the many stakeholders involved and difficulty to prove the causality between your activities and changes in performance. Politicians and other key stakeholders need to understand the need for qualitative research to complement quantitative metrics, such as 'bed nights' or 'visitor spend'. Impact measurement research can be expensive and is often not included in the initial budget—or only insufficiently. The recommendations shared in this chapter will hopefully help you to master those hurdles and make your sustainability endeavour a success.

3.4   Bits of Advice from Leaders and Changemakers

Sustainable tourism is on everybody's mind now. Whether you are a graduate fresh out of college or an experienced tourism manager, the advice shared here will help you to make the gradual shift towards sustainability. Our panel members—academics, entrepreneurs and organisational leaders—provide valuable tips based on their own experience and observations. As Darrell Wade of Intrepid Travel pointed out, now is the perfect time to get one's foot in the sustainability door. How? Here our panel's bits of advice for aspiring tourism changemakers and future sustainability leaders:

- Be passionate
- Build a good team
- Find a mentor
- Be open to continuous learning
- Never give up, be persistent and patient
- Understand local issues
- Understand the market
- Do not expect to get rich

Be Passionate
A strong desire to make this world a better place through sustainability is a good kickstart to a career in tourism. If tourism is your passion, working towards making it better will be so much easier.

There will be lots of ups and downs, small victories and failed attempts, encounters with supporters and naysayers, but what remains consistent is the passion to achieve your objectives. Believing in oneself is important, and staying true to one's convictions is key, suggested Rachel Dodds—a 'pracademic' with broad experience in teaching and advocating for sustainability in Canada and beyond.

Passion tends to have a positive cascading effect on others as well. Winning their attention, convincing and including them in the sustainability journey is easier when you are passionate about what you do. Staying passionate is especially crucial when the sustainability road gets bumpy—which it most likely will at some stage. In these instances, it is key to listen to your own intuition and instinct and to follow your heart.

Build a Good Team

A group of like-minded people can together make a greater difference and cover more ground in implementing business ideas or sustainability initiatives than a solo achiever. Also, a team brings diverse ideas and strategies to the table. As Christian Baumgartner mentioned, sustainable development needs a small group of committed people who have the motivation and energy for long-term work.

A team keeps the show running despite the hurdles. Rebecca Hawkins recommends taking others with you as it can be a lonely path to travel alone. Shannon Stowell and Darrell Wade too recommend rallying others to join you in your mission.

Find a Mentor

Seasoned leaders achieve their expertise after having spent many years picking themselves up every time they failed. Frankie Hobro advices seeking out those who have succeeded and letting them mentor you, to learn from them and their mistakes. Marcus Cotton reflected this opinion by suggesting newbies to sustainability to tap into the wisdom of those with many years of experience.

No matter how brilliant the idea, how strong the team, or how large the funding, guidance and advice from long-serving or accomplished individuals go a long way in achieving long-lasting results. In this vein, Brian Mullis also recommended securing a mentor.

People from different schools of thought put forward different suggestions. Academics have a deeper understanding of the industry on the whole, whereas practitioners have more hands-on experience in the practical setup. In this situation, Peter Richards advises finding great teachers in both areas.

Antonis Petropoulos: 'Listen carefully to "three bits of advice" from older colleagues and then proceed with your original plans, slightly modified to accommodate some of their more interesting, but usually conservative, views. But do not let anyone put you down with phrases like "this is already being done" because you know you will do it better!'

Be Open to Continuous Learning

Linked to the above—an eagerness to continuously learn new skills is a healthy habit to have. Learning is not restricted to knowledge resources alone but also learning from peers and picking skills along the way. There is no such thing as a 3-month course on becoming a tourism changemaker or sustainability leader which will deliver all the know-how: rather, it takes time to grow professionally.

Peter Richards therefore suggests committing to educating yourself both academically and in day-to-day operations. Though books have a wealth of knowledge, getting your hands dirty as a tourism practitioner makes the real difference in succeeding. As Sonja Gottlebe emphasised: stay active and learn on the ground, not only through zoom sessions. Explore and experiment on your own, and do not fear losing, as this is an important part of becoming successful.

Like a sponge absorbing knowledge, come into a new situation with your eyes open as a learner and not as the person who already has the expertise, suggested Christian Baumgartner. Frankie Hobro: do not be afraid to try things out and to step outside of your comfort zone. Mistakes should be seen as progress as long as you learn from them and use them to find solutions.

When starting out on your sustainability journey, you will want to get the foundation right. Anna Spenceley and Lisa Choegyal stressed the importance of understanding the subject matter of sustainable tourism and how it might apply to your operation. Anna recommends looking out for examples of good work and resources on sites like Google Scholar or ResearchGate. Lisa Choegyal recommends keeping an eye on insights shared by https://sustainability-leaders.com/, as well as examining successful models.

Moreover, never stick to just one skill or find yourself trapped in a mental silo, as experience gained in different professional areas—within an organisation or in other industries—can go a long way, Elisa Spampinato finds.

And once you have learned new things, share them! Share and collaborate, so others in tourism can benefit from your new knowledge.

Never Give Up, Be Persistent and Patient
Be aware of the uphill battle which challenging the status quo or wanting to change a system almost always is. Acknowledge that the initial struggle is difficult but well worth the effort. The golden rule before embarking on your changemaker mission is to be prepared for headwinds, for example when your recommendations are ignored, no matter how sensible and logical they are, reflected award-winning Richard Butler in our panel conversation.

The secret is to never give up but to make amends along the way to fine-tune your objectives. Peter Richards encourages seeking ways to adapt, adjust, and tailor your knowledge to the real situation, rather than trying to push stakeholders and situations to adjust to your tools and knowledge. James Crockett likewise recommends being flexible. As Rebecca Hawkins put it, aim big but remember that every practice you change is a victory.

Reto Fry of Laax Destination in Switzerland urged to look for your supporters in your immediate working environment, instead of wasting too much time trying to convince the wrong people. Anne De Jong shares this thought, warning that some people simply do not care. Accept it and find ways to stimulate them to make changes. If at first you do not succeed, find another way. As Rachel Dodds put it, sustainability is like sales—everyone buys into it for a different reason.

This, however, without putting at risk your integrity or sacrificing your own sustainable tourism principles, Lisa Choegyal warns. Above all, stay humble and grounded. Be confident without being arrogant—important advice shared by several of our panel respondents.

Understand Local Issues

Understanding a destination's spirit, its identity, the woes of its communities, environmental condition, or its current state of affairs requires physical presence. To be effective as a tourism changemaker or destination sustainability leader, Beatriz Barreal finds it important to know and understand the value of all its asscts—historical, cultural, social and economic: to make it your treasure, perhaps even falling in love with the destination.

Elizabeth Becker, too, recommends immersing yourself in one locale and getting to know the issues intimately and how tourism really affects people, places, culture and wildlife. Then continue learning by visiting successful places as well as destinations that are battling challenges like overtourism, income disparity, lack of stakeholder engagement, and so forth.

For this to work, it helps to speak different languages, Elisa Spampinato noted: the language of the politician, the entrepreneur, the tourist, the traditional community and the local citizen. Understand their world, their fears and their needs and try to build bridges between them all. Open dialogues and create innovative solutions to continue paving the road for sustainability.

Kelly Bricker recommends listening carefully to the challenges of stakeholders, and to consult a wide variety of them. The more diverse and inclusive the stakeholders involved, the better usually the outcomes. Gianna Moscardo suggests focusing on the people with the least power in any tourism setting and how tourism will impact them. If you always start with the most vulnerable stakeholders and work backwards, she finds that you are more likely to identify leverage points that could make real change.

Hence, as Shivya Nath offered, deep understanding is essential before taking action. Having good intentions alone does not suffice, which leads us to the next key bit of advice shared by the panel: understanding the market.

Understand the Market

Business is business and therefore a good understanding of how the tourism market works is key to succeeding as a changemaker or leader. Testing the waters is important before diving into any new venture, so best to always check the effects of sustainability measures on the economic development of a business or destination, advises Christof Burgbacher.

Further, how do your ideas or objectives meet the market needs? Is your sustainability solution financially viable in the long term? Can you withstand pressure to lower prices in the face of competition or crisis?

Amine Ahlafi encourages us to undertake a sound analysis of the environmental, economic and socio-cultural ecosystems of the tourism activity, before getting started.

Once your idea is out of the box, Jonathan Tourtellot sees the next task to create demand for your services and abilities, should it not yet exist.

Do Not Expect to Get Rich

Truth be told, financial incentives are a deciding factor for a variety of reasons and goals in life. Breaking even, improving profit margins, etc., drives business decisions.

In the case of sustainable tourism development, a budding leader who is in it for the long term is best advised not to seek instant financial gratification or a great salary, Frankie Hobro suggested. Greg Bakunzi also warned that if the sole focus is about earning money through sustainable tourism, then it is going to be a challenge.

Frankie provides an honest assessment when she says, 'expect to work hard initially for very little or even as a volunteer to get your foot in the door and to gain respect as someone who really is serious and focused about what you want to do'. Vicky Smith likewise advises working from the bottom up to survive financially but prioritising practical experience over money.

Along similar lines, Peter Richards' recommendation is not to go straight into well-paid, well-managed jobs. Rather, to push your limits by working under very challenging conditions, with insufficient resources, time, people, money etc., since this is how most organisations in the area of tourism and sustainability function.

Key Takeaways Follow these eight key bits of advice and you are well prepared for a career as a tourism changemaker or sustainability leader. Committing yourself to supporting sustainable tourism is not the easiest career option out there—it might be one of the hardest. Yet, it can be hugely rewarding, so long as you build resilience in the face of opposition and other challenges, keep your spirits high when things get tough and are driven by a strong passion to make a positive change, rather than a hope to get rich. Inspire and lead, but stay humble, and listen. Take risks, stay creative, think out of the box, push your limits and see what lies on the other side of ordinary and conventional ideas.

Do not try to reinvent the wheel: most knowledge to become more sustainable is already available. Find a mentor and never stop learning, to really understand local issues or the tourism industry. This will take time, but through persistence and patience, and smart networking, you will succeed and shine.

Sustainable Tourism: Future Trends and Priorities

4

As a future-focused activity, sustainability management is always about anticipating what comes next. Which trends might influence the competitiveness and ability of businesses and destinations to shine and thrive? Which tendencies will impact sustainable development strategies? What 'hot topics' (should) occupy the minds of sustainability researchers around the world? Here is a snapshot of the state of affairs in late 2021.

4.1 Trends Likely to Impact Sustainable Tourism Success

2020, the year of the coronavirus pandemic, has been truly testing and exceptional for most destinations and tourism businesses. Yet, throughout this difficult time the key purpose of sustainability has essentially remained the same: to make destinations attractive to visit, and businesses resilient. So, what are the sustainability priorities and prospects for tourism professionals in the years to come?

Judging by the panel,[1] one thing is clear: tourism is in a stage of transformation, and a lot of changes are going to sweep through the industry in the coming years. In the future, when we look back at the evolution of this industry, the complete halt in travel and hospitality caused by the pandemic will be a good case study on how to prepare for unforeseen shocks. So, what trends to keep an eye on, what avenues to invest time and energy in, and what to opt out of? Here are my top picks from the panel answers:

- Transformation towards sustainability
- Avoiding overcrowding through smart visitor management

---

[1] This chapter is based on a summary of two panel sessions, with full answers available at https://sustainability-leaders.com/sustainable-tourism-hot-topics-2020/ and https://sustainability-leaders.com/sustainable-tourism-priorities-2021/

© The Author(s), under exclusive license to Springer Nature Switzerland AG 2022   49
F. Kaefer, *Sustainability Leadership in Tourism*, Future of Business and Finance,
https://doi.org/10.1007/978-3-031-05314-6_4

- Focus on regenerative tourism
- Mitigating and adapting to the changing climate
- Move towards digital
- Locals first

Transformation Towards Sustainability

One of the hottest topics currently is how to emerge more sustainably and build-back-better rather than reverting to business as usual once the pandemic ends and travel returns to normal. Global awareness of sustainability in tourism has reached an all-time high. This demand for sustainable tourism is evident by market research data showing an increasing business case for sustainable travel, as Anna Spenceley shared during the panel:

- Air Travel Sustainability survey[2]—A survey of 464 people in April 2020 found that 58% were thinking more about the environment and sustainability now compared to before COVID-19.
- Booking.com[3]—Travellers are becoming even more conscientious of how and why they travel, with over two-thirds (69%) expecting the travel industry to offer more sustainable travel options.

Roi Ariel, General Manager of the Global Sustainable Tourism Council, too sees a silver lining, something good coming out of the pandemic, in that destinations are picking up the pace to becoming more sustainable. He said that in the past year, many DMOs and government tourism departments have been using this pandemic-induced travel pause as an opportunity to work on sustainability.

While the marketing and promotional teams had little work due to restrictions, the development and sustainability teams had time to work on the integration of clear sustainability schemes (such as the GSTC Criteria) into their long-term strategic and recovery planning. This means that sustainable tourism businesses will likely find it easier to receive support and promotion from the governments where they operate.

Roi shared a few examples:

- Turismo de Portugal is launching Tourism Sustainability Plan 2020–2023 that covers many relevant issues including professional training in sustainability.
- Switzerland has just launched its Swisstainable scheme.
- Utah has its Red Emerald Strategic Plan to promote responsible visitation and help preserve our world-class quality of life for future generations.
- Visit València plans to be carbon neutral and has designed its sustainability strategy through a control panel with indicators that unify the SDGs and the GSTC Criteria.

---

[2] https://skift.com/2020/05/13/covid-19-airline-industry-sustainability-climate-change/

[3] https://news.booking.com/smarter-kinder-safer-bookingcom-reveals-five-predictions-for-the-future-of-travel/

The global trend towards a circular economy that focuses on conscious production and consumption is also catching up and becoming more important for tourism. Lucy McCombes suggests taking early advantage of this move and reviewing how we perceive successful growth and development beyond the usual economic measures. This will require innovative planning and design of new tourism products, services and destination strategies that reduce waste and bring in new ways of thinking.

On similar lines, Kevin Teng, the Executive Director of Sustainability at Marina Bay Sands in Singapore, believes the next movement that will define the direction that tourism will take will be food security and resiliency, including local/regional suppliers, diverse supply chains, and more environmentally friendly options for consumers. All of these are part of sustainability.

Avoiding Overcrowding Through Smart Visitor Management
Though affected destinations have acknowledged the overtourism problem, more needs to be done to deter visitors from swarming cities or places of interest. The negative impact of mass tourism is for everybody to see and there is a critical need to orient ourselves towards sustainable tourism.

Jonathon Day suggests that the 'rightsizing' of tourism will be an ongoing theme. In the past, it was hard to imagine a world without tourism and the emphasis was on more passengers. But, as the pandemic wanes, communities know what life without tourism looks like and can make more informed decisions about the right level of tourism for their destination.

Owing to the need for restricted travel and a lesser congregation of people, new business models include decentralised and hybrid events, added Kirsi Hyvaerinen. The old-school fairs where square meters were sold per stall will diversify into 'matchmaking market places'. Spreading tourist activities throughout the year and distributing travellers away from main attractions are some of the solutions. Kirsi recommends dispersing tourism regionally and seasonally with the help of smart technologies (e.g., open data)—especially in those places that are (again) under pressure. She reminded us that the real problem is not overtourism, but lack of sound management.

Focus on Regenerative Tourism
There is a new and growing trend towards regenerative tourism, of which destinations, businesses and investors alike should take note. Jonathon Day feels that regenerative tourism will remain a hot topic for a long time, likely to reinvigorate the sustainability conversation. Professors Willy Legrand and Vik Nair concured that the topic in vogue is regenerative tourism. The tourism sector is a job provider for millions and regenerative tourism is as much about the socioeconomics of tourism as it is about the tourism-nature interaction. The pandemic has been a gentle reminder that we cannot dissociate the social from the environmental and economic pillars. The tourism industry has been shaken to its core and talks of regeneration are heartwarming for many.

Taking the concept of sustainable tourism one step further, regenerative tourism—-as illustrated earlier in the book—basically is about leaving a place in a better condition than you found it, as a traveller—and to make sure destinations benefit from tourism more than it harms them.

Vik continued that before the pandemic, we collectively failed to protect the very resources that tourism relies on—commitment from the host and guest in stewarding the human and natural resources. Regenerative tourism provides an opportunity to get back to the sustainability philosophy.

Before regenerative tourism took shape, staycations and low tourism had been emerging as options for more sustainable travel before COVID-19. Gianna Moscardo feels that one of the challenges for both these trends was to encourage people to rethink their travel and to find value and benefit from travelling closer to home and slowing down. And the COVID lockdown, closed borders, travel restrictions and mandatory quarantines have made these two options the only ones in many places, opening up opportunities to encourage staycations and slow travel beyond COVID-19.

Vik Nair: What we need is more than sustaining but making the destination better than before (net positive). Our aim should not be to do the minimal to maintain the status quo, but we should be focused on restoring and regenerating the capability to live.

Mitigating and Adapting to Climate Change

Climate change is no longer the elephant in the room. Citizens are actively taking measures to reduce their carbon footprint, which is a hint for governments and corporations to make tourism a carbon-neutral industry. To future-proof the industry, it is recommended that businesses change their operational models accordingly. The sooner, the better.

Localising and decarbonising travel should be a top priority given the climate emergency, suggested Albert Salman. Willy Legrand mentioned that citizens across nations see the urgency for climate action, and a tourism industry that has the desire to bounce forward should decouple its development from carbon emissions, since nations are trailing back.

According to Natalia Naranjo Ramos, tourism must return while taking into account the climate emergency context. And this can only be achieved with the public, private and communities coordinated work for sustainable management of destinations and initiatives in the territories.

Move Towards Digital

Automating and streamlining manual processes in tourism makes them faster, more cost-effective and measurable. Destinations that make the most of the digital revolution will strive forward quickly, one step ahead in achieving sustainable tourism. Also, digitalisation is an often-quoted process in managing destination overcrowding.

Masaru Takayama of Asian Ecotourism Network predicts a smart digital transformation in tourism. For Maja Pak of the Slovenian Tourist Board, digitalisation will be one of the pillars that DMOs will address first, post-Pandemic—to improve user experience and also safety standards.

Locals First

And, last but certainly not least, considering the needs and desires of the locals in tourism decision-making will be a priority. Be it sustainable, regenerative, or transformative tourism, locals must have a say. A top-down approach has rarely worked, as observed in most destinations. To be truly sustainable, destinations need to listen, act and ensure the social and economic sustainability of the local communities. Marcus Cotton hopes for a greater focus on putting those at the centre of tourism decision-making.

For that, we need destination management with strong and continuous involvement of the local communities, said Kirsi Hyvaerinen. This approach reaps benefits, as putting communities at the heart of tourism helps improve resident, visitor, and environmental well-being, according to Brian Mullis. Maja Pak concured that local communities and their satisfaction with tourism will play a more central role, as will local experiences.

Key Takeaways With the ongoing transformation towards sustainability in tourism we will see a strong focus on visitor management and digital solutions, for instance to avoid overcrowding. This is one of the keys to happy local communities. It is also essential for enabling regenerative tourism, which together with mitigating and adapting to climate change is part of the main trends our panel foresees for the next years. Although many solutions already exist, some questions and issues are in need of further investigation, as we will explore in the next chapter.

4.2 Research Priorities: Questions to Be Answered

Now that we have almost reached the end of this first part of the book, we can reasonably conclude that there is no silver bullet to solve the current tourism crisis, or to "achieve" sustainability. However, with less than 10 years left to reach the ambitions expressed in the 2030 Sustainable Development Goals, there is a renewed urgency to examine our actions and to balance human aspirations with the planet's ability to sustain them. But which topics are the most burning, in need of further investigation?

Our panel—which includes both researchers and practitioners—has some clear ideas. The following appear to be the most relevant 'hot topics' for researchers to address and for practitioners to keep an eye on as we continue to develop and improve sustainable tourism theory and practice:

- How to implement sustainability policies successfully
- How to adapt to new climate realities
- How to reduce greenhouse gas emissions
- Gender equality: how to empower women

- Understanding travellers better
- How to engage stakeholders in sustainable tourism
- How to communicate sustainability

How to Implement Sustainable Tourism Policies Successfully

Everybody wants sustainable tourism to be implemented, but how to go about it? There is immense curiosity in this regard as most entities are first timers in this transformation to a better form of tourism.

Many questions surround the implementation of sustainable tourism policies over the next 5–10 years. For one, because sustainable tourism is not a one-size-fits-all solution to fix the problems of the industry. Each destination has its unique circumstances that need careful planning and management.

Joanna Van Gruisen seeks more clarity and information on the actual economic benefit to the communities in which the tourism enterprise is situated—not in terms of GDP but locally and specifically.

There are a plethora of declarations and certifications to guide destinations and tourism businesses towards sustainability and to secure strong actions and commitment. But what value do we actually generate by adhering to those, and how do they contribute to the SDGs?

Christian Baumgartner, for one, would like to see more evidence that destinations which focus on sustainability in their policy planning and management are also more resilient. And Kelly Bricker sees a need in identifying and measuring the change within destinations that adopt sustainable practices, especially regarding the reduction of CO_2 emissions, energy use, or better supply chain management.

Joanna Van Gruisen added that the industry requires a measurement of tourism's social/cultural impact—positive and negative. The aim is to see how/where the whole tourism industry can improve to benefit the local economy and positively impact the social fabric of the area in which it occurs.

More importantly, what do the locals want? Elisa Spampinato calls for more research from a local community perspective and their needs.

Along similar lines, Jonathan Tourtellot thinks constant monitoring of impacts is beneficial to establish better measures of success and identify better tourism management practices.

How to Deal with the Climate Emergency

Frequent floods and wildfires show that more than ever healthy and functioning ecosystems are key to most travel experiences—not to mention the well-being of communities. As an industry, tourism is highly dependent on the natural world Climate change negatively affects the very "product" that tourism is trying to sell.

How much industry leaders worry about the climate emergency is apparent by the number of mentions in the panel regarding how tourism should adapt to the climate emergency and help mitigate it.

Questions to tackle include:

- How to establish sustainable tourism amid a climate emergency?
- How does tourism affect the climate crisis and how to remedy those issues?
- How can tourism businesses collaboratively contribute to the international climate goals?

Megan Epler Wood in her panel response outlined a possible direction, starting by looking at the total footprint of the tourism industry, understanding how to measure it, creating highly responsive levels of data analysis at the local level to manage impacts, and trying to use systems of analysis that can be translated internationally but gathered locally. She suggests a focus on lower-impact transportation, global analysis of climate resilience investment and dynamic data gathering to gauge tourism impact and destination well-being.

Greenhouse gas emissions are the—now very visible—elephant in the room with regard to the changing climate. The question is: how can low carbon development work for an industry which is so energy intensive? Further, since we already have the knowledge and many tools, but still lag the required progress towards positive climate impact: how can we solve the apparent logistical and political issues that are slowing progress? And if it is too late to reverse the process, what can we do now to mitigate the negative effects that our future generations will be facing?

How can we tackle rising emissions and decouple the industry from carbon-based energy sources? How can we reduce aviation's significant carbon footprint, especially when travelling to long-haul destinations? The same applies to cruise tourism.

Research which helps to find solutions to these urgent questions will be hugely valuable for making tourism (and human activities in general) more environmentally sustainable.

Gender Equality: How to Empower Women
According to the UNWTO Global Report on Women in Tourism 2010,[4] 54% of people employed in tourism are women, but on average those earn 14.7% less than men. How can tourism offer more opportunities to women, to achieve gender equality—Sustainable Development Goal 5?

Elisa Spampinato expressed the need for more research on gender equality and the impact of tourism on women, both in terms of damage (current situation and past mistakes) as well as potential (what we can do to support and encourage positive change). There is little research on these topics globally, and we need to highlight the situation with better scientific means to understand the real impact and act accordingly.

A pain point that the industry needs to iron out is the lack of funding or financial support for women-powered sustainable tourism start-ups. The issue was highlighted by Vicky Smith, reminding us that women form a minor percentage of investment grant recipients. To have more clarity, she would love to know the tourism-specific statistics on investment in sustainability/applicant gender/value.

---

[4] https://www.e-unwto.org/doi/pdf/10.18111/9789284420384

Understanding Travellers Better

Consumer expectations are often explored through market surveys. Destinations and large corporations rely heavily on them to forecast demand, to make informed decisions. But the downside of this exercise is that there are a plethora of surveys to catch consumer vibes. The real question with regard to sustainability is: how much of the intent expressed in a survey translates into action?

For example, Euromonitor's *Voice of the Industry: Sustainability* survey[5] (shared in the panel by Vicky Smith) found that 64% of global consumers want to travel sustainably and with a purpose. Based on this data, how many of them actually went on a responsible holiday? To address this discrepancy, Xavier Font suggests more experiments about actual behaviour and fewer surveys about intentions.

Glenn Jampol and Richard Hammond also call for better post-consumer data: on how the consumer makes decisions on where to travel, what they expect when they arrive and how much they are willing to pay to achieve these needs. Not just claims, but actual behaviour and decision-making. Richard feels confident that there is an increase in sustainable holiday bookings, but it would be helpful to demonstrate it with real data.

The COVID-19 pandemic interrupted both the tourism industry and many sustainable development projects. The standstill in tourism provides immense insights, for instance the opportunity to compare data from pre-pandemic with post-pandemic travel. Lisa Choegyal recommends post-COVID consumer demand surveys and analysis for assisting governments as they restart travel post-COVID.

How to Engage Stakeholders

Fair and equitable stakeholder participation and engagement is a prerequisite for sustainable tourism development. Yet, decision-making has mostly been a top-down exercise as we have seen over the years with local communities and the environment bearing the brunt of unregulated tourism.

We know that including the perspectives of those being impacted by new strategies and developments is crucial. But how can we assure inclusive decision-making and compromises which work for all involved at the destination level?

How can we obtain greater buy-in and ownership of sustainable and responsible practices? António Abreu suggests that more knowledge is needed about the interactions between tourism, people and nature. This, he reckons, would help investors, governments and professionals to make better decisions. We need a better system in place to generate a shared understanding and dialogue, to prevent conflicts.

How to Communicate Sustainability

A lot has been spoken, discussed, and debated about the responsibility to be shouldered by governments and businesses to make tourism sustainable. But how much of the onus is on the visitors themselves?

---

[5] https://www.euromonitor.com/voice-of-the-industry-sustainability/report

Glenn Jampol finds it problematic how sustainability is currently communicated in the industry. He said, 'We do not know how to inspire travellers to put responsibility and conservation in their planning. They certainly do not want to feel guilty, and the "green" travel movement has taken advantage of that fear. We have not discovered how to make them highly proactive in their decision making'.

This boils down to the question: how can we communicate the role of and need for sustainability in a way which motivates travellers? Linked to this, which channels of communication work best—traditional or social or both? What about timing, and so forth.

Improving behaviour change at the individual level is achievable, and Jonathon Day encourages researchers to adapt behavioural science and decision architecture to the tourism context.

Key Takeaways   The panel highlighted some important questions and areas where research is needed to help us future-proof the tourism industry and enhance its sustainability. Some issues require technical solutions, whereas others are all about people and communication. We will clearly achieve a more sustainable tourism faster if we manage to really involve local communities in destination planning, and there is a good chance that the process will work best when women in particular feel empowered.

A good quote to conclude our expedition into research priorities and questions to be answered comes from Gianna Moscardo: 'We don't really need a great deal of academic research in general—what we need is a lot more action-based research—applying what is already known (especially from beyond tourism) to change practice and then evaluating practice so it can be fine-tuned'.

What Are *Your* Sustainable Tourism Questions?

Having presented you my summary of the SLP panel's take on some of the most relevant topics associated with the art and practice of sustainable tourism development and management, I would love to hear from you: which questions or issues keep you awake at night, linked to making tourism more sustainable? Tell me via LinkedIn,[6] Twitter (@floriankaefer), or by sending me a message through the website, Sustainability-Leaders.com. Let us stay in touch!

---

[6]https://www.linkedin.com/in/floriankaefer/

Part II

Interviews with Sustainability Leaders and Changemakers

Perhaps you are wondering who the professionals are whose ideas, insights and advice have influenced my understanding of sustainable tourism and who continue to inspire my work. Allow me to dedicate this second part of the book to those individuals who have actively contributed to our expert panel, helping me and so many others understand the complexities involved in developing, nurturing and marketing sustainable businesses and destinations.

Some of the interviews are already a few years old, so I have added an up-to-date short description for each person. One of the shortcomings of publishing a book like the one you are reading right now is that it can always 'only' be a snapshot of what has happened until the time when one starts writing the book (and this one took over a year to put together. . .). It is thus quite possible that some of the persons featured in the following (in alphabetical order) will have moved on to different topics, and that new experts will have joined the SLP community. This is therefore by no means a definite list of sustainable tourism experts and changemakers—if you'd like to find out who is currently active in your area (location or area of interest), please visit https://sustainability-leaders.com/.

Abdulla Radaideh on Sustainable Hospitality in Jordan

Jordan | Abdulla Radaideh is a mechanical engineer who started his career at the InterContinental Aqaba Resort in 2008. He has been with IHG [InterContinental Hotel Group] from the junior level to becoming a part of the executive committee. In 2013, he was nominated by the Ministry of Environment & Tourism and by the Jordan Ecolabel operators to speak at the 'Ecolabels for the Tourist Accommodation Service in the Mediterranean' program about ecolabels and how the tourism industry can be a part of it. He has worked towards implementing Green Engage, Green Key, Blue Flag and Eco-school programs. InterContinental Aqaba Resort was the first beach in the Middle East to earn the Blue Flag and Green Key certification in 2010 and has continued to renew these certifications for the last 11 years.

Areas of expertise: hospitality, sustainable business, Middle East, UNSDG 12 (Resource Efficiency)

The following interview with Abdulla was first published in August 2015 on Sustainability-Leaders.com.

F. Kaefer, *Sustainability Leadership in Tourism*, Future of Business and Finance, https://doi.org/10.1007/978-3-031-05314-6_5

Abdulla, a few words about your professional story—why did you choose to work in hospitality?

Before joining the hotel industry, I worked in factories for the timber industry, some of which have a devastating impact on the environment. Although I'm a mechanical engineer and factories offer a very good income, I wasn't comfortable at all and took the decision to leave all kind of factories and start a new career with a focus on conservation, rather than destruction.

I joined the InterContinental Aqaba resort seven years ago, where I noticed a big difference in the business approach, with many initiatives and programs like Green Engage. IHG [InterContinental Hotel Group]'s steering wheel has four major quadrants: financial returns, guest experience, our people and finally the responsible business part, which is about reducing our impact on the environment and supporting the community.

To be honest, I decided to work for IHG because of the hotel group's commitment to responsible business, and over the years worked myself up from junior level to becoming part of the hotel's executive committee.

When did you first discover your passion for sustainability?

Energy conversions—renewable sources—was a subject I was very interested in during my studies at university. Jordan is one of the poorest countries in terms of available water and energy resources, and I always had that dream of making a positive difference and changing people's mentality toward the environment and energy.

My dream is now becoming true at InterContinental Aqaba. Maybe my work doesn't change the whole world, but I have managed to change the mentality of certain groups and colleagues toward sustainability.

As a leading force behind establishing the Green Key ecolabel in Jordan, how has your view on sustainable tourism evolved?

Tourism is one of the most important sectors of Jordan's economy, with about 50 thousand employees, half a million tourists and more than 2 million overnight stays in 2014. Moreover, studies indicate that 75% of consumers want a more responsible holiday and almost half of global consumers are willing to pay more for products from companies that show a commitment to social responsibility.

Five years ago, we started the sustainability journey with JREDS "The Royal Marine Conservation Society" with the Green Key program, then another 20 hotels followed.

The change is noticeable in the change of five-star hotels from heavy consumers of energy and water to green, responsible hospitality businesses with sustainable energy management plans and sustainable tourism programs

Which aspect of implementing sustainability at the InterContinental Aqaba Resort in Jordan do you find the most challenging?

Waste management. We are separating our waste inside the hotel for the recycling but all the waste is going out to the landfill after being collected by the local authority, except for some recyclable material for which we have been able to contract private companies to take them instead of the local authority, like oil, metal cans and some plastics.

Why the focus on sustainability at Aqaba Resort in the first place?

- Being environmentally responsible is part of IHG's vision of providing "Great Hotels that Guest Love", and this is part of our yearly KPO's [Key Performance Objectives]
- Saving money through reduced energy consumption (up to 25%)
- Winning more business through promoting use of Green Engage to corporate and other guests
- Engaging employees in making a difference
- Competitive advantage over other hotels

Which sustainability achievement at Aqaba Resort are you most proud of?

Water: in addition to all the mandatory water criteria from the Green Key and IHG Green Engage programs, we have dug a couple of wells that accumulate the underwater coming from the sea. We installed a RO unit which produces an average of 140 cubic meters per day from these wells, which covers nearly 40% of the hotel's water consumption.

Which have been the main benefits of implementing sustainability initiatives at Aqaba Resort?

In addition to the reputation that we are gaining from being a green hotel, there are clear financial benefits. We have reduced our energy consumption by 29% and are working towards a reduction of 33%.

As a luxury resort, do you communicate your environmental initiatives to guests?

Yes, in many locations: on the guest room TVs and lobby screens, in the guest room directory, on the concierge lounge at Lobby, on the Blue flag board at the beach and at any special environmental occasions that we are participating in.

We are also placing posters, banners and flyers at the lobby for special events like Earth Hour or World Environment Day. And our front of house staff communicates our sustainability initiatives directly to our guests.

Your thoughts on the current state of sustainable tourism in Jordan?

Tourism is Jordan's largest private sector contributor to GDP and foreign exchange, and supports more jobs than any other industry, as a result, a Jordan Tourism Development Strategy was developed under USAID/AMIR 2.0 (launched by HM King Abdullah at the 2005 World Economic Forum), a project known as "Siyaha", which is enhancing Jordan's competitiveness as an international tourism destination.

In Jordan there are many environmentally friendly tourism spots like the Ajloun Forest Reserve, Azraq Wetland Reserve, Shaumari Wildlife Reserve, Mujib Reserve, Dana Biosphere Reserve, Wadi Rum and Feynan Ecolodge, in addition to many NGOs like JREDS who is taking care of enhancing these spots and spreading the awareness toward sustainable tourism which is getting better and better year over year.

Your 3 bits of advice to other sustainability managers?

Start with employee engagement and get them committed, since they are the root of all the changes that we are seeking.

Then get Eco-Labels like Green Key, Blue Flag or Green Globe, and focus on the communications, especially with guests directly through front of house staff and through the media.

Finally, look at ways to reduce energy consumption and link this to your green initiatives and programs, since possibilities to save money are the main reason and factor for the owners and investors to support your ideas, especially if changes need initial investment.

Link to the interview: https://sustainability-leaders.com/interview-abdulla-radaideh/

Adama Bah on Responsible Tourism in The Gambia, Africa

The Gambia | Adama Bah has over 40 years of experience working in the tourism industry in The Gambia and other West African countries. He has spent more than 25 years in operational roles in The Gambia's hotel industry and later as Deputy General Manager responsible for operations and human resources. From 2006 to 2013, Adama became The Gambian coordinator for The Travel Foundation, UK. An associate member of the International Centre for Responsible Tourism and a speaker at destination conferences organised in different countries: India, Brazil, Canada, UK, South Africa, Spain, Belize, etc. In November 2004, he won the international award for The Greatest Contribution to Responsible Tourism at the Responsible Tourism Award Ceremony at World Travel Market in London. In July 2019, Leeds Beckett University awarded him an Honorary Doctorate for dedicating much of his life to responsible tourism, both in The Gambia and internationally.

Areas of expertise: responsible tourism, Africa, destination development, hospitality, UNSDG 8 (Equal & Fair Economic Opportunities)

F. Kaefer, *Sustainability Leadership in Tourism*, Future of Business and Finance,
https://doi.org/10.1007/978-3-031-05314-6_6

The following interview with Adama was first published in May 2018 on Sustainability-Leaders.com.

Adama, do you remember the first time you heard or thought about the sustainability of tourism?

In July 1994, we had a takeover of The Gambia Government by the military, bringing to power President Yaya Jammeh. This was followed by a travel warning by the British Foreign Office, advising its citizens against travelling to The Gambia. Scandinavian countries also issued a warning. This move brought the tourism sector in The Gambia to a complete halt.

I was back then working as a Personnel Manager at the Bungalow Beach Hotel and was faced with the situation of terminating the services of over 100 staff who just reported in October for the beginning of the tourist season. I had sleepless nights!

This experience made me think of issues of sustainability in tourism—if foreign governments can make it possible or impossible for tourists to visit or not to visit our country, then we need to look at ways to make our tourism less vulnerable and more sustainable.

From here I made plans to take voluntary retirement from my work and, in 1995, left to start a local NGO called Gambia Tourism Concern, with the main aim of advocating for sustainable tourism.

Having worked as a hotelier in The Gambia for 25 years—what made you decide to dedicate your career to the development and promotion of responsible tourism?

I left my job mainly to promote sustainable tourism out of frustration that tourism in The Gambia was controlled mainly by foreign powers and investors, with little participation of locals. In 1995, we organized the first sustainable tourism conference in The Gambia, looking at the role of small enterprises and community involvement in tourism.

In 1999, with sponsorship from Vocational Services Overseas (VSO), I attended the first Commission for Sustainable Development session on Sustainable Tourism at the United Nations in New York. Here I met Dr Harold Goodwin and invited him to The Gambia to a conference organized by Gambia Tourism Concern on Community Based Tourism.

It was after this conference when we applied to the UK Department for International Development (DFID) for a project aimed at improving the participation and income of informal, small businesses in tourism. This project made it possible for the participation and income of tourist taxi drivers, craft vendors, juice pressers, fruit vendors and local guides to be improved and recognized by Government and Industry as important partners.

At the end of the project term, we handed it to the Responsible Tourism Partnership (RTP), consisting of government represented by the Gambia Tourism Authority, private sector tourism associations and international tour operators. This partnership also helped the government to come up with a responsible tourism policy for The Gambia.

My motivation comes from seeing different stakeholders playing their part in practice to make local participation and gain come to fruition in The Gambia.

Which are the main topics or concerns linked to tourism sustainability at the moment in West Africa?

Tourism is a business. Our governments invest heavily sometimes at the expense of communities to make it possible for tourists to visit our destinations. Much of the financial resources used by our governments to invest in tourism are loans to build the required infrastructure. Who pays for these loans? It is the citizens of these destinations. As a business, we need to gain more than what is invested. This gain must translate into development and measures that deal with the poverty of our people. If not, what is the point?

Secondly, tourism should not destroy our culture and environment but must do everything to promote our cultural heritage and maintain our environment.

Lastly, the relationship between our tourist guests and locals in destinations must be based on respect, mutual harmony and genuine learning and sharing from each other, rather than a relationship based on exploitation.

Harold Goodwin in his interview made the point that "Knowing what is in the tourism books and journals is not enough and theory needs to be tested against experience." Which businesses, destinations or projects in the West Africa region would you recommend tourism students to visit or check out, to learn about best practice on the ground?

All countries in West Africa are unique in themselves. However, I know and practice in The Gambia and Senegal more than any country in West Africa. So, I will recommend people to contact me for advice if they want to visit The Gambia or Senegal.

We are working on a new project call the Ninkinanka Trail. This is an interesting 5-day excursion where you visit a wide variety of natural and cultural heritage with communities along the river Gambia.

As the founder of the International Centre for Responsible Tourism-West Africa, do you notice a growing interest in tourism sustainability in Africa?

We are trying very hard to get engagement going through the International Centre for Responsible Tourism—West Africa. However, it has not been an easy task. Not much in terms of collaboration between West African countries is realized so far. We will keep on pushing!

Africa has been "fed" for many years with well-meaning development aid offered by governments and charities, a practice often criticized for undermining efforts to foster entrepreneurship and self-sufficiency. To your mind, which kind of actions or support from international organizations would be the most useful to help African communities to benefit from tourism as a source of income and a means of wildlife and nature conservation?

What we need is fair trade and where it is a charity, it should be designed to make us self-reliant economies, not aid or charities that are designed to make us more dependent. Our people must be trained to have the required skills and also make it possible for them to have the markets to sell their products. Tourism, if managed properly, can make this possible.

We should link local production to the tourism market so that tourists will buy, eat and drink what is local.

How important is a destination's sustainability performance nowadays for its competitiveness?

We need to translate who we are, what we want and the better future we need from tourism into simple, attractive messages that make our destinations unique. For example, The Gambia is the people, their cultures and the environment.

We need to be honest with what we present—for example if The Gambia is branded as the "Smiling Coast" the smiles must not be plastic. A real smile means better working conditions for workers; doing something to deal with the negative impacts and so on. Nowadays, with social media, nothing is hidden anymore.

Which trends do you observe in the international sustainable tourism community?

The trend is towards "local", "neighbourhood" and "authentic" experiences. In The Gambia and Africa in general, we do not use this only as a marketing tool. This is who we are for real!

What role does destination branding and marketing play as facilitator (or inhibitor...) of more sustainable tourism?

What we need are simple, honest and unique messages to help tourists and locals play their part to make the destination a better place to live in and a better place to visit.

Looking back at your career so far, which three bits of advice can you share with newcomers to the field of sustainable tourism development in Africa?

Be informed; be persistent and critical where necessary. And be ready to dirty your hands to make a point.

Link to the interview: https://sustainability-leaders.com/adama-bah-interview/

Estonia | Aivar Ruukel is the founder of Soomaa.com and a nature guide in Soomaa National Park, Estonia. He is committed to continuing the culture of building and promoting the traditional dugout canoe in Soomaa National Park. Aivar is a Board Member of the Global Ecotourism Network since 2018, a member of the Estonian Ecotourism Network since 1993, one of the initiators of the national ecotourism quality scheme 'Estonia—the Natural Way' (2000), and the organizer of the first European Ecotourism Conference held in Pärnu in 2010. Aivar is a sustainable tourism auditor at Green Destinations and a board member of EDEN Network AISBL. Aivar has won many accolades, like the Best Promoter of Tourism by the Estonian Association of Travel Agents in 1997, Best of the Year by the Estonian Hotel and Restaurant Association in 2007, Honorary Nature Conservation Badge by the Estonian Ministry of Environment in 2017 and Tourism Developer of Pärnu County in 2020.

Areas of expertise: ecotourism, tourism business, destination development, UNSDG 15 (Forests & Biodiversity)

The following interview with Aivar was first published in June 2019 on Sustainability-Leaders.com.

F. Kaefer, *Sustainability Leadership in Tourism*, Future of Business and Finance, https://doi.org/10.1007/978-3-031-05314-6_7

Aivar, having been involved in ecotourism for many years, do you remember what first got you interested?

The first time I heard about ecotourism was in May 1993 at a seminar organized in my home region, Soomaa, by the Estonian rural development program Kodukant (which translates as "home area"). That time I had just started as a local tourism service provider, offering bed and breakfast in our family house.

The ecotourism idea was introduced to Estonia as a tool for rural economic development, especially suitable for peripheral areas. I was involved as one of the first members in this movement.

Our first mentor became Jan Wigsten, Board Member of The International Ecotourism Society at that time and owner of the consultancy Eco Tour Productions AB and co-owner of the tour operator Nomadic Journeys in Mongolia. We discussed the concepts related to ecotourism, triple bottom line, and the principles of sustainability together with local farmers, park directors, municipality heads, and other stakeholders at many meetings both locally in Soomaa and elsewhere in Estonia. That way we formed a national ecotourism network.

The most important standpoint that I learned from Jan—and which I support—is that ecotourism is NOT a product. It is not a type of nature tourism. It is not a kind of adventure tourism.

Ecotourism is a term that does not say what the activity or content of the trip is, nor where it will take place. Ecotourism is about how tourism is organised and how it impacts the place where the trip happens; how it impacts its natural environment, the people who live there and their way of life.

Ecotourism is a way to see tourism and to do tourism. It is a value system, it is a mindset, of both travellers, tourism businesses, and other stakeholders.

How has your view on the potential of tourism as a facilitator of biodiversity conservation and local economic development changed over the years?

I participated in the process of creating the Estonian Biodiversity Strategy and Action Plan, back in 1999. It was an incredible learning experience. Together with conservation people, after long discussions and consultations, we agreed that the role of ordinary tourism in the conservation of biological diversity tends to be passive, focused on reducing the adverse impacts of tourism activities on biological diversity.

On the contrary, the role of ecotourism is an active one, to influence positively and to directly support the conservation of biodiversity.

Seizing this opportunity and harnessing the power of ecotourism as an instrument of sustainable development requires, above all, the integration of tourism development and nature protection into the economic interests of the local population.

Reflecting on this now, 20 years later, I feel that things are only getting worse. As we know from the latest reports, at least 1 million plant and animal species are at risk of extinction worldwide. In Estonia, for example, ornithologists witness a strong decline in the number of forest birds. However, logging during the spring-summer period is still a common practice today.

To change this, tourism companies have written a public letter[1] to the Minister of the Environment, calling for the establishment of non-disturbance of breeding birdlife.

In such an era of mass extinction, I do believe that tourism businesses have to take a much stronger position and become more ambitious. After all, the loss of biodiversity is not caused by biological factors, but by social and economic ones.

What role do ICT play nowadays in the context of supporting the sustainability of tourism experiences and destinations?

You have to communicate with your (potential) customers if you want to be in the tourism business. 25 years ago, the means of ICT was a landline phone in a kitchen and Motorola one-way pagers in my pocket for receiving messages. Now everyone is using a smartphone in our hyper-connected world, and this is also the main way to communicate with our visitors and to take care of them.

Social media communication works very well for small tourism businesses because it comes naturally. You do not need to learn any tricks, you can be who you are and share honestly your own personal stories—about your passion for your place, and about the people who make your place special.

The transparency of word-of-mouth communication is supporting the sustainability of tourism experiences and destinations—provided that travellers are demanding sustainability.

What motivated you to join the Global Ecotourism Network as a board member?

I believe in lifelong learning. The best way to learn is to connect with people who are smarter than you. GEN[2] is a great network for ecotourism and sustainability. And its board members are undoubted, all of them, smarter than me. I hope that they make me smarter, better, and more generous.

Estonia is mostly known for its progressive digital policies ("e-residency"), not so much for its sustainability credentials. Would you consider it a "green destination"?

Actually, we Estonians love to identify ourselves as a nature-friendly and environmentally conscious nation. Some facts:

- In 1297 the earliest recorded measure for strict nature protection, when Danish King Menved prohibited logging on Northern islands.
- In 1644, farmers fought against the watermill's dam building in Southern Estonia on the holy river of Võhandu.
- In 1910 the designation of the first bird sanctuary, Vaika islands.
- In 1971, Lahemaa, the first national park of the Soviet Union was founded in Estonia.
- In 1995 the Sustainable Development Act was adopted by Parliament.

---

[1] https://www.linkedin.com/pulse/public-letter-minister-environment-support-piece-birdlife-ruukel/

[2] https://www.globalecotourismnetwork.org

- In 2008 the concept of World Cleanup Day was born in Estonia when 50,000 volunteers participated in collecting litter to keep local scenery pristine. Around 16 million people in 113 countries have participated in cleanup days since.

An interesting fact is that already by 1938, the Institute of Nature Preservation and Tourism had been established under the Ministry of Social Affairs in order to address the issues of tourism development, promotion, and conservation of natural resources.

Which destinations within the country would you consider leaders in sustainability?

For tourism destinations that work most systematically with sustainability today, I would name three of our six national parks: Matsalu, Lahemaa, and Soomaa. All of them use the European Charter for Sustainable Tourism in Protected Areas[3] as a practical destination management tool.

Also, there are currently over 20 hotels and other types of accommodations in Estonia with the Green Key label. And since the year 2000, we have a national ecotourism labelling scheme, "Estonia-the Natural Way".

Through your work, you are in close contact with destination managers and marketers. Which trends do you observe in how they approach the development, management, or promotion of their destination?

I can see that DMOs of different regions in Estonia are today shifting the focus of their work from destination marketing towards destination management, pursuing a more integrated approach. This is the right way to go, I believe.

Success of tourism development and management needs stakeholders of the destination to gather together, to discuss and decide what kind of tourism they want—or not—to have, and together take actions to make it become a reality.

Which are the main challenges in Estonia, preventing businesses or destinations from becoming more sustainable?

Lack of vision and understanding where the world is headed.

How do Airbnb Experiences and other apps and portals affect the traditional tour guiding business?

The digital evolution of tourism is here to stay. I do not see a reason why not to adapt our business to the changing online landscape. It means, of course, that businesses have to put some effort into learning about OTAs and apps, how they work, what are the costs and what are the possible benefits. And you need to figure out if you want to collaborate with them.

For the current summer season, we have managed to get a "Book Now" button on our TripAdvisor listings. We are working to have similar cooperation with GetYourGuide and some other sellers. Airbnb Experiences hasn't yet launched in Estonia.

What does it take to become a great tour guide?

---

[3] https://www.europarc.org/library/europarc-events-and-programmes/european-charter-for-sustainable-tourism/

In our small local nature tourism business, tour guides are the most valuable part of our company and the most important part of what we are doing and selling.

A good guide needs a combination of great knowledge and great personality. I think that the second is more crucial. Good guides love their work and to help people. They have a lot of empathy for other people.

Maybe these things can be learned through training, but most of the good guides I know seem to have been born with those traits.

And there is one more thing about great guides: they love their place. This love allows them to share the sense of the place with visitors, and to make their visit so memorable.

Tourism professionals sometimes avoid engaging with "sustainability", since it is not something usually part of their KPIs. Reflecting on your own experience, which advice can you share with tour guides or tour operators in terms of how to deal with sustainability?

In my opinion, this is a language issue. The sustainability lingo is too often misused and overused. So many public sector projects, often EU funded, talk about sustainability, and the outcome seldomly makes a positive contribution to the lives of real people.

At the same time, I know many colleagues, tourism professionals, practitioners, guides, and operators, who practice excellent sustainability, but never use this word.

To be honest, travellers hardly ask about 'sustainability', 'resilience', 'carrying capacity' and the like. Instead, they want authentic experiences. They ask for locally grown, organic food, local guesthouses to stay in, local people to meet, local events to participate in.

Our clients will ask for activities such as canoeing and walking, and of course, they want to explore nature in a way which does not disturb wildlife. But they won't call it 'a low impact' experience. So, while aspects closely linked to sustainability are expected by customers, the word itself is more meaningful as a development and management concept for your business or destination strategy, not when communicating with travellers.

Anything else you'd like to mention?

Part of my work since the 90ies is related to keeping alive the heritage of the local dugout boat of Soomaa. When I started my business, these boats were still in use at some riverside farms. Luckily there were still two boat masters, and together with them, we had a boat-building summer camp during five summers 1996–2000.

Soomaa is the last place in the European Union where the culture of building and using dugout canoes has survived. Some colleagues have said that this is a nice example how tourism businesses can bring new opportunities for preserving cultural traditions, and how it can contribute to knowledge being passed on over the centuries. I do agree.

Now, together with some partners, we have proposed to nominate Soomaa haabjas (dugout canoe) to the UNESCO list of intangible cultural heritage in need of urgent safeguarding. We hope to submit the application to UNESCO in March 2020.

Because the dugout canoe tradition is widespread among many Finno-Ugric indigenous peoples on the territory of Russia, the recognition by UNESCO would also motivate Estonians' kindred peoples to preserve and revitalize their own dugout canoe traditions.

I am looking to establish contacts with indigenous peoples across the world, who have their own ancient dugout canoe traditions and who could be interested in cooperation.

Link to the interview: https://sustainability-leaders.com/aivar-ruukel-interview/

Alexandra Pastollnigg on Sustainable Tourism Standards and Increasing Transparency

Switzerland | Alexandra Pastollnigg is the Founder of Fair Voyage,[1] an online sustainable travel agency for socially and environmentally responsible trips. Making a stand for more ethics in tourism, Alexandra has been serving as a non-profit board member, advisor, speaker, guest lecturer, and awards judge in the field of sustainable tourism. Before venturing into tourism and social entrepreneurship, Alexandra gained over 12 years of experience in the financial services industry. She also authored the book Kilimanjaro Uncovered and cycled 11,000 km+ from Cairo to Cape Town, all the way through Africa. Alexandra holds a Master of Finance from The University of Hong Kong. She has lived in nine countries and speaks five languages. Originally from Austria, Alexandra is now at home near Zurich, Switzerland, and enjoys the Swiss lakes and mountains to regenerate.

---

[1] https://fairvoyage.com

F. Kaefer, *Sustainability Leadership in Tourism*, Future of Business and Finance, https://doi.org/10.1007/978-3-031-05314-6_8

Areas of expertise: entrepreneurship, sustainability, tourism business, UNSDG 8 (Equal & Fair Economic Opportunities)

The following interview with Alexandra was first published in August 2020 on Sustainability-Leaders.com.

Alexandra, with a professional background in finance, what motivated you to shift focus to tourism and sustainable development?

When I climbed Mount Kilimanjaro in 2015, I became aware of the exploitation of porters on the mountain. Many were working in unhealthy conditions and without the right equipment, fair pay, or equal opportunity. I left Tanzania wanting to solve that. I wrote a book, Kilimanjaro Uncovered, sharing lessons I'd learned, such as how to choose a responsible tour operator for a Kili expedition.

Then, while taking a social entrepreneurship program with Philanthropy University through the Haas School of Business at the University of California, Berkeley, I explored solutions for the problems I'd seen. That's when I realized that sustainable tourism could actually be a business.

At the beginning of 2017, I stepped away from my mergers and acquisitions job for four months to bike from Cairo to Cape Town. The adventure opened my eyes to the beauty of Africa and expanded my boundaries of what is possible. That's when I knew I could not go back to finance; it was my calling to help others see the world and do it in a sustainable way.

A lot of travellers book their holiday activities through OTAs (online travel agencies), not least owing to the competitive prices these offer. How can such big corporations commit themselves to make travel more sustainable?

OTAs need to integrate sustainability checks and audits into their rankings. For example, hotel booking platforms should rank accommodation providers with the highest sustainability ratings on top.

OTAs often say that they cannot do this because "there are no good independent sustainability audits yet for tourism." This is not true. The Global Sustainable Tourism Council (GSTC) has set baseline standards for sustainability and accredits certifications for accommodations and tour operators.

When providers have been awarded such a certification, it means that they have passed an independent sustainability audit that has verified their sustainability commitment in line with the GSTC standards. OTAs simply need to start using GSTC accredited certifications as a ranking criterion and eventually make them mandatory throughout their supply chain.

For transportation companies (such as airlines) and the platforms selling them, meaningful carbon offsets should be mandatory—or at least the default option, so that consumers have to actively opt-out rather than find a way to opt-in.

By "meaningful" carbon offsets, I mean fully priced and sustainable solutions that invest in innovative green technologies or community-based forest restoration and conservation projects. The vast majority of currently commercialized carbon offsets are not fully priced or do not invest in sustainable solutions, thus I do not endorse them.

How can we drive consumer demand towards sustainable products, to put pressure on tourism businesses to be ecologically and socially responsible?

Consumers are increasingly interested in choosing sustainable products, but there is no easy way for them to know which products live up to their claims. We need to inform them that the GSTC standards exist and help them see which companies adhere to them.

This is where one single logo or badge would come in handy: consumers could then look for this emblem and feel confident that these companies are vetted.

Of course, being socially and economically conscious—doing things like paying staff living wages, investing in quality equipment, and buying carbon offsets— almost always also increases costs to consumers. And there will likely be many who are unable or unwilling to pay the higher price.

My hope is that more companies will make their operations sustainable and only work with providers in compliance with sustainability standards; that consumers will choose to travel more responsibly, even if it means traveling less frequently; and that consumers who are not willing to pay a fair price will eventually no longer find companies or destinations willing to serve them.

At the GSTC 2019 Global Conference you urged industry leaders to put personal agendas aside and to create an easily recognisable logo to help vacationers around the world relate to sustainability. How has the industry response been so far?

COVID-19 broke out soon after the conference and the tourism industry has been fighting for survival.

It is important to understand that financial and economic viability is a necessity for sustainability. When companies can no longer pay their bills and are forced to lay off staff, environmental and social sustainability becomes secondary.

Many large travel companies have had to file for government bailouts. On the positive side, governments right now have a unique opportunity to attach stringent sustainability criteria to their financial rescue packages. Will they take this opportunity?

Even in these challenging times, we've continued to move forward at Fair Voyage. We are reworking our communication materials to make them resonate more with travellers, to explain the science behind sustainability standards in simple terms.

In the absence of a globally accepted and recognized logo, we have created our own badge (Fair Voyage Curated) for companies that we have approved to use our platform because they are committed to the GSTC standards.

At Fair Voyage you offer responsible tours to conscious travellers. Which are the main challenges in vetting suppliers for their sustainability practices, so you can work with them?

Fair Voyage only works with suppliers that are committed to sustainability and to being independently audited for sustainability practices by GSTC standards. That's the Fair Voyage screening criteria.

Right now, there are simply not enough suppliers that adhere to the standards. Some countries—especially in developing markets—have none. Where some do exist, the companies may just be starting out and have only met the first level of standards.

For cases like this, we have developed a phased approach in collaboration with Travelife,[2] the leading GSTC-accredited certification for tour operators, that enables small companies with proven sustainability commitment to become vetted over time. As their revenues increase, so must their levels of sustainability.

When we need providers for our tours, we first seek out those already certified by GSTC accredited certifications or at least in compliance with local standards for sustainability. If we come up empty but find a provider that seems like a good fit and is interested in becoming certified, we use the phased approach, encouraging them to develop their business in a sustainable way.

Our goal is to make all components of Fair Voyage tours sustainable. We are sharpening our sustainability criteria to go beyond the providers to ensure that every single element of a client's travel experience is handled in accordance with best sustainability practices, from transportation to accommodations and activities.

How has the current pandemic affected businesses working with Fair Voyage? How are you helping them through these testing times?

The pandemic has been especially catastrophic for businesses and local communities in developing markets that rely on tourism.

In Tanzania, for example, tourism generates nearly 30 percent of the export revenues. Fair Voyage has strong ties to East Africa, and so, when the pandemic hit, we partnered with local and international organisations and sustainability leaders to run support groups to assist the local businesses, especially around Mount Kilimanjaro. Through this initiative, we've been able to help facilitate health and training resources, webinars with experts, and making other connections that will hopefully lead to long-term sustainable solutions. You can read more about these efforts in this article.[3]

In your experience with Fair Voyage, which demographics are showing the strongest commitment towards opting for a responsible holiday?

The younger generation seems more committed to sustainability, but as I mentioned earlier, sustainability has a price, and the younger generation often cannot afford it. The travellers that Fair Voyage sees booking trips tend to be 35 years and older, well educated, and affluent.

Awareness of sustainable travel seems to be highest on the west and east coasts of North America, along with Australia, New Zealand, and Western to Northern Europe. These are some of the wealthiest parts of the world, demonstrating the link between education, financial security, and sustainability.

Which aspects of working as a sustainable tourism entrepreneur do you find the most rewarding?

There are three aspects I find rewarding: being in sustainability, being an entrepreneur, and working in tourism.

First of all, it's incredibly rewarding to know that I'm making a positive impact on sustainability with my daily work. Every morning when I'm starting my work, I

---

[2] https://www.travelife.info/index_new.php?menu=home&lang=en

[3] https://amanda-mckee.medium.com/saving-lives-at-kilimanjaro-6bfac642be68

know my mission, my 'why'. And I enjoy the freedom to be 100 percent consistent with my personal values, with what my heart is telling me.

I've also found that being an entrepreneur is perhaps the most effective tool for accelerated personal growth. Even just these past three months, I've learned and grown more than I ever thought possible, and I continue to have major "aha" moments every week. To me, this is hugely exciting.

Speaking specifically about tourism, I believe in the potential of travel as a force for good—to foster intercultural understanding and personal growth. In my mind, travel may be the single most powerful tool to create and maintain lasting world peace. I want to live in a peaceful world.

Knowing that I'm giving my best every day and working to create such a world is the biggest reward for a sustainable tourism entrepreneur.

Anything else you'd like to mention?

Within 20 years, I want to be living in a sustainable and peaceful world. I strongly believe that this new world will be created and led by impact entrepreneurs. We need more impact entrepreneurs, urgently.

I didn't grow up in an entrepreneurial culture, didn't have work experience in tourism before starting Fair Voyage, and don't have any formal education in sustainability. Yet, by following my heart, taking action, and learning by doing, I've become a sustainable tourism entrepreneur and recognized as a sustainability leader.

I want others to know that they can do this, too. They can become sustainability leaders, and together we will build a healthy planet.

Link to the interview: https://sustainability-leaders.com/alexandra-pastollnigg-interview/

Ally Dragozet on Marine Conservation, Plastic Pollution and Tourism Solutions

The Netherlands | Aleksandra (Ally) Dragozet is a marine biologist, social entrepreneur, sustainable tourism expert with 10+ years of experience in the tourism industry as both a professional and an operator. Ally holds a Master of Aquatic Biology/Limnology from the University of Amsterdam and an undergraduate degree in Environmental Biology from the University of Toronto and the National University of Singapore. Ally founded the sustainable tourism consultancy Sea Going Green[1] in 2017 to alleviate the negative impacts of the tourism industry on the marine environment. By measuring the environmental impact (CO_2 and waste flows) of marine tourism operators, coastal businesses and island destinations, Ally has gained deep insights into the changes that need to be made in the tourism industry to enable sustainable tourism in marine-dependent areas, especially SIDS. As a marine biologist, Ally understands that the profitability of coastal and island destinations is linked to the health of the local biodiversity.

---

[1] https://www.seagoinggreen.org

F. Kaefer, *Sustainability Leadership in Tourism*, Future of Business and Finance, https://doi.org/10.1007/978-3-031-05314-6_9

Areas of expertise: conservation, entrepreneurship, responsible tourism, sustainability, tourism business, UNSDG 13 (Fight Climate Change), UNSDG 14 (Oceans & Marine Life)

The following interview with Ally was first published in August 2020 on Sustainability-Leaders.com.

Ally, you advise businesses and destinations on sustainable marine tourism. Why do you care so much about the topic—who or what got you interested?

I can honestly say that I have always felt a strong connection to the marine environment since I was a child. Even though I grew up in Toronto, Canada, being half Croatian meant that I was able to spend my summers with family in a tiny beach town. Interacting with marine life and beautiful coastal landscapes at such a young age inspired me to study to become a marine biologist. At the same time, as tourism in Croatia grew over the years, I saw changes happening to the state of the environment and the direct social impacts of overtourism. It was at this point that I realised that I also wanted to find a way to use my interest in marine biology to make the tourism industry more sustainable, inclusive, and resilient.

I also had many other personal ties to the tourism industry as my parents had careers in the airline industry, so travelling really became second nature to me. I was able to gain hands-on experience in the tourism industry throughout my studies, which I've been able to apply since starting Sea Going Green.

This leads to my story of how I founded Sea Going Green. Following my studies at the University of Amsterdam, I realized that I wanted to offer something different from just consulting services and/or ecotours. I really wanted to help create a formative change in the tourism industry as a whole.

Having traveled to many parts of the world and seeing plastic pollution in what would otherwise be beautiful blue waters, definitely was a wakeup call for action. Then the idea just clicked for me that I could create a company to combine both of my interests and passions—tourism, and the marine environment, hence Sea Going Green was born.

With the current pandemic on everyone's mind—how is it impacting tourism sustainability—and marine tourism in particular?

The current pandemic, which has now touched almost every sector, is impacting tourism's sustainability prospects in a huge way.

While there was a huge push for pivoting tourism towards more sustainable practices pre-pandemic from a business and accountability perspective, now environmental or social sustainability is no longer the industry's biggest priority, as everyone is trying to stay afloat and ensuring their economic survival.

You can see it simply in the newfound plastic pollution issues that have been born out of COVID-19 times. Plastic pollution is, unfortunately, increasing due to consumers and businesses adhering to safety restrictions and precautions, which has led to the rise in the use of single-use items such as masks and gloves.

Tourism companies are trying to reopen in a safe way, but this also, unfortunately, means using more disposable materials such as single-use plastics to meet customers' safety expectations.

The increase in tourism operators using single-use plastics means that in many cases marine tourism destinations are seeing more litter on beaches from disposable masks and other items being left behind. Especially in the U.S. and beyond, many plastic bags are being retracted in light of the virus.

Unfortunately, I have to say that the tourism industry seems to have taken a few steps back in terms of sustainability. On a brighter side, a big takeaway from the pandemic is that without overtourism and unsustainable practices, nature has come back to areas that are no longer under pressure and in terms of pollution, water and air quality are improving. This should be a big sign showing the industry how important sustainability will be as we move forward, COVID having provided somewhat of a blank slate for how tourism can be managed in a sustainable way.

With the climate crisis already lurking around the corner, which topics would you consider the most urgent for us to pay attention to, regarding its potential impact on the health of our oceans?

I do believe the climate crisis is one of the major things we need to focus on in terms of the largest scale of negative impacts putting pressure on our oceans. Biodiversity will become increasingly fragile in the face of global warming, which will have huge consequences on the entire ecosystem and food chains.

This can of course only be slowed and stopped by formative changes by national governments and more accountability for huge CO_2 emitters like airlines, the fast fashion industry, among many others.

Another issue that I also think is in urgent need of tackling is the enforcement of marine protected areas and facilitating the proper management of the oceans.

Enforcing marine protected areas will make or break the way oceans are managed. If done well, issues like overfishing and biodiversity damage can be effectively combated.

Also, putting in place measures to build capacity for tourism destination management is incredibly important, because if business continues as usual and destinations continue to be mismanaged as we saw in Maya Bay (Thailand) and Boracay (The Philippines), the health of our oceans will be more at risk—resulting in bigger and more long-term consequences.

Congratulations for having been featured as one of the 2019 Forbes 30 Under 30 Social Entrepreneurs, for your work to make the tourism industry more sustainable! How has this recognition impacted you and your work?

Thank you! I am really proud to be represented in the category of a social entrepreneur since making an impact is really what my work is about.

The recognition that I have received since has been great for my personal brand and our work at Sea Going Green; a lot of business prospects and speaking engagements have opened up since. Being a part of the Forbes network has also been an amazing way to be able to connect with other entrepreneurs and to share and learn from their experiences as well. Also, personally, it was always an achievement I hoped to reach.

Which projects or initiatives aimed at making marine tourism more sustainable are you currently involved in?

Due to the pandemic, some of our projects that were in the pipeline had to be put on hold, but we, as a result, have been able to get involved with creating a very inspiring grassroots movement workshop series along with WWF-NL on the island of Bonaire. The goal of this series is to stimulate sustainable development among the local community, by helping them form their own initiative while promoting sustainable marine tourism initiatives in Bonaire.

We are also involved in a cross-industry project which is focused on creating a 'smart assessment sustainable tourism destination tool' for destination management, which we hope to test with marine tourism companies as soon as possible.

The cruise industry being one of the most polluting travel activities—how can it be more environmentally sustainable and become part of the solution, rather than adding to our environmental problems?

The cruise industry is definitely one of the biggest polluters in the travel sector, but it also is one of the fastest-growing tourism products. This means that it also has the possibility to become a great example of how a tourism activity can adapt towards incorporating more environmentally sustainable practices.

We need a sustainability pioneer and champion within the cruise industry to show how environmental sustainability can ensure a more long-term profitable business that will appeal to future generations while protecting the marine environment that the industry depends on. The solutions are already out there for the cruise industry, they just need to be adopted by the cruise operators.

Do you see progress made in global efforts to reduce the littering of our oceans?

Absolutely. I think there are amazing global efforts to reduce the littering of our oceans, especially in terms of awareness and research on where the largest leakages are. Yet, I do believe that the majority of progress is happening at the grassroots level. Grassroots organisations really seem to be taking the lead in doing ground-work by engaging the local community and relevant stakeholders to reduce littering in the oceans—from cleanups to promoting reusable items.

In your experience, which regions or destinations around the world are the most innovative or pro-active in how they support a healthy ocean and marine life?

In my experience, I don't think any region or destination is perfect, but of course, there are some that are more ahead of others in terms of destination management. With that said, even for those that aren't quite there yet, I believe that many are moving in the right direction.

I would like to point to Hawaii as a great example of being the first U.S. state to ban coral reef harming sunscreens, as well as putting a plastic bag ban in place.

Costa Rica is also a great example that we use frequently as a case study. The country plans on banning all single-use plastics and becoming carbon neutral by 2021. This is on top of banning all captivity and activities with marine mammals and protecting its Savegre River by making it a UNESCO Biosphere Reserve, not to mention designating 25% of the territory as protected by the National System of Conservation Areas.

We've heard from other interviewees that social entrepreneurship has helped to stop or reduce illegal wildlife trade. Do you know examples where this has been achieved in a marine environment?

What first comes to mind as an example is the shark fin trade, which has now been banned in many countries as a result of grassroots pressure put on governments to take action in the form of petitioning and later, policy-making.

Illegal marine life trade has been reduced and stopped in certain areas thanks to social entrepreneurship. There are many non-profit organizations and nation-wide campaigns that have encouraged locals, expats, and tourists, as individuals, to report if they see illegal trading taking place at markets, which has helped close up shop for some traffickers. These efforts, along with government crackdowns, have definitely helped put a stop to these practices around the world, though there is still a lot of work to do.

Another important, yet linked, topic that has been tackled by social entrepreneurship is targeting marine mammal captivity, including using them for shows and other tourist activities. I believe many enterprises have started a really powerful movement and have had some great success with it, especially in collaboration with booking companies like TripAdvisor, which now no longer sells tickets to marine life entertainment shows.

Tourists are also becoming more educated about their impact on local marine life. For example, if your hotel is on a beach that is a turtle nesting area, it will either be shut down to the public (like in Costa Rica) or you will be asked to close your blinds in your hotel room to not disturb hatching turtles (Florida). Communicating with guests about their impact on wildlife can make a huge difference and set new norms for interactions, which is better for species and more sustainable for the industry itself.

Reflecting on the many projects you have been involved with so far, which parts of advocating for sustainability do you find the most challenging?

I think the part that I find the most challenging is advocating for the long-term commitment and planning that sustainable development needs. This comes from knowing that the tourism industry is very 'short-term minded' so if you start advocating for long term sustainability commitments, they don't see the immediate benefits from it, and it makes the case for sustainability a harder sell.

Your three bits of advice to social entrepreneurs keen to contribute to the sustainable development goals, especially those linked to marine tourism?

If I had to give three bits of advice to social entrepreneurs to contribute to the SDG goals linked to marine tourism, I would urge them to:

- Really take the time to understand the impacts of tourism on the marine environment from both the private and public sector perspective.
- Pick which aspects of marine sustainability (economic, community, environment) you want to focus on early because the world of marine sustainability is vast.
- Make a great network for yourself—there is no need to reinvent the wheel, instead focus on making great partnerships to increase your impact. The more people you have on board, the bigger your chances of success will be.

Anything else you'd like to mention?

My team and I spent a lot of time during the beginning of the quarantine researching how we could support the tourism industry to open up again in a safe and profitable way, while still keeping sustainability at the core of operations. As a result, we were able to launch our custom online guide, which is full of tips and strategies for businesses trying to get back on track.

If anyone has any questions about the guide or our services, please do reach out! We can only build back better if we work together.

Link to the interview: https://sustainability-leaders.com/ally-dragozet-interview/

Morocco | Amine Ahlafi is an Urban Architect with a master's degree in Cities, Spaces and Societies from François Rabelais University, France. He is an expert in tourism and sustainable development, also a coach and auditor in ecolabelling and sustainability. After founding '2D dama' in 2010, his company consults international and national organizations (public, private and NGO) regarding sustainability and tourism. He is part of the Foundation for Environmental Education, as the Executive Board member in 2012 and Vice President in 2016. Amine has accompanied the Moroccan Ministry of Tourism on the sixth and the seventh Moroccan Sustainable Tourism Awards' editions, the Moroccan Charter on Sustainable Tourism, and the African Charter on Sustainable and Responsible Tourism signed in COP 22 at Marrakech, etc. He is currently involved in sustainable tourism regional development supported by the German Cooperation Society and Swiss contact.

Areas of expertise: responsible tourism, sustainable development, destination, Africa

The following interview with Amine was first published in October 2016 on Sustainability-Leaders.com.

F. Kaefer, *Sustainability Leadership in Tourism*, Future of Business and Finance,
https://doi.org/10.1007/978-3-031-05314-6_10

Amine, when did you first learn about sustainable architecture and urban design?

One of the workshops I attended during my studies in architecture at the National School of Architecture of Rabat was entitled "Sens Space," inspired mainly by the design of Sir Frank Lloyd Wright.

It was my first contact with a new approach integrating nature and the man at the heart of the architectural project, while maximizing the functionality and performance of the building. This has changed my way of thinking about architecture.

At that time the concept of sustainability was not used to qualify architecture.

A few years later, I had the chance to work for leading organizations in environmental protection and sustainability areas: The Foundation Mohammed VI for the Protection of the Environment and the international Foundation for Environmental Education. My daily tasks have thus influenced my approaches to architecture, urban design and development.

In 2010, you founded 2d DAMA, an architectural and urban development consultancy focused on the use and promotion of sustainable practices. Can you tell us a bit about the agency?

The agency is based on the belief that the act of planning, design, development and construction can only succeed under the banner of sustainable development.

It reflects my vision of promoting sustainable development in architecture with the advice and support of several stakeholders in Morocco and abroad. Its achievements are the result of the commitment of a dedicated and motivated multidisciplinary team who shares the same values.

2d DAMA also works with sustainable tourism development projects. How did you first get involved in the tourism sector?

One of our favorite areas is sustainable tourism. The inspiration comes from a combination of my previous jobs for the Ministry of Tourism and for the Mohammed VI Foundation for the Protection of the Environment. It was reinforced by my missions for 5 years as national coordinator of the international program Green Key for tourism accommodations, owned by the international Foundation for Environmental Education (FEE).

We have developed proven expertise, and we continue providing technical assistance and studies for the Ministry of Tourism and some tourism operators.

What major challenges does Morocco face in regard to sustainable development, especially as it relates to the tourism industry?

I do believe the major challenge is environmental education and education about sustainable development.

Other measures are to be implemented as well: the integration of sustainability within the life cycle of the products, Corporate Social Responsibility for the tourism companies.

What changes have you noticed in recent years in Morocco's approach to tourism development in regard to sustainability?

The Strategic Vision Tourism 2020 clearly established the objective and the actions to be implemented in order to position Morocco in the list of top sustainable destinations.

A significant number of programs have already been implemented:

- Morocco is co-lead of the international program 10 YFP ST.
- Morocco has developed a National Charter for Sustainable Tourism and has initiated the development of an African Charter for Sustainable Tourism under the auspices of UNWTO.
- Morocco holds the national trophy rewarding the best sustainability initiatives.
- Morocco has institutionalized a national day dedicated to sustainable tourism.
- The National Confederation of Tourism has a unit dedicated to sustainable tourism, and various tourism operators have hired a sustainable development manager.

What role should architects and developers play in sustainable tourism planning and development, at a destination level?
They must integrate sustainability into all stages of the project from the planning. They should start by a value chain analysis before implementing the operational actions.

They have a role to play to convince the stakeholders by showing the benefits of sustainable tourism.

Do you think there is a successful synergy between destination managers, developers and architects in Morocco? If not, what can be done to strengthen cooperation between these stakeholders?
Not yet. Fully achieving the benefits of sustainability practices cannot be done in the short term and sometimes requires investment. These are perceived by tourism operators and developers as real challenges.

Awareness and incentive measures could contribute to strengthen cooperation between the stakeholders.

You are also a consultant at the Mohammed VI Foundation for the Protection of the Environment. Can you tell us a bit about the organization and your role there?
The Foundation Mohammed VI for the Protection of the Environment is an NGO chaired by HRH Princess Lalla Hasnaa.

The Foundation has placed education and awareness-raising at the heart of its mission, thus contributing to the objectives set by the summits in Rio in 1992 and 2012 and in Johannesburg in 2002 in terms of education for sustainable development, to which Morocco has subscribed.

Several programs and projects were initiated, including: Clean Beaches, environmental education and awareness, Flowering Cities, Voluntary Carbon offset, and Qualit'Air. These successfully implemented programs are based on environmental education, coastal protection, responsible tourism, restoration of historic gardens and the preservation and development of palm groves and oases.

The Foundation is the Moroccan representative of the international Foundation for Environmental Education.

I am proud to work as an advisor for the foundation and to contribute to the implementation of several programs and projects. I used to be the national coordinator for the Blue Flag and Green Key programs, and now I am chairing the technical committee for developing the first Environmental Education Center in the country.

Of the projects and initiatives, you have been a part of over the last 6 years, which are you most proud of?

I'm most proud of my team. Thanks to them, we can face various challenges and implement interesting projects and programs:

- Technical assistance and coordination of the Environmental Education Center project
- Design of a primary school with only raw earth material
- Technical assistance and design of the National Sustainable Tourism Charter and the African Charter
- Technical assistance and design of the national label for handcraft in Morocco
- Planning and design of the Green Master Plan for Marrakesh City
- Technical assistance and design of the national Sustainable Building label

I am also very proud to be nominated Vice President of FEE.

Link to the interview: https://sustainability-leaders.com/interview-amine-ahlafi/

Angelique Tonnaer Kırkıl on Hotels and Tourism Sustainability in Turkey

Turkey | Angelique has been an Auditor and Advisor for Green Destinations since 2018. She is trained in sustainable agriculture, like organic farming, apiculture, equestrian management, and mosquito prevention and actively practises them. Her love for nature and its preservation motivates her to fight towards limiting the (mostly government-approved) use of pesticides and chemicals in the hotel sector and municipalities. These entities often 'confuse' the use of pesticides with 'good service' to inhabitants or tourists who are unaware of the damage they can cause to their health and biodiversity. Currently, she is working on raising awareness on alternatives to pesticides.

Areas of expertise: hospitality, tourism business, sustainability, destinations, UNSDG 12 (Resource Efficiency)

The following interview with Angelique was first published in June 2017 on Sustainability-Leaders.com.

Angelique, when did you discover your passion for sustainability and travel?

Actually, I discovered it very early. Since I was young, I was close with nature. My dad used to take me to the wetlands in our village Thorn in the Netherlands, to explain me about the special nature there. And I was already a little activist trying to

© The Author(s), under exclusive license to Springer Nature Switzerland AG 2022   91
F. Kaefer, *Sustainability Leadership in Tourism*, Future of Business and Finance,
https://doi.org/10.1007/978-3-031-05314-6_11

protect insects and animals, where possible, and organizing water clean ups with my friends.

Via the orchestra in our village that I was member of, I discovered my passion for traveling. As we played international music, we had a lot of concerts in and outside of the Netherlands.

So, there it started. Then, during my studies of International Relations I became more interested in the human/social side of sustainability, while studying about world politics, development of countries, human rights, etc.

After some years working in and around EU [European Union] institutions, I coordinated the establishment of the Dutch National Youth Council to promote youth participation in all areas that concern them. There I learned about how to really involve (young) people.

The only way one gets optimal involvement of people is not to talk about or decide for them, but with them.

In 2005 I coordinated the project Cool2Know with the goal to connect local Turkish youth with tourists in mass tourism hotels. The idea: it is 'Cool2Know' each other and the local life. A kind of community involvement, literally.

Recently with TRIADA,[1] my company, we were part of the INTOUR[2] project and coordinated the SUSTOUR project, aimed at introducing Travelife to accommodations, tour operators, travel agencies and educational institutes in Turkey.

Travelife for Hotels & Accommodations is a global sustainability management scheme for the tourism industry run by ABTA (the UK's largest travel association).

Travelife for Tour Operators is run by ECEAT a not-for-profit organisation based in the Netherlands, helping tour operators to improve their social, economic and environmental impacts.

In cooperation with TUI, we have also launched sustainability criteria and a management system for excursion and activities providers.

The biggest trigger for my sustainability passion was the birth of our son in 2010. This—and I am sure that other parents agree—awakes a kind of super awareness about the world and how we leave it for our children.

Since 2011 I have been an auditor for Travelife, auditing hotels and other accommodations against the Travelife criteria mostly in Turkey, the Caribbean, and in the coming months Africa.

How is sustainability integrated in the services of TRIADA, your consulting company in Antalya?

First of all, we try to 'be the change you want to see' by working ethically and with respect for people, environment and community. Almost all projects and activities that we have designed and carried out during our 10 years of existence have had an empowering and idealistic aspect. The focus is always on improving an

---

[1] http://www.triadaconsultancy.eu

[2] https://www.travelife.info/intour/

existing situation, be it in community development, youth participation or international cooperation.

Sometimes our projects are directly related to sustainability in tourism, but we also cover other areas. For instance, we helped promote sustainability of local banana production by exchanging know-how and best practices between the Canary Islands and Alanya. Very recently we designed and coordinated a project to empower social entrepreneurship skills of young people.

Since 2010 we got involved more closely with sustainability linked to tourism, by helping develop Corporate Social Responsibility practices among Turkish accommodations, educational institutes and other tourism companies. First as consortium partners in the INTOUR project (focused on eco innovation) and later as owners of the SUSTOUR project (focused on lifelong learning). One outcome of the latter was that we became the official representative in Turkey for Travelife for Tour Operators, Travel Agencies & Educational Institutes.

In June 2012, together with the English and Dutch Tour Operator Associations ABTA and ANVR, the Ministries responsible for tourism and the tourism NGOs ECEAT and the Travel Foundation, we interviewed 50 stakeholders in the three main touristic areas of Turkey. The purpose was to identify the main challenges for sustainable tourism development in those areas, and to develop common projects to tackle them.

What motivates you to work on the sustainable development of tourism in Turkey?

Let me start with a quote from an astronaut: 'We are all astronauts on spaceship earth...' That is my starting point. We are all inhabitants of this world, which is just a little dot in space that should be treated with care.

I see myself as a global citizen with local roots in Holland, living in Turkey. I feel responsible for making a change where I can, which is mostly here in the Turkish province of Antalya. I contribute through my professional activities, but also through initiatives such as organizing a forest clean up with the local school, for example.

As professional dedicated to the sustainable development of tourism, I see my main role in helping coordinate the different national and international actors and efforts in this field, because working together is essential for the projects to be effective.

Tourism is one of Turkey's main economic sectors. In my opinion, sustainable development in tourism is strongly needed in Turkey, because it is the only way to keep the destination attractive in the long-term.

For a long time, sun, sand, beach and all-inclusive luxury hotels were the main attractions and focus of tourism in Turkey. Those are well developed and well known for their good price/quality and service mindedness.

But Turkey has so much more to offer. There is now a shift toward diversification of tourism—for example development of cultural trekking routes and opening up the national reserves of Antalya province for the public (between 2016-2019). It is very important to do this in a sustainable way, as this kind of tourism takes place in more vulnerable places with a limited carrying capacity.

As member of a committee which assists with this process, my role is to look out for good practice examples abroad, so we don't have to reinvent the wheel!

Which are the main tourism sustainability challenges that Turkey faces today?

Turkey's main challenges in (mass) tourism, in my view, are:

'Betonification': There is no need for more large-scale hotels serving mass tourism. There are enough of those already, and building more would be a threat for the scarce "empty" spaces left, especially along the coast.

However, due to the current structure in government and the huge amounts of money that are earned by project developers, it is a challenge to restrict or stop new buildings.

Fortunately, people's awareness and activism are also increasing. For example, the plans to build a hotel right on top of the ancient Phassilis ruins near Kemer have been stopped, at least for the time being. Hopefully, original places such as Olympos and Çıralı will succeed to remain without concrete buildings forever.

Lack of sustainability awareness: While in tourism awareness about sustainability issues is higher than average because tourism depends on it to flourish, there is still a strong need to move from short-term, money-oriented thinking to a view that only tourism that follows a long-term vision and is developed sustainably can survive. 'Better places to visit, better places to live'. . . is a suitable motto for this.

Lack of control of excursion & activity providers: It is a challenge for the Turkish government to control all the activities and excursion providers that are active in tourism. Especially those working in vulnerable areas should be better monitored.

Need for more involvement and better cooperation of the main stakeholders in tourism: It would be very good if the main stakeholders in tourism would be more involved in national and international cooperation for sustainable tourism.

Geopolitical tensions: The current tensions in the region, but also in Europe, have a strong impact on the international image of Turkey, which has recently suffered a sharp decline in tourist arrivals. This mainly hurts the 'man on the street'.

The positive side of this is that more and more Turkish people now can stay at high-end hotels they would have hardly had a chance to experience before. . .

Your key lessons learned from promoting innovation and sustainability management in Turkey?

First, that it is very important to start with the base—education—so that the future workers are sufficiently equipped to implement the necessary changes in tourism companies!

Second: hotels are much easier to convince because they can earn a lot in a short term by implementing water and energy saving measures. Other tourism companies, such as travel agencies and excursion providers, are harder to convince because in their eyes, sustainability measures take a lot of time and effort with too little financial return, at least in the short term.

Third: During times of economic crisis, tourism companies are more reluctant to work on sustainability. We tried to convince them that this was exactly the time for it, because as they had little work they could devote more time, but they were hard to convince.

Fourth: It is essential to have global sustainability management schemes such as Travelife that help to facilitate sustainability in tourism.

Fifth: I call upon all the big tour operators to use their power to convince their supply chains to adopt sustainability!

Sixth: It is important to prevent greenwashing. The main tools for this are control via standardization and certification systems. However, for the public to understand properly there should be as few of those as possible...

Lastly, working on sustainability in tourism can sometimes be tiring because you start to notice the sustainability side of everything. Being so aware of things like rubbish in the forests, new development projects, motorways, lack of attention to sustainability in the general news, can sometimes be frustrating. But I have to pick my battles.

Where do you see the main priorities for the hospitality industry in Turkey in terms of sustainability? And the challenges?

As a Travelife auditor for hotels and other accommodations, I can see that local and national governments are doing a lot to make mass tourism more environmentally friendly, and push for better social welfare. There are now incentives for hotels to reach the national certification 'Green Star'. There are also strict laws regarding, for example, recycling of (hazardous) waste and treatment of wastewater. In Antalya, hotels are fined if their wastewater is not clean enough, for instance.

Also, there is more control now of how hotels are to treat their employees, with regard to working times, etc.

Many of the larger 5-star hotels that I have worked with have high sustainability standards. This includes not only the environmental part, but also community welfare projects, such as assisting local schools and NGOs.

An interesting cultural footnote here is that in Turkish culture it is not well regarded to show off your positive sustainability actions as a hotel, whereas on behalf of Travelife I tend to encourage hotels to promote their good deeds to a wider public. We usually find a middle way.

Challenges and priorities:

Employees in tourism should be better represented and trained to be able to defend their rights and interests. Lack of a union for tourism employees and the current economic crisis make workers vulnerable and dependent on the goodwill of their bosses.

Inherent to the all-inclusive system is a lot of food waste, which is a big loss in all senses. It is important to find creative ways to reduce it, but the hotels fear complaints from their guests. A friendly way to communicate the message to guests would be a message such as: 'Take all you can, but eat all you take'.

Smaller boutique hotels should also be motivated to join international certification systems as to help them to become more sustainable. Travelife has recently developed a checklist for small hotels, in addition to its checklist for larger accommodations. I hope that it will help small hotels to be included in this growing sector of sustainable accommodations.

Especially for areas with a concentration of mass tourism, hotels that have big gardens (such as Antalya Lara) should be able to use wastewater for their gardens that has been cleaned by the water authority.

Which criteria must hotels fulfil to be certified by Travelife as a sustainable tourism business?

Travelife for Hotels and Accommodations is a certification scheme that helps hotels and accommodations manage their social and environmental impacts and communicate their achievements to customers. Travelife was founded in 2007, and is run by ABTA, the travel association, with a head office team of 8. Travelife works with circa 50 independent freelance auditors and has over 1500 hotel members globally.

Travelife Audits occur every two years. They only take place when a hotel is open. Hotels are audited against 163-point criteria that include environmental management, business policies, labour and human rights and community integration.

It is not easy to achieve Travelife Gold certification. Hotels rarely achieve gold on their first audit, and will usually have to work on improvements. There has been only one occasion for me when this has taken place. But to not reach Gold on your audit is not a failure. Sustainability is not easy.

As an auditor I ask questions, compile evidence, take pictures. If I find non-compliance, the hotel is given time to fix it. Sometimes I need photos showing that issues have been fixed. Other times I need to see a strategy plan.

Some of the questions I will ask during an audit:

- Energy and water consumption records, volume per guest night and targets (if hotels do not comply with flow rates specified in the Travelife criteria and this is their first-year audit). Targets must be set by the hotel and an action plan needs to be put into place and implemented.
- How often backwashing of the swimming pools takes place.
- Whether the water comes from a borehole. If yes, then I want to see the official borehole license.
- I check with workers how they are trained, how their extra hours are paid.
- Whether they store local and/or organic, bulk products, in recyclable packaging.
- What refrigerant they are using.
- How compliance and disciplinary procedures are communicated, and if staff may join a union.

Your 3 bits of advice for hotel managers keen to implement sustainability practices but not sure how to get their property owners/investors on board?

Focus your message on the benefits:

- Sustainability and Corporate Social Responsibility = quality assurance!
- Saving water and energy will automatically save money.
- If you as hotel commit yourself to sustainability, automatically you will help your destination to become a better place to visit and a better place to live

in. Sustainability is the key to being able to enjoy your destination in 20 years as much as you do now.

Link to the interview: https://sustainability-leaders.com/interview-angelique-tonnaer-kirkil/

Papua New Guinea | Dr. Anna Spenceley is an independent consultant who works on sustainable tourism issues. She has been analysing the impact of COVID-19 on the tourism sector in protected areas for the EU,[1] UNEP, UNESCO and Luc Hoffmann Institute, and compiled a blog post of over 1200 materials on COVID-19 and sustainable tourism[2] for public use. She is a signatory to both the Future of Tourism[3] and Tourism Declares a Climate Emergency.[4] Anna is Chair of the IUCN World Commission on Protected Areas (WCPA) Tourism and Protected Areas Specialist Group (TAPAS Group),[5] is on the Board of the Global Sustainable

---

[1] https://op.europa.eu/en/publication-detail/-/publication/bda7e04d-7c9c-11eb-9ac9-01aa75ed71a1/language-en/format-PDF/source-194167807

[2] https://annaspenceley.wordpress.com/2020/04/02/covid-19-and-sustainable-tourism/

[3] https://www.futureoftourism.org

[4] https://www.tourismdeclares.com

[5] https://www.iucn.org/commissions/world-commission-protected-areas/our-work/tourism-tapas

© The Author(s), under exclusive license to Springer Nature Switzerland AG 2022
F. Kaefer, *Sustainability Leadership in Tourism*, Future of Business and Finance,
https://doi.org/10.1007/978-3-031-05314-6_12

Tourism Council[6] and sits on the Independent Advisory Panel of Travalyst.[7] She is editor of the new Handbook for Sustainable Tourism Practitioners[8] from Edward Elgar and co-author of Private Sector Tourism in Conservation Areas in Africa[9] from CABI.

Areas of expertise: community-based tourism, destination sustainability, responsible travel, tourism business, Africa, UNSDG 1 (Eradicate Poverty), UNSDG 8 (Equal & Fair Economic Opportunities)

The following interview with Anna was first published in March 2020 on Sustainability-Leaders.com.

Anna, do you remember what inspired you to dedicate your career towards sustainable tourism and the conservation of protected areas?

I've always been fascinated by nature and wildlife, and have been particularly inspired as a child by documentaries from Sir David Attenborough and the BBC that showcase the beauty and diversity of our planet. Later I had the good fortune to get a place on a Tropical Biology Association[10] course in Uganda's Kibale forest, and then take a backpacking tour around some of Kenya and Tanzania's iconic protected areas—and that really sealed it. I knew then that I wanted to dedicate myself to working on using tourism as a way of generating benefits for local people while financing conservation of wildlife and protected areas.

A masters project in Zimbabwe's Hwange National Park, a PhD in Kruger National Park and its private game reserves, and a couple of postdocs on top really cemented this. This academic grounding allowed me to move into more of a practitioner role, and I now have incredible opportunities to work with clients and colleagues on a variety of sustainable tourism initiatives across the world.

Now I feel really privileged that I have the opportunity to work on sustainable tourism issues at a local and international level. For example, in my voluntary capacity, I have the honour of being the Chair of the IUCN World Commission on Protected Areas' (WCPA) Tourism and Protected Areas Specialist Group (TAPAS Group).[11] I sit on the Board of the Global Sustainable Tourism Council (GSTC), and also am part of the Travalyst Independent Advisory Group.[12]

Africa has been the focus of much of your work. Which market trends do you observe right now which might influence responsible tourism practice and the sustainability performance of destinations there?

---

[6] https://www.gstcouncil.org

[7] https://travalyst.org

[8] https://www.e-elgar.com/shop/gbp/handbook-for-sustainable-tourism-practitioners-9781839100888.html

[9] https://www.cabi.org/bookshop/book/9781786393555/

[10] http://www.tropical-biology.org

[11] https://www.iucn.org/commissions/world-commission-protected-areas/our-work/tourism-tapas

[12] https://globalnews.booking.com/travalyst-coalition-announces-development-of-new-sustainability-frameworks-to-help-travelers-find-sustainable-travel-and-tourism-options/

Last year Sue Snyman and I published a book called Private sector tourism in conservation areas in Africa[13] which included 32 case studies of accommodation operations. Collaborating with this suite of enterprises—some of which we had worked with for many years—this publication allowed us to look at the business models, environmental, social and economic benefits generated and changes over time.

For example, we found that 26 of the lodges collectively employed over 1600 people, of whom 77% were from local communities, and the lodges were spending USD 16.3 million each year on local wages.

The trend towards employing local people, and buying goods and services from them to support tourism, coupled with supporting rural entrepreneurs and small enterprises can really bolster rural economies.

The World Travel and Tourism Council recently released a study on The Economic Impact of Global Wildlife Tourism,[14] and highlighted that wildlife tourism contributes USD 343.6 billion (or the equivalent of the entire GDP of South Africa), and 21.8 million jobs (6.8% of all travel and tourism jobs). The impacts of this sector in Africa and worldwide are really substantial.

There is an increasing trend of protected areas seeking partnerships with private sector operators to develop and operate sustainable tourism services, such as accommodation and restaurants.

With support from the Convention on Biological Diversity, myself and co-authors Sue Snyman and Paul Eagles produced the 'Guidelines for tourism partnerships and concessions for protected areas' report[15] to assist them. These guidelines provide guidance to support protected areas that want to forge commercial relationships with tourism investors and operators, in a way that supports their conservation objectives while benefiting local people. Their development and testing engaged professionals from 10 African countries and protected areas to ensure that they were practical and relevant.

These and a range of other issues are also covered in the IUCN Best Practice Guidelines on Tourism and visitor management in protected areas[16] that I co-edited with Yu-Fai Leung, Glen Hvenegaard and Ralf Buckley, and complementary guidance from the United Nations Development Program[17] and International Finance Corporation.[18]

Then there is also the change in the growth of a market demand for sustainable tourism. A while back Andrew Rylance and I wrote 'The Responsible Tourist', as a

---

[13] https://www.cabi.org/bookshop/book/9781786393555/

[14] https://www.atta.travel/news/2019/08/the-economic-impact-of-global-wildlife-tourism-wttc/

[15] https://www.cbd.int/tourism/doc/tourism-partnerships-protected-areas-web.pdf

[16] https://portals.iucn.org/library/node/47918

[17] https://www.undp.org/publications/tourism-concessions-protected-natural-areas-guidelines-managers

[18] https://documents1.worldbank.org/curated/en/459431467995814879/pdf/105316-WP-PUBLIC-Tourism-Toolkit-19-4-16.pdf

modest ebook that aimed to help travellers make informed decisions about destinations, online booking platforms, and hotels and tour operators that operate sustainably. The book was an effort to separate 'greenwash' from actors that were reporting valid claims of sustainable practices. It is fantastic to see how the Travalyst coalition is taking these issues—of connecting travellers to good actors—up to a whole new level.

You are part of the Travalyst advisory group—what is this new initiative about? How does it add value to the already existing organizations dedicated to facilitating a more sustainable tourism?

The Travalyst coalition is a global partnership founded by The Duke of Sussex together with leading brands Booking.com, Skyscanner, TripAdvisor, Trip.com and Visa. The overall aim is to "change the impact of travel for good," and the first steps have been to develop sustainability frameworks to serve as a cross-channel guide for scoring sustainability practices across the travel and tourism industry.

I am a member of the independent advisory group that has been brought in to help advise, check, and scrutinize their work, and while it is only early days, I am enthusiastic for the impact the organisation can have.

I think that Travalyst will add value through its leadership in drawing together these major players who've agreed to collaborate on this collective interest to innovate sustainable tourism tools. The partners represent some of the most influential companies in the travel and tourism industry, and their digital platforms have collective access to the majority of the world's travellers.

I am excited to play my part, and believe that these efforts have the potential—if done correctly—to really accelerate the mainstreaming of sustainable tourism practices globally.

What criteria will tourism businesses or destinations have to meet, to be part of this new initiative?

The ambition is to ensure everyone within the travel and tourism industry can play a role, and can ultimately benefit.

Recently, a summit took place in Scotland to discuss the draft tools with tourism stakeholders, and there will be more opportunities for tourism business and engagement like this in the future. If people want to get involved at this stage, they can sign up at the Travalyst website.

Reducing poverty through conservation and tourism: which are the common challenges locals face in terms of benefiting from revenue generated through protected areas, or in their destination?

Common challenges that I've observed include a lack of access to markets, and information and training about how to engage in tourism, and the 'build and they will come' fallacy.

In my earlier studies, I believed that creating community-based tourism (CBT) enterprises could solve this, but the more I read, learned, and saw, changed my perception. CBT is difficult to establish, and sometimes even more challenging to sustain over time as a commercially credible concern.

While joint-venture operations allow partnerships between communities and the private sector to counter capacity constraints of rural people in developing countries, it can take many years and technical inputs to establish the correct governance conditions to make them work.

In 2016 I co-authored Operational guidelines for community-based tourism in South Africa[19] with Andrew Rylance, Sadia Nanabhay and Heidi van der Watt for the International Labour Organisation. These guidelines attempted to overcome the challenges faced in CBT and joint-ventures. We suggested steps that could be taken not only to develop new ventures, but also how to improve CBTs that are not succeeding, or modify tourism enterprises that want to engage more with local communities.

One of the important things we can do is to quantify the economic effects that tourism has in local economies. In the past I've used value chain analysis tools, case studies, and book compilations such as Tourism and poverty reduction: Impacts and principles in developing countries[20]—co-edited with the late Dorothea Meyer—to examine the benefits that local people see from tourism.

Now, together with fifteen contributors drawn from the TAPAS Group, I'm co-editing 'Visitors count!', which will provide practical guidance for protected areas on the economic analysis of visitation. This will help destinations to understand how to assess their economic contributions and impacts, and also how to communicate the results to decision makers and the public.

Which KPI would you recommend destinations to use, to measure and monitor the success of their sustainability?

I think that the GSTC Destination Criteria[21] (version 2.0) are a great tool for this purpose. Tourism destinations can become certified by independent bodies as sustainable—including those that use standards equivalent[22] to the GSTC Destination Criteria such as EarthCheck and Biosphere Responsible Tourism.

I've also been supporting UNESCO to develop a Visitor Management Assessment Tool (VMAT), which will provide tools for World Heritage Site managers to evaluate their sustainability, and make decisions on which areas need to be improved. This tool combines GSTC criteria and also elements relating to the Outstanding Universal Values of the sites, and will be available in the near future.

Also, I'm in the process of editing a new Handbook on applied research tools for sustainable tourism,[23] which is due to be published by Edward Elgar in 2021. This will contain a series of 'how to' papers by experienced sustainable tourism practitioners, consultants and academics that will help others measure and monitor sustainability in practice. The book will cover issues relating to planning and

---

[19] https://tkp.tourism.gov.za/Documents/Community%20Based%20Tourism%20Operational%20 Guidelines.pdf

[20] https://www.routledge.com/Tourism-and-Poverty-Reduction-Principles-and-impacts-in-develop ing-countries/Spenceley-Meyer/p/book/9781138936089

[21] https://www.gstcouncil.org/gstc-criteria/gstc-destination-criteria/

[22] https://www.gstcouncil.org/gstc-criteria/gstc-recognized-standards-for-destinations/

[23] https://annaspenceley.wordpress.com/2019/07/08/in-development-handbook-of-applied- research-tools-for-sustainable-tourism-a-guide-for-practitioners/

designing sustainable tourism; enhancing the sustainability of existing tourism facilities; and also balancing overtourism and undertourism (i.e., visitor management in practice).

So, the Handbook will bring together practical advice from leading international practitioners in sustainable tourism. I am truly humbled that so many people I admire have agreed to contribute to the volume with their insights and expertise.

To your mind, what role does tourism play in response to the global climate crisis?

I think tourism has a crucial role to play in responding to the climate crisis. This is not only in looking at the transport, and how we travel, but also where we stay, what we consume and buy during those trips.

The UN Intergovernmental Panel on Climate Change[24] (IPCC) says that we have just over a decade to restrict global warming to 1.5 °C above pre-industrial levels, if we are to avoid extreme floods, droughts and heat. And the World Economic Forum suggests several ways that we can all reduce our carbon footprint through the way we travel. These include travelling light by reducing consumption while we travel; travelling with trust, but using accommodation that is certified under standards equivalent to the GSTC Industry criteria, and which use renewable energy, and by travelling slowly and carefully.

With the immensity of the challenge, partnerships addressing climate change are crucial. Tourism Declares a Climate Emergency[25] established by Jeremy Smith is a new collective of travel companies, organisations and professionals that recognize the urgency of climate change, and are accepting responsibility, and aim to act and collaborate to find solutions.

I recognize my own responsibility here too, and have signed up to the collective, and have published a declaration and plan that I'm implementing. I'd strongly encourage all those reading this to consider doing the same.

Another complementary initiative is the Strong Universal Network[26] (SUNx) founded by Geoffrey Lipman, which provides a system for tourism destinations and stakeholders to build climate resilience through climate-friendly travel.

Which aspects of working as a sustainable tourism consultant do you find the most rewarding?

Most rewarding aspects include that I'm learning all the time. I get the opportunity to work on a complex array of challenges and collaborate with others to identify the best solutions in specific settings.

I enjoy linking people together who have good things to share, or who can benefit from collaborating with one another, including through the TAPAS Group. I like working on a range of projects—whether as paid or as a volunteer—from the level of international organisations and networks, to national governments, private business and with local community members.

---

[24] https://www.ipcc.ch/sr15/

[25] https://www.tourismdeclares.com

[26] https://www.thesunprogram.com

Being adaptable and flexible during assignments is critical. The diversity of projects keeps me engaged and busy, whether it is working on the design of new projects and initiatives, the evaluation of ongoing or completed programs, synthesising analyses and compiling technical tools or case studies.

And which aspects do you find the most difficult?

Most difficult is not having enough time to do all of the things that I want to do, or that need to be done. For example, I make my living as a consultant, but I want to balance this with the voluntary activities (e.g., the TAPAS Group, GSTC and Travalyst) and family life.

I've found it challenging to translate complex technical or academic terms and sector jargon into easily understandable messages—and this is an area where I am still trying to improve!

Then, sustainable tourism work often requires travel to destinations, to gather information and local stakeholder perceptions, to do field work, and to validate options. However, this is going to be more challenging in the coming months with the coronavirus COVID 19 pandemic. I'm already adapting with an increasing emphasis on desk-based support and teleconferences. While this will reduce my carbon emissions, it is a challenge to really understand a place if you haven't been there, experienced it, and had face-to-face conversations with people that live there.

I'm also concerned about the impact it will have on tourism businesses, the livelihoods of people they support, and the natural and cultural assets that tourism financially supports. Still, we all need to adapt and be pragmatic in the face of this challenge.

Link to the interview: https://sustainability-leaders.com/anna-spenceley-interview/

Anne de Jong on Responsible Tour Operators and Sustainable Travel

The Netherlands | Anne de Jong is a sustainable tourism consultant and a passionate changemaker, fascinated by the complexity of the tourism industry. Anne supports tour operators and destinations on their journey to become more sustainable. She aims to contribute to future-proofing the tourism industry. Through her consultancy Fair Sayari[1] and as the co-owner of the Good Tourism Institute, she supports and teaches tour operators how to combine running a successful business and sustainability. Her main focus is to build a sustainable and profitable business while making a positive impact on local communities and the environment. She emphasises making sustainability easier, more accessible and attractive. Anne also works for sustainability certification schemes like Travelife for Tour Operators and

---

[1] https://fairsayari.com

Green Destinations, where she coaches, supports and audits tour operators and destinations to comply with those sustainability standards.

Areas of expertise: tourism business, destination sustainability, responsible tourism, UNSDG 11 (Sustainable Human Settlements), UNSDG 8 (Equal & Fair Economic Opportunities)

The following interview with Anne was first published in May 2020 on Sustainability-Leaders.com.

Anne, you are relatively young compared to most of the sustainable tourism changemakers whose story we have shared so far. Tell us about your journey: how did you get here?

I graduated almost three years ago, when I finished my master's studies in sustainable tourism development (MLE at Wageningen University). I became first interested in sustainable tourism during a minor as part of my bachelor studies (international tourism management). We were working on a case study where we had to sustainably develop the destination of Zambia. This was the moment I knew I wanted to continue this career path and it made me choose a master's in sustainable tourism development.

During this two-year master's, I did my internship at Travelife for Tour Operators, a certification scheme for sustainable tour operators. I learned all about sustainability standards, the implementation, and most importantly, how to train tour operators in developing their business. After graduating, I applied for a variety of jobs but none of them really fit me. I wanted to work in the tourism industry, preferably with a focus on Africa and sustainability.

This combination turned out hard to find for a recent graduate. I ended up continuing to work for Travelife and was offered another freelance job. In order to accept, I had to register myself as a company and this got the ball rolling! Besides Travelife, I also worked for a Dutch tour operator for a while, but since a year I am fully independent.

It wasn't my intention to start my own business but now I wouldn't want it any other way!

What inspired you to work as a sustainability consultant?

During my studies, I already knew I wanted to work in the tourism industry, hence the choice for studying international tourism management.

While travelling the world, I found that there are many negative consequences connected to tourism and that sustainable tourism is key to changing that. I wanted to contribute to this solution and personally support tour operators and destinations in this change towards sustainable practices, also to make it easier and accessible for them to contribute to future-proofing the tourism industry.

I stand for a fair tourism industry which balances all the components, such as animal welfare, human rights, fair practices, and collaboration. My consultancy is called Fair Sayari and for a good reason! Sayari means planet in Swahili, so Fair Sayari = Fair Planet.

What is your view on responsible travel—and has it changed since you started being active in the field?

In my opinion, responsible travel is the only way tourism can exist in the future. It's simply taking care of the environment, communities, and wildlife—and protecting it. It's about making sure local communities and nature benefit from tourism, leaving a positive footprint.

I believe that tour operators that are not working sustainably now, are basically sabotaging their future self.

And yes, my opinion has definitely changed, but so has the tourism industry. When I was still studying, responsible tourism was really a niche. Only a select group was fully committed, and tour operators were either working responsibly or they weren't.

Over the last years I have seen the change towards sustainability becoming a standard for more and more tour operators: something which is now integrated into their business strategies, rather than a side project. Tour operators have realized that they need to protect tourism assets to be successful in business.

How can we make responsible tourism more accessible for tour operators?

I have noticed that most tour operators starting out with sustainability are often overwhelmed by all the possible practices and standards to comply with. I believe that we can only make a sustainable tourism industry possible if every tour operator out there is equipped to do their part and take their own responsibility. That would really change the pace of development, and I have the knowledge to get them started.

On my website, I share blogs to make responsible tourism more accessible, easier, and fun for tour operators, by sharing practical tips, tricks, and best practices. I am also hoping to inspire them to start implementing sustainability.

I have recently launched my new project: the Good Tourism Institute,[2] where we help tour operators become the best version of themselves. To get them started, we have written an eBook[3] that can be downloaded for free, about the basics of becoming a better tour operator. From strategy development, to supply chain management, communication, and online marketing—all in a sustainable way.

By supporting tour operators to take small and structured steps, it is more likely that they'll invest time and money in sustainability and become better tour operators.

To create an overall sustainable tourism industry, we don't need a handful of tour operators practicing sustainability perfectly; we need thousands of tour operators doing it—even if imperfectly.

As an entrepreneur yourself, which aspects of getting a foot into sustainable tourism consulting did you find the most challenging? And how did you overcome such hurdles?

When I started as a sustainable tourism consultant, having only recently graduated, I had the knowledge but lacked the experience. In the beginning, I experienced difficulties in being taken seriously. It really helped to be working for a company such as Travelife for Tour Operators, before taking on individual assignments. It gave me time to grow, learn, and to make a name for myself.

---

[2] https://goodtourisminstitute.com
[3] https://goodtourisminstitute.com/ebook

I've now been working for Travelife for almost four years, and looking back I can see how much I have grown into the role of a sustainable tourism consultant. Gaining such experience has opened great doors for me, among others becoming a representative for Green Destinations for Africa, hosting workshops for Dutch tour operators in collaboration with ANVR, and providing sustainability training abroad in countries such as Uganda, Kenya, Ghana, and Sweden.

Which trends will shape the tour operator business in the next years?

Well, this is extremely difficult to forecast at the moment, due to the coronavirus pandemic. I am expecting hard times for the tourism industry and it's unclear when we are allowed to freely travel again. I am predicting an increase in domestic travel at first, which will slowly expand to destinations farther away.

I notice that more tour operators are interested in sustainable tourism now, due to COVID-19, as they have realised how important it is to protect our planet and how much we need it for the tourism industry. Travellers are yet to follow this trend, but as long as tour operators are taking the lead and showing travellers what it means to travel responsibly, we are on the right track to making tourism more responsible and sustainable.

Is there a specific travel business or destination which has impressed you recently for its innovative approach to sustainability?

There are numerous businesses and destinations that have impressed me over the years, but one of the most significant ones is Kara Tunga[4] in the Karamoja region, Northern Uganda. The Karamoja region was unthinkable as a tourism destination due to decades of isolation and conflict. Peace has returned and Kara Tunga now develops community tourism in the region. They focus on improving community livelihoods and have started a tourism academy for locals. They collaborate with local communities and shepherds, and have developed community tours to share this unique region with visitors.

I coached Kara Tunga during the process of becoming a Travelife Partner, and visited them last year. I was impressed by their mission and strength at implementation. They really are a best-practice example of community-based tourism.

Much of your consulting focuses on Africa: which African destinations would you consider leaders in implementing sustainable practices or champions of promoting responsible travel?

This is difficult to say, as I don't feel one African destination can be considered a champion in promoting responsible travel, as they are all doing very well. I mostly work with destinations in Sub-Saharan Africa and each of these destinations has great best practices, extremely sustainable tour operators, and protected areas doing their best in terms of sustainable tourism.

However, the same destinations also have tour operators and parks that are still working with captive animals, without following animal welfare guidelines, are involved in trophy hunting or exploiting local communities for the sake of tourism.

---

[4]https://www.kara-tunga.com

What I like about African destinations, in general, is that they show resilience, passion, and genuine commitment. They care to protect their destination and their tourism product and are very much aware that they need to protect it, to be able to work in tourism in the future.

Link to the interview: https://sustainability-leaders.com/anne-de-jong-interview/

Antje Monshausen on How to Reduce Inequalities in Tourism

14

Germany | Antje Monshausen holds a Diploma in Geography. After gaining her first working experience in Guatemala and Bolivia, she started working with Tourism Watch at Church Development Service (former EED, now Bread for the World) in 2008 and was later responsible for financial support of international organizations. Since 2012 she is Head of Tourism Watch[1] at Bread for the World. Antje is Chairwoman of the international multi-stakeholder initiative Roundtable Human Rights in Tourism e.V, vice-chair of the children's rights organization ECPAT Germany, and a member of the certification council of TourCert. Furthermore, she is active in the Transforming Tourism initiative, where civil society organizations from all around the world demand the transformation of tourism in line with the 2030 Agenda for Sustainable Development.

Areas of expertise: destination sustainability, overtourism, responsible tourism, sustainability challenges, UNSDG 10 (Reduce Inequality), UNSDG 8 (Equal & Fair Economic Opportunities)

---

[1] www.tourism-watch.de/en

© The Author(s), under exclusive license to Springer Nature Switzerland AG 2022
F. Kaefer, *Sustainability Leadership in Tourism*, Future of Business and Finance,
https://doi.org/10.1007/978-3-031-05314-6_14

The following interview with Antje was first published in January 2020 on Sustainability-Leaders.com.

Antje, what inspired you to work with Tourism Watch, focusing on sustainability and the betterment of communities and destinations?

German author Hans-Magnus Enzensberger wrote in his theory of tourism: "The tourist destroys what he seeks by finding it." This quote always challenged me, since I started to travel myself. The answer is more than just adding some sustainable practices to existing tourism. A radical shift in perspective is necessary.

Tourists should only be able to find what people living in destinations are willing to show. Participation and self-determination of local communities are key in any development which aims to benefit people.

This change of perspective, putting the communities at the centre, inspires me.

Tourism Watch is a tourism-critical civil society organization. We believe that another tourism is possible and urgently needed. Developing tourism is not a goal in itself. Instead, tourism must be an instrument that contributes to sustainable development. If tourism is not benefiting local people, for example by increasing their quality of life and environmental integrity, it is not worth it.

Tourism Watch is currently analyzing how booking platforms are changing tourism. How do they benefit small businesses and help us fight inequality in destinations of the Global South?

Digital travel and booking platforms are challenging and changing the way tourism is organized, with a severe impact on the entire supply chain. The main question we pursue in this context is whether this reduces or increases inequalities in the Global South.

Theoretically, the integration of local tourism businesses into digital booking platforms could democratize and balance the (global) economic power structures by making intermediaries between producers and clients obsolete.

Practically, the opposite is the case, because large national or international booking platforms are too often dominating the market, by determining the rules of access through opaque algorithms and imposing price models that are far too competitive for small, often family-run businesses. While the platforms maximize their profit, they outsource the risks to stakeholders along the supply chain.

On the other hand, we see that the possibilities to sell products directly are an important precondition for the economic empowerment of small businesses. Some of them make good experiences by selling their sustainable or community-based tourism experiences on specialized digital platforms.

But these platforms themselves compete with larger, often venture-capitalized and global platforms. In most cases, they are not able to attract a critical mass of tourists and tourism businesses, in order to build a counterweight and to foster sustainable tourism development in destinations.

Is digitalization only changing the product side?

The digitalization of tourism also changes how we travel: for a large number of tourists the smartphone is their most important travel equipment. Nearly all services—from international to local transport, from accommodation to excursions and catering—can be booked in advance or on the go—without talking to a single person.

All those providers who lack access to the booking platforms—for example because of language or technical barriers—face severe challenges and can lose their market access completely.

What do you expect from booking platforms?

Booking platforms are companies, just like all other companies. Therefore, according to the UN Guiding Principles on Business and Human Rights,[2] they have a responsibility to respect human rights along their supply chain. This means that they have to make sure that they respect the right to decent work (e.g., concerning the working conditions of staff that are cleaning private apartments).

Employees and contracted self-employers need easy access to remedy, which is far from reality. For example, Uber drivers in South Africa would have to address their complaint to a Dutch Court, where the corporation is registered.

Booking platforms should stop using oppressive contracts forcing suppliers to offer the lowest price exclusively to one platform. Some countries do not allow such business practices—but we see that governments in many countries are still not well equipped to legally and administratively address the challenges that digital companies are causing (e.g., taxation of transnational digital companies).

Tourism Watch is part of civil society networks from the Global South and North. What are the most common challenges that come with such interactions? And how can they benefit from each other?

Tourism is perceived as a "white industry" which creates jobs, revenues, and economic development, with nearly no downsides. This narrative is very strong.

Many human rights defenders and civil society organizations that are protesting against specific tourism developments (such as the expansion of airports, large-scale resort developments) or consequences of tourism (such as water shortage or unmanaged waste) are facing criminalization by governments, or intimidation by powerful business stakeholders. Their protests are de-legitimized as being against any form of economic development.

Therefore, it is extremely important to create space for an exchange of civil society groups and to learn from experiences and develop joint strategies.

The issues that we observe as the most urgent from a civil society perspective are: pressure on land and violation of land rights, lack of participation in decision-making processes in the destinations, as well as concerns regarding severe violations of workers' rights and sexual exploitation of children.

The Transforming Tourism Initiative[3] observes worrying trends in tourism and demands bold counteraction by politicians and industry. How proactive have both been in promoting sustainable travel, e.g., in Germany?

---

[2] https://www.business-humanrights.org/en/big-issues/un-guiding-principles-on-business-human-rights/

[3] http://www.transforming-tourism.org

We see that more and more tourism companies are addressing some issues around sustainability. Yet, this is often not systematic, but the result of public pressure or progress in legislation (think of the modern slavery act in the UK). More holistic approaches, like human rights impact assessments that include stakeholder consultations in destinations, are still uncommon.

On the public level, we see that local communities and civil society are not yet accepted as important stakeholders which must be consulted in advance of any tourism development. Politicians are declaring areas as special economic zones for tourism, often in closed meetings with investors, without any consultation of civil society stakeholders.

While the rhetoric of business stakeholders and politicians on sustainability and human rights is getting stronger, we see worrying counter-developments, for example with the UNWTO convention on tourism ethics, which lacks an obligatory complaint mechanism.

Can you share examples of DMOs that planned strategies with local stakeholders in decision-making, with a real, positive impact on the destination?

We see some DMOs—for example in Peru, Ecuador or Uganda—that have integrated sustainability into their tourism development strategies and are undergoing the TourCert[4] sustainability certification for destinations. This includes local consultations. However, those destinations are still a minority.

Even though many destinations suffer from overtourism, the central indicator of success for most DMOs is still the number of visitor arrivals and their spend, and not the well-being of the destination's inhabitants.

What measures should destinations take with respect to sustainability, in the wake of the climate crisis?

In the context of climate change and tourism, flight mobility is still the most concerning issue. Even though aviation is the most climate-damaging form of transportation, the number of flights is growing severely. Every avoided flight is an active contribution against global warming.

Destinations can do a lot to reduce their own air-dependency by focusing on land-based mobility. Additionally, they should try to increase the length of stay of their guests through innovative product developments that include the hinterlands of the destinations.

A tourist visiting Barcelona by train, staying for two weeks instead of just one weekend, not only produces less carbon dioxide but also more revenues in the region, and more chances for interaction with local people on eye level.

Link to the interview: https://sustainability-leaders.com/antje-monshausen-interview/

---

[4]https://www.tourcert.org/en/

Antonio Abreu on Developing Sustainable Island Destinations 15

Portugal | Antonio Abreu is a biologist with a PhD in marine biology, a researcher and manager of the UNESCO Chair in Biodiversity and Conservation for Sustainable Development at the University of Coimbra, Portugal. He has extensive international work experience, having worked in the last 20 years for UNESCO, African Development Bank, World Bank, IFAD among others, as well as governments of different countries in Africa, Asia, South America and Europe. He has worked at UNESCO as a Programme Specialist in the Division of Ecological and Earth Sciences and was a member of the National MAB Committee. Antonio has coordinated several applications, periodic reviews and evaluation processes of Biosphere Reserves in Portugal and abroad and was part of the Expert Group responsible for the MAB Programme Strategy Review. He is the scientific coordinator of Príncipe Island UNESCO Biosphere Reserve. Currently he is also Vice President of the European Network of Environment Councils, representing the National Council for Environment and Sustainable Development.

Areas of expertise: ecotourism, island destinations, UNSDG 15 (Forests & Biodiversity), UNSDG 8 (Equal & Fair Economic Opportunities)

© The Author(s), under exclusive license to Springer Nature Switzerland AG 2022   117
F. Kaefer, *Sustainability Leadership in Tourism*, Future of Business and Finance,
https://doi.org/10.1007/978-3-031-05314-6_15

The following interview with Antonio was first published in October 2015 on Sustainability-Leaders.com.

Antonio, a few words on your professional work and responsibilities?

I am a biologist holding a PhD in marine biology but during the last years have worked on environmental management, biodiversity conservation, protected areas and in particular with UNESCO Biosphere Reserves. Climate change, environmental impact and integrated coastal zone management are also among my current fields of work.

I am a frequent university guest lecturer on tourism and environment. So, I would say that I am a highly versatile natural science professional with expertise in bringing scientific and technical research into international policy, management and governance.

Do you remember the first time you heard about sustainability in tourism, and your initial thoughts?

I have been following the relations between tourism and environment for more than 20 years, both as teacher and as manager. For example, some years ago I was coordinator of a Laboratory of Environment, Tourism and Sustainability.

As a biologist I understand the close relations between tourism, the natural and social environment and the need to appreciate and communicate these complex connections, and to use this knowledge in favour of the sustainable development of touristic destinations and their local communities.

Now at the end of 2015, has your view changed?

No. But the industry has changed a lot.

Tourism is now a massive and fast-growing sector with impacts (positive and negative) that most of the times are not taken into consideration by the specific sector actors. Landscape management, water, soil, pollution, biodiversity, climate change, IT, environmental management are just a few examples of things that now are part of the tourism business, requiring a totally different capacity from the professionals in the sector.

Then there is the interaction between hosts and visitors and tourists' impact on cultural heritage and the lifestyle of the communities where tourism activities occur. This is a part of the considerations and challenges that are shaping the concept of sustainability nowadays.

As Scientific Advisor for Príncipe Island Biosphere Reserve, which have been your main lessons, or insights?

My current role as advisor evolved from my former responsibility as technical/scientific leader of the application of Príncipe Island to be nominated as UNESCO Biosphere Reserve (BR). I was asked to continue providing support and to help build a local team that in the mid-term will be able to manage the BR and to run projects on conservation, development and awareness-raising.

Being involved in this process allowed me to gain unique insights as I had to engage with Príncipe's people and to learn their views about nature and people.

Most rewarding from a personal perspective has been to be recognized as a son of the Island. This has a special meaning as it puts me in a position totally different from the usual advisor. I can gain a deeper understanding of the real nature of these people and how they connect with their environment.

Which was the process involved for Príncipe Island to become a UNESCO Biosphere Reserve?

It was a real surprise to me when I received a phone call from the Minister of Foreign Affairs of Portugal at that time, asking me to have a meeting in Lisbon with someone from Príncipe Island who wanted to propose it to be nominated by UNESCO.

Then I met with His Excellency Eng. José Cassandra, the President of the Autonomous Government of Príncipe and a few months later I was preparing the application to UNESCO. I had to gather a team of experts from abroad covering all technical issues, together with a local team that helped us with the field work and initial contacts with the communities and all relevant stakeholders.

Then there was a lot of technical work collecting and producing base line information and at the same time explaining and promoting the concept of Biosphere Reserve among the communities.

From the very first phone call to submitting the application dossier for evaluation by UNESCO, this process took nearly 3 years of work, but with a very happy end: in July 2012 the small island of Príncipe obtained its title as UNESCO Biosphere Reserve.

Obtaining the Biosphere status was a very special and emotional moment, which I had the chance to experience in Paris together with Príncipe's delegation. Hearing President Cassandra address the plenary at UNESCO was a tremendous moment. But a few minutes later he was already thinking about what to do next, so here we are.

In which way was the local community of Principe involved in this initiative?

Totally and all the time. During the preparation of the application, all were informed and had the chance to participate. Not only during the usual conferences and training workshops but discussing potential initiatives and proposing activities.

The application file included the usual support letters and declarations from local, national and international institutions but, also, individual signatures from hundreds of Príncipe's citizens who didn't miss the opportunity to formally support their own Biosphere Reserve.

This close connection very much reflects the awareness that Príncipe's residents have about the Biosphere Reserve, which is also reflected in the success of initiatives such as the Water & Recycle Challenge project, aimed at removing plastic from the island, and other activities, such as the upcoming Capture Zero project for Marine Turtles, the Biosphere Run with over 300 runners, etc.

What does it mean for Príncipe to be a UNESCO Biosphere Reserve?

First of all, it means a clear awareness about the need to find ways for the sustainable development of the island and an effective understanding of the role of nature and biodiversity in particular in this process.

The UNESCO Biosphere Reserve is an effective platform for experimenting initiatives and ideas demonstrating that it is possible to find solutions, to learn from others' experiences, through the thematic and geographic networks of Biosphere Reserves.

The Biosphere Reserve can be considered as the vehicle that Príncipe uses heading towards sustainability and which brings Príncipe to the global cooperation for sustainable development, not only to learn but also to teach and share.

Your advice to governments of small island destinations eager to embrace sustainability?

Being small means, most of the times, difficulties. If you are small and insular, then we need to add distance, lack of critical mass, vulnerability, and many other conditions that can only be addressed through sustainability, efficiency and not following the usual path.

In such situations, sustainable development becomes a decisive tool for ensuring that decision-makers take into account tourism impacts and benefits from a multi-sector perspective, and not simply by way of cost-benefit analysis.

Tourism investments should be tested and shaped under a sustainability vision, as tourism leads to several and significant impacts which most of the times are not taken into consideration sufficiently at the time of decision-making.

For instance, the use of water, landscape, waste production, energy, impacts on biodiversity should be considered together with the usual indicators such as employment and income. But even these classic topics should also be assessed differently. For instance, number of jobs tells nothing about the quality of these jobs, if they are just and fair.

How to measure the success of sustainable tourism at destination level?

There are many potential indicators but, as we are just starting this new Post Rio +20 period, I would suggest that we should build a system based on the recently approved Sustainable Development Goals. They are universal, comparable, easy to understand and motivate all social actors to work together under a common set of goals.

Link to the interview: https://sustainability-leaders.com/interview-antonio-abreu-biosphere-reserve-principe-island/

Antonis Petropoulos on the State of Ecotourism in Greece

Greece | Antonis is the Founder and Editor of Ecoclub.com.[1] He holds a BSc in Economics and an MSc in Economic History from the London School of Economics (LSE), and an MSc in Shipping, Trade & Finance from the City University of London. Before founding Ecoclub in 1999, Antonis worked in the shipping sector. Following a couple of abortive forays into politics, his current favourite hobby is being a highly rated short-term rentals host.

Areas of expertise: ecotourism

The following interview with Antonis was first published in July 2015 on Sustainability-Leaders.com.

Antonis, why did you start the International Ecotourism Club (ECOCLUB. com) in 1999?

After a rather brief career as an employee in the Shipping industry, a trip to Belize and a lot of soul-searching I decided not to wait for a mid-life crisis and do what I really wanted to do, and play my little part in the broad global movement for environmental and social justice. Not through politics, but in a quiet, practical and

---

[1] https://ecoclub.com/

© The Author(s), under exclusive license to Springer Nature Switzerland AG 2022 121
F. Kaefer, *Sustainability Leadership in Tourism*, Future of Business and Finance,
https://doi.org/10.1007/978-3-031-05314-6_16

pleasant way, through appropriate forms of tourism, such as ecotourism and through the new, revolutionary medium that was then the Internet.

Do you remember the first time you heard about ecotourism? What got you interested?

First of all, I would like to thank you for this great opportunity, it feels funny being on the receiving end after having interviewed a number of people myself. I also feel I must clarify that I do not really see or wish to see myself as a "leader".

From a social ecology and mutualist perspective, I distrust leaders and hierarchies and believe everyone must lead themselves—especially in a sector and a movement like ecotourism/sustainable tourism, which is all about the grassroots.

I do not exactly remember when I first heard the very word "ecotourism"—it was certainly during the last century (!) perhaps in an academic article. But I do remember the first time I practiced ecological tourism, without realising it at the time, when—like many other young students of my generation—I travelled around Europe by rail (with an Interail pass). This was back in the summer of 1988.

Some forget that in ecotourism, as in Cavafy's famous poem (Ithaka), it is the journey that matters, how you reach your destination, along, of course, with the destination itself, the communities and the businesses and all tangible and intangible things that constitute a destination. I was interested in environmental and social issues from a young age, thanks to my parents and my teachers, and also greatly enjoyed travel, so ecotourism seemed a natural choice, the perfect combination.

As ecotourism community manager and online publisher, which have been the most significant changes over the last 15 years?

I would say the first dot-com boom and bust, the advent of Google search which killed the old directories, and finally, the meteoric rise of the social media which is currently challenging Google and old-style media alike.

Somehow travel booking and review sites have survived by adapting and adopting, but also by being immune to the noise. We, as in us the 99%, have been fortunate that despite concentration and oligopoly tendencies, the Internet largely continues to offer a level-playing-field with low barriers to entry, with increasingly important open-source, creative-commons and crowd-funding options.

Despite the best efforts of some corporates and governments, it has not fragmented, it is not fully monitored, and free speech and activism are still possible in most countries.

Which have been the main lessons for you?

The main lessons involved learning to live by the consequences of one's life choices and how to be flexible, realistic and open-minded without compromising one's ideals, which is, after all, the most practical way "to be the change".

Being based in Athens, your thoughts on the current state of sustainable tourism in Greece?

Well, Athens, a major tourist destination since at least the Roman times (and in a way sustainable as the world's oldest continuously inhabited capital city) despite its notorious smog (nefos) and 1960s "sea" of concrete has become more tourist-friendly and eco-friendly in the past decade, peculiarly thanks to an unsustainable event like the Olympics, which facilitated the construction of, an ever-growing since, subway and tram system.

There are many paradoxes and contradictions and conflicts in Greece, a result of its situation in geopolitical and cultural (and economic!) crossroads. One of these contradictions is the fact that it combines first-world tourism infrastructure with third-world environmental practices. Perhaps I am being too harsh, but you get the picture.

Is Greek tourism as a whole environmentally, socially, economically and politically sustainable? I am not sure. But equally, I am not sure about international tourism as a whole in an early 21st century with climate change, recurring economic crises, wars, refugees and destitute migrants.

We have to realise that apart from sustainable tourism development, there is also the option of sustainable tourism degrowth, especially in developed and unsustainable destinations. I would classify some parts of Greece in that category also.

Most Greek tourism commentators consider Greece unlucky due to its high seasonality (a direct effect of 5 months of bad weather and 3 months of unpredictable weather), but I believe this is a silver lining in that it prevents us from a tourism monoculture.

In which countries/regions do you observe most interest in sustainable tourism?

There is interest worldwide, both in countries that are more or less environmentally aware, egalitarian, free and rich in tourist attractions and those that lack any of these characteristics.

But there is usually a long distance to be covered from mere "interest" to full "success". One also has to separate source markets and destinations, although some countries are both.

Many of the destinations usually identified with ecotourism/sustainable/green/ responsible tourism are in the global south and include Costa Rica, Belize, Brazil (Amazon), Ecuador (Galapagos), Tanzania, South Africa (Cape Town), India (Kerala), Seychelles, and many Caribbean islands (from Dominica to the lesser-known Barbuda). But you can find isolated best practices in nearly every country. Connecting these isolated efforts is crucial to moving forward.

It would also be fair to include all countries (and regions) in the global north (including the Antipodes) that have sound environmental policies in all economic sectors and well-managed protected areas as well all cities that have green infrastructure (transport, housing etc.) waste management policies and combine multiculturalism with social cohesion and integration. These urban centers are usually major producers of green tourists.

From your experience, which issues do ecotourism operators struggle most with?

A common issue for all genuine green operators is "unfair" competition from all those who cut corners, by not even meeting legislation (such as environmental, safety and tax), let alone high standards. In this world of actually-existing-capitalism, the truth is that few consumers (around 10%) would choose the more expensive but greener holiday option if they could find a cheaper, less sustainable or greenwashed one.

This race to the bottom inevitably takes place on the backs of exploited tourism employees (and trainees, and interns) and small owners blackmailed by multinational tour operators.

I long for a world where quality tourism is affordable for all and accessible to all, but it just is not possible under the current global system.

Which sustainable tourism stories, written or personally witnessed by you, have been the most inspiring and memorable?

When it comes to stories and inspiration, I prefer non-fiction and things that can and have been done, to eloquent fiction and utopia. For this reason, I consider that the most inspiring sustainable tourism stories that I have been involved with are community micro-projects decided by local communities. Some of these were facilitated by our members and co-funded by ECOCLUB.com between 2004-2011. I hope that we can relaunch this activity soon in an improved and expanded framework, with the cooperation of our members.

Which achievements at ECOCLUB.com are you most proud of? And what keeps you motivated to continue?

I try to avoid pride, one of the "seven sins" (Superbia). My motivation is an optimistic—perhaps—belief that I can put my ideas into practice through ecotourism. And my son—as Marx & Engels first wrote, "like boni patres familias, we must hand the world down to succeeding generations in an improved condition".

The main achievement of ECOCLUB.com—International Ecotourism Club so far is to have attracted a number of people who truly care for the greater good as members, and as a result, to have survived since 1999, unlike a number of other well-intentioned but less fortunate sustainable tourism initiatives.

The goal during this, our second, decade is to improve mutual support and deepen cooperation between our members, so that we can jointly help local communities meet their real needs and aspirations through genuine ecotourism, ecological & equitable tourism.

What does your normal working day look like?

Currently, as exciting or as boring as that of most other people working in an office, I am afraid, if a bit quieter as there are far more emails than phone calls. But I enjoy and often get enjoyable disruptions in the form of visits by friends and colleagues from afar. Before my son was born, I used to travel a lot more, but I am already "catechising" him in the ways of ecotourism, starting from our local (but world-renown) Athenian attractions. The more local the more eco.

Your advice to ecotourism businesses and destinations on how to tell compelling sustainability stories?

In relation to ecotourism businesses, always tell the truth, be yourselves (human not hero, unless you are one), avoid hype, stick to the facts, do not go into minutiae, once more do not forget the humans, the hosts, your employees—even better let them tell the story. Be competitive without being antagonistic, cooperate. Know your audience, it is discerning and will see through any fake.

In relation to ecotourism destinations, I respect those destinations that decide how they wish to both develop and present their tourism image in a direct democratic fashion, aiming for maximum involvement of all their citizens, and not just tourism "stakeholders"—certainly not just relying on experts and advice—including this one!

Link to the interview: https://sustainability-leaders.com/interview-antonis-petropoulos/

Ariane Janér on Ecotourism and Sustainable Tourism in Brazil

Brazil | Ariane Janér is a Dutch zoologist with an MBA who has been working with sustainable development and ecotourism in Brazil since 1991. She is active in the field both as a consultant and through NGOs. Her projects range from developing community tourism projects to implementing sustainable procurement in large companies, designing tours, setting up websites, business plans, and marketing plans. In 1993 she co-founded the NGO EcoBrasil to help develop ecotourism in Brazil. Ariane is a former advisory board member for TIES, ABETA and currently the Executive Secretary of the Global Ecotourism Network (GEN), Board Member of the Golden Lion Tamarin Association, Technical Director of the Instituto Homo Caballus (iHOCA). Part of the Climate Reality Leadership Program in Brazil, she is a regular speaker in Brazil and at international events and has contributed to various publications.

Areas of expertise: ecotourism, tourism business, sustainability, UNSDG 15 (Forests & Biodiversity), UNSDG 8 (Equal & Fair Economic Opportunities)

The following interview with Ariane was first published in August 2016 on Sustainability-Leaders.com.

Ariane, as a biologist, you were probably aware of the concept of conservation and sustainability early on. When did you first learn about sustainability in relation to tourism?

First, I learned about "the dots", and then I managed to connect them.

Ever since I was a child, I was interested in animals and nature. My aunt gave me a gift membership to WWF Holland. My parents traveled to far-off places when we were very young and always came back with lots of stories. We avidly read National Geographic, learning about places all around the world.

When we started to travel as a family, there was always an itinerary with a narrative and a learning experience, and it combined nature, culture and relaxing at the beach. My parents planned everything in detail. In those days, you did this by letter, telegram, a fixed phone and a guidebook. Sustainability was not an issue then, there seemed to be a lot of space in the world.

During high school, I became increasingly aware of environmental issues. Things like Limits to Growth, the oil crisis of 1973 and the car-less Sundays, acid rain, a heavily polluted river Rhine and a major toxic waste scandal come to mind.

My "sustainability in tourism moment" probably came in Mexico. I did my final thesis for my MSc in biology on sea turtles and spent a year in Mexico. We patrolled and collected data on many beautiful beaches. When I visited Cancun in 1981, it was still under construction.

It was horrifying to see the great contrast of a mega resort with the (then) relatively untouched neighboring places on the Yucatan peninsula. Big resorts also do not combine well with sea turtle nesting beaches. In Mexico, I also became much more aware of the importance of the local community for both conservation and in tourism.

What made you decide to make the professional transition into the sustainable tourism industry?

First, I decided I needed to learn more about business in order to help nature. For my MBA thesis, I was lucky enough to be able to study a long-term WHO project in West Africa on the management of the control of River Blindness (Onchocerciasis). This terrible disease is transmitted by blackflies, who breed in flowing water and thus affects those living and farming close to rivers. This opened my eyes to another reality: the importance of very basic health and security conditions for communities.

I started working at Shell, hoping to learn about management and also travel to other countries. At the time, I thought that there could also be a future there in alternative energy. After about three years working in Holland, they sent me to Brazil.

After a year, I met my husband there. Next to running his own business, he was supporting some pioneer conservation initiatives in Brazil. At the same time, I realized that I would not have a career in alternative energy, as it was not really a priority for Shell. If I wanted to stay in Brazil, I needed to do something else.

Friends of mine had started one of the first ecotourism operators in Brazil. One of the founding partners and an ecotourism pioneer was Silvana Campello, a biologist who had organized the first official ecotourism guide course in Brazil for Embratur in 1987. She had just been offered a dream job at the World Bank and the partners were looking for a replacement. So, I switched.

You have been the owner of Bromelia Consult for a quarter of a century. The business started out as an ecotourism operator and has expanded into a well-respected consultancy for sustainable development in the tourism industry. Can you tell us more about the first incarnation of Bromelia and how it has evolved over the years?

It was a lot of fun running a small ecotourism operator in Brazil, researching, setting up and operating tours, being a guide, training guides and marketing to international tour operators. Just when we started out, we were asked to set up a day tour to visit the Golden Lion Tamarin Conservation Program near Rio. This beautiful little monkey was in danger of going extinct in the wild. During Rio 92, this was quite a hit with the delegates.

But despite these small successes, business-wise, the timing was a little off. Brazil was in an economic crisis; international tourism was down by 50% and the country was not considered a reliable destination for ecotourism by international tour operators. On top of that, airline travel to Brazil was more expensive than to competing destinations. So, demand was low and you had to compete with conventional operators, who had much more marketing clout. Unfortunately, my partners did not have enough capital to hold out for better times and, though we got close, we did not manage to raise venture capital. My two children were born in those years. I decided I could not do it alone; tourists and toddlers demand a lot of attention.

Bromelia went into hibernation, and I made money being a freelance consumer market researcher for Euromonitor and later added being a financial analyst at a boutique investment bank called Baxter Straub. These experiences were key for reinventing Bromelia as a consultancy for sustainable businesses (not only tourism related). The focus was on tailor-made business plans, marketing research, strategy and training. I also worked on larger projects as part of a team. A good example of the latter is the Brazilian Program for Certification in Sustainable Tourism (PCTS in Portuguese), which was coordinated by the Hospitality Institute and financed by the IADB, Ministry of Tourism, APEX (Brazilian Export Agency) and SEBRAE (Small Business Support System).

The main objective was improving on the quality and sustainability of the medium and small tourism businesses in Brazil, through standardization, training and technical assistance, certification and marketing. I was part of a core team of quality, standardization and sustainable tourism experts. The program was targeted at hotels at up to 50 rooms and reached out to stakeholders in key destinations in Brazil.

You co-founded EcoBrasil (the Brazilian Ecotourism Association) in 1993 and stayed on as Technical Coordinator until 2014. What was the creation process of the organization, and what were some major projects or initiatives that took place during your time there?

Many of the Brazilian ecotourism pioneers at the time met at an Ecotourism Congress in Ilheus in 1993. The organizers used a pretty girl on a donkey at the beach as the cover of their promotional pamphlet. They didn't understand what ecotourism was, but they knew it was trendy. We put our heads together and concluded we needed to organize ourselves better and formally founded EcoBrasil at the Adventure Travel Congress (organized by the Adventure Travel Society, which later became ATTA) in Manaus. Our inspiration was The Ecotourism Society (TES not TIES at the time) and Megan Epler Wood and many other international pioneers were there at the conference. Most of the original founders of EcoBrasil are still leaders in ecotourism and sustainable tourism in Brazil and internationally.

In the 90s, the emphasis was on investing in people by spreading knowledge and providing training about ecotourism. We partnered with the government, experts in different areas and major NGOs on projects like Ecotourism Guidelines for Brazil (1994), WWF Brazil's Community Based Ecotourism Program (1996 – 1999), Pilot Program for Ecotourism in Indigenous Areas (1997), Funbio's Best Practices in Ecotourism (2000 –2003), Braztoa's International Benchmarking in Ecotourism (2004 – 2005).

Gradually, the focus of interest of the government shifted more towards sustainability in tourism, adventure tourism, social inclusion and community-based tourism and using standardization and certification as tools for promoting quality, safety and sustainability.

New organizations have been created. ABETA (Brazilian Trade Association for Adventure and Ecotourism) reflected a growing necessity to help businesses become more professional. ABETA also extended the work of the PCTS into safety and adventure tourism activities standardization and certification and has brought this to the international stage.

The NGO Semeia is now promoting Public-Private Partnerships to support tourism in protected areas and has managed to push the public sector.

What challenges did EcoBrasil face in regard to ecotourism development in Brazil and positioning the country as an eco-destination?

It is important to realize that EcoBrasil was just one of many organizations working with tourism, conservation and communities. On the national level IEB (the Brazilian Institute for Ecotourism), was also doing good work. For instance, they did an extensive inventory of potential ecotourism destinations in Brazil.

Brazil is a huge country with a large domestic market. Though it was often seen as one destination in the eyes of the international market, it always was a country of many destinations. To get the word out to the field, you faced logistic problems or, simply put, needed to spend money on travel.

Partnerships with the public sector, tourism trade, conservation NGOs and development agencies were essential. WWF Brazil, Conservation International, The Nature Conservancy, SOS Mata Atlantica were some of the NGOs supporting ecotourism.

The word "ecotourism" had immediate marketing appeal. The conventional tourism trade and governments from national to destination level saw it more as some kind of fairy dust you could sprinkle on nature tourism in general. The much-vaunted growth numbers of 15-20% per year seemed especially appetizing to the short term thinkers. So, when partnering with government, tourism trade and conservation NGOs, you had to be careful to manage expectations, without losing interest.

Serious private investors were not that gung-ho on the tourism sector in general and, even if they were, they were more interested in large projects than risky small-scale ecotourism ventures. Many investments funds, even those with an environmental focus, were looking for high IRRs (over 20%). In Brazil, interest rates have always been high and just the transaction cost of getting a loan is a barrier for small business. So, many ecotourism ventures were self-financed and/or received seed money or support from NGOs or government programs, like the Proecotur (IADB funded Ecotourism in the Amazon Program). Not all of these investments were wise, but many of Brazil's top ecotourism products established themselves in the 90s.

Efforts were also made to implement a concession system in National Parks, but despite setting the example of Foz de Iguaçu, the combination of bureaucracy and barriers to entry for small businesses meant that this is still waiting to take off.

It is challenging to maintain the appeal of a good ecotourism product in a destination that is not sustainably managed. And destinations are always in danger when short term economic interests like mining, oil and, sadly, big tourism arrive with big promises and then leave the local community with empty hands. Or—in the case of the Samarco mining dam disaster—with a terrible legacy.

You are a founding member of the Global Ecotourism Network (GEN). What prompted the organization's creation?

EcoBrasil had been working with TIES [The International Ecotourism Society] since 1993. In 2010, I was asked to be part of what proved to be the last Advisory Board of TIES. It was a great group of people from all over the world. All of us got increasingly frustrated with the lack of financial transparency, the management and also with the direction TIES was taking. We tried and tried to resolve the problems internally. After the successful ESTC 2014 in Bonito, the Advisory Board came to the conclusion that it could not support future ESTCs without a deeper understanding of the financial risks.

In response, TIES quietly removed the Advisory Board from their website without communicating this to us first, and we therefore publicly resigned as a group. We then discussed if it was worth setting up a new organisation and contribute to carrying the flame for ecotourism and be a beacon for sustainable tourism.

What makes GEN different is the many years of practical experience in all aspects of ecotourism.

We have seen products, destinations and tourism organizations start up, grow, try, fail and innovate. There is a lot of reinventing the wheel and lost opportunities in ecotourism and we think that we can help focus on the issues that make a difference. GEN aims to be a global network that is very hands on.

Anything else you'd like to mention?

One of the advantages of getting older is that you can see how things turn out. I tend to follow projects and initiatives I have been part of.

When I researched the Kemp's Ridley Turtle in Mexico it was on the brink of extinction with less than a 1000 nests per year, less than 1% of the 1947 level. But thanks to long term cooperation between Mexico and the US, numbers rebounded to over 20,000 in 2012. At the moment, researchers are worried again as numbers are falling, which might be an effect of the Deep Water Horizon disaster, but also a decline in their favorite food, the blue crab. Today turtles are an important tourism attraction in many places. Tamar in Brazil is a great example of that.

The Golden Lion Tamarin is also an interesting example. The population that was on the brink of extinction in the wild in the 60s, is now already over the target set for 2025, and a key problem now is assuring enough protected rainforest and ecological corridors. The protected areas, which are usually private, offer new opportunities for tourism.

It is also important to remember that we are all "dwarfs standing on the shoulders of giants".

The Golden Lion Tamarin is now a tourism attraction, because a Brazilian primatologist, Adelmar Coimbra Filho, managed to rally both the Brazilian and the international conservation community to save it. The Uakari Floating Lodge at Mamirauá, for which I did the original business plan, was built to support economic alternatives in a sustainable development reserve. The reserve was created through the efforts of Marcio Ayres, who did his thesis on the rare and endemic Uakari Monkey. And he was inspired by 19th century explorer Henry Bates, who wrote The Naturalist on the River Amazons.

Link to the interview: https://sustainability-leaders.com/interview-ariane-janer/

Audrey Scott on Sustainable Tourism Strategies and Social Media

Germany | Audrey Scott is the Co-founder of Uncornered Market,[1] a tourism development consultancy working with sustainable tourism organizations, travel brands and destinations to create, position and market community-driven, sustainable travel experiences and culturally responsible destinations. Together with Daniel Noll, her business partner and husband, she transformed their around-the-world creative sabbatical and journey into a business of exploring, sharing stories, and advocating for travel as a force for good—for local communities, travellers and businesses—through their work as advisors, writers and speakers worldwide. The award-winning Uncornered Market blog advocates for responsible, respectful travel that cares for our planet and its people.

---

[1] https://uncorneredmarket.com

Areas of expertise: destination sustainability, marketing, sustainable development, tourism business, UNSDG 11 (Sustainable Human Settlements), UNSDG 12 (Resource Efficiency)

The following interview with Audrey was first published in May 2020 on Sustainability-Leaders.com.

Audrey, together with your partner Daniel you advise tourism businesses and destinations on sustainability. Do you remember what first got you interested in sustainable tourism?

Oddly enough, we didn't really encounter the phrase 'sustainable tourism' until a few years into our work and the around-the-world journey that evolved into Uncornered Market.

At the end of 2006, we set off on a creative sabbatical to travel around the world, at a time when travel blogs were a novelty and social media barely existed. Our goal was to tell the stories of people and places that we felt didn't have a voice, so as to break down fears and stereotypes and to humanize the places we visited. We articulated and advocated for travel grounded in respect—for local people and culture, socio-economy, and the environment. We believed that travel could enrich travellers just as it benefitted the people and communities they encountered. And to us, those multiple layers of respect are what underpin sustainable tourism.

A few years in, a board member from the Global Sustainable Tourism Council (GSTC) reached out to engage us in a communications and social media project, characterizing our work and advocacy as sustainable tourism, noting that the art of our work was that we didn't use the term 'sustainable tourism'. We just lived it, experienced it, articulated it, and advocated for it.

We found many organizations, travel companies and destinations doing impactful work—yet unable to articulate what made them different, bogging down in technical terminology and certifications, and losing the through-line of why travellers ought to care about sustainability. In other words, how they could make decisions in line with their values and have a positive impact on their travels that they wanted.

For clients and our work with them, the implications were not only in marketing, communications, and storytelling but also upstream in product development, operations, and strategy.

As tourism development strategists, what areas do you specialize in? What approaches or models define your work?

We specialize in destination strategies, tourism product development, and sustainability marketing, applying this model of the three layers of positive impact between communities, travel industry, and travel consumers. We first presented this model at the WTTC Global Summit in Sevilla last year (2019) together with Tourism Cares,[2] to illustrate how tour companies could integrate purpose directly into their product development. The model also serves to articulate the essence of our work in tourism development.

---

[2] https://www.tourismcares.org

Our work is community driven. We ask a lot of questions, engage local stakeholders across the spectrum. We bring to surface all the raw materials of a destination—culture, cuisine, history, nature, architecture, and essence—much of which may be hidden or poorly articulated before we arrived. These characteristics are often either taken for granted or at the other extreme, overstated with respect to their importance in the traveller experience.

Then, we facilitate, helping clients and destinations balance it all in the creation, positioning, and delivery of experiences. Our product development process is iterative and rapid; the way we help build incubators is entrepreneurial. Along the way, we keep in mind traveller trends, trade links, market access, and commercial viability—and how the client can position it all and deliver by connecting the operational dots.

We adapt constantly to new data, from when we construct, plan field visits and research itineraries to when we're on the ground in destinations. We constantly uncover opportunities and connections which may inform and help destinations or companies create, articulate, and deliver better travel experiences that also benefit the local community.

There are many (sustainable) tourism development consultants out there. What differentiates your work and approach?

We apply an entrepreneurial and private sector lens to all our projects, even those with NGOs and donor organisations. This means we are always building for commercial viability, market access, and sustainability, also in the financial sense. For projects in the developing world, that implies the core objective of viability, of survival after outside funding ends.

We've seen too many tourism development projects fail around the world because they were executed in a vacuum, sealed off from the real world of how the tourism industry actually operates and how travel consumers really make decisions.

Furthermore, our work underscores that sustainable tourism need not be exclusive for the rich or well-off, but instead be available across budget levels. The success and breadth of the sustainability movement, we believe, depend on this.

To projects, we not only bring extensive experience in the tourism sector—including the various professional hats we've worn over the last 13 years in positioning, product development, marketing, sustainability, and training but also our vast knowledge from visiting over 100 countries and seeing what works and what doesn't, in terms of sustainable and commercial success. We are also entrepreneurs ourselves, so we understand first-hand the challenges of starting and running a small business, as well as how to be agile, lean, and creative to get things moving with very few resources.

Results-orientation and commercial viability seem to drive your work in sustainable tourism development. Can you talk about this in the context of recent projects?

As senior advisors on a USAID BGI tourism development project in Kyrgyzstan, we were asked to develop and position four regional destinations and DMOs, including developing new local tourism products with the goal of increasing local income and employment. Within one season, we worked with the regional DMOs and local SMEs to create over 20 new adventure, cultural, and culinary tourism products. Within two seasons, this helped boost visitor numbers up by 45%, extended stays to increase the depth of connection, and local revenue from the newly created tours increased by around 75% across the destinations.

Most recently in the Alay region of southern Kyrgyzstan, we have worked with Helvetas[3] and the local CBT provider (Visit Alay)[4] to help develop tourism in a remote region that is considered 'offbeat' even for Kyrgyzstan. Our role included a regional marketing strategy combined with conceptualising, testing, packaging, positioning, and marketing adventure tourism products that feature the beauty of the Pamir-Alay Mountains and the unique nomadic culture of this region.

We know that 'build it and they will come' just doesn't work, no matter how endowed a destination may be. This is particularly true in remote regions that appear difficult to reach, in particular, because there is such limited information available.

So, we worked actively with the local CBT director to not only improve the tours to meet adventure travel and consumer trends, but also on partnerships with OTAs and related travel websites, SEO/content marketing, and packaging.

In the last 3 years, we've seen some inspiring changes. Not only have the visitor numbers increased in the region by almost 60% (3500–5500), but more importantly the number of tourists booking local tours has doubled. Due to this, the average daily spend has almost tripled ($45–$120) and more than doubled the number of local people (from 130 to 300) involved in and earning an income from this community-based tourism.

For an area where income opportunities are very limited outside of the coal mines, subsistence farming/shepherding and going to Russia for work, these positive results from the Alay Region's tourism development really have a strong impact on so many families and local people.

In addition to your consultancy work you still run a successful travel blog. What role has your blog and its community played in informing your work?

Our travel blog and the community around it has been invaluable. It enables us to stay directly connected to and continually engage with travelers—their desires, goals, and experiences—and changing travel trends. This helps ground us and our work. We can see firsthand from our blog posts and social media activity what works and gets traction from advocacy, marketing, and even conversion perspective. Even more importantly, we can see what doesn't work.

We also have used our blog to help train our clients as to what destination strategy, positioning, and product development we've advised and trained on, it could look like 'in action' when marketing to and connecting with prospective travellers.

---

[3] https://www.helvetas.org/en/kyrgyzstan/what-we-do/how-we-work/our-projects/Asia/Kyrgyzstan/kyrgyzstan-small-business

[4] https://visitalay.com

For example, the Experiential Travel Guide for Karakol, Kyrgyzstan illustrated the destination positioning we'd proposed and brought to life through the cultural and culinary tourism products we'd worked on into something that travelers could connect with, engage and eventually book.

After we'd published this, the DMO and local community members we spoke to said that our guide made real what we'd spoken of during our capacity building and training exercises. Not only that, but it also helped them to see how their destination and their culture could be proudly shared with engaging prospective travelers.

Finally, we love our community. We want to continue sharing stories from around the world that inspire travellers to travel more sustainably, to places they may not otherwise have considered, and to connect with local people in impactful and meaningful ways.

What can sustainable tourism brands and destinations do to share their efforts more effectively with travellers?

First and foremost, eliminate the jargon. From there, focus on how sustainable tourism experiences do two things:

- How they connect travellers more deeply to what they desire, like interactions with local people, cultures, nature, and destinations;
- How the structure and economics of these meaningful travel experiences deliver benefits and positive impact on local communities and their environments.

We believe the two are linked, but it isn't always communicated to prospective travellers.

Sustainable tourism brands, including those with sustainability certification, should be transparent about their sustainability journey, that they are always working to improve and to optimize their impact. Bring the traveller along on this journey and be sure to illustrate the important role they play. And don't be afraid to talk about the difficulties.

The journey is long, evolving, and ever-changing.

What do you think will be the role of niche bloggers and social media influencers in travel and tourism, following the coronavirus pandemic? Do you think that we will see a change from quantity to quality?

I hope that niche bloggers and social media influencers will use their platforms for awareness, education, and advocacy. In other words, for 'good'. This means sharing stories and information that help their audiences travel with more care, so they better understand the impact—good and bad—of their travel decisions and how they spend their money.

I believe there will be a focus on supporting local businesses in the travel recovery process. Bloggers and influencers can play a role in helping raise awareness of small businesses and their importance.

Ironically, many of the same things we advised travel bloggers and influencers to help combat overtourism also apply for how they can help the travel industry recover from the pandemic in a healthy and responsible way.

I always hope there will be a shift towards more quality content vs quantity.

Not all bloggers and social media influencers will survive the economic impact of the coronavirus pandemic. It remains possible that, as travellers become more discerning and careful about their travels post-COVID-19 crisis, they will seek out more authentic and quality content to inspire them and help them make travel decisions.

Any thoughts on how the COVID-19 crisis will impact or change sustainable tourism?

This remains to be seen.

We're cautiously optimistic that the sustainable tourism movement can gain a foothold amidst the 'Great Reset'. Destinations and travel companies have a unique opportunity to rebuild a better-managed tourism industry that minds the environment and serves the local community just as it tends to the needs of travellers. Travelers, too, have an opportunity as never before to make choices according to their values.

We are not naïve, however. There are a myriad forces working against this vision, including an economic fallout greater than our current imagination, the temptation of destinations and companies to recoup losses by seeking volume and quantity over quality, and the inclinations of some travellers to tick off bucket list items at any cost.

In this respect, today's moment is as every moment has been in the history of travel: of challenges and opportunities in the face of competing for short- and long-term views.

Link to the interview: https://sustainability-leaders.com/audrey-scott-interview/

Mexico | Beatriz Barreal is an accomplished executive with a domestic and international presence as a key speaker, trainer and inspiring storyteller. Passionate about sustainability, a results-oriented, decisive leader with proven success in creating sustainable practices and multi-stakeholder councils based on collaboration. Creator of the SDSF methodology—Single Differentiating Success Factors, an alternative to the SWOT matrix, to help find and engage with your purpose in life from your heart towards purpose-oriented marketing for destinations, companies, families or people. Natural entrepreneur, founder of Neographika Inc and Sustainable Riviera Maya.

Areas of expertise: destination management, marketing, tourism business, destination sustainability, UNSDG 8 (Equal & Fair Economic Opportunities)

© The Author(s), under exclusive license to Springer Nature Switzerland AG 2022
F. Kaefer, *Sustainability Leadership in Tourism*, Future of Business and Finance,
https://doi.org/10.1007/978-3-031-05314-6_19

The following interview with Beatriz was first published in June 2016 on Sustainability-Leaders.com.

Beatriz, when did you discover your passion for sustainability?

Being an entrepreneur in marketing and graphic communication, with a bachelor's degree in graphic design and communication, with post-graduate diplomas in marketing, corporate communication and strategic communication, for more than 12 years, something inside of my heart started asking for something more to be achieved with my daily job.

In 1999, I became a board member for Amigos de Sian Ka'an, NGO. From there, I started my immersion in the topic of conservation and the awareness of the needs in the surrounding communities and the urban population within the destination. I continued as a board member for almost 11 years, following the desire of my heart to do something along the lines of the three pillars of sustainability.

In 2001, I set out with a friend to start an NGO, Toma el control (take control), for teenagers in high-risk situations. In 2009 I left the Amigos de Sian Ka'an board to start Sustainable Riviera Maya.

I was visualizing an NGO that would be able to join all the efforts in sustainability, in order to give full potential to their strengths and create the needed awareness. When we can see all the efforts together, sustainable tourism becomes an engine for development.

My abilities in communications, and good relationship with the entrepreneurs and authorities in the region, allowed me to be the creator of the first Sustainable Realty Investment Forum in 2011. For the very first time in the Riviera Maya, this forum brought together realtors, developers, United Nations agencies, investors, federal authorities from tourism, energy, environment, education, NGO's, academics and civil society. Over 3 days we shared ideas and worked towards a Sustainable Riviera Maya.

What was your view of responsible tourism when you first started your professional career?

I was a believer that responsible tourism could become the engine of balanced development, the creator of real opportunities, better cities, protected ecosystems and empowered people.

How has your view of sustainable tourism changed since then?

Now, I see how difficult it is to achieve common visions and goals over personal or selfish interests. My work with global authorities on the topic, federal and local authorities and entrepreneurs allowed me to see that revenue and shortsightedness are often the engines for those decision-makers. On the other hand, there are more and more dreamers that give heart and passion every day to make sustainable tourism a reality.

When and why did you establish Sustainable Riviera Maya?

Our objectives are aligned to the UN Sustainable Development Goals, and we work in different paths to achieve solutions on each of them.

I fell in love with the Riviera Maya region in 1992. I left Mexico City and, with no job or starting point, soon I created a marketing and communication firm, Neographika Inc., with 42 employees.

After 16 years of witnessing how development was being done in the area, in 2009 I started Sustainable Riviera Maya envisioning a large NGO that would combine all sustainability efforts in the region, small or large, and which would unite stakeholders through a shared vision. From there, the goal was to be able as a destination to co-create a new future, under the lights of a global vision towards a sustainable world. I wanted Sustainable Riviera Maya to become a powerful asset for all stakeholders in the region.

What are the main challenges Sustainable Riviera Maya faces today?

To create the financial resources for Sustainable Riviera Maya to be able to pay professionals in sustainable tourism so that the different programs can be led by someone else and not only me.

To be able to bring together all stakeholders, despite the selfishness and short-sightedness of some key players in the history of this destination, and to have all agree on the right priorities for a vital, sustainable destination.

To be able to change the view of tourism as being separate from local development, and to avoid the migration of Mayan Communities by creating local sourcing chains under fair trade principles.

On a national level, a key challenge here in Mexico is to become stewards of our natural capital and to foster the commitment our country has with the UN as a signatory of the natural capital declaration.

What does it mean for Riviera Maya to be listed as one of 14 Early Adopters recognized for their commitment to sustainability by the Global Sustainable Tourism Council (GSTC)? Did you notice increased booking numbers, for instance?

Actually, my goal was never to have more bookings, but better bookings, better clients, better entrepreneurs, better citizens, better visitors, better communities.

For Riviera Maya, becoming one of the only 14 EA-D meant being able to bring together all the professional efforts done over years in one platform. And now, with the Council for Sustainability—created and lead by Sustainable Riviera Maya so far—we are making decisions towards the better stewardship of our destination, such as:

- Signing with EarthCheck[1] to start the certification process as a destination, with the important participation of the key stakeholders of the Council,
- Improving and fostering the Ambassadors Program—created by Sustainable Riviera Maya—as a tool for better interaction and commitment from key stakeholders.

Through this, we achieve the baby steps toward a formal stewardship multi-stakeholder body, with a multi-sector vision and coherent actions regarding sustainability planning.

3 books linked to sustainability that every tourism professional should read?

---

[1] https://earthcheck.org

- La Revolución Generosa (Generous Revolution) by Stefan Klein
- Es Posible Otro Turismo? (Another Tourism, is it possible?) Facultad Latinoamericana de Ciencias Sociales, Sede Costa Rica
- EMPODERAMIENTO, Un camino para luchar contra la pobreza. Empowerment, a path to fight against poverty

If you had to start your professional journey all over again knowing what you know now about sustainability, what would you do differently?

Maybe I would start a marketing business but with the three-axis of sustainability in the core of the business, and focusing on corporations committed to sustainable tourism.

But, perhaps, I would have started Sustainable Riviera Maya earlier! Like the very first business of my life!

In your view, what characterizes a sustainability leader in tourism? What kind of personality or skills are needed?

A sustainability leader needs to be: generous, visionary, flexible, adaptable, very ambitious on long-term changes, with great certainty in life itself, able to work as a team, good communicator and conciliator; adapted to wealth and prosperity, with a long-term vision, and... IN LOVE WITH LIFE! In all its manifestations: people, planet and profit! And a sustainability leader needs to think of everything in terms of sustainability, even using specialist search engines... like Ecosia.[2]

Link to the interview: https://sustainability-leaders.com/interview-beatriz-barreal/

---

[2] https://info.ecosia.org

Benjamin Lephilibert on Hotels, Food Waste and Other Sustainability Challenges

Thailand | Benjamin Lephilibert is the Founder and Managing Director of LightBlue Environmental Consulting,[1] a UN awarded Best Small Business and social enterprise working predominantly on food waste prevention. He has worked across Asia and Europe with hotel groups (Marriott, Accor, Hyatt), restaurants (Cofoco Denmark, Bo.lan, J'AIME by Jean-Michel Lorain), international organizations (Michelin Guide Thailand, GIZ Laos, WWF), and government agencies (TCEB). He is an international speaker (50+ international conferences and webinars), guest-lecturer in culinary and business schools (Institut Paul Bocuse, FERRANDI Paris, ESCP Europe), a counsellor at Franco-Thai Chamber of Commerce, solution designer of FIT Food Waste Monitoring Tech, co-designer of The PLEDGE™ on Food Waste certification and benchmarking system, and a certified trainer for the Global Sustainable Tourism Council.

Areas of expertise: green hotel, hospitality, tourism business, Asia, UNSDG 12 (Resource Efficiency)

---

[1] https://www.lightblueconsulting.com

F. Kaefer, *Sustainability Leadership in Tourism*, Future of Business and Finance, https://doi.org/10.1007/978-3-031-05314-6_20

The following interview with Benjamin was first published in March 2019 on Sustainability-Leaders.com.

Benjamin, sustainability is now a key topic for most hotels, but it wasn't always that way. Do you remember the first time you came across sustainability in connection with tourism? What triggered your interest?

Yes, I remember this very well. I actually randomly ended up in Thailand looking for a job, knocking on companies' doors. I also went to ACCOR, which was at their regional office here in Bangkok. During the conversation with the HR director, we talked about my working experience, which was actually working with my dad, who distributed agrochemicals. Then I said to the guy, I was feeling a bit more eco-conscious and then he said: 'Oh that's interesting'.

And there was a position that opened a couple of days before our interview [related to sustainability]. He said: 'this is what it is, are you interested?' And I said, 'Yes that's great'. So that's how I got into it.

Initially I wasn't really into it. I wasn't born into sustainability from day one, but it definitely grew on me big time.

Looking back, how have your views on tourism and sustainability changed over time?

Well, dramatically. I think it started with a very naïve perspective on the sustainability agenda. I thought things would move forward very quickly. The more I was confronted with the reality of the situation in the field, specifically hotels, the more I realized we are evolving in a very cynical kind of world and the industry in particular. There are very few players who are actually doing it in a genuine way.

It is sad and discouraging to a certain extent, and so my views have changed a lot on tourism and sustainability. At the same time, I see so many opportunities that remain untapped, that are not done properly, done halfway or being stopped, which is sad. So, it's a mix of feelings between despair and hope.

As Director of Light Blue Environmental Consulting, in a nutshell, how does the company support tourism businesses?

Within the tourism industry, we are directly working with hotels and restaurants. We have several aspects of operations we can help improve—from support to certification (Green Globe, Travelife, Green Key), to more specific energy and water audits, and waste management. We do more and more practical workshops, training and campaigns related to specific SDG too.

Our strongest skills are on food waste and food waste prevention, and in the restaurant industry as well where we support leading restaurants achieve ThePLEDEonFoodWaste.[2] This is what we do, in a nutshell.

To your mind, which issues or challenges do hotels struggle most with, linked to tourism sustainability—for example in Thailand or the wider Asia region?

The first thing that comes to mind is, waste, waste, waste. It's crazy how neglected the issue is, waste and food waste. For food waste I have great hope

---

[2]https://www.thepledgeonfoodwaste.org

because the savings we experience are dramatic. Seriously, positively impacting the bottom line. But other waste is a concern, there is a low incentive and again we would rely on those who are really into it, before governments and strong regulations. Or the public.

And I have faith that the public may force hotels and stimulate them through travel agents and tour operators to be more responsible on this aspect.

Plastic has made some noise; it does help but it is only scratching the surface. Hopefully it will help trigger the discussion on this.

Tourism is sometimes criticized for being a very fragmented industry, dominated by short-term, silo thinking and lack of collaboration across destinations, agencies and institutions. From your experience, which are the main barriers that might prevent hoteliers from fully embracing the many solutions now available for more sustainable business operations?

There are several main barriers. I think there has been a lack of clear communication from players, like us for instance, but also in other sustainability-related fields like energy, waste and water, alternatives for chemicals, about business cases. This is central in our communication, and it is all about the bottom line, bottom line, bottom line. 'How much can we save?' or 'How can we differentiate your product from the competition?' or 'How can we help retain talent?'

I think that one of the key aspects with sustainability in hotels is the lack of clear KPIs related to efficiency in terms of energy, food waste, and a lack of clear benchmarks. Hoteliers love to compare themselves, it's one of the things we should take into account more.

Another barrier is a clear push from authorities for clear, independent checks of performance. How are hotels currently impacting the destinations they are involved in? I think that is a major aspect.

The third barrier is the urgent need for consumers and tourists to ask, to request, to demand responsible practices. Every tourist is voting with their wallet, and I think it's a matter of time before we can federate and put together those tourists that are more concerned to put pressure on actors to be more responsible.

Are there any differences in how professionals in Asia approach sustainability, compared to Europe?

We have experience in South East Asia, the Indian ocean and Northern Europe. I would say there are different levels of initial awareness. You wouldn't start from the same basis. In terms of approach, I guess it would be slightly different. When we worked with our customers in Europe, like a restaurant group in Denmark, we would start from a more advanced level, for sure. But there is still some nice work to be done.

How important is a hotel's sustainability performance nowadays for its competitiveness?

It's so low, because sustainability performance is unfortunately now self-reported and self-claimed, which is so sad. This was happening like 6 – 7 years ago, where the main players pulled out from certification schemes, claiming it was costing them too much, there was overlapping in reporting.

But my assumption is that they were very happy to invest this cash into their own communication initiatives, instead of investing in the sustainability initiatives itself. So, we moved away from genuine changes, in most of the cases, to more surface-visible sustainability that would help polish their image and green credentials.

The reality is that sustainability is not visible to any external parties, to customers who are just visiting the hotel. It is something that should be in the culture, something that is in the back of the house, not visible to the naked eye.

So, how important is sustainability performance for hotel competitiveness: zero. I would say it is not to be seen, except for those, the tip of the iceberg, the initiators of #skipthisstraw. Who is requesting to get records of the average amount of waste raided per head or per head night; or the quantity of water that is used per hotel room? If you knew the figures, you would have nightmares.

Which trends or changes do you observe in the international sustainable tourism community right now?

Well, to be very honest, I stepped back from what you would call the international sustainable tourism community over the last years, focusing on my mission to help reduce food waste.

What I can tell, however, what we've seen recently, directly related to this question: the movement around plastic is gaining momentum. So, as it is something that is easy to communicate, like #skipthestraw, hotels and restaurants love to do that.

So, I would say this is a positive change, but it is scratching the surface because all the very important matters are unfortunately remaining unchecked, and that's a bit sad. So, I would say the plastic trend is growing, which is great.

Food waste prevention is also growing, also great. I would say it's positive, but the negative side is the step backwards we are taking, to withdraw from international independent verified certification programs, and that's very sad.

Which three bits of advice can you share with hoteliers eager to make their operations more sustainable, but not sure how to get started?

The first one is about picking your fight. Define where you want to start: is it energy, is it water, is it waste, is it local communities, is it plastic pollution, is it chemicals. Pick one as a start, that will give you focus.

The second advice is: make sure that you don't give that in the hand of one person and hope for the best. Make sure that your team (or department) will understand why you are doing it.

If your team doesn't understand why, they won't see it as something nice or something that they are eager to do. Start with something very simple. Show them videos, about the impact of plastic pollution on the ocean, show them videos of food waste, on global warming, on local communities. Show them short videos on topics like that and then you can even get feedback from the team and see which topic they are most interested in seeing in the hotel improving on.

And that's the best way because they feel involved from the beginning. You might be surprised and find superstars going that extra mile and bringing the energy extra high.

The third one is, once you get started make sure you are using Key Performance Indicators. If you do not have any monitoring system on the performance of the initiative you are trying to take, then it becomes just like communication: 'Are we doing good?'—Yes. It's just a gut feeling.

You need to be able to define 'what is the performance in terms of energy efficiency.' That would be kilowatt-hour per occupied room, for example. Or what is the performance of waste? The amount of waste raided per number of people on site + number of customers per month. That's how you need to do that.

What are the main trends in hotels now, regarding sustainability?

I would say that, and this is wishful thinking, more use of technology. There are entire hotel operations that are still managed like 10–15 years ago, and there is so much technology now that could be used as leverage to use the right information and to improve on performance, energy, waste—food waste of course. That's our key expertise. I think that is crucial.

The hospitality industry has run with the same KPIs for ages and in order to bring new ideas, we need to bring new information to help them understand and see new opportunities. It's a business, if they are not strong in positioning themselves as sustainable, they have to at least see the benefits in terms of utility. So, technology can certainly help collect the right information and be used to improve the operation.

I also see another big way forward for sustainability in the hotel industry: consumer demand is gaining more grounds. It still needs to be higher; pressure needs to be stronger. There are many hotels that are not transparent, not disclosing.

When you take the hotel industry leaders, you have on one hand the corporate social strategy when it comes to sustainability, and on the other hand the operations where they are like 'huh?'. They have heard about the program once, but no one is in charge, there is no communication, there are no KPIs, it's sad. So, I am hoping the public can be more aware about that and be more demanding.

And finally, I see a huge trend coming up where hotels are really shifting—food waste is coming up, there is a shift in the way they are traditionally serving food and growing trends on food waste prevention.

Do buffets still have to be part of the way we serve food? Not to mention responsible, local, organic, seasonal sourcing. It's big and these trends are here. It's easier said than done to integrate those into operations, of course. But it's gaining ground and I am hoping it will become mainstream. And not because it's a requirement that is good to have, but a must-have.

Link to the interview: https://sustainability-leaders.com/benjamin-lephilibert-interview/

USA | Brett is Chief Executive of The Travel Corporation (TTC), a 100+ year strong family-owned and run international hospitality and travel company. TTC has an award-winning portfolio of 40 niche brands that operate on all 7 continents in over 70 countries. They specialize in immersive, experiential guided journeys and tours; luxury hotels and river-cruising; independent and youth travel brands. With over 35 years of experience, Brett is also an enthusiastic advocate of sustainable and meaningful travel. He is the founder of the TreadRight Foundation, a non-profit funded by TTC, which supports projects around the world that focus on empowering women in the communities their brands travel to, protecting wildlife and actively reducing their carbon and plastic footprints. He is included in the book 'I am Eco-Warrior' by Roger Moenks and helps drive the industry's sustainability efforts as a long-standing Executive Committee member of the World Travel and Tourism Council and their sustainability committee.

Areas of expertise: responsible travel, tourism business, UNSDG 13 (Fight Climate Change), UNSDG 15 (Forests & Biodiversity)

The following interview with Brett was first published in April 2020 on Sustainability-Leaders.com.

F. Kaefer, *Sustainability Leadership in Tourism*, Future of Business and Finance,
https://doi.org/10.1007/978-3-031-05314-6_21

Brett, you are an industry veteran with three decades of experience in the leisure travel industry. How has your journey been so far?

For the past three decades, it has been an incredible privilege working in our constantly evolving and dynamic industry. As the Chief Executive of the Travel Corporation (TTC), we are a fourth-generation family owned and run business. We are driven by service and we put the guest at the heart of everything we do.

One of my most fulfilling endeavours has been creating our not-for-profit TreadRight Foundation in 2008. I continue to fully support and engage with various tourism organizations, including as an executive member of the World Travel and Tourism Council (WTTC) and an associate member of the World Tourism Organization (UNWTO).

In 2017, TTC was also a Diamond Sponsor for the United Nations' International Year of Sustainable Tourism for Development.

TTC is celebrating its 100th anniversary this year. How do you think has the industry evolved over the past century?

Travel has become a global force for good. WTTC acknowledged that in 2018, tourism contributed US$8.8 trillion to the global economy and contributed 319 million jobs to the world economy.

Travellers are now more interested in conscious and sustainable travel. They are seeking immersive and local experiences. We are committed to working on the UNWTO's 17 Sustainable Development Goals (SDG's), including supporting more women's empowerment and WE Charity projects around the world, helping to protect endangered wildlife initiatives, and eliminating single-use plastics.

With the current global pandemic crisis, I think it will be fascinating to see how both consumer behavior and the industry change over the next decade. There will likely be many changes, as we've seen over the past century, where leisure and business travel exploded in volume and diversity.

Just imagine, only 25 million people travelled internationally in 1950, and that number exploded to 1.5 billion less than 70 years later.

With 50+ projects worldwide helping local communities, protecting wildlife and the environment—what is the future vision of the TreadRight Foundation?

TreadRight's vision is simple, it's to Make Travel Matter. What's important, and where we see our greatest opportunity for impact is the way in which each of our core business functions interpret that vision. That means:

- how we build experiences for our 42 travel brands
- how we support the destinations we travel to through our operations team
- how we tell stories to educate our guests and trade partners about how we should behave while travelling

How do you implement and scale sustainability initiatives across 42 brands under the TTC umbrella?

TreadRight projects are identified in partnership with the product and experience teams at each of our travel brands, including Trafalgar, Contiki, Insight Vacations & Luxury Gold.

Our TreadRight team, based in Toronto, coordinate with our brand teams to identify projects that fall into our three pillars of Planet, People and Wildlife and, in addition to offering an opportunity for guests to interact, experience and learn, which is critical.

We have scaled sustainability within our business, having been on this path for over a dozen years. So, all of our team understand and embrace our vision and our sustainability initiatives. These have included reducing our carbon footprint within our offices by installing solar and other investments, eliminating all unnecessary single-use plastics, and working with third parties to get them involved on a sustainable footprint journey as well.

In one of your interviews with Skift,[1] you mention that the younger demographic is not as committed to sustainability as previously thought. How can companies such as TTC encourage them to travel responsibly?

I believe that there is a clear intention on behalf of the youth market to choose responsible businesses. However, we lack the evidence that actual purchasing decisions are being influenced by questions of sustainability.

That said, it's unrealistic that any market can fully appreciate the intricacies and potential impacts of all industries, tourism included. That's why we feel it is our responsibility to explain the impacts of our business, how we are addressing them and what their choices are.

In your recent interview with Forbes,[2] you stress that each company within TTC has a Head of Sustainability. What are the benefits of creating such a role in senior management?

At the TTC level, we have a Chief TreadRight & Sustainability Officer, who is a member of our global Executive Committee. This role sets the direction on sustainability at the group level and works personally with each brand head to interpret and integrate our direction at a brand level.

This role is critical in a number of ways; it's a clear indicator to all existing and new team members the level of commitment and the value that we at TTC place on sustainability.

From a progress perspective, it also ensures accountability and provides me with a clear and direct link on our progress and efforts.

Ownership on sustainability must extend far beyond marketing teams in 2020.

Which of your sustainability initiatives across TTC could be easily replicated in other destinations or by other businesses?

The setup and integration of our TreadRight Foundation is certainly unique, however the guidance and leadership that it takes on all matters of corporate social responsibility is not.

Our TreadRight team has evolved to a point where it oversees TTC's CSR strategy, entitled How We TreadRight as well as our foundation's funding efforts.

---

[1] https://skift.com/2019/05/09/skift-forum-asia-the-travel-corporation-maps-plan-to-sustainability/

[2] https://www.forbes.com/sites/douggollan/2020/01/16/how-one-company-is-scaling-sustainable-and-meaningful-travel/

Other destinations and businesses can absolutely set themselves goals that directly address the areas of their business where they can make a material impact, ideally against the Sustainable Development Goals.

The UN Sustainable Development Goals have quickly become the language of sustainability, certainly for the business community, and we've embraced them at TTC.

Is there any initiative you'd like to highlight?

One to highlight would be Iraq al Amir. This is a great example of a private business working with government, other tourism organizations and local NGOs to get projects realized and underway.

I highly commend the Jordanian Tourist Board, Tourism Cares, and the founder of Baraka (Muna Haddad) for helping us bring this project to fruition, as well as help maintain its continuity.

What challenges should DMOs foresee and prepare for over the next few years, as the number of tourists (post COVID-19) is likely to continue to rise?

COVID-19 will pass, travel will endure, and DMOs should be focused on protecting the fabric of what makes their community unique and set KPIs that go beyond arrival numbers.

From where we stand, many DMOs have limited interaction with the travel trade, who have a considerable opportunity to encourage greater sustainability within our industry.

Travel is coming to a standstill due to travel restrictions related to the Corona-virus outbreak. How can tourism businesses and destinations prepare for such crises?

Our vibrant travel and tourism industry is extremely resilient, which we have demonstrated during past global crises including SARS, Ebola, MERS and unfortunate terrorism attacks.

Tourism businesses and destinations can prepare for such crises with continued cooperation and engaged international dialogue. After the travel restrictions are lifted globally in collaboration with the World Health Organization (WHO), we are fully committed and united to support the recovery across our 39 offices worldwide.

It is difficult to know at this time what will be the fallout and the changes that will come from this global pandemic crisis. I don't think anyone could have anticipated—not to mention prepared for—this global shutdown. Like all other companies in our great industry, we have been deeply and significantly affected.

It is so upsetting to see how many people have been affected across the world by this crisis and the shutdown. We have responded by trying to keep as many of our team employed as possible, by remaining positive and optimistic while dealing with the issues as responsibly and as timely as possible. We are also planning what to do when this is over, and to ensure we are more responsible and ready to do the right thing once this is over.

Imagine you could turn back time and start all over again. Knowing what you know now about business, travel and sustainability, what would you do differently?

What I would do differently is to make TreadRight and its core principles a foundation of our company from day one. But otherwise, I wouldn't do anything differently, frankly. I do believe, fundamentally, that one has to do well in business in order to do good in the world.

Link to the interview: https://sustainability-leaders.com/brett-tollman-interview/

Brian Mullis on How Guyana Promotes Sustainable Tourism

22

USA I Brian is a destination management, development and marketing specialist passionate about community-led conservation and regenerative tourism. As the Director of the Guyana Tourism Authority (GTA) between 2018 and 2020, Guyana became globally recognized as a leading sustainable destination and tourism became its second-largest export sector. Before leading the GTA, he founded and led Sustainable Travel International for 14 years. Brian is currently working with the World Bank and Rwanda Development Board as a Tourism & Conservation Technical Advisor and with Planeterra to scale up community tourism globally. Leading initiatives and delivering solutions within governmental agencies, multinationals, MSMEs, and community leaders in 70+ countries has given Brian a unique ability to foster multi-stakeholder collaboration, bridge communication divides, and generate tangible results at scale. It stems from 25+ years of experience in CEO positions in the private, public and civil sectors and a long track record of generating positive socio-economic and conservation outcomes through tourism.

Areas of expertise: destination branding, destination marketing, sustainable tourism, UNSDG 8 (Equal & Fair Economic Opportunities)

F. Kaefer, *Sustainability Leadership in Tourism*, Future of Business and Finance, https://doi.org/10.1007/978-3-031-05314-6_22

The following interview with Brian was first published in January 2020 on Sustainability-Leaders.com.

Brian, prior to joining the Guyana Tourism Authority as its Director, you led Sustainable Travel International, a non-profit focused on sustainable tourism. What inspired you to move to Guyana to lead its national tourism board?

After landing my first dream and becoming the owner/operator of an international ecotourism and adventure travel company focused on providing guests with once-in-a-lifetime experiences, I quickly discovered that it was relatively easy to make money but somewhat difficult to make a difference in the communities we visited.

So, I made a shift that has affected my career path ever since; I founded Sustainable Travel International with a colleague in 2002 with the aim of taking the wonder and thrill of travel and making it better by using tourism as a means to improve the livelihoods of locals, and protect the environments they depend upon.

Little did I know I was embarking on the most challenging undertaking of my life, yet I felt empowered by our vision to mainstream the concept of sustainability in tourism.

Fourteen years in, I was happy with our progress and the level of impact we were having at scale within multinational corporations and supporting governments worldwide. At the same time, I realised I needed a change. Years of 60+ hour work weeks, extended travel and constant stress associated with running dozens of projects had taken its toll.

After becoming a father of two beautiful children from Ethiopia, I knew I had to find better work/life balance. So, I took a step back. I passed the reigns to a successor, reassessed what I wanted to achieve in the next stage of my career, and then set out to secure my next dream job.

It was paramount that my next role enabled me to have a deep, meaningful, and long-term impact. Little did I know that the small South American country of Guyana would soon become my next destination and family's home. Leading the Guyana Tourism Authority (GTA) has been an honour and a privilege.

How have your views on Guyana as destination changed since then?

Guyana, as a relatively recent entrant into tourism, presents an opportunity to leap-frog tourism development, to avoid the mistakes many other destinations have made, and to become a model sustainable destination.

Despite having a long-standing sustainability agenda and some of the most intact and spectacular natural landscapes in South America, the country remains unvisited by outsiders. With approximately 90% of the country's population concentrated along the coast, Guyana has remained remarkably free of large-scale development and pollution.

More than 80% of the country's forest and vegetation remains in a natural state. This vast expanse is one of Earth's last great regions of tropical wilderness, home to thousands of plant and animal species, many of them found nowhere else. Jaguars, Harpy Eagles, arapaimas, giant anteaters, giant otters, anacondas, and more still thrive in this interconnected ecosystem and can be encountered with relative ease if you know where, when and how to spot them.

Sustainability is a way of life in Guyana. Nine indigenous nations have occupied Guyana's interior regions for thousands of years, cultivating a close relationship with the forests and savannahs that sustain them. They were Guyana's first scientists, whose wisdom long preceded modern medicine and ecology.

As might be expected, indigenous people have extraordinarily keen senses and possess an unsurpassed knowledge of the plants and animals around them. With no need for a translator, visitors are able to really get to know their indigenous hosts, and gain true insight into their lives. It is far more of a genuine human connection than most travel experiences provide.

Sustainability is an important success indicator now for any destination. How do you master this balancing act between wanting more attention and visitors, and at the same time having to protect the natural environment and ensuring the wellbeing of local communities?

Guyana is not a mass tourism destination, nor is it 'touristy'. We are pursuing a progressive path to tourism development focused on maximising the local socio-economic and conservation outcomes.

There is a proven formula, involving inter-ministerial, multi-stakeholder collaboration through structured partnerships and integrating sustainable destination management and development best practice into all aspects of our strategy planning, policy, product, and promotions.

While we want to incrementally increase the number of travellers in the areas of Guyana that will accommodate more volume, our primary focus is on increasing the value that each traveller represents.

Our destination marketing strategy is predominantly centered on attracting travellers who seek out authentic nature, culture and adventure experiences. These travellers tend to stay longer and spend more during their vacations, travel with a lighter environmental footprint, and many want to leave a positive impact on the people and places they visit.

We are also making a concerted effort to scale up community-led tourism. To my knowledge, Guyana is the only country in the world where tourism that is led by indigenous communities is the primary focus. It is home to some of the world's best examples of community driven, owned and led tourism. The host indigenous communities own and manage the enterprises, which results in all of the residents receiving economic benefits.

Furthermore, due to the market demand for birding and wildlife spotting, host communities have an incentive for conservation and habitat protection, and due to demand for cultural experiences, these communities determine what is sacred and what they want to share, which fosters cultural pride and the protection of culture and heritage.

To put things into perspective from an environmental standpoint: if, for example, each European visitor to Guyana generates 2.79 tons of CO_2 and Rewa Village alone is protecting primary rainforest that absorbs approximately 70,000 tons per year, the potential for scaling up community-led and owned tourism enterprises country-wide becomes clear.

This defines conservation travel—tourism that creates a net positive gain in ecosystem services and exemplifies why we're supporting an increasing number of communities to establish tourism enterprises and community conservation areas.

If you had to summarize Destination Guyana in the length of a tweet—what would you say?

Are you ready to embark on an epic adventure? Experience diverse landscapes, exotic wildlife, warm hospitality, a mix of cultures and authentic community-led and owned tourism. Travel how it used to be and let Guyana help you to rediscover the meaning of a five-star experience.

As demand for a more responsible tourism is growing, do you think the success of DMOs should be measured linked to the UN sustainable development goals?

Absolutely. To help maintain the ecological health of the planet, the UN Sustainable Development Goals (SDGs) should be adopted by all industries, including the tourism industry at large and related sectors.

The United Nations World Tourism Organisation (UNWTO) and its member states have formally recognised the actual and potential contribution of tourism to all 17 SDGs. That being the case, the tourism sector has an opportunity to take a leadership role, and in the process protect the assets upon which it depends.

Well-designed and managed tourism is renowned for its potential to contribute to the preservation of the natural and cultural heritage upon which it depends, empower host communities, generate trade opportunities and foster peace and intercultural understanding.

We need more governments engaging in destination stewardship and the tourism private sector demonstrating global corporate citizenship. This involves, for example, investing in World Heritage Sites and national and community protected areas and ensuring that local communities benefit more from tourism than international investors.

Travel and tourism should be driven less by short-termism of the markets and more by the longer-term protection of our shared natural and cultural heritage.

Few emerging destinations and tourism benefits have sophisticated means like Tourism Satellite Accounting for measuring the impacts and benefits of tourism. But now that myriad proven sustainable tourism solutions are readily accessible, everyone can make more of a concerted effort to maximise their positive and minimise their negative impacts.

Some of the outcomes will be difficult to quantify, yet many of the qualitative results will be apparent. What's important is sustainability needs to be embedded in strategy and driven by a more holistic set of metrics.

Which main trends do you observe right now, likely to impact the work of destination leaders in the years ahead?

There are a few key trends that are transforming destination management and marketing generally, and more so for tourism in Guyana.

Conservation tourism is a big one: tourism businesses, communities, donors, and government working together to make net positive contributions to the conservation of biodiversity and ecosystem services.

As previously mentioned, we are responding to this trend by helping more communities establish their own tourism enterprises to incentivise the establishment of community conservation areas. This has several key benefits, including improving wildlife spotting and birding.

Furthermore, with an increase in flying shame among environmental conscious travellers, visitors to Guyana can support communities that protect forests that sequester far more CO_2 than is generated from their travels.

Traveling in Guyana is experiential in that all of the senses are engaged through many of the experiences that are available.

While this trend continues to grow, few destinations have integrated into their strategies to support MSMSE development. In an effort to build on the trend, Guyana was among the first countries in the world to launch a national initiative to scale up experiential travel and join the Transformational Travel Council.[1] We believe in transformation travel and that it happens from the inside out, often inspired by exploring a new destination and experiencing the extraordinary. New products and experiences are being developed with this mindset in mind.

As your readers well know, placemaking inspires people to collectively reimagine and reinvent public spaces as the heart of every community, strengthening the connection between people and the places they share. This collaborative process is beginning to gain momentum in Guyana's capital city of Georgetown, where business and community leaders are increasingly coming together to revitalise public spaces.

Which cities, countries or regions have served you as source of inspiration, in how they approach destination branding?

I consistently seek out examples of best practices. The destinations that are balancing sustainable destination management, development and marketing are the most interesting. Through the work of groups like The Travel Foundation,[2] Green Destinations,[3] Swisscontact[4] and Sustainable Travel International,[5] for example, destination managers can identify and learn about all types of destinations that are on the cutting edge in varying ways and means.

From the Azores, Barcelona, and Bhutan to Namibia, Scandinavia and Thompson Okanagan, there are many multi-national, national, regional and community level examples to follow for inspiration.

The Guyana Tourism Authority is interested in sharing lessons learned and success stories with other destinations that are doing all they can to benefit residents, visitors and society as a whole.

Link to the interview: https://placebrandobserver.com/how-guyana-promotes-sustainable-tourism-through-destination-marketing/

---

[1] https://www.transformational.travel

[2] https://www.thetravelfoundation.org.uk

[3] https://greendestinations.org

[4] https://www.swisscontact.org/en

[5] https://sustainabletravel.org

Carlos Sandoval on Building an Ecolodge in the Middle of Nowhere

Argentina | Carlos Sandoval has dedicated more than three decades of his professional life to ecotourism development. As the Director of Yacutinga Lodge and its Private Environmental Reserve, he provides ecotourism advice and consultancy to the private sector, promoting the importance of environmental consciousness and respect for host cultures. To live a life consistent with the best sustainability criteria that drove his professional life for so long, he now lives on a farm in Patagonia, where he grows his own food. Carlos strongly supports environmental education and building awareness in society and is optimistic about the future of the planet.

Areas of expertise: ecotourism, ecolodge, tourism business, South America UNSDG 12 (Resource Efficiency)

The following interview with Carlos was first published in April 2016 on Sustainability-Leaders.com.

Carlos, why did you build the Yacutinga Lodge in Argentina?

F. Kaefer, *Sustainability Leadership in Tourism*, Future of Business and Finance,
https://doi.org/10.1007/978-3-031-05314-6_23

In 1997 I worked in the Republic of Paraguay, developing an incoming tourism company specialized in ecotourism. I realized that in the zone around the Iguazu Falls (the border between Brazil and Argentina), there weren't any established hotels dedicated to providing guests with an authentic experience and intense contact with the exuberant nature that characterizes the region.

I realized that this was the perfect opportunity to build Yacutinga as a company governed by modern concepts of sustainable tourism and functioning as a symbol of social responsibility and environmental protection for the region.

How has your business (and life) changed since then?

After 1998, I left my activities in Paraguay to move to Argentina and began construction of the Yacutinga Lodge. The place I chose to build it created a lot of problems due to its distance from any urban centres.

The Lodge's pristine location required a very strong environmentalist conviction in order to minimize the negative impact that the construction of the lodge would have on an area of jungle practically untouched by humans.

I decided to live in the middle of the jungle, taking on many of the restrictions of that way of life. I spent two years living together with the people that were building the lodge; two years of great personal growth.

To achieve this, I had to leave my family and children, somehow entranced by the need to be constantly present during the construction and implementation of the Yacutinga Lodge.

The two years we needed to build the Lodge from scratch were very intense: working from sunrise to sunset, living in camps, at first even without paths or potable water, not to mention electricity.

I felt like a pioneer at the end of the 20th century. I learned so much about the local people that accompanied me on this construction adventure. Observing and interpreting natural processes that weren't in the biology book, I shared meals with representatives of the aboriginal communities.

They taught me to look at the jungle from a different point of view. It was a time when we truly practised the strictest form of sustainability: living in a pristine environment, reverting to a more primitive lifestyle, discovering the essence of nature, living with the local population and feeling like you are part of the natural environment.

Building Yacutinga Lodge has been undoubtedly one of the most spectacular chapters in my life.

Has your understanding of sustainability in tourism changed since?

I began my career in tourism in the 70s. In those days, no one spoke about sustainability; in fact, we had just begun to speak about ecotourism. Most tourism companies and tourist agencies weren't yet aware of the need to regulate touristic activity to avoid resource depletion and to minimize the negative impact on the host communities.

Many years have passed, and the criteria for sustainable tourism have transformed into a conceptual necessity for every tourism stakeholder, whether it be an employer, an employee, a public institution, an NGO or the tourists themselves.

Today, due to population growth and the massiveness that characterizes tourism in the 21st century, it is impracticable to have a tourism business without responsible practices.

It is not only because of a situation of tourism resource depletion or the cultural alienation that communities potentially suffer because of tourism development but rather, the consciousness that our planet provides us with finite resources. Therefore, providing a tourism service that is properly managed and functions as consciousness-raising has become a social, environmental and corporate duty.

Since the beginning of our ecotourism operation in 2000, many things have changed in our region: some good and, unfortunately, many bad. The negative impact of agricultural expansion and the need for excessive consumption affect natural areas, where biological ecosystems take refuge and are conserved. Not to mention uncontrolled population growth and the penetration of exotic ways of life into host communities. These adverse effects result in environmental impoverishment and loss of cultural diversity.

The good news is that many tourism entrepreneurs in the Misiones Province, and also the region's Ministry of Tourism, have understood the situation and begun to take adequate actions to reverse these negative impacts. Since early 2015 we have a tourism policy at a regional level.

I am confident that this new public-private sector initiative will succeed in achieving the desired balance, and that we can achieve the sustainability that our industry urgently needs.

In your experience, what are the priorities for sustainability in tourism?

A major challenge lies in successfully applying the concept of sustainable tourism to all levels of the commercial value chain. Sustainability should not merely consist of pompous statements and should be used less as a marketing angle.

I believe that the great challenge is to effectively create awareness among consumers, providers and regulatory agencies, to develop an industry that goes beyond the economic and the commercial, making a tangible contribution to environmental education and the acquisition of social commitment.

How does an ecolodge differ from other kinds of accommodation?

An ecolodge is an establishment with soul. It needs to be business-focused in order to sustain its financial feasibility, but it doesn't forget its commitment to the environment that shelters it or the community that hosts it. This commitment should be reflected in the reality of the business and give the ecolodge its unique personality.

An ecolodge should educate through tourism and be beneficial to the place in which it operates, therefore creating awareness and responsibility for the future.

Carlos, why did you certify Yacutinga Lodge with the Rainforest Alliance?

Yacutinga Lodge has been certified by Rainforest Alliance for the second consecutive year now. Initially, the Misiones Provincial Government invited us to become certified, which is when I realized the immediate benefits that the certification created for our business.

We decided to pursue the Rainforest Alliance certification for a second year as a way to measure and evaluate our sustainability performance internally.

The rating achieved in the last certification was very high, to a large part thanks to the behaviour of the Yacutinga Lodge staff members, who firmly push the company toward perfection in the realm of sustainability.

The Rainforest Alliance certification process has helped us understand what is balance and what is luxury. It has helped unite our staff behind a common cause, together tackling the many challenges. And it has ingrained in all of us an environmental and social commitment.

Which part of the certification process did you find the most difficult?

The certification process of the Rainforest Alliance is based on the completion of various objectives in three different categories: environmental conservation, social commitment and the internal organization of the company.

Yacutinga began as an Environmental Conservation Project: protecting 570 hectares of jungle by performing low-impact ecotourism activities and promoting scientific research. For that reason, we initially thought that due to our undertaking of actions favoring environmental conservation, this part of the certification would be easy to complete.

Surprisingly, it turned out to be the opposite, because the environmental requirements of the Rainforest Alliance Certification included concepts that unfortunately aren't dealt with in our region and, therefore, made them almost impossible to achieve. Among those were the control and monitoring of carbon emissions and efficient recycling (not only at the internal level of the business) of waste generated by tourist visits (batteries for example).

Other aspects that weakened our certification was adapting our services to guests with reduced mobility and the prioritization of purchasing goods from sustainable providers since they are practically non-existent in our region.

To understand these limitations, we needed to communicate that Yacutinga Lodge is in a truly pristine and undeveloped area.

However, without a doubt, what constantly threatens our performance is the lack of vocational training on conservation in our municipality. Many times, we feel defenceless against poaching and environmental deprecation.

Our ecotourism efforts are sometimes slowed down by municipal authorities stuck in a short-sighted mentality of extracting natural resources rather than conserving them. In other words, we find ourselves in a mildly hostile environment due to our strong convictions about environmental conservation.

However, I am optimistic that things will change. The Rainforest Alliance certification has begun to have a strong effect in the province. More and more accommodation providers want to gain certification and act in favour of sustainable tourism.

I am confident that soon the negative impacts that exist in our region will be drastically reduced, thanks to the awareness that this certification has brought. For us, this is without a doubt the greatest benefit of the Rainforest Alliance certification.

Which have been the greatest challenges in terms of operating the lodge?

Undoubtedly, the greatest challenge that we have had, and continue to have, is to meet and exceed the expectations of Yacutinga Lodge's guests during their stay with us. Even if we did the impossible to achieve it, there are external factors that are outside our control and work against customer satisfaction.

Sustainable tourism in our region is still a new concept and not well understood. Not by those that offer the services or the official agencies, and not by the visitors either, some of whom don't realize that sustainable tourism requires a different perspective on vacationing.

In my view, sustainable tourism means enjoying our leisure time in a conscientious and balanced way, respecting the local culture and interpreting the environment through experiences. Fundamental for this to work is the support on the part of the tourist in adapting to these concepts, enjoying them and giving them value.

At Yacutinga Lodge, it is often difficult to satisfy guests who consider themselves responsible travellers, but in reality, are not. Their demand for comfort is overwhelming for the environment in which we are immersed, and which is fundamental for the concept of sustainability that governs my company.

Link to the interview: https://sustainability-leaders.com/interview-carlos-sandoval/

Christian Baumgartner on Sustainable Tourism in the European Alps

Austria | Christian Baumgartner studied Landscape Ecology and currently chairs the Sustainable Tourism and International Development programme at the University of Applied Sciences Graubünden (CH). In addition, he is the owner of Response & Ability. Christian specializes in the development and implementation of sustainable tourism and sustainable regional development. He has led tourism development projects in Europe, Central and Southeast Asia. Christian is the Vice President of the International Commission for the Protection of the Alps CIPRA and—due to his vast experience in labelling and monitoring—a member of the advisory board of some national and international eco-labels. Christian was a member of the (former) Tourism Sustainability Group of the EU Commission in DG Enterprise and of the Multi-Stakeholder Advisory Committee (MAC) of the UNEP 10YFP Sustainable Tourism Program and a member of several national and international advisory bodies for tourism.

Areas of expertise: destination sustainability, mountain tourism, responsible tourism, sustainable tourism, Austria, Europe, UNSDG 13 (Fight Climate Change)

The following interview with Christian was first published in July 2020 on Sustainability-Leaders.com.

© The Author(s), under exclusive license to Springer Nature Switzerland AG 2022
F. Kaefer, *Sustainability Leadership in Tourism*, Future of Business and Finance,
https://doi.org/10.1007/978-3-031-05314-6_24

Christian, having been involved in the consulting and teaching of tourism and sustainable development for many years, do you remember what first got you interested in the topic?

Well, this is a nearly life-long story. I got involved in the youth organisation of the Austrian branch of an International NGO—Naturefriends.[1] I learned to ski there when I was six. Later on, started to organise youth camps and international seminars. And as the Naturefriends organisation was involved in the alpine approach of 'gentle tourism' already in the 80s, it was easy to come in touch with this topic. It was at that time that I was also very much influenced by the work of Jost Krippendorf and Robert Jungk, arguing against the destruction of the Alps by mass tourism, and Dieter Kramer's work on social tourism.

My studies of landscape ecology gave me a professional background, and after my studies, I founded an NGO: respect—Institute for Integrative Tourism and Development. Respect started with a critical but constructive approach to tourism in Europe—one of the few NGOs that tried to cooperate with the tourism industry to develop things—but soon widened its focus to development cooperation and global tourism development.

How has your view on sustainability and tourism changed over the years?

When I started to work in that field—before the Rio conference in 1992—we neither talked about sustainable development nor sustainable tourism. The term 'sustainable tourism' developed mainly at the Lanzarote conference of UNEP and UNWTO in 1995.

As far as I can remember, my personal view in those days included both environmental and social issues, and economic impacts as well.

The discussion within the tourism industry has changed a lot. In the beginning, many practitioners misunderstood sustainability as a new word for environmental protection. Today, this view is widened and now also includes regional products and cultural aspects in our common understanding of tourism sustainability. But still, good working conditions, human rights, and governance as inherent parts of sustainable tourism have to be explained to many.

Personally, I learned to be more pragmatic when it comes to step-by-step approaches, instead of big solutions, which is especially important if you work with large tour operators. It is important not to be satisfied with only small steps forward, but to always keep the long-term goal of transforming tourism in mind.

You joined the Institute for Tourism and Leisure at the University of Applied Sciences of the Grisons (FHGR), Switzerland, as Professor for Sustainable Tourism. Why? How is tourism sustainability approached at the university?

When joining, I was very impressed that sustainability is not only important at the FHGR Institute for Tourism and Leisure, but an important value for the whole university, which it tries to implement in different areas and on different levels. The Institute wants to position itself as a leading Swiss academic institute in

---

[1] https://www.nf-int.org/en

sustainable tourism—having both a regional, alpine, and a global view on tourism development and its role as an implementer of the Sustainable Development Goals.

That means we have a teaching focus on sustainability in tourism: the 'Sustainable Tourism and International Development' major, since its inception four years ago has created huge interest among the students.

But sustainability is also one of the four strategic core areas of the Institute, which means that we focus our research and services in this area. Those projects have a rather wide approach—both concerning the content but also the geographical coverage. For example:

- We cooperate with an important Swiss winter sports destination, aiming to transform it as a model region for sustainable tourism.
- Together with UNWTO, WEF, and Global Compact—financed by SECO, the Swiss state secretariat for tourism—we are in the development phase of a high-level executive course focusing on using tourism as an instrument for the implementation of the SDGs. This course will start with a series of virtual 'resilience in tourism' workshops, to use the COVID-19 break for more sustainable tourism.
- And, to illustrate our development approach, we are currently evaluating the success factors of community-based Tourism (CBT) in Kyrgyzstan and are preparing a know-how transfer from there—together with Swiss knowledge—to other mountain areas in Georgia and Iran.

Over the last 15 years, you have gained considerable expertise through your consultancy Response & Ability. How has the profile of clients and projects changed during this time?

When we started our efforts in sustainable tourism, both as an NGO and with my company, the clients were mainly from the development sector. In Austria and the Alps, that was the boom period of independent regional development, which is often linked to tourism. The governmental and civil society organisations dedicated to development cooperation (EZA) also discovered tourism as an instrument of sustainable development at that time, something that has been reflected in recent years in the SDGs and the UN declaring 2017 as the International Year of Tourism for Sustainable Development.

And it was the time for awareness-raising campaigns, both among travellers and among tour operators and destinations.

Especially the latter has clearly changed: today there are many more tourism stakeholders, accommodation providers, tour operators, but also destinations that are seeking support in implementing their sustainability efforts. However, tourism is still a major theme in development cooperation.

Communication methods have become more professional—or even more subtle. From the wooden hammer of the 1990s, we have now moved on to nudging.

You also teach at FH IMC Krems. Are there any differences between Austria and Switzerland in how sustainable tourism is approached in research—and especially in practice?

In research and teaching, I cannot identify any general differences; it is rather a matter of how much importance different colleges and universities attach to the topic of sustainability. At the moment, I see the momentum here, more at Swiss universities, but perhaps this is also because I live very close to Switzerland.

Surprisingly, in practice, apart from obvious similarities, there are also considerable differences in tourism between the two neighbouring countries, which is also reflected in the implementation of sustainability.

What is similar, however, is that the really good examples tend to be at the level of a few destinations and businesses, rather than being reflected in the large governmental framework.

In Austria and Switzerland, tourism subsidies tend to be counterproductive in terms of sustainability, and demonstrated creativity with regards to new winter scenarios—keyword climate change—is lacking in both countries.

One clear sustainability advantage of Switzerland is the fact that travel to holiday destinations is often more climate-friendly. The reason for this, however, lies less in 'tourism' than in the almost perfect public transport system in Switzerland.

Which part of making a tourism business more sustainable is the most difficult, in your experience?

The larger a company is, the more difficult it generally is to take drastic measures. If shareholder value is the main focus, it becomes more difficult than with smaller family-owned companies.

Money, of course, plays a major role. Once accommodation providers understood that operational costs could be reduced by saving energy, water, and waste, environmental protection measures were no longer a major issue.

Much more difficult—but still feasible—is to make a financial argument around working conditions, such as offering year-long contracts rather than just seasonal. This for me belongs to the most underestimated area in the context of sustainable tourism.

And it becomes very difficult when the common good stands against individual financial or emotional gain, that is when well-established business models should be changed to increase the good of all. We all know that flying is one of the biggest sustainability problems of tourism. Does this change supply or demand? Only marginally.

The biggest challenge with such issues is to get politicians to create framework conditions that are supposedly (!) contrary to economic and consumer interests but in the interests of the long-term well-being of Nature and communities.

Which are the main sustainability trends right now, likely to impact tourism professionals in the years ahead?

I am quite sure that tourism will change dramatically within the next 20 years—due to climate change and its impacts, to the point of increased political unrest, for example in the arid regions of the world, but also through returning crises and imponderables as we are currently experiencing with COVID-19.

Tourism professionals must be experts in resilience. They must be transformation managers and, above all, good communicators—beyond marketing expertise. The challenges of future work lie in multi-stakeholder management, process moderation, and intercultural work. Soft skills will, therefore, become increasingly important.

You just attended a roundtable discussion in Berlin on Germany's tourism strategy. What advice did you share with members of parliament and industry leaders?

- More clarity is needed. The anchoring of sustainability in existing tourism strategies is superficial and unspecific.
- More effective government climate protection measures are needed—not only, but above all, in air travel.
- Regional climate change adaptation strategies are needed.
- More customer transparency is needed—such as through a climate or sustainability footprint on travel packages.
- Booking platforms are needed as partners. Their rankings should primarily be based on sustainability criteria, rather than price.
- More knowledge and comparable monitoring are needed, e.g., through a comprehensive implementation of the EU Tourism Indicator Scheme ETIS.
- More intensive and priority marketing for sustainable tourism is needed.
- There is a need for the participation of all stakeholders in the decision-making process for tourism development or tourism projects.
- Financial steering instruments are needed—and a promotion policy that is clearly oriented towards sustainability.
- We need an agreement on the belief in growth and new indicators for assessing tourism success.

Which tourism businesses or destinations have inspired you within the last years for their innovative or exemplary approach to sustainability?

At the level of accommodation facilities, apart from good environmental protection measures (which should be the expected standard nowadays), I can only think of Austrian examples:

- Hotel Hochschober and Alpenresort Schwarz for their employee programmes
- Hotel Magdas for its social approach
- Boutiquehotel Stadthalle for its energy programme

On a regional level, one of my favourites is certainly the Val Poschiavo in Switzerland, with the 100% Val Poschiavo project, which shows how cooperation between tourism and local production can function at a high level.

Many tour operators have begun to take a closer look at the human rights impact of their activities. Kuoni, and now DER Touristik play a pioneering role in this regard. Also, the members of Forum Anders Reisen who have begun to include CO_2 compensation for their air travel in the price of their holiday—exemplary in terms of climate protection.

Link to the interview: https://sustainability-leaders.com/christian-baumgartner-interview/

Morocco | Christof Burgbacher is the Founder of Consulting-Elementerre and the Director of Sustainable Tourism at HAM. He has worked in the international hotel industry for many years in Germany, France, Costa Rica, Tunisia and Morocco. He formed a connection to sustainable tourism in Costa Rica while working for a collection of upscale hotels. This experience opened his eyes to a new way of working: economic, efficient and always oriented towards the future. He is involved in several sustainable tourism development projects and advises international organizations and companies. He is referenced by the program One Planet—Travel with Care of the UNWTO and is a member of the Global Leaders Network of Green Destinations, chosen as a preferred partner. Christof has won awards on several occasions as the Best Sustainable Tourism Consultant in Morocco/North Africa.

Areas of expertise: hospitality, sustainable development, tourism business, UNSDG 12 (Resource Efficiency), UNSDG 8 (Equal & Fair Economic Opportunities)

© The Author(s), under exclusive license to Springer Nature Switzerland AG 2022   173
F. Kaefer, *Sustainability Leadership in Tourism*, Future of Business and Finance,
https://doi.org/10.1007/978-3-031-05314-6_25

The following interview with Christof was first published in November 2019 on Sustainability-Leaders.com.

Christof, on your website you mention that Costa Rica's approach to sustainable tourism was an eye-opening experience which had a lasting impact on how you see tourism now. What made your visit such a transformative experience?

That was in 2006. I got the chance to do a management traineeship at Cayuga Collection. Cayuga operates luxury hotels in a comprehensively sustainable way. Until then, sustainable tourism for me was limited to small lodges in rural areas, with the consequence that guests would have to put up with a lot of restrictions in order to spend a sustainable holiday.

Cayuga has shown me the opposite, that luxury and sustainability are not mutually exclusive. Natural products, natural air conditioning, employees who feel comfortable always contribute to a positive guest experience.

I was particularly impressed by two things:

- Cayuga always tries to find a sustainable alternative to classic products and services.
- The motivation of the employees to contribute to the sustainable development of the hotel. I have never experienced such motivated employees anywhere else, and that's because they are integrated in the sustainable development of the hotels.

I was also able to witness fascinating life stories and careers. Cayuga invests a lot in the education and training of its employees, enabling them to pursue a career that offers them and their families prospects.

Before experiencing Cayuga, the hotel industry for me was about providing people with a pleasant moment. I come from the luxury hotel business, and often had the feeling that my work made life more pleasant for people who already have everything they need.

My internship at Cayuga gave me a new perspective and showed me that the hotel business can be much more.

Having worked in renowned hotel groups in France, Germany and Costa Rica, what inspired you to choose Morocco as base for your consulting business?

I have to admit, I didn't really choose Morocco. I was offered the position as reservations and yield manager for the opening of a luxury hotel in Marrakech.

The idea of starting my own business in sustainable tourism came when I discovered the potential of the country.

Tourism has played a major role in Morocco for a long time. In addition, Morocco has an extreme wealth of natural products, in terms of food, beauty and health products.

Certainly, my decision was also influenced by the fact that already in 2012 Moroccan politics documented the will to attach greater importance to sustainability in tourism.

At the same time, I also felt that there was a lack of knowledge and solutions to actually put sustainability into practice.

To this day, many people tell me that they consider this to be a very courageous step. I would find it crazy not to have tried.

Are there significant differences or similarities in how sustainability is approached in Costa Rica, compared to Morocco?

That's difficult to answer. There are already very big differences within these regions. Between the sustainable development of Costa Rica and Nicaragua, for example, there are extreme differences. Or in Europe between countries in Scandinavia and Eastern Europe.

It's therefore best to look at individual destinations to see how they approach the issue. In Europe, leading cities in sustainability would certainly be Copenhagen or Edinburgh.

With the exception of Costa Rica and now also Slovenia, hardly any destination has succeeded in positioning itself as a whole, as a sustainable destination.

Morocco would certainly have the potential for this and would benefit extremely from it economically. However, here too sustainability is almost exclusively limited to rural areas.

What does social sustainability mean for you? What role does it play nowadays in the context of sustainable tourism?

Social sustainability means dealing with people. In the tourism sector we talk first about employees and the local population. This should be a matter of course in tourism, as it is one of the economic sectors in which the human factor will not be replaceable in the long run. The employees always make the decisive contribution to a successful guest experience.

If you ask 100 people today what they mean by sustainable tourism, I am sure that at least 80 will equate it with ecological tourism.

Also in the media, and even in the professional world, the terms are handled very uncleanly and thus contribute to a very vague picture. This clear distinction is important to me because it would help to attach more importance to social sustainability.

A company or a destination can only say it is 'sustainable' if it actually takes measures in the area of ecological and social sustainability.

Particularly in the hotel and restaurant sector, we have extreme deficits in staff management and training, and this worldwide. Although tourism is one of the largest employers in the world, only a small part of those jobs is actually stable and good enough to open up a real perspective for young people.

In Germany, for example, the industry is urgently looking for workers in the housekeeping, service and kitchen departments. These departments are very unattractive for young people and in times of very low unemployment they prefer to search for a job in other sectors.

In this context, I find it unfair to accuse the young generation of convenience or even laziness. The sector should be more honest with itself and needs to work hard to become more attractive for young people.

In countries like Morocco, this misguided personnel policy has the consequence that the sector suffers from a strong staff fluctuation, and tourism has not the best reputation as employer here either.

Which are the main barriers slowing down progress towards social sustainability?

The biggest problem I see is that it is simply forgotten. There have been a few surveys recently that have shown that social aspects are becoming more important for guests, but the focus is still on meeting the demands of ecological sustainability. So, the pressure on businesses and destinations to become active in this area is not yet very high.

Another obstacle is that social sustainability appears to be more abstract than ecological sustainability, not as easy to measure or to 'fix' through technical solutions, for instance. Cost-benefit calculations are more difficult to establish and measures which can be taken are often very complex and require long-term monitoring, which means much more time-consuming.

To make matters worse, we have too many hotel managers and owners, especially when it comes to SMEs, who do not have a professional hotel background. This lack of professional experience has little effect on the design or equipment of a hotel, but impacts the staff management and communication strategy. Necessary knowledge is often missing, and too little importance is attached to the topic.

How can local communities be better integrated into tourism development?

Here, too, we must certainly first look at the respective cultural context and the circumstances. The integration of the population in Berlin will be different from Marrakech, or a Berber village in the Atlas Mountains.

Regarding tourism development and sustainability, we should stop assuming that we always know what the local population really wants or needs. The actual needs and demands of the local population can only be identified and met through participation processes. These must of course be adapted to the respective circumstances and problems, and have to be permanently installed.

Without this participation, the acceptance of tourism and visitors is at risk. In Germany and Europe more broadly, conflicts between tourism and the local population are increasing, especially in the cities.

By now, many regions worldwide depend on tourism; yet, many of these regions also suffer from its consequences. In both cases, only the involvement of the local and regional population can mitigate these conflicts and add value for the stakeholder groups.

In Morocco there are different tour operators who try to establish a direct contact between tourists and the local population. In most cases the intentions are good, but often the actual needs of the people involved are not taken into account.

If we design these participation processes professionally and handle the results correctly, we can put tourism development in a positive light again. This seems to me to be crucial, especially in view of the importance of the sector.

I am currently founding a company together with my father and my brother, to fill this gap. My father has a background in tourism policy and my brother is an expert in very innovative participation processes. By combining our skills, we want to offer companies and destinations the opportunity to develop sustainability strategies that are based on the needs and demands of the population, employees and guests.

What opportunities does social sustainability offer to tourism companies?

From a business point of view, an investment in social sustainability is an investment in employees and thus in the service quality of the company. In today's hunt for 'likes' and positive comments, employees are the decisive factor. Properly trained, they can solve many problems that guests have during their stay and sometimes even turn bad moments into a positive experience.

In Morocco I am often asked by hotel managers: "What if I train my employees and they then leave the company?" I always answer: "And what if you do not train them, and they stay?"

In Morocco, I see the greatest potential in middle management. I consider the role of department heads in imparting competencies to be very important. They should also play an important role in the sustainable development of a hotel. In smaller hotels and riads where you won't find departments as such, the manager should take on this role.

Unfortunately, this is rarely the case, and these people are often only entrusted with control tasks.

A solution could be to offer training to managers and heads of departments in hotels, in which the aspects of 'total quality'—i.e., the combination of sustainability and service quality—are conveyed. These trainings must, however, be adapted to different company sizes and types.

In times of overtourism and flight shame, I see a bigger investment in social sustainability, not only as an opportunity but as a necessity. Destinations like Morocco are dependent on air travel. They should therefore give potential guests the opportunity to generate a positive impact with their trip, not only ecologically but also socially.

We should not forget that traveling allows people to know each other and helps to develop a better understanding for other cultures. Protecting our environment is surely our biggest challenge, but I really don't want to imagine a world in which people stop traveling to other countries.

Your thoughts on the sustainability of Morocco as a destination? Which challenges does the country face and which initiatives are adding to its sustainability profile?

Morocco has had the topic of sustainable tourism on its agenda since 2010, but very much limited to rural areas. There are also many projects and programs in the area of development aid. Unfortunately, they often don't have any impact as there is only little or no competence in the field of tourism within the active organizations, and because communication and marketing are not considered in the conception of the projects.

In addition, there are some scattered initiatives and private projects, but there is a lack of a coherent mission statement.

Morocco has certainly the will to position itself as a sustainable destination, but if they want to be a golf destination at the same time and become more attractive for cruise shipping, we cannot speak of a clear orientation.

Companies are not really being helped with this issue either. Some years ago, a cooperation was concluded with a well-known sustainability label that has already certified several hotels in Morocco. However, there is a lack of professional structures, both in terms of advice and audits and, above all, in terms of communication. In many cases, the result is a sign on the reception desk, but nothing more.

I think that the young generation sees the topic of sustainability not only as a necessity but also as an opportunity. Perhaps not yet so much in tourism, but certainly in culture and the creative industries. I really hope that this generation will be given a chance and that they will make use of it.

Anything else you'd like to mention?

There's a subject which I consider as a problem for sustainable tourism in general. In tourism, there is a strong black-and-white thinking about sustainability: either a company is sustainable or not. Apart from the fact that this contradicts the actual principle of sustainable development as an on-going process, it has the consequence that it excludes a large number of businesses from becoming more active.

Certain companies cannot be described as sustainable on the basis of their conception—for example, a hotel with an adjoining water fun park. The water and energy consumption would be so enormous that it is impossible to speak of sustainability in this area. In all other areas (waste reduction, staff management, local integration), however, that same hotel might be a champion in sustainability.

Other companies, especially small and medium-sized ones, are often afraid of taking the step towards greater sustainability because they know they cannot live up to this demand for perfection.

I estimate that the 'Generation Fridays for Future' will deal very radically with the topic of green-washing in tourism, and will reward honesty and gradual development. So, we should get away from judging what is sustainable and what is not, and instead focus on what a hotel actually does for sustainability.

And: we should leave the assessment to the customer. In the end, they are the ones to decide whether the commitment of a company meets their demands and ideas of sustainability, or not.

Link to the interview: https://sustainability-leaders.com/christof-burgbacher-interview/

Australia | Craig Wickham is the owner and Managing Director of Exceptional Kangaroo Island and Chair of Australian Wildlife Journeys. Craig lives and works on Kangaroo Island, where he grew up in a business family with a diversity of tourism ventures. He studied wildlife management and worked with the South Australian National Parks Service before getting into a private enterprise. Craig has been travelling internationally promoting his touring business, Kangaroo Island and Australia for over 20 years. Craig, his wife Janet, their family and their local team have been in business since 1990. Leadership and sustainability are two primary tenets Craig follows, and he has been a consistent contributor to community and industry. He has held board positions with the South Australian Tourism

Commission, SA National Parks Council, and a range of community responsibilities, including Local Government, with 4 years as Deputy Mayor.

Areas of expertise: ecotourism, tourism business, wildlife conservation, Australia, UNSDG 12 (Resource Efficiency), UNSDG 15 (Forests & Biodiversity)

The following interview with Craig was first published in July 2019 on Sustainability-Leaders.com.

Craig, do you remember the first time you heard about sustainability in connection with tourism? What got you interested in the topic?

I grew up on Kangaroo Island with a strong interest in the environment which surrounded me, and my family invested in the tourism industry in the mid-70s. As a young fellow, I worked for the National Parks and Wildlife Service, followed by a 12-month stint living and travelling in Southern Africa. I saw some excellent examples of sustainable tourism and active conservation in Namibia and elsewhere in South Africa—both public and private. I experienced some extraordinary landscapes in the Cedarburg Mountains in the Western Cape, and walking the Otter Trail in Tsitsikamma National Park.

On my return to Australia, I studied Wildlife and Park Management and then had a series of roles with National Parks and Wildlife Service—largely focused on visitor management and environmental education.

I loved the visitor and environmental interaction but the bureaucracy—less so. I decided my future would be better served in the private sector and ended up in a business that blended my experience in travel, environment, and hospitality in my home community of Kangaroo Island.

From the outset, I knew that to offer a diversity of really engaging and immersive wildlife encounters and have an impact on how people see the natural world, I needed to focus on small scale tourism. So, from 1990 we really had a strong focus on sustainability. And in order to ensure that I was pursuing what was 'best practice' at the time, I got involved in the Ecotourism Association of Australia and have retained an active membership from that time.

How has your view on tourism and sustainability changed over time?

One really strong 'wake up' was realizing that no matter my values, the rest of the world will continue to head in a specific direction, regardless of my activities within my business.

An example of this is recycling—we were doing everything we could possibly do to 'reduce, reuse, recycle'. Yet the 'recycle' piece was really difficult, living on an Island and the waste stream that was outside the 3 R's that went to landfill—well that was an issue. We had a very poorly run landfill and I decided to get involved to change that, and some other things, by running for Local Council.

I was elected to Council and at my first Council meeting was presented with the fact that we (our community) were being fined many tens of thousands of dollars for numerous breaches of our license conditions, due to poor management of the landfill. Long story short—the landfill was closed, and we now have kerbside recycling serving every home on the Island—exported off-Island and managed in a co-operative facility with 3 neighbouring mainland Councils. I ended up serving

10 years in Local Government, the last four as Deputy Mayor, and we got some good management plans in place.

My view is still that tourism can be a powerful force—both for change and resisting it.

I recently had the good fortune to spend the Norwegian National Day with a range of community leaders in Lillehammer and we had a wide-ranging conversation about the value of tourism in the face of globalisation diluting culture. My argument was (and is):

Tourism, if driven by communities and not being allowed to be imposed on communities, can be the thing that maintains diversity. That's because cultural (and natural) diversity is a strong reason for travel. If everywhere is the same as home, why not stay home?

I am hopeful that strong local networks can join up with others to develop a powerful coalition of shared values. For example, I see so many parallels in what the SlowFood movement is trying to achieve with what those of us with passion for retaining healthy wildlife populations—they are complementary movements and combined I think there is a great opportunity to retain nature and culture.

Tony Charters in his interview mentioned that, fortunately, the consumer is becoming more knowledgeable about sustainability. Does this match your experience? How important are sustainability credentials for the success of a tour business nowadays?

I would like to say I can see great commercial benefits from a market demand perspective, but it is difficult to see this in action. It is quite market dependent—and within markets, there is often polarisation. For some, sustainability is very important and for others, it is less of a factor, if at all relevant for their decision-making.

During my recent trip to Scandinavia, the Swedish tour wholesalers were particularly interested given the emergence of the flight shame concept, the social pressure to not take long-haul flights given the carbon dioxide emissions. This has quite some momentum in Sweden, I noticed. The train conductor, as we arrived in Gøteborg, stated plainly "thank you for travelling in such an environmentally responsible manner as our train."

My presentation, which always includes mention of our Advanced Ecotourism Accreditation, was suddenly the source of much interest and probing as to what stands behind it. So, there are positive signs!

What role does sustainability play for the small group and private wildlife tours which you offer through your company, Exceptional Kangaroo Island?

Sustainability is at the core of what we do—but of course, there are tensions. It would be far more environmentally responsible to walk or ride bicycles, or travel in much larger vehicles. However, larger groups are incompatible with the immersive wildlife experience.

When discussing sustainability, it is useful to identify issues in 5 key areas—all of which are vital and work hand-in-hand.

The first is the guest service and interaction elements—those things our guests see and experience.

The second is the behind-the-scenes elements that we can influence at the enterprise level: purchasing decisions; service design; energy procurement; handling of waste; water management; insulation; landscaping and so on.

The third element is the community infrastructure and service piece—waste management; water supply; external energy supply and renewable choices; co-operative activities around land management—pest and weed reporting and control programmes; co-ordinated revegetation programmes and the like.

A fourth piece is the establishment and marketing of destination culture, which can be either deliberative or 'random'. What I mean here is that the deliberative setting of destination culture is where there is a conscious planned effort to identify and define what the place stands for and what expectations are set in the minds of potential travellers. This includes branding, choice of destination descriptors, images used, and the tone of marketing material which industry and agency partners create.

The 'random' (for want of a better word) elements of destination culture influence are the social media posts, the broader media stories about the place, the word of mouth and the destination reputation—only some of which can or arguably should be influenced by anything other than the experience offered.

The final element is the regulatory framework, which is pretty broad—Federal and State legislation; license and permitting; development planning frameworks; codes of practice, and the like.

Our approach has been to try to participate in every one of these five areas. If we are only active in the first two then with a small-scale business catering for less than 2% of visitors to the Island, we can have very little influence over things which ultimately impact our visitor's experience, our commercial success, and our long-term sustainability.

We try hard to influence the direction of our industry and the behaviour of both the operators with whom we share the destination and that of our guests—whether they be on a guided experience or self-driving, and happen to be at the same place at the same time.

From a best practice point of view, what makes Kangaroo Island exceptional as a destination?

At present there is significant alignment between community values and those of the visitor population—that is, people come to Kangaroo Island for the same reasons why we as locals choose to live here. This has been recognised in the design of our development plan, the content of strategic tourism plans, and those of key agencies. It is also reflected in the establishment—15 years ago—of the Tourism Optimisation Management Model (TOMM).

TOMM is a multi-agency and community process to identify our desired future state and compare it with the current status by way of regular resident and visitor surveys, and the collection of data on a specific range of indicators covering our social, economic, environmental and visitor satisfaction performance.

Whilst this programme is not resourced at the same level as it has been in the past, and is in need of a refresh and rejuvenation, it still has influence. It's been critical in establishing a forum for regular and thoughtful discussion between key players (agencies, land managers, tourism businesses) focused around the question of sustainability.

A discussion around Kangaroo Island would be incomplete if it did not recognise two core elements of our community. We are very fortunate to have a legacy of a large portion of the Island that has retained its natural vegetation, which is the habitat for our wildlife (on private and public lands).

The natural resource management approach has been positive and engaging—especially in comparison to most other places. I think we are closer to achieving a balance between room for nature and production on our farmland. Is it perfect? No. Can we do better? Of course—but we are continuing a journey from a good starting point.

The other element is the fact that the tourism industry recognised the value of observation of wildlife in the wild, thanks to the work of a tour operator called Don Dixon, who has only just passed away as an old man. Don pioneered the establishment of the Australian Sea lion experience at Seal Bay. Without feeding the animals he was able to develop a trust with this population. And that approach of benign habituation without feeding or handling the animals is largely the modus operandi now for wildlife observation across the Island.

Through Australian Wildlife Journeys,[1] ***you actively network with other independent tour operators across Australia. What led to this collaboration, and how have you benefited from it?***

There were several drivers that led to the establishment of Australian Wildlife Journeys. One was the observation that there appears to be a gap in the market in educating the travelling public and the traditional travel trade distribution system about the opportunities for wildlife observation in Australia.

The National and State Tourism Organisations present at a high level and seem to have been running a succession of campaigns that, whilst they are based on market research, are quite superficial in terms of the exposure of wildlife and the truly unique nature our long period of isolation from the rest of the world has given us.

In the minds of the more serious nature traveller, Australia just does not rate. East Africa, Antarctica, Komodo, Costa Rica, the Pantanal, Alaska—these are all considered the hot wildlife places and Australia is rare amongst them. We believe that we have much to offer given the number of endemic animals—mammals, reptiles, birds, and even some of our marine organisms.

Another contributor was that much of the 'consumption' of wildlife experiences was and still is, captive. Whilst well-designed and operated captive animal experiences can play a positive role in environmental education and visitor experiences, we cannot hope to conserve wildlife in captivity. Without habitat, we have no hope.

Our members believe that by sharing immersive wildlife experiences in the natural habitat, we can give our guests an understanding of how critically important it is to protect natural habitat.

---

[1] https://australianwildlifejourneys.com

We also believe there is great benefit in sharing our guests between us—it is likely that if someone is visiting me on Kangaroo Island for a wildlife experience, they would also enjoy joining up with Janine and Roger at Echidna Walkabout in Melbourne, or exploring the rainforest with James at Far North Queensland Nature Tours[2] in Cairns.

In terms of lessons learned, it is still early days—we are only now going into our 3rd year and since we are located right across the country, spending time together face to face is infrequent. When we do get together, it is amazing, as there is great enthusiasm and energy, and a great sharing of ideas.

We are getting greater consistency in how we approach things—from the transactional 'booking terms and conditions' type of activity, to shared image libraries, market knowledge, and the latest research to share with guides and, ultimately, with our guests.

Destination development was the focus of your work as chair of the Kangaroo Island Steering Committee. To your mind, which are the main issues right now which could potentially threaten the sustainability of tourism on the island?

We are vulnerable to many of the same stresses as other destinations—not all of these are current issues, but they are certainly visible on the horizon:

'Overtourism'—which has many elements, some of which are infrastructure stresses, availability, volunteer exhaustion, crowding. Issues that diminish the quality of the experience for visitors, and locals losing access to the very things that they value about living in a place.

Globalisation—a generic sameness of experience—same food, same clothing stores, same vehicles, same tacky souvenirs. Note that these are issues largely because they are 'contrived' rather than 'derived'. A mentor of mine, David Crinion who worked in planning at the South Australian Tourism Commission, impressed on me the need to focus on the derived rather than the contrived as a key differentiator for developing tourism in alignment with community values—and it remains an important concept in this discussion.

The external imposition of developments—one which has been imposed here is the cruise market. I believe the cost of this outweighs the benefits—the numbers of arrivals in one pulse are larger than the original facility design and certainly larger than the experience design parameters. Cruise arrivals coincide with an already busy season, so all resources are stretched.

With cruise arrivals, I feel that we are unable to provide an experience which is on-brand and we risk diluting the reputation of the destination. It is also pushing the bounds on acceptance of tourism by those in the community who do not derive any direct benefit from it.

The attraction of visitors for whom the community and natural values are less important—once we lose the 'shared values' experience, tensions will become more apparent.

---

[2] https://fnqnaturetours.com.au

Seasonality—We have established a programme 'Open All Year' in an attempt to get the industry focused on a complete four-season approach. We really cannot cater for more visitors when we are already busy. Forward bookings for key periods in Spring and Summer for late 2019 are very strong, yet in the cooler months, there is considerable underutilisation of resources. We need to act strategically to get this message across.

One of the challenges is that tourism is multi-channel and the providers are a mix of a plethora of small businesses and a few very large ones. For completely different reasons there is inertia in both, which is proving difficult to influence.

Keeping decision-making local is very important—otherwise, we are condemned to suffer from what is often referred to as 'unintended consequences'—outcomes which were unforeseen largely due to a lack of understanding of social and natural relationships and not understanding the importance of an empathetic, positive and enthusiastic host community for the successful tourism industry.

Which are the keys to a successful responsible tour business?

- Clear identification of an offer that an identified market is prepared to pay for
- Consistency of service
- Thoughtful planning of the experience and the systems which deliver it
- Consistency in marketing
- A global view to understanding what we are competing with and how to position ourselves

Where do you see opportunities to be seized in Australia right now, in terms of making its tourism offerings more sustainable and encouraging responsible travel?

I think there is a need to really understand the value of our natural experiences here in Australia, and to have a strategic approach to developing and presenting this to the world.

The strategic planning that I have discussed for Kangaroo Island has not occurred at a national level. We started down a very strategic path with the development of a programme called 'Australia's National Landscapes' which articulated around the nexus between Conservation and Tourism and was a tenure-blind approach led by Parks Australia and Tourism Australia. Sadly, this programme was abandoned just as it was starting to gain some real traction at the grassroots level.

In the absence of a strategic approach, we will continue to lurch from chasing one new bright shiny thing to the next.

There is a great opportunity for Protected Areas Management Agencies to offer more incentives for achievement of a higher level of sustainability. We have excellent, world-class accreditation systems which are externally audited and already in place. There can be significant financial, or efficiency bonuses provided for those operators who reach higher levels of accreditation—making licenses cheaper—or valid for longer, than those operators who choose not to participate.

In the accommodation sector, there is a great opportunity in at least a couple of areas. One is to showcase energy-efficient, beautiful design with passive solar architecture and completely 'off-grid' with complete reliance on renewable power.

One example is a property developed by one of our guides, Tim Wendt, Oceanview Eco Villas.[3] I can see this property having real influence over future design choices for guests who stay and experience the place.

Another area of opportunity in accommodation in Australia is beautifully designed places at a lower price and service point. It seems that no-one goes out deliberately to create a two-star property—they build a four-star and as the place gets run down it ends up being two stars by default! As a result, we (Australians) associate places with few stars with bad experiences. I have stayed in beautiful hotels in Europe rated two stars and they were superb. No restaurant, no room service, no porters—but great linen, comfortable beds, simple timber furniture, and a great sense of place.

Recent changes to development planning regulations—at least in South Australia, will result in far better energy efficiency of new builds, as there is a requirement for a much higher performance rating, so this will have a clear benefit.

If you could turn back time, is there anything you'd do differently—lessons learned?

I would definitely start working with private property owners from the outset— the ability to participate in revegetation and have access to private areas and guaranteed solitude—that is really well received by guests.

The experience of each enterprise is always so different—access to start-up capital makes an enormous difference and it has taken years to get to the point where we can start to invest in areas that have previously been out of reach.

It would be nice to think I could have developed and sustained networks with other operators earlier, but the 365-day nature of our businesses means that we are all time-poor.

Link to the interview: https://sustainability-leaders.com/craig-wickham-interview/

---

[3] https://www.oceanviewkangarooisland.com.au

Dagmar Lund-Durlacher on Sustainable Tourism in Austria and Germany

Austria | Dagmar Lund-Durlacher is the founder and CEO of the Institute for Tourism Sustainability and Senior Research Associate at the Centre for Sustainable Tourism at Eberswalde University for Sustainable Development. Dagmar completed her doctoral studies at the Vienna University of Economics and Business and was later the Dean and Head of the Department of Tourism and Service Management at Modul University Vienna. Between 2010 and 2014, Dagmar chaired the BEST (Building Excellence in Sustainable Tourism) Education Network. She is also a technical expert for UNEP for their Transforming Tourism Value Chains in Developing Countries and Small Island Developing States project and part of the UNWTO working group 'Measuring the Sustainability of Tourism', aiming to develop an international Statistical Framework for Measuring the Sustainability of Tourism. Since 2010 she is co-chairing TourCert, an international certification organization for tourism providers and destinations and part of the Scientific Advisory Council of Futouris e.V., a sustainability initiative of the German travel industry.

© The Author(s), under exclusive license to Springer Nature Switzerland AG 2022   187
F. Kaefer, *Sustainability Leadership in Tourism*, Future of Business and Finance,
https://doi.org/10.1007/978-3-031-05314-6_27

Areas of expertise: tourism research, Germany, Austria, Europe, UNSDG 11 (Sustainable Human Settlements)

The following interview with Dagmar was first published in August 2015 on Sustainability-Leaders.com.

Dagmar, when did you first hear about sustainable tourism?

Having joined the anti-nuclear and peace movements as a high school student in the late 1970s, I got interested already in sustainable tourism during my graduate studies of business administration in the mid-1980s when the term sustainability had not been born yet.

It was the time when the first criticism of mass tourism arose because the social and environmental impacts of mass tourism became visible also in the small and beautiful Austrian alpine destinations. Alternative forms of tourism were being discussed, called soft tourism or alternative tourism.

During a conference in Lloret de Mar, a mass tourism destination in Catalonia (Spain), I got to know my Swiss colleagues Jost Krippendorf and Hansruedi Müller and our discussions about the potential risks and negative impacts of mass tourism reinforced my interest in sustainable tourism and inspired my future research. I stayed connected with my Swiss colleagues who at this time were leading the discussion about alternative forms of tourism in Europe.

How has your view of sustainable tourism changed over time?

In the 80s and early 90s, I was mainly concerned with the negative ecological and social impacts of tourism and exploring ways to avoid or combat these.

This has constantly developed into a more comprehensive and holistic view of sustainable tourism, aiming at improving society's well-being and maintaining the natural resources for future generations.

As a leading sustainable tourism researcher, which have been the main lesson/insights for you personally?

It takes a long time to fully grasp the complex concept of sustainability and there is nothing like a 100% right solution for sustainable tourism businesses or destinations.

Sustainable tourism development is a process of learning. It is about involving all stakeholders, developing mutual beneficial relationships, embracing principles such as transparency, honesty, trust, inclusiveness, information sharing, consensus decisions and shared responsibilities.

Engaging all stakeholders in the developing process and transparent and inclusive communication is maybe the most important lesson which I have learned.

Has sustainable tourism awareness changed in recent years?

The awareness among policymakers, the industry and the consumer has definitely increased in recent years, but we are still far from sustainability becoming mainstream. A major part of the business is still going as usual, despite the growing environmental concern due to the global warming discussion, natural disasters and food scandals.

But sustainability has made it at least on the agenda of major industry associations and governmental bodies and there are many great initiatives to encourage sustainable tourism.

Why did you decide to lead BEST EN (Building Excellence in Sustainable Tourism)?

To my knowledge, BEST EN is the only international network of academics who are focused on research AND teaching in sustainable tourism. When I got again more engaged in teaching, I was searching for appropriate teaching resources and found the BEST EN website and their resources.

I was very impressed by their open access policy and joined the network in 2006 and became an executive member in 2009. From the beginning, I was impressed by the inclusive and collaborative working atmosphere of the network, the people who all were driven by the desire to develop and strengthen the concept of sustainable tourism.

It was an honour and a pleasure to accept the chair position knowing that I was supported by an active and committed executive committee and many supporters within the network.

We all had the same goal, to 'create and disseminate knowledge to support education and practice in the field of sustainable tourism'. And I enjoyed every minute working with a dedicated team of very knowledgeable and engaged academics.

The main insights from your four years at BEST EN?

The field of sustainable tourism calls even for more interdisciplinary research than tourism itself already does. Working with colleagues from different disciplines such as economics, psychology, anthropology, geography and many more provides the opportunity to look at issues from different perspectives.

There is definitely still a need for profound research in the field of sustainability, but not only creating knowledge through research is important. Emphasis also has to be placed on disseminating this knowledge to our students, the industry and the wider public.

One important task of academics is the transfer of knowledge, and BEST EN always aimed to develop and provide educational resources in different formats which are accessible to all.

One of BEST EN's major strategic decisions was to provide research and educational resources free of charge because we wanted to provide these resources especially to students and colleagues who did not have the financial means to buy them.

Therefore, most of the resources produced by BEST EN 'friends' are downloadable from the BEST EN website for free. There is also no membership fee for joining the network and for me BEST EN is a very good example for a network which is pulled together by a common goal and not by formal structures.

How do you assess the current situation in Austria in terms of sustainable tourism? And in Germany?

The topic of sustainable tourism is not very prominently on the agenda of the tourism industry in Austria. I have been thinking a lot about the reasons for this absence of public discussion and in my opinion, this is because tourism in Austria is still taking place in a quite healthy environment and that sustainable tourism topics are not a major concern for Austrian tourism businesses.

On the other hand, sustainability has certainly reached the public discussion in Germany, especially the large tour operators and cruise lines are affected, but also destinations in Germany are preparing for more sustainable tourism offers.

In Germany, politics and industry associations show a clear commitment to sustainable tourism development and Germany-based initiatives such as FUTOURIS or TourCert underline the importance placed on this topic.

Your advice to graduate students and emerging researchers interested in sustainable tourism?

Don't lose your passion. It is not easy to find jobs in sustainable tourism, especially well-paid ones. Get engaged with networks and stay close to decision-makers. If a sustainability approach is not supported by the decision-makers of an organization or business, it most likely will fail.

One of the key skills necessary when working in/for sustainable tourism are excellent communication skills, because communication with stakeholders within and outside the organization or business is crucial for a successful sustainable tourism development.

Your favourite book on sustainable tourism in 2014/2015?

Nachhaltiger Tourismus (2015), edited by Hartmut Rein and Wolfgang Strasdas. This is the first comprehensive book in the German language which addresses the phenomenon of sustainable tourism. It was high time that this book was published because the majority of tourism programs in the German-speaking world are still taught in German and with this book many student cohorts will be reached with this important topic.

This leads me to the remark that we still face a language barrier in our globalized world and that we have a lack of knowledge about the research of our colleagues in regions such as Latin America, France, China etc.

In your view, where is sustainable tourism research and teaching headed in Austria and Europe?

Besides environmental aspects, there are more and more social aspects included in the sustainable tourism research agenda. Still, climate change, negative environmental impacts, resource scarcity, and mobility issues are seen as important topics, but social issues such as integration, poverty alleviation, workforce-related aspects, and social well-being gain importance. In recent years also the concept of social entrepreneurship entered the research agenda.

A positive development are the increased efforts to measure sustainable tourism by developing and applying sustainability performance indicators as well as to measuring the social impacts.

Which is the best way to connect researchers and professionals in the sustainable tourism field? In which way can they help each other?

Through common goals and joint projects, be it in joint research activities or knowledge transfer. Researchers and industry must communicate with each other. Researchers should know firsthand about the questions and problems the industry is facing and expert panels can be a good way to exchange this information.

On the other hand, it is also important that researcher transfer their research results to the industry and often this is not very successful due to the lack of common platforms and language.

Research results are often presented in a very sophisticated manner exclusively in scientific journals which are not accessible or read by the industry.

This would call for a 'translation' medium which filters the relevant information and provides it in an applicable format to the industry.

Link to the interview: https://sustainability-leaders.com/interview-dagmar-lund-durlacher/

Portugal | Daniel Frey has 20 years of experience with reputed international hotel chains working on operational and corporate levels, which have taken him around the world. He studied at the Lausanne Hotel School and furthered his studies at Harvard Business School (AMP159). In 2018, he completed the Business Sustainability Management course with CISL (University of Cambridge, Institute for Sustainability Leadership). His recent work includes destination work at different levels (strategy & concept definition, working with government agencies in aligning policy, stakeholder engagement & implementation of destination certification framework). Daniel offers strategic advice, consultancy, training and coaching to various tourism players and project consultancy in sustainable design and construction of buildings. He is also an independent auditor for reputed and officially recognized certifications for the tourism industry—destinations, hotels and other tourism operators. He has helped conceive, establish and implement the vision and sustainability strategy for tourism players aligned with UNSDGs.

Areas of expertise: hospitality, tourism business, destinations, Portugal, Europe, UNSDG 12 (Resource Efficiency)

© The Author(s), under exclusive license to Springer Nature Switzerland AG 2022
F. Kaefer, *Sustainability Leadership in Tourism*, Future of Business and Finance,
https://doi.org/10.1007/978-3-031-05314-6_28

193

The following interview with Daniel was first published in December 2015 on Sustainability-Leaders.com.

Daniel, was sustainability a topic when you started your professional career? What got you interested?

The spirit of respect, preservation and biodiversity was a gift that our parents shared naturally with all of us children, but it was not much in evidence when I started my career.

My experiences in different parts of the world made me aware that there is more to life and work than earning money—I saw things happening around me: deforestation, slash and burn, inequalities and other things that made it clear to me that it was time for change, and that 'business as usual' was not good enough.

I wanted to consolidate these experiences and show young people that it is possible to be sustainable and build a business case around this—a win-win situation.

Has your view on sustainability and sustainable tourism changed in recent years?

Hospitality leaders are increasingly aware and knowledgeable about the issues facing our industry, and this in turn results in them embracing sustainability as an integral part of their strategy.

The leadership challenge is to embrace sustainability both top down and bottom up: to empower staff to generate change and grassroots sustainable practices.

For too long sustainable development has been a synonym for cost savings, energy efficiency and other actions that do not constitute a strategy but serve one purpose only—to reduce outgoings.

Why did you start Green Growth, and which achievements so far are you most proud of?

I had a vision that led to the creation of a sustainable hotel chain and I launched its first property (Inspira Santa Marta) in Lisbon some years ago. The continued success of that brand made me want to use and share what I had learned and the competencies I had acquired with other professionals that needed guidance in that field.

Green Growth has clients in various points of the globe, and we are proud that all of them have embraced a sustainable approach to their businesses. However, there is one place that definitely stands out: Chepu Adventures in Chile—their vision, concept, commitment is exceptional, and they have won several awards—only a visit can convey the real spirit of their venture.

Your key professional insights from your time as CEO of award-winning Inspira Hotel in Lisbon?

Very clearly the importance of empowering and motivating the team. Having sustainability embedded in the strategy and everybody's day-to-day realities from the very first day made it possible to have the team carry these values and excel in their practical implementation.

The enthusiasm and commitment of all staff members were key to the hotel's success; this translated into motivation, quality, differentiation. The hotel remains a showcase for sustainable development and a leader in the Portuguese capital.

In your view, which are the most important ingredients for a sustainable hotel?
The first ingredient must be top-down leadership, following a clear strategy with direction and objectives for this sustainable approach, enabling culture change in the organization and engaging all stakeholders across the board.

How to measure the success of sustainability initiatives in hospitality?
The way to measure is to establish a corporate scorecard with clear KPIs (key performance indicators) per section, as part of executives' objectives and incentive schemes. These KPIs include energy efficiency, operational indicators, HR indicators, etc.

3 books on sustainability which you'd highly recommend?
Hot, Flat, and Crowded: Why We Need a Green Revolution—and How It Can Renew America, Release 2.0 by Thomas L. Friedman
Cradle to Cradle: Remaking the Way We Make Things by Michael Braungart and William McDonough
Leading Change by John P. Kotter, and with it *Our Iceberg Is Melting* by John P. Kotter and Holger Rathgeber

We witness growing confusion and concern about eco-labels and sustainability certifications in hospitality and tourism. Which would be your advice to business owners and destination managers regarding how to choose the right certification?
Yes, greenwashing, tick box certifications—quick fixes have invaded our world and turned sustainable development into an excel sheet formula, when what is really necessary is genuine change, vision and conviction.

Firstly, a certification needs to be recognized by the GSTC. Then there is a handful of renowned certifications that themselves are there to ensure their members instigate and generate change—their criteria are verified by specialized professionals who understand the businesses they certify and the sustainability aspects.

Measurement-based certifications are important: these are quantifiable, based on KPIs, benchmarks and not just statements.

An interesting question to ask certification schemes is how many projects that applied for certification didn't make it, or what number of certification requests were rejected. Most of them will tell you zero, as they are interested in receiving membership fees first.

Your advice for tourism businesses or destinations eager to gain recognition for their sustainability efforts through sustainable tourism awards? Which ones do you consider most reputable?
Before hunting awards, there is a great deal of groundwork to be done requiring internal work and dedication.

People should talk about you not because of your marketing efforts but because they feel there is a difference, a spirit, a drive—these are the elements that attract nominations or put businesses forward for these awards.

There are interesting awards sponsored by the UN, and the WTM Responsible Day is a great forum with an interesting approach. But again, what really matters is what you do, not the labels and awards you have got on your wall—credibility stands and falls with your integrity.

If you had to start your professional journey all over again, knowing what you know now about sustainability, what would you do differently?

I would firstly look for a university or hotel school that integrates these subjects in the curriculum and then seek to work for employers that are known for their endeavours in sustainable development.

Link to the interview: https://sustainability-leaders.com/interview-daniel-frey/

Darrell Wade on How Intrepid Champions Carbon Management and Sustainable Travel

Australia | Darrell Wade is the Co-founder and Chairman of Intrepid Travel, the world's largest certified travel B Corp. A highly regarded entrepreneur and sustainability advocate, Darrell and Geoff ('Manch') Manchester founded Intrepid in the late 1980s as a new way for people to explore the world that was immersive, sustainable and gave back to the communities they were visiting. Darrell has been an active advocate on climate change since 2005. He was one of the first Australians to complete Al Gore's Climate Reality Project training and is Vice Chair of the World Travel and Tourism Council, where he leads the development of its sustainability agenda. Darrell is part of the advisory group for Travalyst, a global partnership founded by The Duke of Sussex, and is the Chairman of The Intrepid Foundation as well as the director of two philanthropic foundations.

Areas of expertise: destination sustainability, entrepreneurship, responsible tourism, tourism business, Australia, UNSDG 13 (Fight Climate Change)

F. Kaefer, *Sustainability Leadership in Tourism*, Future of Business and Finance,
https://doi.org/10.1007/978-3-031-05314-6_29

The following interview with Darrell was first published in May 2021 on Sustainability-Leaders.com.

Darrell, you have three decades of experience in the travel industry. When did you first discover your passion for sustainable tourism?

I've been lucky enough to have been a regular traveller since a very young age, but it wasn't until I was 21 that I headed off for a year backpacking through Asia that my travel paradigm changed. Then in 1988, I lead a group of 14 friends across Africa in a converted rubbish truck. The group included my fellow Intrepid co-founder Geoff 'Manch' Manchester and my new wife Anna.

That was the first time Manch and I talked about creating a style of travel that gave travellers the opportunity to enjoy local experiences and stay in communities. It was a very connected, fun way to travel to be a part of the destination, rather than removed from it. We didn't use the word sustainability those days—no one did—but before long, we realised that it had the huge fringe benefit of being a highly sustainable and responsible way to travel.

We started Intrepid the following year—I was running the business from my kitchen table in Melbourne and Manch was our first tour leader in Thailand. We had 47 customers that first year, so we certainly were not a success from day one! It all grew from there, and in 2019 we carried 460,000 passengers.

Responsible and sustainable travel has always been at the heart of our business— we realised that not only did our customers love the respectful way we travelled, but local communities were really supportive too. Everyone was a winner! Realising this was a bit of a hidden competitive advantage, we employed a responsible travel manager very early on. Everyone thought we were some kind of crazy left-wing hippy outfit!! But actually, we knew it was the key to our success and so wanted to build sustainability as a part of our business model.

In 2002, we established The Intrepid Foundation so travellers could give back to the communities they visited—again a great way to reinforce our values to our travellers, whilst also getting them to actively support our work. Once again, everyone won out of the arrangement—our travellers, our staff, the local communities we relied on and us as a business. It was a unifier.

In 2005, I read the book The Weather Makers by Tim Flannery and I experienced a serious existential crisis. I realised that the travel industry—and our business—was directly contributing to climate change. We were a part of the problem! I asked the whole executive team to read the book and I became one of the first Australians to complete training with Al Gore and the Climate Reality Project in 2007.

From there, we set a goal for Intrepid to become carbon neutral, which we achieved in 2010. Once again some thought we were a bit crazy taking such a stance and going out on an expensive limb like that. But actually, we were just putting our values into action—and our customers really appreciated us for doing that. It won us business rather than costing us business.

Intrepid Travel is the world's largest carbon-neutral travel company since 2010. How are you planning to achieve Intrepid's next ambitious goal of becoming a climate-positive company?

We realised in 2019 that being carbon neutral simply was no longer enough. The world wasn't changing fast enough, emissions were still rising, as were global temperatures. Carbon offsets play a role, but they are not enough alone to address the climate crisis. Real action and commitment are needed to actually reduce the amount of carbon we're all producing, not simply offset it (although we continue to offset, of course).

Around this time, it was clear that global warming was having ever greater impacts—and this wasn't only hurting people, communities and wildlife around the world, it was also affecting our business. We were having to change itineraries due to extreme weather and our office in Bangkok experienced flooding.

At the start of 2020, we declared a Climate Emergency[1] as a founding member of Tourism Declares, a collective of businesses, organisations and individuals who are concerned about the climate crisis.

As part of that, we adopted our seven-point climate commitment plan, which includes setting science-based carbon emissions targets and becoming climate positive by actually creating an environmental benefit by removing additional carbon dioxide from the atmosphere. That includes offsetting more carbon than we produced in 2020 (25% extra), which is our first step towards our commitment to being climate positive. We use gold-standard carbon offsets that produce a number of other social and environmental benefits, such as the Savannah Burning project in Arnhem Land in Australia's Northern Territory.

Despite 2020 being a devastating year for travel, Intrepid became the first global tour operator with verifiable science-based targets through the Science Based Targets Initiative.[2] This independently assesses corporate emissions reduction targets in line with what climate scientists say is needed to meet the goals of the Paris Agreement for a 1.5 °C future.

From a practical perspective, that means we will transform our business for a low-carbon future by reducing emissions across our operations and trips, over the next 15 years.

This is a science-led approach, which will see us reduce absolute scope 1 and 2 greenhouse gas emissions by 71% by 2035 from a 2018 base year. We'll also reduce scope 3 greenhouse gas emissions from our offices by 34% per full-time employee equivalent, and from its trips by 56% per passenger day over the same period.

Congratulations on Intrepid Travel being named as one of the World's Most Innovative Companies in 2021, according to Fast Company. What impact does this recognition have on your business?

Thank you! This is indeed a fantastic recognition—to be named alongside travel brands such as Airbnb, particularly after the year that was 2020.

Intrepid was recognised for our commitment and action to address climate change, our product innovation with the development of a range of 100+ new

---

[1] https://www.tourismdeclares.com

[2] https://sciencebasedtargets.org

local and domestic trips, and our ongoing commitment to purpose initiatives, such as our not-for-profit The Intrepid Foundation.

This recognition helps to demonstrate that standing for something and leading with purpose is the best way to differentiate your business and brand.

We know that our people and prospective employees want to work for innovative organisations that reflect their own values, so this helps us attract great people. We also know that customers want to support businesses that support causes that are important to them.

Intrepid is the world's largest certified travel B Corp[3] and we've seen how other B Corps, in many different industries, have been able to grow their business and brand by living their values and leading on social or environmental issues.

Of course, having our name in one of North America's leading business publications helps to raise awareness too, especially since we've just launched a new range of trips for Americans in the US. So, it's another example of creating a positive feedback loop: values-based decisions lead to good outcomes lead to better customer and staff experience leads to further growth. And on the cycle goes! It's not rocket science so I really hope other companies can learn from our experience and start to put values at the heart of what they do.

What tips or suggestions do you have for smaller travel establishments that want to engage with carbon offsetting, but don't know where to start?

Carbon offsetting is a good start, but it certainly isn't the only thing a travel business—whatever its size—can or should do. Look to reducing emissions where you can, then offset the balance that is difficult or impossible to reduce.

Not all carbon offsets are created equally, so the first thing is to do some more research and understand a little bit more. Gold Standard[4] is a good place to start! They are what it says on the tin—the gold standard.

We focus on measurement, reducing emissions and reporting because this is the only way businesses can really start to take meaningful action. I suggest that all travel leaders download Intrepid's free Decarbonisation Guide,[5] which is available on our website. This provides a step-by-step guide and explanation about how to start your carbon journey. It is authored by Intrepid's Environmental Impact Specialist Dr Susanne Etti, who is an expert in the field. It includes resources, links and information to help other travel businesses start to measure your business's carbon emissions. Dr Etti has even conducted numerous one on one discussions with companies who've downloaded the guide to help them get started, just because she's personally so passionate about climate change and the UNWTO has written an article about this collaborative approach.

Finally, I would recommend that businesses reach out to organisations like Tourism Declares to find out more. Tourism Declares a Climate Emergency is a

---

[3] https://bcorporation.net

[4] https://www.goldstandard.org

[5] https://www.intrepidtravel.com/au/download-our-quick-start-guide-decarbonise-your-travel-business

community of 230+ travel organisations, companies and professionals who have declared a climate emergency and are coming together to find solutions.

Tourism Declares is really mobilising the industry, including organisations of all sizes, and they have ambitious plans ahead of COP26 later in 2021. The travel and tourism sector needs to more or less halve its carbon impact by 2030. We can only achieve this goal if we all work together, sharing learnings, supporting each other and advocating for innovation.

Which achievements are you most proud of as the Co-Founder & Chairman of Intrepid Group so far?

- I guess ultimately, it's about providing extraordinary holiday experiences to millions of people, and doing that in a way where we all grow and learn things. It's incredibly satisfying.
- Becoming carbon-neutral all the way back in 2010 was a big achievement, since many companies are only thinking about it now, and more recently becoming the first global tour operator with verified carbon emissions targets by the Science Based Targets initiative in 2020.
- Achieving B Corp certification in 2018, which included becoming the first certified B Corps in Vietnam, Cambodia and Sri Lanka. B Corp is a validation tool for yourself, your staff and your customers that you are heading in the right direction and how to improve that direction in the future.
- Our decision to ban elephant rides on our trips back in 2014, which was a catalyst for the rest of the industry and many companies followed our lead.
- Achieving our goal to double our number of female tour leaders globally—we did this in 2019, a year earlier than we planned—and it was an important milestone as tour leader has often been viewed as an unsuitable job for women in some countries around the world.
- In 2020, I was named elected as a Vice-Chair of the WTTC and Chair of its Sustainability Committee. This is recognition of the decades of work that Intrepid has put into sustainability and I'm proud that Intrepid is seen as a leader and that I'm in the position to help influence our industry globally.

Following the coronavirus pandemic, what trends will we observe once travel restrictions are eased?

The pandemic has clearly had a devastating impact on Intrepid and all travel businesses. But I really believe that we will see a strong upsurge in demand for leisure travel and particularly our style of travel when tourism can restart again safely.

The days of mass tourism and people rushing through countries to tick off 'the sights and overwhelming cities with overtourism are hopefully long gone. The pandemic will change our societies and I believe people will choose to travel more mindfully, slower and stay longer in destinations, especially if they have the option with their jobs to work remotely for part of the time.

We're certainly seeing people travelling more closer to home already. We actually think this will be a long-term trend too as people increasingly factor the carbon emissions into their lifestyles and personal choices. Intrepid has already launched more than 100 new local trips in countries including Australia, the UK and most recently the United States. This is the first time we've offered trips to people in their home countries. In the long term, I think over half of our business will be in domestic travel.

And, travellers will want to be in nature and do active trips, such as walking, hiking and cycling. We're already seeing strong interest in these sorts of trips in places like Maine in the US, Cornwall in the UK and Northern Territory in Australia. This has a lot of benefits for local communities and, of course, more local travel and fewer flights mean fewer carbon emissions, which is a win-win.

The pandemic has set many destinations back, with no hints of a full recovery anytime soon. To what extent will the current situation impact the commitment of cities, regions, or countries to sustainable tourism?

This is a real risk and a serious concern, especially for us as the largest global adventure travel company, operating in more than 100 countries.

In some ways, the pandemic has prompted some cities, regions or countries to commit more deeply to sustainability, as they look for new innovative ways to reinvigorate their economies and reshape their workforces. More than 230 organisations have now declared a Climate Emergency, including destinations such as Visit Scotland and the Oregan Coast Visitors Association.

But that is certainly not true everywhere.

One of the challenges is that sustainability is sometimes viewed as expensive and difficult—and changing the way we live, work and travel will be hard—but I don't believe we have any other choice. The pandemic has devastated many destinations but the impacts of climate change will be much worse, so we all need to commit to building back better and helping others to do so.

The pandemic has highlighted the huge social, health and economic gaps that exist around the world. I'm worried about the impact on children and education, that there could be a rise in child labour and an increase in orphanage tourism. We've worked with our partners Rethink Orphanages[6] in the past to raise awareness about this and together we've advocated for orphanage tourism to be included in the Modern Slavery Act, making it illegal for operators to include it in their supply chain. But according to UNICEF, millions of children will be pushed into poverty due to Covid and orphanage tourism is likely to return, given the pattern of demand following other disasters.

These are the sort of things that businesses can do to help ensure tourism doesn't just return after Covid—but that it restarts in a responsible and sustainable way.

Anything else you would like to mention?

Science-based targets are the most effective way to tackle carbon emissions and I really encourage every business leader to learn more about them. The sooner businesses respond, the greater the benefits—not only for the planet but also

---

[6]https://rethinkorphanages.org

businesses that are first movers in this space will have the most opportunities to create value.

We've also seen that measurement is a real struggle for many businesses, but this is a really essential step. Otherwise, businesses risk 'greenwashing' and not actually taking meaningful action.

Finally, I'd remind other leaders that they aren't alone. We're all on this journey together—so reach out to your network, get involved in organisations and really approach sustainability as an urgent business priority. And reach out to me or the leadership team of Intrepid—we're here to help if we can!

Link to the interview: https://sustainability-leaders.com/darrell-wade-interview/

New Zealand | David Simmons is an Emeritus Professor of Tourism at Lincoln University (New Zealand). At his retirement, he was also principal research strategist and University research chair. Internationally, he was the Director of Research for Australia's Sustainable Tourism Cooperative Research Centre (2008–2010) and is the chair of EarthCheck Global Research Institute.

His expertise is in destination governance, planning and management, and enhancing the financial, economic and sustainable yield from tourism. He continues to offer advice, keynote presentations and consultancy to the tourism sector. Internationally he had the opportunity to work on international tourism (planning and governance) programmes in Cambodia (WWF), Mauritius (UNDP), Niue, Vanuatu (UNWTO/UNDP), Nepal (ITTO, MFAT and WWF), India (WWF), Sarawak (E. Malaysia), and North Korea (UNWTO) and more recently the sustainability platform for the Azores. He is a fellow of the International Academy for the Study of Tourism and the Council of Australasian Tourism and Hospitality Educators.

Areas of expertise: tourism research, destinations, sustainable development, Australasia, UNSDG 12 (Resource Efficiency)

F. Kaefer, *Sustainability Leadership in Tourism*, Future of Business and Finance, https://doi.org/10.1007/978-3-031-05314-6_30

The following interview with David was first published in May 2018 on Sustainability-Leaders.com.

David, when did you discover your passion for sustainability linked to tourism? Who or what stirred your interest?

Not surprisingly, my interest in nature and natural history was inspired by my father, an elementary school teacher from a farming background. During the longer school holidays we were always active in the outdoors—picnics, walks and longer treks, as the years progressed. In my late teens, I was very fortunate to travel the world via sport and soon realised how very privileged we were in New Zealand, in an increasingly crowded world.

My first degree is in ecology (botany and zoology) and it was natural to migrate to resource management, and onto tourism via human geography and planning. Tourism was in its early boom and I wrote a piece for the then government (1986) on the future management of tourism in New Zealand and its effects on New Zealanders.

How have your views on tourism sustainability changed over the years?

My views have certainly continued to evolve. This is especially so as new understandings and metrics become developed and socialised.

In short, we need to 'measure to manage' at all levels in tourism production and consumption: business, destinations, itineraries (land, sea, and air)—and of course via tourist behaviour/consumption.

As a highly respected academic researcher in the field—are there any topics you think should be addressed more by the sustainable tourism research community?

Indeed. In many respects, we need to re-think the consumption paradox in tourism. We might ask what 'high yield' tourism is in terms that reach beyond financial metrics—and reference regional and local economies, alongside the sector's sustainable draw on resources?

Does tourism always need to be consumptive of common property resources? Could, for example, tourism contribute to social and natural capitals by enhancing the concepts of restoration and regeneration?

To your mind, which are the main topics and concerns linked to tourism sustainability at the moment, especially for destinations?

Mindless consumption of common property resources, environments, and cultures. How might they be valued (priced?) in the tourism system.

For us, deep in the southern hemisphere, the carbon footprint of aviation (and cruise) is a very serious concern, as it is for the many developing countries for which tourism is a primary source of foreign exchange.

The coronavirus pandemic has hit the travel sector hard, yet may also lead to opportunities in that it forces tourism to rethink status quo operations. In your view, how can we emerge from the crisis in a way which facilitates a more sustainable future for travel and tourism?

For New Zealand, this is domestic tourism which generates 58% of all in-country tourism expenditure, while New Zealanders themselves (a population of 4.8 million) made 3 million trips overseas last year and in so doing took $NZ 6.5 bn out of our economy.

Elsewhere the highest order challenge appears as 'distance = (carbon) costs'.

A new paradigm has to emerge—one where tourism gives back more than it draws down on all the resources that support it. Slow travel, localism, and mindful long-haul travel (which may well be shaped by post-COVID pricing) will need to become part of this new paradigm.

What tips do you have for professionals who want to deepen their knowledge about sustainable tourism, but don't know how to start?

For anyone choosing a course to study, I would look first at the credentials of the instructors (ahead of the institutions themselves).

Learning is best achieved in a mixture of reading and discussion. Courses need not be standard 'degree' type courses but might include micro-credentials where like-minds can meet and be challenged to new horizons.

Link to the interview: https://sustainability-leaders.com/david-simmons-interview/

Puerto Rico | Eddie Ramirez graduated from Fairleigh Dickinson University in 1986 with a BS in Hotel Management. He was awarded the Green Globe of the month for outstanding performance in Sustainable Tourism in 2009 by Green Globe. Eddie started Casa Sol Bed and Breakfast in 2013, which currently has four eco-labels—Green Key, Sustainable Tourism Facility by the Puerto Rico Tourism Company, Green Leader Platinum by Trip Advisor and STEPS by PRHTA. In 2015, Casa Sol B & B was recognized as Small Green Hotel of the Year 2015 by the PRHTA, an organization that focuses on helping communities and individuals in need, providing essentials during emergencies like Hurricane Maria. He is certified under the Green Business League as a Consultant—Certified Sustainability Officer since 2011. In 2015, Eddie was the Chairman of the Conservation Committee of PRHTA and implemented the Sustainable Tourism Education Program, a practical tool for assessment, benchmarking and education.

Areas of expertise: hospitality, tourism business, entrepreneurship, sustainability, Caribbean, UNSDG 12 (Resource Efficiency)

© The Author(s), under exclusive license to Springer Nature Switzerland AG 2022   209
F. Kaefer, *Sustainability Leadership in Tourism*, Future of Business and Finance,
https://doi.org/10.1007/978-3-031-05314-6_31

The following interview with Eddie was first published in June 2015 on Sustainability-Leaders.com.

Eddie, why did you choose to work in tourism?

It all started when I was 10 years old. My father owned a restaurant and I worked with him during the weekends. Later on, he bought an old building, renovated it and opened a theatre café restaurant. On the second floor of the building, he had eight apartments that he rented out.

As I grew up and continued working for him, the interest in hospitality grew. I saw a future potential of turning the business into a small boutique hotel.

In 1980-81 I took a summer job with the Puerto Rico Tourism Company in New York City. A great experience, which gave me even more reassurance that hospitality, restaurant- and tourism management was my future.

During my college years at Fairleigh Dickinson University in New Jersey (USA), I had the opportunity to attend three European Hospitality Seminars, each year in a different country (Austria, Switzerland and Spain).

I also held part-time jobs in the industry working for hotel chains like Marriott, Holiday Inn, Sheraton, Hilton, and even at a Horse Race Track where I was dishwasher, waiter, runner, bellman, room service attendant and kitchen helper.

All these opportunities just gave me a strong background in the field. I graduated in 1986 and went on a corporate management training program at the Grand Hyatt in New York City, a hotel with 1,500 rooms. Later on, I transferred back to Puerto Rico to work at the Hyatt Dorado Beach and moved to the Caribe Hilton some years after.

What does sustainable tourism mean for you?

I find this definition the most useful:

Sustainable tourism, in its purest sense, is an industry which attempts to make a low impact on the environment and local culture, while helping to generate income, employment, and the conservation of local ecosystems. It is responsible tourism that is both ecologically and culturally sensitive.

Did sustainability play a role during your time at Caribe Hilton in San Juan?

During my 20 years at the Caribe Hilton, I held positions such as Room Service Manager, Minibar Manager, Restaurant Manager, Banquet Manager, Assistant F&B Manager, and Procurement Manager with responsibility for environmental sustainability.

As an Environmental Sustainability Manager, I was nominated Green Globe of the month for outstanding performance on sustainable tourism in 2009. I also did an online course to become a CSO (Certified Sustainability Officer).

When I was asked to spearhead Caribe Hilton hotel's sustainability program, I just knew the basics on sustainability. I asked myself, can I really make this happen? Lack of knowledge was a true challenge, but I had the passion and thirst to make Caribe Hilton the first hotel in Puerto Rico to get an eco-label.

The rest is history. Caribe Hilton, now an icon in sustainability matters, became the first hotel to receive not just one but two eco-labels, Green Globe and Green Key. It was also awarded Green Hotel of the Year twice by the Puerto Rico Hotel and Tourism Association.

Last year I was nominated Chairman for the Conservation Committee of the Puerto Rico Hotel and Tourism Association. I am currently working on an endorsement program called STEPS (Sustainable Tourism and Environmental Performance Stewardship).

Through orientation sessions, trainings and support, we want to encourage hotels in Puerto Rico to start an environmental program and to get certified. 20 properties are already working towards STEPS endorsement.

Which sustainability aspect at Caribe Hilton did you find the most challenging?

Caribe Hilton was very challenging. Even with the blessing of management, the investment issue always generates friction. Accountants are keen on short-term ROI [return on investment], whereas savings through sustainability initiatives are mid-to-long-term.

Just like in agriculture, you first need to prepare the ground, then plant the seed, maintain and care, and eventually you get to enjoy the crop. Sustainability isn't a low-hanging fruit you can just pick and enjoy.

Changing corporate culture and habits is a real challenge, too. At Caribe, I needed to change routines and the way the team members thought, which was complicated by the still wide-spread mentality in Puerto Rico that conservation/recycling is too much work.

Why the focus on sustainability at your latest venture, Casa Sol B&B?

Because it's just the right thing to do morally. It also fits our lifestyle, and together with my wife and our children, we all believe in the value of protecting our natural resources and helping our local community prosper.

Sustainability also makes sense financially, of course. The savings and low operating costs can help you keep in business even during hard times.

Recognition is another benefit of focusing on sustainability. But ultimately, that feeling of satisfaction from knowing that you are doing the right thing is probably the greatest benefit.

Which sustainability aspect did you find most challenging?

At Casa Sol, the most challenging aspect was to get the permits to install the solar panels. Being located in the historical area of Old San Juan, it felt like every agency had to put in their two cents.

Do you communicate your environmental initiatives to guests?

Yes, we do this as part of welcoming our guests, and they just love it. New arrivals always ask: can we go see your rooftop garden, can we see the solar panels, how does a solar water heater work. . .

Which part of your Green Key certification did you find the most challenging?

Challenging was the uncertainty that one had when getting audited, asking yourself "did I do everything the right way", or "did I miss something"? Implementing initiatives was just great, fun and rewarding as you see installations taking shape and becoming effective and productive.

For us, to have been awarded the Green Key certification is truly an honour and a seal for our commitment to helping preserve our natural resources for generations to come.

We are true believers in that little by little we can make a change, and our goal is to make this change a very contagious one, until all of humankind contribute their share to helping our planet stay alive.

Your 3 bits of advice to managers of small hotels eager to improve their environmental performance/sustainability?

Go for it! It's not as hard, difficult or expensive as it is said to be. If you want to be successful long-term, embracing sustainability is the way to go. The savings are there, plus you have a social responsibility as a business owner and/or administrator to be an agent of change for your community.

Your thoughts on the current state of sustainable tourism in Puerto Rico and the Caribbean region?

Regarding sustainable tourism in Puerto Rico, we have a long way to go. As I mentioned earlier, this is a cultural thing, we need to enforce and promote sustainability principles through education. This needs to be part of the curriculum in all elementary schools.

The central government definitely needs to step up to the plate, they play a key role. We have laws but no one enforces them, it's just not on their list of priorities, a real shame.

In Puerto Rico, we only have seven hotels with eco-labels, out of a total of 106 properties. As part of the Board of the Puerto Rico Tourism Association, I am making it a priority to promote sustainable tourism by creating awareness and educating through STEPS.

As for the Caribbean region, they are a few steps ahead thanks to a less US-based and more European culture. You can see governments are more involved and promoting sustainability by making and enforcing laws.

Link to the interview: https://sustainability-leaders.com/interview-eddie-ramirez/

UK | Elisa Spampinato is the CEO and Founder of Traveller Storyteller.[1] As a travel writer, she recognises herself as a Community Storyteller since she focuses primarily on the stories of indigenous and rural communities around the world. She designs and delivers workshops on Storytelling that are tailor-made for Community-Based Tourism (CBT) projects. A passionate and responsible tourism professional and a wholehearted and committed global Community-Based Tourism expert, Elisa is the CBT Ambassador for the Transformational Travel Council. For almost two decades, she has been a writer, speaker and researcher—in NGO and civil society initiatives, academic research, local development projects, international campaigns and advocacy for human and cultural rights, and ethical tourism. Elisa is particularly committed to raising awareness about gender equality and diversity in the travel industry. She is an Associate of Equality in Tourism and a Mentor for Women in Travel CIC.

---

[1] https://travellerstoryteller.com

F. Kaefer, *Sustainability Leadership in Tourism*, Future of Business and Finance,
https://doi.org/10.1007/978-3-031-05314-6_32

Areas of expertise: community-based tourism, responsible tourism, UNSDG 8 (Equal & Fair Economic Opportunities)

The following interview with Elisa was first published in July 2020 on Sustainability-Leaders.com.

Elisa, you wear many hats—Ambassador at Araribá Turismo e Cultura DMC and Travolution Travel, editor at Travindy.Brasil, community storyteller, among other roles. What keeps you motivated to dedicate your time and energy to supporting and promoting sustainable tourism?

My professional journey has given me the opportunity to be actively engaged in different projects and activities over the years, and I wear many hats. But all my hats are telling the same story, just from different angles and to a different audience.

As an Ambassador of Araribá Turismo e Cultura DMC and Travolution.Travel, I am working on giving visibility to a hidden universe of traditional communities immersed in a variety of biospheres and natural environments.

Travindy.Brasil is my personal project, born from an idea to create a digital community where multiple actors of the tourism industry could converge, get inspired, learn from each other, and potentially create partnerships. But above all, I wanted to create a space where the local communities could be seen and recognised.

As a Community Storyteller, my mission is to narrate the stories of hidden communities and create content—with words and images—to build a bridge between travellers and communities.

Travelling and visiting traditional communities opened my heart and it helped me reconnect with myself and the world around me. I would like to allow others to experience the same, starting with a story.

I like to look at sustainability as a dynamic balance that needs to be constantly maintained and which is never achieved, once and for all. It is without a doubt a collective journey. Nevertheless, supporting and promoting it should be everyone's constant responsibility. I am doing my part, with my professional choices and through my words.

Slum tourism is one of the initiatives you are especially interested in. How would you advise tour operators to handle fair sharing of revenue with the community, generated through slum tourism?

My research on slum tourism gave me the chance to observe CBT in an urban area, for the first time. I come from the premise that the community should be at the centre of the way we design tourism.

In urban reality in general, and in slums in particular, the socio-economic and cultural complexity is such that talking about 'community' represents a huge challenge in itself. Communities are everything but 'uniform' realities, especially in a slum.

Nevertheless, I still believe that urban communities should also have a central role if tourism exists within their territory.

If the community does not have formally organised tourism activities, any external actors that would like to offer tours there should take some measures that guarantee the participation of its members and a local positive impact.

Employing community guides is a basic option that an external agency could implement. However, supporting the training and professional development of the guides would be a much more responsible form of giving back to the community. It will, in fact, contribute directly towards local development, social innovation, and inclusiveness.

My advice is to invest in building relationships with the local community. A regular dialogue with the resident association, local business organisations and cultural entities should guide the product development process.

Especially in the case where a local agency or tour operator already exists, the design of the tour—the what, how, where and when a tour happens—should be the result of a co-creation process with their representatives. Because the community should never be the object of a transaction, but always one of the protagonists of the process.

Also, I strongly believe that co-design is the key to sustainability in the tourism sector, and not only in CBT.

Amran Hamzah in his interview mentioned that less than 10 per cent of community-based tourism (CBT) worldwide can be considered successful, most suffering from lack of local capacity, lack of leadership, poor understanding of the market, among other reasons. In your experience, how can CBT be more efficient?

I am aware that CBT means different things in different international contexts. For clarity, the definition that I am referring to here is the one accepted in the Latin American continent.

Brazil probably guides the continent's academic reflections on the subject and recognises the role of the protagonist—the community as the foundation principle of CBT. Other defining principles are cooperation, enhancement of local culture and history, and the protection and conservation of the environment. I totally agree with that.

It is not easy to establish if Latin American CBT is successful, even when considered exclusively from a financial perspective. Based on my experience, though, if we adopt a social perspective, the 'success' of these experiences is undeniable.

Latin American CBT experiences show clear empowerment of the local communities and the strengthening of internal communitarian connections. Also, it is unquestionable that there is an increase in both, individual and collective self-esteem and the impact on gender equality.

The process of rediscovering and reconnecting with their own traditional culture and history through tourism-related activities naturally lead their members to value their own material and intangible heritage. Also, their ancestral knowledge, such as traditional medicine, directly linked to their relationship with the natural environment. Not to mention the impact that it has on the local economy.

CBT projects are not always economically stable. What suggestions do you have for such projects to be profitable and hence become financially independent?

The CBT experiences in Brazil have grown a lot in terms of internal organisation, but they still face limitations in terms of commercialisation and marketing, which continue to be the biggest problems.

In the model chosen by 'Route to Freedom', a CBT experience near Salvador in the state of Bahia, tourism has been consciously included in the local development production chain. Here, and in many other CBT cases, this has been possible because all the economic activities work collectively, according to the principles of the solidarity economy. A grassroots alternative way or organizing the local economy which has been growing in economic terms in the last decades in Brazil.

In any case, even if tourism is practised as a complementary economic activity, the question of its financial independence is a journey that the community should confront, and this would start by recognising it as a business, with its needs and priorities. These include establishing an international presence and their place in the supply chain, in a dialogue with other stakeholders. In this journey, the steps should include professionalism and an improvement in their capacity to negotiate.

The participation in CBT networks is essential to gain strength and power and to learn directly from good practices and success stories. On the other side, the alliance with external strategic partners, such as universities, non-profit organisations, trusted agencies, and tour operators are crucial and needs to be addressed more incisively, in my opinion.

Brazil is currently reeling from the coronavirus pandemic. How is the current crisis affecting Brazilian destination communities? How resilient are they?

The traditional communities in Brazil have been severely affected by the current crisis, especially those who depend strongly on tourism activities for their survival. What has emerged—and continues to emerge—is a kaleidoscope of solidarity initiatives.

I believe that the endemic absence of government intervention has made resiliency a part of the Brazilian traditional communities' DNA, who are so accustomed to facing traumatic moments and uncertainty.

To protect their ethnic groups, all the traditional communities of the country had to strictly isolate themselves and many resources have been raised to support them.

The immediate response comes from social movements, civil society organisations, as well as simple volunteers. This army of solidarity has been able to provide the communities in the lockdown with food and items to guarantee not only their subsistence but also the basic, individual, and collective hygiene practices. Stock items such as cleaning products and hand sanitizers have been distributed with handmade fabric face masks, especially to indigenous communities.

What raised to the occasion has been the intricate network of organisations that surround CBT, and which overlap constantly with their economic, non-profit activities and daily actions.

As a CBT expert, how do you see the Brazilian indigenous communities coping economically due to the travel restrictions owing to the Coronavirus pandemic?

As a clarification, indigenous communities are only one of the roughly thirty traditional communities officially recognised in Brazil with whom I collaborate, the others are mainly quilombola, caiçara, and ribeirinhas.

The way the different traditional communities are currently coping with the COVID-19 pandemic depends mainly on the kind of relationship they have with tourism itself.

The quilombola communities of the Ribeira Valley in the State of São Paulo, are essentially self-sufficient in food production and they are coping well during the pandemic. They have maintained tourism as a complementary source of income, and they have been able to protect their land and maintain a strong relationship with the natural environment. So, nowadays, they continue to produce all they need.

During the current pandemic, we have observed that the most affected have been those communities whose traditional activities are no longer carried out, either because they have been completely abandoned in favour of tourism activities, or because it was no longer possible to continue them.

What have been your top priorities as a Community Storyteller, during the pandemic and what will you focus on with respect to reviving travel, once the travel restrictions are lifted and life goes back to 'normal'?

Since the beginning of the pandemic, I have maintained contact with the communities I know and constantly reaching out to others using different means. I found that telling their stories in this particular moment of forced immobility could be a way to support and re-connect with them, as a person, as a tourist and as a professional.

So, rather than wait for a moment when I would be able to travel again, I started to travel through stories, and that is the message of my writing project, StoryTravelling.

I have recently launched a new campaign called Destination Community. The purpose of the campaign is to highlight people and the different aspects of their culture and nature. It is an opportunity for tourists to learn about communities, the variety of ethnicity, culture, tradition, knowledge, and also, how unique each one of them is.

For this campaign, I am collaborating with Travolution Travel and Araribá Turismo e Cultura, but I plan to extend the campaign to other countries and continents beyond Latin America if the opportunity arises.

The Amazon jungle is under constant threat from logging, cattle grazing, poaching, and mining. Are there any CBT success stories from this region that are braving the odds and bringing a positive difference?

Well, I dare to say that in the Amazon, as well as in many other regions of the country, all the CBT initiatives have done an incredible job in maintaining their livelihood, culture and protect their natural environment.

Let me explain how.

The richness and diversity of resources have always put the Amazon at the centre of big economic interests. Constant threats, an alarming increase since the beginning of 2019, have been the norm but the control of the land always has been a key issue.

The law in Brazil recognises the existence of traditional communities, but at the same time, it does not automatically acknowledge them the right to live on the territories they occupy, which must be legally recognised first. This, unfortunately, does not always happen.

The unsolved question of the rights of the land attracts vested interests and it is a common scenario where, in different degrees, all the Brazilian traditional communities are trapped.

For a CBT project, tourism is a means to end these threats. It is an instrument to improve their livelihood but, first and foremost, to preserve their culture, ancestral knowledge, and at the same time, to guarantee their right to stay on their land.

CBT is not only an opportunity for tourists to dive into an astonishing variety of the natural environment, traditional knowledge, and cultural activities, but it is also an opportunity to directly and continuously support those communities that have contributed to protect and preserve the natural environment, so they can continue to do their remarkable work.

Would you consider tourism in its current state in Brazil sustainable? Which are the main issues?

Unfortunately, no. I would not define tourism in Brazil as sustainable, on a large scale as yet.

I would say that the national discussion on sustainability is still relatively young. However, there are several organisations such as BRAZTOA[2] (The Brazilian Association of Tour Operators), ABETA[3] (Brazilian Ecotourism and Adventure Tourism Trade Association), SESC São Paulo,[4] and SEBRAE[5] in some of the States.

Also, in many other non-governmental organisations, such as Projeto Bagagem,[6] the national reference for CBT, the persistent efforts by the public university and initiatives such as MUDA![7] (CHANGE! Brazilian Collective for Responsible Tourism), have greatly influenced the discussions and reflections on the theme.

A lack of dialogue among diverse communities, being unaware of local realities and the lack of plurality of stakeholders in the decision-making process are the main obstacles in the journey towards a sustainable tourism sector in Brazil—and a more responsible society.

I reaffirm my belief that to make tourism sustainable, we should change the way we perceive the actors in our society and recognise the roles and responsibilities of each one of them in the collective orchestra.

I continue to be optimistic and focus on innovative and inspiring stories that I relentlessly look for. I am also determined to make Travindy.Brasil an instrument of this collaborative journey.

Anything else you'd like to mention?

Tourism impacts not only the economic activity and the environment, but also—and greatly—the relationship between the human, social, cultural, and economic aspects of the destination.

---

[2] https://www.braztoa.com.br

[3] http://abeta.tur.br/en/the-abeta/

[4] https://www.sescsp.org.br/online/revistas/tag/3098_TURISMO+COMUNITARIO

[5] https://www.sebrae.com.br/sites/PortalSebrae/canais_adicionais/sebrae_english

[6] https://projetobagagem.org/site/pt/

[7] http://www.coletivomuda.tur.br

I suggest, as an industry, to ask ourselves what kind of relationships we want to create through tourism, then start redesigning those relationships.

Link to the interview: https://sustainability-leaders.com/elisa-spampinato-interview/

Spain | Evarist March Sarlat is the CEO of Naturalwalks, whose goal is to make tourism an enjoyable experience to explore the beauty of nature and culture, improve the health of the planet and make a positive difference to the locals. For more than 20 years, he has been guiding, training and advising professionals in guidance and tourism associated with nature and culture through his company Naturalwalks. He believes that the natural world we are blessed with is a good place to start living better.

Areas of expertise: ecotourism, tourism business, entrepreneurship, Spain, Europe, UNSDG 15 (Forests & Biodiversity)

The following interview with Evarist was first published in August 2019 on Sustainability-Leaders.com.

Evarist, having been involved in ecotourism for many years, do you remember what first got you interested?

I'm not sure when I began to feel that what I had been doing as a guide and as a company was ecotourism. It became clear to me when I met a group of small-scale entrepreneurs in Barcelona, and we wanted to bring to light our work in responsible tourism. There I realized that the tourism that interested me was called ecotourism:

based on people who love natural resources and where the client takes an attitude of responsibility to sustain the place for future generations.

Today these debates are more necessary than ever, and it seems unfeasible to me that there would be tourism without this basic attitude if we're talking about quality.

'Quality' is an overused word in tourism, but I think that it is also seldom applied in its most literal meaning. I work quite a lot as a trainer in Latin America and its use there often has little to do with its meaning.

How has your view on the potential of tourism as a facilitator of biodiversity conservation changed over the years?

Without having so much experience in this subject, I would say that tourism originated as a way to get to know the planet. Today it is more necessary than ever that tourism becomes an opportunity to get closer to nature and to the people that connect with it on a daily basis.

Humans live with more and more urban patterns and that means that we have lost, in many places, direct contact with nature's actors that used to connect us to nature: farmers, fishers, shepherds, or others that work with the land and the sea.

Today that natural bridge is getting lost and leisure—which we call tourism—is the way that many of us interact with the most natural territory.

Connecting people firsthand through interpretation allows us to enjoy, learn, and value what it is that gives us the most complete form of health—nature—and at the same time raising awareness of the consequences of our way of living.

In addition to your work as an award-winning tour guide, you also advise El Celler de Can Roca—one of the world's best restaurants—in the use of edible, healthy plants. Are algae, fungi & co in haute cuisine becoming more popular now—perhaps moving on from ultra-processed food?

Without a doubt, the world of cooking has taken a turn in recent decades. We live in a globalized world and we have the luck of being able to choose among many styles of cuisine: from raw food to molecular gastronomy.

For all types of cuisine, the foundation should always be the use of local and natural products. In that sense, we are living a real revolution with regards to the inclusion of many ingredients from both the sea and the land—frequently forgotten.

In addition, modern cuisine should contribute to the awareness and sensitivity of our current environmental problems!

In my work with El Celler de Can Roca,[1] we are also doing our part in making the connection between food and wild plants—today commercialized—that we have saved from being lost. As an example, 15% of the plants that we use are invasive: the second most serious problem affecting the planet's diversity following climate change.

In 2018, the US World Food Travel Association recognized your company NaturalWalks as the second-best tour operator of food & wine tours. What is your secret to success—what makes your tours so popular and unique?

---

[1] https://cellercanroca.com

I think that we are constantly looking for quality and that is why we work with the ecotourism and Premium sectors, with very demanding clients.

Since the very beginning, we have understood that our activities should be unique, which means that they have to make sense in the place where they are being carried out. Furthermore, we have sought to be original and stay away from 'what the market wants', which I honestly continue to not understand.

I believe much more in offering activities that come out of their own desires— genuine activities beyond fashion and trend, and what the marketing dictates which will result in seeing the same thing in different places on the planet.

I believe that tourism continues to be based on principles from the twentieth century, where information prevailed and where places were shown without communicating their essence in a clear way and without a powerful and conscious activity behind every activity.

Like the Basque Country, Catalonia is well known among connoisseurs for its great cuisine, natural beauty and the entrepreneurial spirit of its people. In your view, how is it developing as a destination—is it on the right track or at risk of overcrowding?

I have always thought that the problems with tourism in many places are based on the fact that the same things are offered, they are done in the same way, and they are offered at the same time—without thinking about the consequences. Obviously, that can generate a disaster for the future of tourism and the planet.

Imagination is lacking in the creation of products that are usually based on what is in the land/territory rather than on a specific look at it—that is, transmitting a message or the essence of a specific aspect of the place.

We continue encouraging competition unconsciously because tourism today is not based on differentiating between what you know and what you feel like doing but rather on what the tourist is supposedly interested in, and I think nobody knows what that is.

On the other hand, quality is confused with quantity, and quality with high prices, which remains incorrect.

We need to attract and enhance a cultured, educated, and sensitive audience and that only happens by improving the standards of the product and the professionals who carry it out.

What makes a great tour guide?

There are three essential elements that we should consider:

- The passion and love for the profession. That's to say, it's so important that the guides feel that they are part of a future profession, not a temporary job.
- We need educated guides, with education and values beyond transmitting a professional image. It's impossible to be a great guide without being a good person who sets an example. Remember that there are guides who work 18 h a day during travel weeks and that requires a very deep contact with customers.
- We need guides who want to constantly improve because this is a very demanding job. Guiding puts us to the test every minute because guiding, especially in natural environments, is the game of life: constantly learn and adapt.

Although not everything is the responsibility of the guides and for that reason, it's important that they form part of the decision-making bodies: in public administrations, companies... which is not very common.

As coordinator of the postgraduate degree in ecotourism and nature guiding at the Autonomous University of Barcelona, do you witness a growing interest in the topic, among students?

Well, surprisingly, yes, quite a lot. At the start in the middle of the economic crisis, it was difficult, with students only essentially from our surrounding area. This year more than half of the students are from different places around the world: from China to Patagonia.

And increasingly more students with a lot of interest and from places with immense potential for the development of quality in ecotourism—like Colombia, Brazil, or Chile.

I think that ecotourism is going to keep growing every day, because there is a social demand from the public to be closer to nature and to do so in a more respectful and conscious way.

We need good professionals and for that, we also need good training.

Where do you see the link between the preservation of natural diversity and cultural identity, for example in Catalonia?

Throughout the world, local cultures are also an expression of physical diversity. Catalonia is a melting pot of the diversity of many people and cultures with whom we have lived and exchanged over centuries and continue to do so today.

We forget that strengthening minority cultures—especially ones with native languages—is a way of maintaining that necessary diversity.

In Catalonia, we have a great tradition of using wild ingredients, especially plants and fungi. Take, for example, a traditional liquor such as Ratafia—with more than 40 species of herbs—or how the same plants of the past are now used in high cuisine, it's a way of showing the diversity of ecosystems, habitats, and species of a place and also how the locals have related with nature over centuries in many areas: as a source of health, spirituality, identification with the land, etc.

Tourism professionals sometimes avoid engaging with 'sustainability', since it is not something usually part of their KPIs. Reflecting on your own experience, which advice can you share with tour guides or tour operators in terms of how to deal with sustainability?

Today sustainability is no longer negotiable and those of us that work in nature know that firsthand.

To be sustainable means taking into account how our actions impact the planet and other beings. I invite you to:

• Focus on the quality of services instead of quantity.
• Balance the number of clients per guide according to each place and each activity.
• Ensure the quality of guides by improving working conditions: fees, working hours, participation in decision-making.
• Create a midterm and long-term strategy for the involvement of guiding services to grow your relationship with the company.

- Make sure that the profile of the guides corresponds to the profile of services and quality that you want to offer. Not everyone is meant to be a guide!
- Look for specialized certifications that guarantee the quality of your guides.
- Invest in selfless public actions that support sustainability.

Anything else you'd like to mention?

I would simply like to reinforce that in the world of tourism, it has been forgotten how vital guides are. They don't often appear on stage in debates, and they are an important element of the chain because in the end, they are the last and most important link in guaranteeing the quality of services.

Nature doesn't lie, and those of us that work in it know that well. Humans are doing something wrong, and tourism can be a great opportunity to get closer to and learn from nature and, therefore, improve our quality of life.

Link to the interview: https://sustainability-leaders.com/evarist-march-sarlat-interview/

Chile | Felipe Vera Soto is a specialist in sustainable tourism. For more than a decade, he has worked towards developing tourism with a vision of responsibility and strategic planning to improve people's living conditions, conserve their heritage and diversify the local economy. He is a member of the Panel of Experts in Tourism of the United Nations World Tourism Organization, representative of Green Destinations in Chile, founding member of Global Leaders Network and expert member in sustainable tourism of One Planet—Travel With Care. He has given conference presentations in Latin America, Asia and Europe, thanks to his in-depth knowledge of the tourism sector. Felipe has contributed to the training of hundreds of new professionals in the tourism sector, instilling a core value: the common good must be the key to a more responsible tomorrow.

Areas of expertise: destination sustainability, ecotourism, sustainable development, tourism business, Chile, South America, UNSDG 8 (Equal & Fair Economic Opportunities)

The following interview with Felipe was first published in March 2019 on Sustainability-Leaders.com.

Felipe, on behalf of the Chilean city of Natales you just received recognition for its sustainability initiatives, at the Green Destinations awards in Berlin. How does it feel to be on stage with the world's most proactive destinations focused on more sustainable tourism?

I am very grateful for the opportunity that the city of Natales has given me to hear my opinions and views regarding tourism sustainability, which has meant that in less than one year we are listed among Green Destinations' TOP100 Sustainable Destinations worldwide and the 3rd best in America, after Galapagos in Ecuador and Thompson-Okanagan in Canada.

In your view, what makes Natales a leading, 'green' destination—and different from other destinations in Chile?

Due to its geographical location, Natales has a vocation of nature thanks to its proximity to the Torres del Paine National Park, which has allowed the development of the tourism value chain.

The municipality of Natales has been a pioneer in the elimination of single-use plastic and the creation of a committee of tourism sustainability, a public-private partnership. It includes the academy with a focus on improving the living conditions of the people, thanks to the conservation of heritage.

Having worked for accommodation, tour operators, the educational system and as an advisor to destinations, which would you consider the main challenges right now in Chile, in terms of advancing its sustainability as a destination?

The main problem that Chile faces in tourism is the lack of professionalism in the industry. The lack of technical knowledge means that many people talk about sustainability but in reality, apply practically nothing.

The lack of practical and local-scale policies is also a gap that must be minimized, to really start the journey towards tourism sustainability in Chilean destinations.

Judging from your experience, which factors or trends within the country might support or hinder a better, more responsible tourism in Chile?

Currently, the tourism industry in Chile is not making the necessary efforts in terms of sustainability. There are just a few low-impact actions. We need to improve those and move faster, to make sure local communities benefit from tourism.

Political decision-makers need to work with well-trained tourism professionals and support the creation of destination management organizations.

Reflecting on your work so far, also as a member of the UNWTO Panel of Tourism Experts, which are your key insights or lessons learned?

First, how quickly the tourism industry has changed in such a short time. More and more visitors are coming to Chile, but we have no adequate destination planning.

When talking about sustainability 10 years ago, few understood the importance. But now there is a base of consciousness that grows every day.

Another very important element is to have data that allows improving the management of the impacts of tourist activities in the destinations. I believe that the UNWTO International Network of Sustainable Tourism Observatories initiative (INSTO) is very important in this regard, especially for emerging destinations.

How can tourism help conserve natural and cultural heritage, and support the diversification of local economies?

Tourism, if managed correctly and with the active participation of the local community—supported by professionals—can be a valuable source of income for local economies. It can empower minorities and help to make a community's economy more resilient, and less reliant on revenues generated through farming or mining, for example.

One community which applies the principles of tourism sustainability very well, in my opinion, is Lake Budi in the Araucania Region, where Mapuches Lafkenches did an excellent job many years ago.

As a country ambassador for the Green Destinations initiative, which destinations around the world would you consider the most innovative right now in how they approach sustainability issues and opportunities in tourism?

Many European destinations are leaders in various aspects. However, I really love how Slovenia has evolved as a country and destination, with a very clear strategy. This is a good example for us to learn from in South America.

Costa Rica, of course, has been a leader in ecotourism for many years and New Zealand with its proposal of community work is something that I am very passionate about.

You are also a founding member of the Global Leaders Network. Why did you join—and what is this about?

The network of global leaders was founded in September 2018 in the Netherlands with the mission of promoting the Sustainable Development Goals of the United Nations linked to tourism. It is a network of professionals and businesses around the world.

I joined because I want to share my experiences and learn from other leaders about good practices in terms of tourism sustainability.

Community-based tourism products and experiences are in high demand. Less known are destination management organizations run by communities, bottom-up.

It is very complex that a destination can be developed sustainably if the local community and its heritage have not been considered in the initial designs.

In October 2009, after living one year in New Zealand, I began to promote this type of initiatives in the Pan de Azúcar National Park in northern Chile.

Then, in 2011, I designed for another national park, Llanos de Challe, a proposal based on the community, which could not advance due to lack of funds.

From those experiences, I learned a lot, which I can now apply in Natales and soon in other destinations in Chile. I am promoting proposals that are administered by the municipalities, adapted to the Chilean reality.

Your 3 bits of advice to destination developers and managers in Chile who'd like to improve the sustainability of their destination?

Understand that tourism is not an industry free of social, environmental and economic impacts.

Understand that tourism cannot by itself maintain the entire economy of a community.

And, sustainability is a journey, and not a goal.

Link to the interview: https://sustainability-leaders.com/felipe-vera-soto-interview/

UK | Born in Scotland and a graduate of Edinburgh University, Fiona Jeffery has had a long and accomplished career in the travel industry. Fiona created and launched the World Responsible Tourism Day, in association with the UNWTO. In 1998, Fiona founded Just a Drop, bringing safe water, sanitation and hygiene education to communities across Asia, Africa and Latin America. As a member of the UNWTO World Committee for Tourism Ethics, Fiona successfully got an agreement at the UNGA to convert the UNWTO Global Code of Ethics into a legally binding Convention. In 2012, Fiona was awarded an OBE by Her Majesty the Queen for services to travel and tourism and the World Tourism Award for her philanthropic vision in creating Just a Drop. She was recognised in the Directory of Social Change Influencer Awards in 2015 and inducted into the Travel and Hospitality Industry Hall of Fame in 2020.

Areas of expertise: sustainable tourism, UNSDG 12 (Resource Efficiency)

The following interview with Fiona was first published in April 2015 on Sustainability-Leaders.com.

Fiona, as a businessperson, what was your view on the importance of sustainability in travel and tourism when you started your professional career?

© The Author(s), under exclusive license to Springer Nature Switzerland AG 2022   231
F. Kaefer, *Sustainability Leadership in Tourism*, Future of Business and Finance,
https://doi.org/10.1007/978-3-031-05314-6_35

When I started in the industry, it wasn't really on my radar, but when I took over World Travel Market in 1994, I simply felt that unless as an industry we protected the very product we were seeking to promote, we'd eventually destroy our own business model. So, I created Environmental Awareness Day in 1996, which attracted about 30 people and was a side-show in a room.

Now called World Responsible Tourism Day, it has become a core pillar of the World Travel Market and attracts the largest gathering of responsible tourism professionals in the world, with over 3,000 attending its 4-day educational programme.

From your experience as a tourism business and marketing executive, where do you see the priorities in terms of sustainability?

There are many but community wellbeing, wildlife protection, environmental protection and climate change, and its impact on water resources all feature. More education is needed for things to change.

Sustainability has to be something that should be built into tourism businesses' training programmes. It should be seen as that important.

In the same way, as you receive sales or marketing training on the job, you should understand what is meant by sustainable tourism practices, what this involves, and have updates on new strategies and developments to improve your personal and business impact.

Do you share the view that sustainability has become mainstream?

No, I don't, not enough mainstream operators take it seriously enough. More companies in my view have to walk the walk for it to become mainstream. I wish it was.

Did sustainable tourism play any role during your time as Managing Director and later Chairperson of the World Travel Market?

For me, it was a key strategic pillar of my business. WTM's Environmental Awareness Day was designed to help raise awareness of the issues and educate the industry. It wasn't a revenue stream, but a commitment to our industry to find better ways of doing things, to get smarter and develop a more acute social, environmental and cultural conscience. It evolved then into the Responsible Tourism Day with a full education programme.

Why did you launch the World Responsible Tourism Day in 2007?

It was a natural development and evolution from its previous incarnation, but this time I had the full backing of the UNWTO.

Do you think being a mother of two kids has influenced your view regarding the necessity for sustainability in travel and tourism?

Most certainly. We all have a greater sense of responsibility when we become parents. But I was also very aware of my responsibilities to the industry when running World Travel Market, and I always wanted what we did to count and add value in a way that went beyond the important commercial benefit of being at World Travel Market.

I've always believed that doing business is one thing, but we should also contribute in a broader way, where we can. I was lucky to be in a position where I could influence, and focusing on the sustainability agenda was a way of making people think differently.

Why did you decide to found Just a Drop?

I created Just a Drop for two reasons. Firstly, I became a Mum and wanted to encourage my industry—travel and tourism—to give back to children and their communities across the world. Secondly, too many people were ignoring the sustainability agenda, finding it difficult to get their minds and business operations around, so I decided to create something that people could still make a contribution towards and find that as a consequence they were able to transform people's lives by donating to our projects in the field.

So, if I wasn't able to engage them on the sustainability agenda, I hoped to appeal to them via corporate social responsibility and philanthropy.

The important thing was finding a way to genuinely improve people's lives in a way that was sensitive to the environment, and making a positive contribution and difference. The fact that a child dies every 20 seconds due to the impact of either dirty or lack of water, both entirely preventable, seemed to me a strong enough cause.

To date, Just a Drop has helped 1.5 million people across 31 countries but there is still a lot to do.

Your (career) advice for newcomers to the sustainable tourism business field?

Listen to your conscience and ensure that, whatever job you take, you put sustainability on the agenda as part of its DNA. Balance short-term issues with long-term goals.

Why did you join UNWTO's World Committee on Tourism Ethics in 2013?

I was honoured to be invited onto the committee and started officially in 2014. We have two meetings a year. The Chairman is a very insightful Frenchman, Pascal Lamy, who was once head of the World Trade Organisation.

Your observations as Chair of the Selection Committee of the WTTC Tourism for Tomorrow Awards?

The entries this year were inspiring as always and exciting because they prove that running successful businesses in a sustainable way is absolutely achievable, but it takes vision, commitment, leadership and hard work.

The diversity of entries was particularly pleasing, suggesting that sustainability is potentially becoming more mainstream and attracting commitment across a broader reach of businesses.

So, we've had a range of applicants across small, medium and large commercial operations all making a big impact, which is very encouraging.

I have been very ably supported by the international team of expert judges led by Graham Miller, Professor of Sustainability and Head of the Tourism and Hospitality School at Surrey University.

I have huge confidence in the team of judges selected, and we share knowledge and insights, and have detailed and involved discussions.

What's always interesting is you can think one thing, believe it's quite clear cut and then someone can put forward a very different perspective, which has to be taken into account. It's not an exact science.

What makes the Tourism for Tomorrow Awards such a special venue?

What makes the Tourism for Tomorrow Awards special is that they really are the leading accolade in responsible and sustainable tourism awards. The judging criteria are extremely rigorous and thorough, with every finalist visited personally by an expert judge able to undertake a root to branch review and ensure the authenticity and veracity of the submission.

How do you keep up to date on sustainable tourism developments?

I learn from the engagement with professional colleagues active in this space. I sit on a number of industry boards and work with a number of organisations, and I learn hugely from all of them. If I'm honest, due to a lack of time I don't spend much time reading specific articles or blogs unless they come in online.

How does the WTTC Tourism for Tomorrow Award help tourism businesses and destinations?

It helps, first of all, by becoming a first-class accolade they can use to effectively market their business or destination. But in addition, they become part of the Tourism for Tomorrow network and I want to increasingly be able to harness the hard-earned knowledge and respect these individuals and companies have garnered, and use them to educate others in the industry on best practice.

To encourage and share learnings more, so people don't have to reinvent the wheel each time, but can tap into this hugely valuable insight these organisations now have and be able to share with the industry at large. We need to find the right mechanic to do this, but it's very important for me to do this.

Link to the interview: https://sustainability-leaders.com/interview-fiona-jeffery/

Florie Thielin on Hopineo and Responsible Tourism in Latin America

France | Florie has travelled across Latin America for 2 years, swapping her web marketing skills in exchange for a room and boarding with sustainable tourism initiatives. She also made videos to showcase good practices that she discovered along the way. Florie is now back in France and part of journalist collective Voyageons Autrement. She also teaches and shares all her training support for free (all in French) on her website 4R Tourisme.[1]

Areas of expertise: ecotourism, responsible tourism, sustainability communication, France, Latin America, UNSDG 12 (Resource Efficiency)

The following interview with Florie was first published in March 2015 on Sustainability-Leaders.com.

Florie, what first triggered your interest in sustainable tourism?

---

[1] https://4rtourisme.fr/tourisme-durable/

© The Author(s), under exclusive license to Springer Nature Switzerland AG 2022
F. Kaefer, *Sustainability Leadership in Tourism*, Future of Business and Finance,
https://doi.org/10.1007/978-3-031-05314-6_36

My interest began when I was a student and decided to write my thesis on the implementation of sustainable policies in hotels, and how to successfully implement the necessary changes to allow the introduction of more ecological practices.

I did my degree in International Business and, in the last year, decided to specialize in tourism and hospitality, mainly because I had always wanted to travel. I loved the idea of selling 'experiences' and thought tourism was an important source of local development worldwide.

Could you briefly describe the Hopineo project, and why you decided to join the adventure?

Hopineo's main objective is to support and promote sustainable tourism, mostly through two methods: HopSolutions and HopTrips. The first harvests the best practices of hotels (the HopSolutions) that participants in the project visit to create a library of sustainability best practices, containing videos and articles that hotels around the world can implement.

The second is the creation of a platform that allows travellers (HopTrippers) to contact sustainable hotels and offer to help them with problems they may have developed or implementing the aforementioned best practices (depending on the specific competencies of the traveller), in exchange for food and accommodation during the mission (HopTrip). At the same time, travellers can take advantage of their stay to enrich the HopSolutions library and give more visibility to the hotels.

For example, in my case, my competency is marketing, and this is what I bring to the hotels that I visit. Another person may be a professional photographer, or maybe they can help with the translation of a web page, etc. The traveller and the hotel together decide which tasks the traveller will complete during their stay.

During this journey, what countries have you visited so far, and which ones are still to come?

My journey began last July (2014), and I travelled through all of Central America. In the past seven months, I have visited Mexico, Belize, Guatemala, El Salvador, Honduras, Nicaragua, Costa Rica, Panama, and I am currently in Colombia. My next destination is Ecuador, where I will attend and participate as a volunteer at the GSTC (Global Sustainable Tourism Council) conference, where tourism professionals meet to share their experiences in sustainable tourism. In the next months, I will also visit Peru, Bolivia, northern Chile and Argentina. If time allows, I will try to make it to Paraguay and Uruguay. The goal is to return to France by next Christmas.

What has been, up until now, your biggest contribution to the hotels that you have visited?

With the hotels, I always discuss the different points that are important to their marketing strategy: the concept of the hotel, their webpage, their pricing strategy, their online visibility, their e-distribution and their reservation management tools.

In the last hotel that I visited, La Cachamera in Colombia, I did a workshop with the staff to discuss sustainability, talk about their current practices and get new ideas. I have various videos that I have collected during my trip, and I showed them to the staff so that they could see for themselves the best practices of other hotels.

Is there a hotel or destination that stands out in particular?

It's very difficult to choose just one hotel or destination, I encounter human experiences and professional marvels every day. I suppose if I had to choose, it would be Hotel Selva Negra in Nicaragua, which in addition to being an eco-hotel, is an organic farm, fair-trade coffee plantation and foundation. I stayed with them for a week and learned so much during that time.

Visiting Costa Rica was also an amazing experience, where I met with the Rainforest Alliance and had an interview with CST, the national organization for sustainable tourism certification. In general, in Central America, there are many private initiatives, but Costa Rica has had a national program of certification that is free and voluntary and a very advanced sustainable tourism policy in place for many years.

How has this trip changed your view of the world and tourism in particular?

I have learned a lot about tourism spending every day with the owners of hotels. I can see the concrete problems they face, whether it is implementing sustainable practices or regarding marketing and operations. However, I have also seen the many things that can be successfully achieved. At the beginning of the trip, I had a lot of theory in my head, and this journey allowed me to see how this theory could be put into action.

On a personal level, I don't think I want to return to work for large hotel chains, I would rather be part of a larger sustainable tourism project working with Hopineo and maybe, for example, organizations like Rainforest Alliance. I would like to use my new skills—especially in marketing—to continue on this path of helping the development of sustainable tourism.

What are the biggest barriers or challenges to the development of a more responsible tourism industry in Latin America, based on what you have seen up until now?

Each country must have a complete program for sustainable tourism. There are many small and private initiatives, but it would help a lot if the Tourism Board of each country supported the hotels with tools and training. This would incentivize the private sector to adapt its offer to satisfy the new demand. Costa Rica is a good example.

Another challenge is communication: many hotels implement sustainable practices but don't know how to communicate this to their clients. It was fairly difficult for me to find sustainable hotels on the internet, because, in general, they don't have good visibility. The government should also help in this concern by, for example, teaching hotels how to make a professional-looking web page.

Another idea is something similar to HopSolutions, a program that collects the best practices of all the sustainable hotels in the country. The hotels, in general, are isolated, each of them in their own regions with their own responsibilities, leaving the owners of small hotels without time to travel and share experiences. Exchanging solutions would save hotels time since the solutions to their problems probably already exist.

What advice would you give to those that are beginning a career in responsible tourism?

I would tell them to begin by doing a HopTrip in their own country—they can contact me, and I will be happy to speak with them and give them advice. Doing this will allow them to understand the real issues and learn directly from professionals in the industry. It's like doing an internship, but instead of only seeing one place, they will visit different companies, allowing them to learn a lot more.

Link to the interview: https://sustainability-leaders.com/interview-florie-thielin/

Frankie Hobro on Marine Wildlife Protection and Business Resilience

UK | Frankie Hobro is the Director and Owner of Anglesey Sea Zoo and Marine Resource Centre. Her vast experience working on hands-on conservation projects with critically endangered species abroad in terrestrial and marine environments makes her a passionate advocate for conservation and sustainability. Frankie spent much of her childhood in North Wales, returned to do postgraduate studies at Bangor University School of Ocean Sciences in 2001, and then settled on Anglesey Island in 2007 when she bought the Anglesey Sea Zoo. Frankie has transformed the business into the only exclusively British aquarium, housing native species with natural seawater and specialising in the captive breeding of endangered species like seahorses and lobsters for reintroduction into the wild. Being a major tourist attraction and local employer, the Anglesey Sea Zoo is a much-needed community, research and environmental hub, focussing on sustainability, education and marine conservation and a marine animal rescue centre.

F. Kaefer, *Sustainability Leadership in Tourism*, Future of Business and Finance, https://doi.org/10.1007/978-3-031-05314-6_37

Areas of expertise: wildlife conservation, entrepreneurship, tourism business, UK, Europe, UNSDG 14 (Oceans & Marine Life)

The following interview with Frankie was first published in October 2020 on Sustainability-Leaders.com.

Frankie, we received an overwhelming number of responses on LinkedIn, nominating you as a sustainability champion for the amazing work you have been doing at the Anglesey Sea Zoo. What motivates you to push boundaries—and how do you stay motivated and committed during the current uncertain, difficult times?

Pushing boundaries is the scientist in me, although I usually call myself a natural historian and conservationist. I have a curious mind. I like to know everything, and everything about what works and what does not work, and why. This curiosity leads me to constantly push boundaries and explore options until one day it works and the results pay off.

My hands-on practical scientific background taught me to observe and record everything very closely and this leads to a strong gut feeling for what works and what does not, and what is worth trying. Then this is applied, not as a risk as such but a well-calculated decision that is likely to pay off. And if it doesn't, at least you know what doesn't work and why; which is productive going forward.

I see failure as a positive thing as it shows you what doesn't work and allows you to test an alternative route towards the solution—as long as you learn and progress from your mistakes. Anyone who has not made a mistake has never tried anything new. Pushing boundaries is the desire to know, a curiosity to find out and the confidence to be the person who takes the plunge. Somebody has to be the first to attempt everything new and I like to be that person. The worst thing that can happen is that it does not work. If nobody ever pushes the boundaries there will never be any progress with anything.

I am very passionate about what I do, and I constantly aim to do my very best and make a real difference, which is my motivation. I have never been motivated by money or materialism—if I was doing what I am doing for the money I would have given up before I even started! I want to influence others in a positive environmental way, through leading by example and laying a legacy of sustainability and positive change.

When you have love and passion for what you do every day, there is no question of losing motivation or dropping commitments. When you are investing in environmental projects and long-term conservation programmes you are signed up for the long haul and you have to accept the hiccups along the way. Your commitment has to be 24/7 especially when you are the owner or driving force as I am here, you can't just quit or walk away, and this is something I accepted from the day I started.

There is no question of losing years of sweat, tears, and investment by stopping a research programme in its tracks or putting conservation work on hold. The animals carry on doing their thing, as does nature, endangered species remain endangered and the protection they need does not become any less just because your time or resources are reduced. It has to continue at any cost and from the day you take that commitment on, your job is to find a way to ensure that this work continues, come what may.

I believe that difficult times, like mistakes, are an essential part of the learning experience and personal growth. Being resilient and resourceful is essential for sustainability and often through the need to adapt to get through the toughest times of absolute desperation, come the most ingenious and creative long-term solutions.

What inspired you to dedicate your professional career to protecting the marine environment?

I have always been environmentally minded and a nature lover—my parents were science teachers in the early days of promoting sustainability, recycling, and renewable energy so I grew up seeing that as the norm.

Although I grew up in the Midlands in England far from the sea, we spent all our holidays in North Wales, based on Anglesey Island which was a perfect playground for my passion for nature. Once I left city life as a teenager I never returned—ever since then, it's been an island life for me!

I had a year out before going to university and decided to travel the world on my own—the best decision I ever made! In Australia, my passion for conservation was properly sparked for the first time while spending time walking in the bush with an old family friend who was a biologist. When I came back to study, I knew that wildlife conservation was what I wanted to do. My final year dissertation for my Bachelor's degree was on conservation and invasive species control in Australia. When I left university, I started to apply for projects abroad—it was competitive work, poorly paid, and I had to work hard to save up enough to cover my own living costs. I managed to secure a place on a very competitive project funded by the Gerald Durrell Wildlife Trust[1] (Jersey Zoo) with the Mauritian Wildlife Foundation.[2]

It was hands-on immersion from day one and I loved it, living in basic conditions in field stations and completely deserted areas with no running water or electricity but surrounded by nature. We were taught the skills we needed in the field and quickly moved on to monitoring and ringing critically endangered bird species. I was in my element, learning abseiling and climbing and other skills to carry out the work and I ended up staying on for several seasons, training other volunteers and managing projects for just over 4 years in total, so I became a specialist in conservation and management of island ecosystems.

I also spent several months volunteering on other endangered species projects on Indian Ocean islands including the Seychelles and Madagascar as the skills I had picked up were easily transferable, and 6 weeks in New Zealand during which I volunteered on monitoring and conservation projects on offshore islands.

While I was the island manager on Ile Aux Aigrettes Nature Reserve in Mauritius, I gained my Divemaster qualification. I adapted my skills for SCUBA-plumbing to maintain the constantly damaged freshwater supply pipe which ran 1 km across the

---

[1] https://www.durrell.org/wildlife/

[2] https://www.mauritian-wildlife.org

sea bed to the island, and I volunteered with the Mauritius Marine Conservation Society[3] doing marine monitoring and underwater clean-ups. I started monitoring the marine reserve around the island with the information provided through a professor at Bangor University Ocean Sciences[4] in North Wales.

When I was ready to move on, I contacted the same Professor to ask about any marine research projects I could apply for to increase my experience in this area. His response was to offer me a fully-funded place on his MSc course in Marine Environmental Protection at Bangor University off the back of my conservation experience.

So that is how I expanded into marine conservation and ended up setting roots back in North Wales!

For my Master's thesis project, I was a research assistant studying the foraging behaviour of manta rays in Baja California, in the Sea of Cortez in Mexico which took me to Mexico and the USA for 6 months. During this time, I was fortunate enough to be an assistant on several other marine projects including marine mammal monitoring in Monterey Bay. Then I spent a year working in the Maldives as a marine eco guide and teaching and running local conservation projects before returning to Seychelles for 3 years where I was an Island Manager on 2 different islands covering both marine and terrestrial conservation work, including the monitoring and tagging of seabirds and turtles, giant tortoises and several critically endangered species of land birds.

These incredible experiences on global hands-on conservation projects involved training and educating local communities and individuals to ensure they understood the need to protect nature and how to do this effectively. It gave me a firm grounding in collaboration and what can be gained through sheer determination, resilience, and hard work. If you were good at your job, you were out of a job and leaving the project in the safe hands of the local community. This solved an environmental problem by giving an alternative solution to local people with a positive result for them and the environment, the common good.

In March 2007, I returned to North Wales and bought the Anglesey Sea Zoo, which was an 'old and tired' visitor attraction much in need of investment and a new direction. I saw the potential to fill an essential niche whilst indulging my own passions and living in a place I love.

From the day I bought the business, I started on my vision of converting it to the unique product that it is today—an all British aquarium focusing on sustainability and environmental protection, carrying out vital endangered species conservation programmes for native species to be re-released into the wild, whilst working closely with local communities and fisheries to educate and involve the public, enhancing their knowledge and love for marine life and thus encouraging them to help protect it.

---

[3] http://www.mmcs-ngo.org/en/

[4] https://www.bangor.ac.uk/oceansciences/

This is a long-term vision, a legacy that needs building. Something like this doesn't happen overnight, it takes years of commitment, dedication, and building bridges within the community and further afield. To be successful, it needs the support and participation of the local community and the public, it needs the involvement of others, for people to buy into it and believe in it and share my passion.

I saw huge potential in the unique location here and the ability to create the first-ever all British aquarium, with water pumped straight from the sea. I realised I had finished chasing my dream, now it was time to start building it.

How is the Anglesey Sea Zoo reducing its environmental impact and inspiring visitors to appreciate nature and protect wildlife?

The very nature of the Anglesey Sea Zoo is sustainable and ethical, it runs as a natural open system, with seawater pumped directly from the sea on the doorstep, through the exhibits, and then back to the sea. This means that all our animals experience natural changes in sea temperature and seasonality such as plankton blooms just as they would in the wild. Happy animals mean healthy animals that display natural behaviour—breeding and feeding well on their own.

There is nowhere in the UK that holds only British marine species and despite a great education system here, the UK is full of people who know nothing about the seas surrounding our island nations, what is in them, or how to protect them. Having lived and worked in developing countries, I have seen the inhabitants of islands with incredible knowledge about the sea, the species there, and the marine environment, but here in the UK nobody was aware of any of these things.

Enabling understanding through education encourages people to care and protect and the Anglesey Sea Zoo does just that—allowing people to see, learn and understand all the amazing species we have in our British seas and the importance of protecting them.

When we first started the British Seahorse Breeding and conservation programme here, I was amazed at how few people were even aware that there are native seahorses in the UK, let alone the fact that they are now critically endangered and on the verge of extinction in the wild. We are the only place in the UK to house and breed both species of native seahorse and we are working with Project Seagrass,[5] Seasearch,[6] and other organisations to re-release these amazing creatures back into the wild where they were once relatively common.

Here we have basking sharks, leatherback turtles, and other amazing ocean giants in our waters but most people never see them, so they are oblivious to their existence. Once they know about them and the threats they face, they want to help to protect them. It is a similar story for many of the more common creatures here which people have seen or heard of—and often eaten—but know nothing about.

I have noticed that once somebody has volunteered on a beach clean-up, where they have to see and deal with the results of others not disposing of their rubbish

---

[5] https://www.projectseagrass.org

[6] http://www.seasearch.org.uk

properly and the damage caused by debris in the environment, their own attitude towards rubbish disposal changes completely and wherever they go, they start to pick up and rethink their own habits. It all comes down to instilling a sense of ownership through involvement and encouragement.

I believe in leading by example through encouraging, enthusing, or even shaming others into doing the same. I have invested in 50 KW of solar panels here—possibly making us the first solar-powered aquarium? We practise regular stock rotation and regular re-release programmes for common lobsters which we breed in our hatchery facility here. We are an Environmental Hub for beach clean-ups with our own voluntary local community group and I am a local coordinator for emergency call-outs for all kinds of marine animal rescue. And we have ongoing research programmes dedicated to the captive breeding and eventual reintroduction of both species of British seahorse and spiny lobsters into our seas, both of which are now almost extinct in the wild having once been commonplace.

I believe in leading by example and sacrificing short-term profits to promote positive environmental change, in people's habits, attitudes, and lifestyles as well as actively protecting our marine environment and the creatures within it.

Anglesey Sea Zoo won Green Key's 'Ethical Green Business of the Year' award last year. How has this recognition impacted you and the work you do at the zoo?

I heard about the Green Key Award 5 years ago through the community work I was doing with Keep Wales Tidy. This appealed as a progressive award with new targets which must be met every year to show a further reduction in carbon footprint and increase in sustainable practise, not just a one-off award which can be won through box-ticking then shelved.

The international recognition of the Green Key Award appealed too, as sustainability is very much a global concept, not just a local one. The Anglesey Sea Zoo was the first business on Anglesey and only the second hospitality business in the whole of North Wales to gain the green key accreditation in 2016 and we have continued to gain it every year since, for 5 consecutive years now, through constantly meeting new sustainability targets. We have also won several Green Business of the Year awards, twice for the Daily Post Business Awards and also through Go North Wales.

These accolades are the most incredible recognition for myself and the team here as the core aim of the work we do is to be green, sustainable, environmentally sound, to pushing boundaries in conservation, education, and research.

It helps us to stand out and be recognised as a centre for positive environmental change and to get others on board with the concept of making a positive difference, and it makes our tireless efforts seem worthwhile and appreciated. We know we are meeting and achieving our goals.

My most prestigious award was in July 2019, when I had the immense honour of being awarded Alumnus of the Year for Bangor University off the back of my ongoing contribution to research and conservation projects, promoting positive environmental change, working with the community, and as an employer to university graduates and volunteers.

Zoos around the world are getting creative to keep funds flowing in, coping with the loss of income during the pandemic. What measures have you taken at Anglesey Sea Zoo and Marine Resource Centre, to ensure financial sustainability?

The pandemic caused extreme stress with the challenge of juggling optimum animal welfare, essential systems maintenance, and ongoing vital conservation and community programmes whilst meeting running costs of £20,000 a month with no income.

This was compounded by a lack of sufficient Government clarity or support whilst we were closed and towards reopening, both in terms of guidance and funding.

Upon lockdown, we dropped to skeleton staffing with 80% of staff furloughed. The remaining staff and volunteers adapted their roles to cover the essential work—there were just 4 of us remaining across the whole site including myself, along with a handful of volunteers, keeping everything running for almost 4 months. It was full-on and extremely stressful and exhausting but fortunately, we came through it!

With animal welfare always a priority, we had to explore new and creative ways to keep the community engaged with our animals and conservation efforts and to raise funds whilst closed.

We posted regular videos over social media with updates on our animals and also did live feeds showing novel behind-the-scenes aspects of the aquarium.

We engaged the public in an emergency funding campaign to help raise funds and the response was fantastic! Without these donations, we may not have managed to keep afloat, so I am extremely grateful to the public for this support at such a difficult time for many people. And it was very reassuring, having spent so much time investing in the local community and environment, to get that support back at a time of such need.

I continued the beach clean-ups and animal rescue work myself with my kids throughout lockdown. Sadly, there was more litter being dropped than ever, but we also had many local people wanting to borrow the kit from our Environmental Hub to do their own clean-ups as they hated seeing rubbish everywhere during their daily exercise, so this role became more important than ever and I believe it positively changed many people's attitude and made them more proactive. I also spent many hours with my kids on the beach collecting food for our animals such as crabs and prawns to ensure that we could keep them healthy and well-fed.

We started a phased reopening in July, strictly following government and industry-specific guidelines to ensure the safety of our visitors, staff, and community and gained the 'Good to go' accreditation through Visit Britain. The extra costs of reopening under COVID guidelines were high and there was no funding for this. We fitted screens at every till, bought extra PPE for staff, set up a fully signposted one-way system across the site with fun, bespoke floor graphics, and fitted numerous hand sanitizer stations across the whole site made from EcoPly, which look great and are environmentally friendly too!

Faced with a reduced capacity of 50% in the main aquarium building to maintain adequate social distancing, I had to come up with ways to increase visitor footfall in the outdoor areas and to accommodate a large number of queuing visitors on busy days.

We put on extra free outdoor activities for visitors, with staff positioned to give hands-on sessions and talks and invested in more covered and sheltered outdoor areas, so bringing all our outreach and visitor interaction sessions outside. Guided family seashore safaris and other free activities offer great value for visitors and they stay on-site longer and learn and participate more while they are here.

We have a great new circus-style events tent to allow activities in a sheltered outdoor environment at a safe social distance and this will be available throughout the year to guarantee that we can offer events and group bookings with ample space in a COVID safe environment. It has also introduced a great new feeling of family fun to the site.

What is the future of marine conservation when there is no relief in sight from rising sea temperature, ocean acidification, plastic pollution, and more recently, deep-sea mining?

It is very easy to be defeatist about the global climate emergency and to feel powerless against the looming crisis with so few positive solutions currently. But a negative attitude will get us nowhere so we must come up with a formula to ensure the survival of humanity and this planet which sustains us.

I am a great believer in the power of positivity—always wear positive pants not negative knickers! There is always a solution, but you have to persevere to find it and work with others who want to achieve the same goal.

I believe there are amazing technologies and discoveries on the horizon which can help us combat and even reverse impending environmental disaster. But to succeed we need to overcome the current lack of recognition of the issues so that we open doors for funding and maximise efforts towards harnessing it. We need absolute recognition and action from the right people in positions of power to bring about immediate change and for that, they all need to look beyond instant gratification and personal gain and learn to work collaboratively for the greater good. Sadly, this seems like an unrealistic request—but we all share one planet, and we are the guardians of our earth so we need to unite in one goal and the responsibility of caring for it.

I see many people working hard, caring a lot, and in their own way making a huge difference but until we can have the same response from the large multinational organisations and collaboration from the global government, we will not see sufficient change. I can see why it is so easy for people who are doing so much to feel defeatist about the bigger picture. But it is important for everyone to play their part in a positive way, however small it may be, as the overall whole result will be a very significant positive change.

Extremely challenging situations often lead to the most successful and extreme solutions. I truly believe that a complete turnaround is not only possible but imminent. Humans can be incredibly resourceful and innovative and when that energy is correctly channeled and combined it can lead to incredible things. We can turn the tide on the impending environmental disaster if we work together.

The state of our oceans is deteriorating at an alarming level. What must the global community do immediately to slow down, if not reverse the damage that is being done?

Many minds are better than one, so collaboration is essential.

There are many of us doing our utmost already to make a difference but as we are doing it alone and it is not enough. If we find a way to join as one large global force, together we can change the world and turn the tide, one wave at a time.

Everybody is connected to the environment in some way and lockdown here restored that link to nature and the outdoors for many people who had lost it. This is a perfect opportunity to rekindle the underlying connection that everyone has for nature and our planet. This pandemic made many people realise how fragile our existence is and how essential it is that we preserve our environment if only to ensure our own health and wellbeing.

We are rapidly nosediving into a global catastrophe. Human impacts are destroying nature and our planet faster than ever daily, this is not only harming wildlife populations and our environment but also our own health and lifestyles.

We need a united global leadership that not only recognises but tackles this issue by pulling together those of us who are desperate to help, and stopping in their tracks those who continue negative exploitation be it through financial or legal means or both. This can be done but it requires strong leadership focused on sustainability and the common good for a united global conservation movement. Currently, we do have nations with the capacity to do this but sadly we also have many who do not have this kind of leadership and that needs to change.

Nobody can do this alone, no single leader or nation or superpower. It has to be a focused, collaborative effort putting competition and financial gain aside and working together for the greater good. If the global pandemic has taught us only one thing, then surely it is that we have to be united as a global community, as nature has no boundaries. Our damaging activities and habits will inevitably bounce back to affect us all negatively, so we all have to collectively work towards a remedy.

Anything else you'd like to mention?

Everyone can do their part and every part is incredibly important.

People often say to me that it feels pointless trying so hard to be environmentally friendly when whole governments and nations are not, so their own contribution makes no difference. But it does. It is important to encourage everyone to continue to make a positive difference, however small—after all, great oak trees from little acorns grow.

Link to the interview: https://sustainability-leaders.com/frankie-hobro-interview/

Belgium | Geoffrey Lipman is currently the President of the SUNx Program, a global initiative to support Climate Resilience, related SDGs and Emergency Response through Climate Friendly Travel. He has played a significant role in the emergence of tourism as a relevant socio-economic sector. As the Executive Director at IATA in the 1970s, he helped drive a new liberalisation agenda, responding to airline deregulation. As the first President of WTTC, he worked to pioneer new systems of measuring the sector, creating CSR Certification and supporting China's efforts to open tourism markets. As Assistant Secretary-General of UNWTO, he spearheaded new development support systems, including the ST-EP Program, led the Davos Climate Summit and launched the G20 Summit recognition program. He has written/ lectured widely on tourism strategy, is the co-author/editor of two books and numerous journal articles on Green Growth and Travelism, and two major EIU studies on airline liberalisation.

Areas of expertise: climate change, sustainable development, tourism research, business, UNSDG 13 (Fight Climate Change)

The following interview with Geoffrey was first published in December 2015 on Sustainability-Leaders.com.

© The Author(s), under exclusive license to Springer Nature Switzerland AG 2022   249
F. Kaefer, *Sustainability Leadership in Tourism*, Future of Business and Finance,
https://doi.org/10.1007/978-3-031-05314-6_38

Geoffrey, when and where did your sustainable tourism journey begin—when did you discover your passion for sustainability?

In Geneva in 1991—prior to the Rio Earth Summit the following year—I had a number of meetings with the late Maurice Strong, Secretary-General of the Summit, to discuss the place of Travel & Tourism in the meeting. That began a friendship and, on my part, learning experience. It's not only what Maurice taught me—simple, obvious things—but his steadfast belief in the principles of planetary co-existence, rights and duties, as well as the role that Travelism (Travel & Tourism) must play.

As the President (and only employee) of the then newly created World Travel & Tourism Council, I wanted to get my head around the sustainability issues. Back then we were preaching the importance of Travel & Tourism as a socio-economic agent, so sustainability was a crucial aspect.

Later at the Rio conference [Earth Summit], I found myself among a handful of tourism people who were thinking about sustainability. It was the moment when some of the most thoughtful and committed people in the world joined together to make our future sustainable. I just wanted to be a part of it and believed that WTTC had a possibility and obligation to be in a leadership role.

During the meeting, I met many leading thinkers—most notably Jonathon Porritt (Founder of Forum for the Future), who convinced me that we needed to go further, faster. So, I asked Maurice Strong what he would advise and he said to do an Agenda 21-focused study of the tourism sector. We did that and shared it with UNWTO (WTO as it then was), and together held regional educational sessions on all continents to spread the findings and the sustainability message.

My then Chairman at the WTTC, James Robinson (who was Chair of American Express), was incredibly supportive of this work, as were others in the Board, like Paul Dubrule and Gerard Pellisson of Accor. Without that early support, nothing could have been achieved in WTTC because sustainable development was not the raison d'etre of the organization.

We created a research centre, launched one of the first industry certification programs—Green Globe, which later helped to spawn Earthcheck and CAST Caribbean. We also started some of the first Sustainable Travel & Tourism events and Awards with World Travel Market.

What was your view of sustainable and responsible tourism when you started your professional career?

I hadn't thought about it—girls, rugby and beer seemed to be the key drivers—I joined IATA in the 1960s and began a 20-year love affair with aviation (and the organization where I ended up as Executive Director) and that sector was not, in those days, focused on such issues.

I considered the Club of Rome—where caring for the Planet emerged as a major discipline—to be somewhat academic and the limits of growth message to be out on a tree-hugger limb.

I guess my first touchpoint with sustainability was in connection with competitiveness issues in aviation, and its growth potential. At that stage, the sustainability dimension began to emerge in the form of concerns about noise and consumer representation—not so much greenhouse gas emissions. I usually found myself with sympathies for the greens, but with my focus on progressively opening markets and expanding the aviation system.

How has your view of sustainable tourism changed over time?

My understanding of sustainability has moved from a detail to a key issue and now to a fundamental part of the future of Travel & Tourism: during the last ten years it has become my central reality.

People have been a big factor—not just Eco-warriors like Maurice Strong and Jonathon Porritt, but colleagues in the sector—public and private. Those include my friend Professor Terry de Lacy from Victoria University, Australia, with whom I've worked in many countries and collaborated on developing green growth research, road maps and models, as well as co-authoring a couple of books and numerous journal articles.

Of course, as you get older you hopefully get wiser—and I think the major shift has been since attending the Copenhagen Climate Summit in 2009, where I suddenly realized that existential means exactly that, and that it has to become my driving reality.

Your thoughts on the Paris Summit on Climate Change?

Bravo, Paris! Bravo Christiana Figueres for managing the process over so many countries and so many years. A big first step to set us on a new course, which of course will need ramping up in the decades ahead. You did what Copenhagen was planned to do in 2009. The 2015 Paris Climate Summit is not the end game, but a clear sign that global leaders are ready to adopt a new low-carbon, renewable energy-based model.

The key will be in implementation, which will need action from the community level up—regions, cities, coastal zones, rural areas, parks etc.—delivering on this nationally committed global framework. And it's essential that travel and tourism leaders and stakeholders sign on with comprehensive, targeted and measurable actions, not speeches and declarations.

Your main insights as a leader in tourism?

Above all, if Climate Change is existential—and I strongly believe it is—then we have to be at the leading edge of carbon reduction, renewables acceleration, biodiversity conservation and impact management. We have to meet the basic climate and sustainability tests of the international community. We have to support and create mechanisms that focus on this.

Second, the more I read and listened, the more I learned (and hopefully understood) that as a minimum, we need green growth: growth for wealth creation, poverty reduction, jobs and trade, but green to minimize the negative socio-environmental impact on people and places. I realised that this is not a message that mainstream 'leaders' will truly advance until it becomes popular.

Third, we need to do more to show what travel and tourism as an economic sector really embraces—the good and the bad, and that it cuts across many industries and government agencies, including transport, hospitality, events, as well as the infrastructure (both soft- people- and hard- structures-) without which the service industry wouldn't function.

Numbers are key.

We all parrot the same growth and jobs data—that's what we refined in WTTC and UNWTO—but we don't have the granular level environment accounting (which exists) integrated with the tourism accounting—that is a fundamental requirement.

Anyone who pushes around data of tourism growth should be obliged to show credible impact data. We must put as much resource into the green part as we put into the growth part, and get a balanced accounting scheme.

And here's an important issue—we need to share more and open up more—much of the mantra about growth is the same as we started in the 1990s. Institutional leaders often repeat the same old PR and lobbying stuff, and the public sessions are often choreographed this way. Look at the trade show 'debates'. Some of the work is puffed out of all reality or is self-perpetuating, some quite basic over-hype.

We create speaking and soundbite events to satisfy our egos, not to keep up with the dynamics or make the tough changes—budget impacting changes. We call them Summits even when they are local. . .we call them High Level and too often they are not. The same people invite the same people, and the media obligingly go along with the game.

Looking ahead—which projects are you currently involved in?

Here are some examples of projects that we have been developing recently, under the umbrella of greenearth.travel:

With colleagues at Victoria University, we undertook an 18-month holistic Green Growth & Travelism study of Bali for the Government of Indonesia—it produced a long-term development roadmap for the island, but also tested interesting analytical models of sustainable development that engaged the local community and university from the get-go. We advised the Government of Egypt on similar issues at Sharm el Sheikh.

With the same colleagues, we published a couple of books on Green Growth & Travelism, because we felt the need for some underpinning to this work. Subsequently, we applied these principles in a Summer School model that we launched at the University of Hasselt in Belgium and will trial in China next year.

With Wayne McKinnon we have developed a new form of transformation support for companies, called Green Growth 2050,[1] that adds the post-2015 framework issues to traditional certification approaches. Green Growth 2050 accommodates aspects of many United Nations principles and declarations linked to responsible practices and sustainability, as well as more tangible metrics on carbon, waste and water. It adheres to the Global Sustainable Tourism Criteria for Hotels and Tour Operators and works in partnership with ICTP, EQi and the Long Run to create a low-cost entry system for small and medium-sized companies, as well as easy online measurement and management.

With others—and for me, this is the most exciting—we have initiated the SUN Program, to create a connected, global community network of solar-powered prefabricated learning and capacity building centres that can operate off-grid. They will focus on Climate Resilience, SDG Support and Emergency Response.

---

[1] https://www.greengrowth2050.com

Can you tell us more about the SUN Program?

SUN stands for Strong Universal Network because the Centres will promote the visionary approach to Sustainable Development.

- Inspired by Maurice Strong, the architect of Sustainable Development—Secretary-General of the 1992 Rio Earth Summit and its Agenda 21 implementation framework
- A unique global post-2015 Agenda support network for Communities. Supporting Climate Resilience, the SDGs and rapid Emergency Response
- Linking dedicated solar-powered SUN-Ark centres, connected through the internet cloud & manned by trained local graduates
- Each SUN-Ark Centre will be small, nimble and low cost/high value & 'Smart Community' enabled.
- Engaging public, private, academic & civil society partners in market responsive, inclusionary, long-term change

The challenge as always is to find supporters and sponsors—and to make the initiatives financially self-sustaining. That's why I put so much faith in Green Growth 2050, which hopefully delivers valuable transformation services, and SUN which will hopefully be an essential, low-cost, support tool for Smart Communities.

3 books on sustainability and tourism which you recommend?

- Obviously, the *Green Growth & Travelism* series—the first of which gathered views of some 50 industry leaders and the second focusing more on concepts and practices
- *The Carbon War* by Jeremy Leggett—a really readable book on climate change
- *From Rio+20 to a New Development Agenda* by Felix Dodds, Jorge Laguna-Celis and Liz Thompson

Finally, looking back at your professional journey, what would you do differently?

I would have stuck with some of the initiatives I launched, like Green Globe and the ST-EP Program, which frankly were taken over by different agenda and re-engineered, with the wrong focus. Likewise, we simply failed to capitalize on the 2008 Davos Climate Declaration. I just moved on and in retrospect, it would have been better to fight. I think it's because I don't have the stomach for the bureaucratic games that go on.

Link to the interview: https://sustainability-leaders.com/interview-geoffrey-lipman/

Geoffrey Wall on the Realities of Sustainable Tourism Research

Canada | Geoffrey Wall examines the implications of different types of tourism on destinations with varied characteristics in the belief that such information is relevant to planning and has implications for sustainability. Much of this work has been conducted in Asia, initially in Indonesia and recently in China. He has also explored aspects of the socio-economic implications of climate change. Recent activities address tourism and sustainable livelihoods, particularly in rural settings, involving the examination of both natural and cultural heritage, links between tourism and agriculture, the involvement of indigenous people in tourism, wildlife tourism and responses to extreme events.

Areas of expertise: tourism research, China, Asia, North America, UNSDG 11 (Sustainable Human Settlements)

© The Author(s), under exclusive license to Springer Nature Switzerland AG 2022
F. Kaefer, *Sustainability Leadership in Tourism*, Future of Business and Finance,
https://doi.org/10.1007/978-3-031-05314-6_39

The following interview with Geoffrey was first published in June 2016 on Sustainability-Leaders.com.

Geoffrey, looking back at your distinguished career in tourism studies, which have been your main professional insights?

My early work on the impacts of tourism predates the wave of interest in sustainable development, although I think it prepared me well to engage with the latter topic. The same is true of climate change.

Although primarily an academic, I have always thought it important to ask the 'so what?' question, and I have had a variety of opportunities to be involved in planning exercises at a variety of scales, from multi-national to local. Thus, I have spent my career trying to understand the implications of different types of tourism for destinations with different characteristics in the belief that the resulting insights can be used to inform tourism planning.

I have generally adopted a broad perspective, involving economic, environmental and socio-cultural dimensions.

Much of my work has been conducted in the developing world, particularly Indonesia and China, where residents are often disadvantaged by so-called development. I acknowledge that tourism is a business but, accepting this, I have tried to argue and illustrate that it is necessary to take a long-term perspective and address environmental and cultural dimensions to move the system in the direction of sustainability.

Based on those insights, where do you see the priorities for (sustainable) tourism research in the near future?

I have found sustainable development to be a difficult concept to deal with because it can mean different things to different people. To guide research or planning it requires further refinement, for example, the development of indicators.

I seldom use the term 'sustainable tourism' because I believe that sustainability demands a multi-sectoral perspective—tourism is not independent but engages and competes with other sectors for the use of scarce resources such as land, labour, capital, water, energy and waste assimilation capacity. I have borrowed the notion of sustainable livelihoods from rural studies to guide much of my recent work. I find this to be a more tangible concept to work with.

Much of my work has focused on destinations where the imprint of tourism is concentrated. However, tourism requires travel, and the impacts of the travel phase have received much less attention, by myself as well as by others.

The travel phase is only now receiving the scrutiny that it deserves, and consumption of energy, especially in long-haul travel, should force one to question if tourism, in its contemporary forms, can ever be sustainable. Even ecotourists may travel long distances, use transportation and other infrastructure, and visit fragile areas so that, somewhat paradoxically, they may even have greater impacts per head than mass tourists.

Thus, a fuller accounting of the impacts of different types of tourism is required and, if done, may give rise to some surprises. Perhaps, fortunately, most tourists are still mass tourists, and more thought needs to be given to ways of making mass tourism more sustainable.

What motivated you to publish 'Tourism: Change, Impacts and Opportunities'
in 2006? Which are the book's key messages?

The 2006 manuscript was a replacement for 'Tourism: Economic, Physical and Social Impacts' which was published in 1982 and reprinted numerous times without modification. Much had happened since that time, including the rise to prominence of sustainable development, the proliferation of niche forms of tourism, such as ecotourism, as well as a massive increase in the relevant literature. Thus, re-writing rather than updating was required.

The title was modified to draw attention to the positive as well as the negative consequences of tourism (the word 'impact' often having a negative connotation) and to emphasize that residents and destinations commonly seek tourism development in the belief that it will improve the lives of residents who respond in a variety of ways and are not simply impacted.

It is my belief that a thorough understanding of impacts is necessary to inform planning and management which, at root, are about managing change. However, I have come to understand that while tourism planning might address the viability of attractions, the desires of tourists and the like, it also should give considerable attention to the desires of residents.

Thus, a substantial part of tourism planning should address the needs of residents. For this to occur, it is important that clear and appropriate goals and objectives should be specified for tourism plans. The development of tourism should not be an end in itself but should be a means of addressing more fundamental societal problems, whether these be jobs and incomes, protecting the environment and heritage, celebrating culture, supporting national identity or whatever.

Much of your research, reviewing and writing has involved projects in Asia
(particularly mainland China and Taiwan)—do you see growing interest and
momentum for sustainability and sustainable tourism in this region?

There are substantial differences between Taiwan and mainland China, not only in political systems but also in environmental awareness and the effective protection and interpretation of special environments in reserves, such as national parks. Tourism has grown rapidly, in line with China's economy, and it is viewed in China as a 'pillar industry'. China is now a major player in international tourism and also has massive domestic tourist flows.

Sustainability rhetoric is widespread and enshrined in legislation at all levels in China, and much of the rest of the world, but implementation is another matter. There is still a strong emphasis on short-term economic gains that override longer-term environmental and cultural perspectives. Massive air pollution as well as issues of water quantity and quality, and less discussed land degradation and pollution, are attracting increasing public attention. Fortunately, many educated young people in China are becoming increasingly aware and concerned about these things, although most are not in a position to make their feelings known or to initiate change.

Which organizations or persons have served you as inspiration?

I was fortunate to receive a good education in strong universities in several countries, as well as encouragement to pursue my interests in tourism (as differentiated from hospitality) at a time when little academic attention was being given to this topic. My home discipline, geography, supported the adoption of broad perspectives that spanned physical, human, technological and other dimensions of real-world problems in diverse parts of the world.

Within tourism, I have great admiration for the work of my good friend Richard Butler. I have also benefited greatly from exposure to the ideas and occasional collaborations with colleagues at the University of Waterloo who have encouraged me to place tourism in a broader context. In particular, I mention Geoff McBoyle, a climatologist, who encouraged me to publish my research on tourism and recreation, and later collaborated with me on research on tourism and climate change; and Bruce Mitchell, a water resources specialist who introduced me to many aspects of resource management. Ian Burton, a leading scholar on hazards and, more recently, climate change, emphasized the human dimensions of environmental problems and put a young academic in touch with novel ideas on human-environment relationships and an international network of scholars.

It is also important to emphasize that I have learned a great deal from my graduate students. It is not possible for a person with limited language skills to work successfully in other cultures in the absence of the inputs and insights of intermediaries.

I do not teach graduate students: rather we explore together and learn from each other. If I had my time again, I would put more effort into learning other languages, for I believe that you cannot truly understand another culture if you do not speak the language.

In your view, what characterizes a sustainability leader?

Leadership in any area requires a willingness to empower others, and ability to listen as well as suggest what should be done, early recognition of needs/ opportunities and a perspective that turns problems into opportunities, commitment and diligence but time for reflection, respect that has been earned and so on.

In the context of sustainability, both depth and breadth are required. It is desirable to be an expert in something so that one has insights pertaining to a specific field than can be offered and shared. At the same time, it is important to have a broad perspective and to be able to put one's knowledge in context. For example, although a tourism specialist, my most interesting and probably useful work has been in the exploration of such themes as coastal zone management, community development, biodiversity strategy, eco planning, heritage planning, and indigenous economies.

Academics in countries such as the UK witness growing pressure to demonstrate the relevance and ROI—societal impact—of academic research. In your view, which is the best way to do this?

I view myself as being a traditional academic whose main aims are to generate and share knowledge. I have not felt pressure to be relevant, although I have always felt it important to point out the implications of my research. However, I have been involved in many planning exercises and international projects where recommendations must be made.

Unfortunately, many tourism researchers feel that they do not get the respect which they deserve.

Tourism education and research has become more and more inward looking, as young scholars read and are encouraged to publish in specialized journals to the neglect of engagement with other disciplines and the place of tourism in societal issues. Having a degree in tourism may or may not mean you know something about tourism, but it increasingly guarantees you know nothing about anything else!'

Link to the interview: https://sustainability-leaders.com/interview-geoffrey-wall/

Australia | Dr. Gianna Moscardo is qualified in applied psychology and sociology and joined the School of Business at James Cook University in 2002. Before joining JCU, Gianna was the Tourism Research project leader for the CRC Reef Research for 8 years, researching how to support making human uses of the Great Barrier Reef World Heritage Area more sustainable. Her qualifications support her research in understanding how communities and organisations perceive, plan and manage tourism to enhance their sustainability. Her work[1] also includes understanding how tourists travel and learn from their experiences, to design more sustainable tourism experiences.

Areas of expertise: overtourism, tourism research, Australia, UNSDG 11 (Sustainable Human Settlements), UNSDG 8 (Equal & Fair Economic Opportunities)

The following interview with Gianna was first published in September 2017 on Sustainability-Leaders.com.

---

[1] https://research.jcu.edu.au/portfolio/gianna.moscardo/

© The Author(s), under exclusive license to Springer Nature Switzerland AG 2022
F. Kaefer, *Sustainability Leadership in Tourism*, Future of Business and Finance,
https://doi.org/10.1007/978-3-031-05314-6_40

Gianna, what was your view of sustainability and tourism when you first started your professional career?

My first serious job after graduating was working in a government research center to support sustainable human use of the Great Barrier Reef in Australia. In this setting tourism was typically portrayed as the main pathway to sustainability, being the economic alternative for commercial fishing and mining. This was in some ways a very gentle introduction to tourism and sustainability, as the tourist numbers at that time were quite small on a global scale, despite having grown quite a lot and quite quickly.

Tourism to the Great Barrier Reef was strictly controlled and so direct impacts of tourism in this setting were very low and we were able to focus much of our attention on how to use tourism as a tool for education about sustainability and marine conservation. This was great in terms of learning about how to use tourism to support local communities and to teach tourists about the environments that they depend on for life, but it wasn't typical of the rest of the world.

When I changed over to being an academic, teaching and researching tourism beyond the local region, I realized that there were major challenges in just managing the sheer volume of tourism, especially in places with far fewer resources.

How has your view of sustainable tourism changed since then?

When I became an academic, I began teaching students from a wide range of different places who had many different and often quite challenging stories of tourism and sustainability from their experiences. I also began attending conferences in new locations, focused on things other than the environmental impacts of tourism and interpretation for tourists.

It was during a visit to Africa to attend an ATLAS Africa conference and meet up with a doctoral graduate that I was more directly exposed to the dark side of tourism. Different people there told me about forcible relocation of residents for resorts, including ecolodges; the consequences of high levels of economic leakage from foreign-owned tourism ventures; and the potential for political corruption in tourism development.

It was this visit that made tourism and sustainability become a much more serious issue for me. It also shifted my thinking away from making tourism itself more sustainable, to rethinking how tourism could work as a tool for sustainability. It also made me question the models and approaches to tourism development and marketing that we were teaching our students, and even whether or not we were teaching tourism and sustainability to the right students.

Where do you see the priorities in sustainable tourism for the near future?

The biggest and fairly immediate priority is managing large numbers of tourists. With more budget travel options available, we see much more international tourism everywhere, especially from China and India because of the potentially huge numbers of tourists.

We have to change the way destinations think about and measure tourism success. We have to change the way policymakers think about and plan for tourism. And we have to change why people travel.

For destinations and policymakers, we have to get them to see tourism as a tool for regional well-being, not an end in itself. We have to replace tourist numbers as the major measure of success with systems that measure tourism's net contributions to things like local business development, resident physical health, the development of social capital, and support for sustainability infrastructure.

In short, we have to ask how does tourism make this a better place to live for residents and keeps long-term residents staying there—and happy about staying there.

We have to change the social narratives about the value of travel for the individual tourist—we have to start seeing travel as a privilege and a responsibility, not a right. As issues of overtourism and tourists behaving badly get more and more mainstream media attention, tourism will be under pressure to justify its existence. We will need to have much better answers than we currently have to the question why have tourism at all?

Your thoughts on the current state of sustainability in tourism in Australia?

Overall, I think tourism and sustainability in Australia are currently in a very good situation. We have a large, affluent country with good levels of regulations around things like working conditions and environmental impacts. And we still do not have large numbers of tourists when compared to other places on the planet.

This doesn't mean that we don't have some examples of unsustainable tourism developments, but in general, there are no major national issues currently. But Australia does face some future challenges, particularly around climate change and tolerance of cultural diversity. In terms of climate change, many of Australia's major tourist destinations are coastal and thus threatened by rising sea levels and more extreme weather events.

Australia's tourism is also vulnerable if tourists start flying less because of climate change pressures. Australia is a long-haul destination internationally and within the country, many destinations are long haul even for domestic travellers.

Like many countries, Australia is currently seeing a rise in the visibility of more extreme political groups opposed to migration and cultural diversity. This is both a threat to tourism and an opportunity for tourism to demonstrate the value of embracing tolerance and diversity.

Your advice to emerging researchers interested in sustainable tourism?

It is currently a difficult time to be an emerging researcher in sustainable tourism, as we have a research world obsessed with prestige rankings and assessments. Those are all based on performance in English language academic journals linked to established disciplines and which are not at all interested in the extent to which research changes the world for the better.

Any researcher in sustainable tourism is by definition focused on applied issues and seeking to change the way tourism is practised. This means partnerships and collaborations with destination community groups, tourism businesses and associations, government departments, academics from other disciplines, and tourists themselves.

Doing this type of research is not currently held in high esteem in the academic world. But this is changing, with an increasing concern from funding bodies about being able to demonstrate that research, especially publicly funded research, is making a difference to policy and practice.

So, the key is to find those partnerships and collaborations locally to do the research and connect to groups of established researchers who share an interest in changing practice—groups like BEST EN (Building Excellence in Sustainable Tourism Education Network), and TEFI (Tourism Education Futures Initiative)—to develop academic publications.

Can you tell us more about BEST EN and how your previous experiences have led you to the role of co-chair?

BEST EN is an inclusive and collaborative network of tourism academics and practitioners committed to furthering the creation and dissemination of knowledge within the field of sustainable tourism. It is a group of people who are passionate about the need for tourism to contribute more to sustainability and who are committed to making change.

I went to my first BEST EN Think Tank in Arizona in 2007. A colleague who knew my interest in sustainable tourism suggested that I might find it interesting, and I enjoyed it very much. The focus on sustainable tourism and applying research to practise appealed to me, but also the Think Tank style was something I found to be very refreshing and rewarding.

BEST EN Think Tanks are very different from traditional academic conferences. They are very focused on the theme and organized to encourage all the participants to work together to generate information and resources on one aspect of sustainable tourism. It is a small meeting (limited to 70 participants) but provides a great opportunity to harness the wisdom and experience of all participants, including local tourism and community representatives, and create resources to support sustainable tourism. Most of these resources are then made available for free on the BEST EN website.

I enjoyed my first BEST EN event so much that I have attended nearly every Think Tank since. As a regular attendee, I was asked to lead some of the collaborations that emerge from the Think Tanks, which connected me to more of the Executive Committee members. Then they invited me to join that group and then to be Co-Chair. I've found the whole experience of working at all levels with the BEST EN group very rewarding.

Best EN's 'Think Tank' is an incredible platform for research-driven collaboration—what is the theme for 2018 and how was this chosen?

We have just recently made the full proceedings available on our website for the 2017 Think Tank on Innovation and Progress in Sustainable Tourism, hosted by the International Centre for Sustainable Tourism and Hospitality at the University of Mauritius.

Our next Think Tank will be held in Lucerne, Switzerland, 19–22 June 2018, hosted by The Institute of Tourism at the Lucerne University of Applied Sciences. The broad theme for 2018 is 'Development and marketing of sustainable tourism products'.

Each year we select the theme in cooperation with our hosts and with input from our network members. It reflects either a sustainable tourism issue of concern to our hosts or an aspect of sustainable tourism that our network thinks is important and that needs further attention.

Which 3 books would you recommend emerging tourism researchers or practitioners?

International Cases in Sustainable Travel and Tourism. Edited by Pierre Benckendorff and Dagmar Lund-Durlacher.

This book was a result of a collaboration between BEST EN and the WTTC's Tourism for Tomorrow Awards. Finalists from the awards were connected to academics through the BEST EN network to jointly write case studies showing how sustainable tourism works in practice.

Education for Sustainability in Tourism. Edited by Gianna Moscardo.

This book emerged from one of the BEST EN Think Tanks and incorporates both resources to support sustainable tourism in general and specific examples and ideas about to educate different stakeholder groups about tourism and sustainability. A wide range of tourism experts and practitioners contributed chapters. Practitioners might find chapters about different dimensions of sustainable tourism and the material on how to educate destination communities and tourists about sustainability particularly useful.

Tourist Season by Carl Hiaasen.

Carl Hiaasen writes crime novels all set in Florida. His stated aim in writing this novel was to discourage tourists from coming to Florida. A consistent theme in this and many of his other novels is that tourism is associated with corruption, crowding, and environmental damage. It is interesting to read how people outside of tourism academia and policy see tourism. A reminder about how bad tourism can be.

Link to the interview: https://sustainability-leaders.com/interview-gianna-moscardo/

Glenn Jampol on Tourism Business and Ecotourism in Costa Rica

Costa Rica | Glenn Jampol has been a tourism pioneer in Costa Rica for the last 30 years. Along with his wife, Teresa Osman, Glenn has been the owner and developer of two award-winning five-level certified sustainable hotels in Costa Rica: Finca Rosa Blanca Coffee Plantation Resort and Arenas Del Mar Beachfront and Nature Resort. In 2010, his coffee plantation won the Rainforest Alliance International Sustainable Standard Setter Award. He has worked as a liaison to the Costa Rican Tourism Board (ICT) and was the vice-president of the National Accreditation Commission accrediting sustainable practices and sustainable tourism certification. He has also worked as a consultant on national and global sustainability policy for the tourism ministry in Costa Rica. Glenn was the Director of the Board at The International Ecotourism Society (TIES) and is currently the

© The Author(s), under exclusive license to Springer Nature Switzerland AG 2022   267
F. Kaefer, *Sustainability Leadership in Tourism*, Future of Business and Finance,
https://doi.org/10.1007/978-3-031-05314-6_41

Chair of the Global Ecotourism Network. Glenn is also a founding member of the newly formed LACEN, the Latin America & Caribbean Ecotourism Network.

Areas of expertise: ecotourism, tourism business, Latin America, UNSDG 11 (Sustainable Human Settlements), UNSDG 12 (Resource Efficiency), UNSDG 15 (Forests & Biodiversity), UNSDG 8 (Equal & Fair Economic Opportunities)

The following interview with Glenn was first published in July 2016 on Sustainability-Leaders.com.

Glenn, when did the concept of sustainable tourism first show up on your radar?

Well, as you know, I attended UC Berkeley in the late '60s and early '70s, and most of us were what they used to call 'tree huggers' with a sense of responsibility for protecting our planet and exposing corporate greed. From this, we developed a strong focus on changing our lifestyle and turning to what we thought was 'natural' or 'organic' products and unique global experiences. This incipient search for authenticity and understanding had—in all its manifestations—health at its core: We were intent on treating our bodies well, fulfilling our newly acquired goals of changing the global consciousness which had permeated and moulded our perception of the world at large.

Additionally, many of us were infatuated with the quest for new travel experiences, and this longing manifested itself into the search for inexpensive and undiscovered destinations. I think it is safe to say that these global travels opened our eyes and hearts to the idea of amplifying our personal, cultural and environmental responsibility. During my travels, I became aware of the extra efforts some owners had put into waste reduction, local products, altruism in the community and a sense of long-term protection.

What was your view of sustainable tourism when you first started your professional career?

That term really didn't exist on any mass level in 1985 when we began our Finca Rosa Blanca project. It wasn't until the mid-90's that it became a more widely used term, and we began to explore it and use it as a descriptor for certifying our own responsible tourism and ecotourism businesses here in Costa Rica.

We had woven this idea of ecotourism into the fabric of all of our business practices and during the development of the Certification for Sustainable Tourism program in 1994/5, we were able to dissect it and clarify what the goals were: the need to expand the meaning of this concept from just an environmental or conservation-oriented focus by adding a social component and to use it to guide our everyday business practices.

How has your view of sustainable tourism changed over the years?

My perception of the definition of sustainable tourism has been radically changed many times over the last 30 years. For instance, many years ago I was asked in an interview on a television show about the ionization system we use for cleaning our pool water at our hotel, and I proudly shared that it only cost me $175 per year to keep it functioning well thereby saving more than $25,000 over the last many years by not purchasing pool chemicals. When the show finally aired, I received many emails from viewers asking me where they could buy this system.

If it hadn't dawned on me before, it certainly made an impression on me then: We are best served when we provide solutions to create a means to financial sustainability. Along with inspirational storytellers and the ethical aspects of good practices, this is what will ultimately and efficiently convert the disinterested into strong advocates.

Since your studies at Berkeley, you have been combining your love of art and your passion for conservation and responsible travel. How have these two career paths intersected? What influence have they had on each other?

Actually, my painting and art career was always parallel to my environmental interests and sense of responsibility. I got my BA in Environmental Design, and as an artist, I was able to incorporate my artistic sense of experimentation and design to the concept of designing the look of our hotel.

Also, over the last 40 years of being an artist, I have collected, traded and made art expressly with the idea of using it as a global exploration of visual experiences. Along with the personalized attention to detail, we see our hotel as a work of artistic creation and many of our guests also comment on this.

Together with your wife, Teresa, you opened the award-winning boutique hotel and restaurant Finca Rosa Blanca Coffee Plantation Resort in Costa Rica. Can you tell us a little about this venture? What are its origins, and how has it evolved over the past three decades?

We found ourselves in the tourism industry in 1989 by sheer accident, having built an experimental home (in the design sense) in an abandoned motocross field in the coffee highlands of Costa Rica and later realizing we needed to find a way to pay for the upkeep. We had no experience in tourism and weren't business-oriented, but we were looking for an artistic and responsible way to create a niche.

We began to incorporate a new tourism model into the one that existed here in Costa Rica. This combined the varied but basic cultural and biodiverse experiences available in Costa Rica with the more upscale and comfort-based attributes of luxury hotels.

Later, we converted a conventional and treeless coffee field into an organic and shade-grown plantation. We funded the initial start-up costs of this conversion with personalized coffee tours from the hotel and realized that this 'agro eco-tourism' offer was not only lucrative but educational as well.

The biggest change in the last ten years has been the huge increase in mainstream tourists who have come to expect good practices or sustainability as part of their hotel's DNA, yet they still do not make choices based solely on this aspect of our business.

In 2008, you and Teresa designed and opened Arenas del Mar Beachfront and Nature Resort, the first five-star and five-leaf resort in Costa Rica. Which have been the main challenges and 'Eureka moments' during the process of building the resort?

Arenas Del Mar was built on one of the most beautiful beaches of Costa Rica. We do not own the land, it is leased as a 20-year renewable 'concession'. We began this project in 1990 but could not open the hotel until 2007! We spent the first 12 years reforesting and planning the sustainable model and trying to work within the difficult restrictions from the government.

Our funding for Arenas Del Mar came from other shareholders who had many traditional, and often conflicting, ideas about the long-term value of sustainability choices when it came to making financial decisions. I believe it is still the only tourism project in Costa Rica that has 100% native species of plants as we started the project by building a nursery for them years before we began construction.

The 'Eureka' moment was the realization that we could and had to become more creative and look for solutions that would offer what the financiers wanted while still minimizing our footprint. With this model we have achieved the highest occupancy in the area.

But certainly, for us, the most difficult and debilitating impediments were the bureaucracy and multi-layered influences that permeated the building on our property, which continues to this day.

You are a founding member and Chair of the Global Ecotourism Network (GEN). Where did the idea to start GEN come from, and what was the creation process?

GEN is made up of the entire Advisory Board of TIES, who as a group simultaneously resigned from TIES in 2015 due to the inability to access critical financial information, despite our numerous requests over the years as the Advisory Committee for this information.

The total lack of financial transparency at TIES and the inability for us to ethically guide the organization forced us to create a new umbrella organization for the representation of global ecotourism: the Global Ecotourism Network (GEN). We urge all the members of the ecotourism world to read our recommendation for the importance of due diligence.

What makes GEN different from all the other sustainable/responsible tourism organizations out there, and how do you see it developing in the future?

GEN is undoubtedly the most experienced organization in the world of sustainable tourism. GEN's board of directors have a long track record of being global leaders in ecotourism—only our name has changed.

GEN brings together the world's national and regional ecotourism associations and networks, destinations, indigenous peoples, global operators, professionals and academicians who will grow the industry, provide advocacy and thought leadership, and who will encourage innovation and authenticity in ecotourism. We are collaborating with the world's community of national and regional ecotourism bodies as an international ecotourism umbrella organization which will provide an easily accessible platform and voice for indigenous peoples and rural communities involved in the ecotourism industry.

Also, we are a think-tank for sustainable development and the evolution of an industry which should be centred on authenticity. We want to disseminate authentic ecotourism trends, innovations, applied research and case studies and in doing so, we hope to be viewed as the authority for ecotourism practices and sustainable tourism among consumers by providing thought leadership and influencing market demand.

Prior to founding GEN, you spent six years as President of CANAECO— National Association of Ecotourism Costa Rica. What were the major achievements and challenges during your time there?

CANAECO is a very important organization for Costa Rica. It has represented the ecotourism and sustainable tourism private sector in this country since 2003 and in my time as president, we were able to multiply our membership. In 2009, we created the now highly recognized International Conference on Sustainable Tourism and Ecotourism; Planet, People, Peace (P3). This conference was officially declared by the government of Costa Rica as being of 'National Interest' in 2012. We also achieved two positions of importance: a permanent seat in the National Accreditation Commission for approving levels of sustainable certification and creating new norms for the CST, and we obtained a permanent seat on the Board of the National Association of Tourism (CANATUR).

Most importantly, we were able to convince the Costa Rican government that sustainable tourism should be the country's official tourism policy.

Our goal was that eventually, the majority of members of the Board of Directors of the Tourism Board (ICT) would be members of CANAECO, and when I stepped down as president in 2014, we achieved this goal with ALL of the board being members of CANAECO including the then newly appointed Minister of Tourism.

Link to the interview: https://sustainability-leaders.com/interview-glenn-jampol/

Gopinath Parayil on Being a Responsible Tour Operator and Voluntourism

India | Gopi is the founder of The Blue Yonder, a pioneering responsible travel company. He is the Co-founder of Chekutty, a livelihood project that became a beacon of resilience during the 2018 Kerala floods. He is currently co-creating resilient destinations bringing the development, humanitarian and tourism sectors together. Resilient destinations focus on building capacities to overcome challenges raised by disasters caused by the climate crisis.

Areas of expertise: responsible travel, tourism business, destinations, entrepreneurship, India, Asia, UNSDG 13 (Fight Climate Change)

The following interview with Gopinath was first published in June 2016 on Sustainability-Leaders.com.

Gopi, what was your view of sustainability and tourism when you started your professional career?

I had nothing to do with tourism business other than being an avid traveller myself, when I started The Blue Yonder. However, even as a teenager, I was already

F. Kaefer, *Sustainability Leadership in Tourism*, Future of Business and Finance,
https://doi.org/10.1007/978-3-031-05314-6_42

involved in development issues because of the left-leaning politics of my home state of Kerala. I also had volunteering experience in community-based palliative care from '93 onwards, which exposed me to another world and possibilities of how issues can be handled through innovative, crowd-sourced, and collaborative ways.

I later began working with UNDP [United Nations Development Programme] funded projects on environmental sustainability in New Delhi as well as being part of Charities Aid Foundation—India in New Delhi and Bangalore. Working on ICTs [Information and Communications Technology] for Development had helped me formulate my own thoughts and insights on how 'development' should be approached.

In my opinion, reducing dependency on 'aid'—whether international or local—was a key driver in these thoughts.

It was clear to me that the values of dignity, sense of belonging and ownership had to drive the journey towards sustainability. Locally driven wealth generation was the key. These thoughts laid the foundation for the growth of The Blue Yonder as a pioneering Responsible Tourism business in India.

What inspired you to set up The Blue Yonder as responsible tour operator company in India in 2004?

It was when I had just returned back from the UK after completing my studies and assignments in Disaster Management at the Royal Military College of Science, that my father passed away. As part of our tradition, there are rituals we do, including requiem for the dead in the holy rivers.

Bharatapuzha, (River Nila) a 250 km long river revered by locals and part of many legends and folklore was one such and I was lucky to have been born in a village nearby that had all the influences of this river. There was hardly any water in the river (it wasn't even beginning of summer then), when we were doing the ritual for my father and the struggle in which I had to take a dip in the river probably became the tipping point for The Blue Yonder.

A river that was so much part of my childhood, romances and growing up was in a decaying state. It wasn't polluted, but for the first time in any recorded history, this river which is so important to our culture apparently stopped flowing during the summer due to sand mining and dams.

This incident resulted in a couple of us setting up a not-for-profit foundation supported by some of the passionate locals to come out with a book on the status of rivers in Kerala.

Funds from personal pockets soon depleted and it was while searching for regular funding sources that I decided to come up with a for-profit travel company that would become the ambassador of this small river and its rich socio-cultural heritage.

We didn't want to be an NGO, dependent on donations and contributions where our dreams would be dictated by donors. We decided to create wealth which can then be used to support various initiatives that will preserve the cultural, natural heritage of the region while ensuring sustained livelihood and sense of respect for the local people.

The result was the formation of The Blue Yonder, the first company in India to position its business on the triple bottom line of sustainability.

Funny enough, we had never heard the word 'Responsible Tourism' then, and were calling it Responsible Tourism, purely because we ended up in this space because of what we believed was our 'responsibility' to 'respond' to our surroundings for a better future.

It was only later that we became aware that there were many initiatives like ours hatching in different parts of the world.

If you would have to start your company all over again, what would you do differently?

Quite a lot.

But deep inside I have no idea how else would we have started. This was a company that was born out of a passion and not out of a business plan. Many of our decisions were impromptu (and it worked!) as we jumped on opportunities that came our way. Striking a balance between passion, professionalism and tricks of the 'game' was always a challenge.

I think we would have been far better off if we had a well thought out business plan from the beginning and not only in the later stages, and had business partners including social investors who understood our kind of business.

Stricter financial management would be definitely a priority. Engaging in too much activism can also scare away investors who saw in us development activists rather than businesspeople. Also, finding the right team which does not compromise on the company's beliefs and values continues to be a challenge and high priority.

How has your education and experience in Disaster Management impacted your approach to sustainable tourism development?

It goes both ways.

In March 2016, during ITB Berlin, The Blue Yonder ran a clinic with the Chief of Disaster Risk Reduction, UNEP (United Nations Environment Programme), focusing on the role of Tourism and Hospitality in relation to disaster preparedness, resilience, management and response.

The objective of the session was to explore the possibilities of how the tourism industry can add value to a destination beyond being a service provider. The clinic focused on the situation of three disasters—the earthquake in Nepal (2015), the flood in Chennai (2015), and Cyclone Aila in Bangladesh (2006).

In Sundarbans, West Bengal (India), our company The Blue Yonder and other like-minded tourism companies were the first responders, even before local NGOs or Government could activate their resources. I was able to be of assistance to the community because of my background in disaster management and understanding of how to utilize our own resources and supply chain effectively.

In the wake of the Chennai floods, we are now working on a manual on how the travel industry can collaborate with civilian society and government agencies. Strategies for mitigation, response, and rehabilitation are more effective when planned, as opposed to knee-jerk reactions like many have been doing.

We must ask ourselves, "How can a tour operator in the UK survive if its destinations like the Maldives or Kerala are underwater or hit by an earthquake?" There is an urgency to recognize the role of travel and tourism industry in disaster management relief for the long-term sustainability of destinations.

Training and motivating employees to implement and 'live' a company's responsible tourism values is often a key issue, together with equal opportunities for men and women. What has been your experience at The Blue Yonder?

Motivation has never been a challenge for us. Employees have always joined us because of the uniqueness of the company. However, sustaining the motivation has been a challenge for multiple reasons. The unique positioning of the company as a combination of development projects and travel experiences is an asset as well as liability.

Over a period of time, our not-for-profit activities totally ceased as the for-profit company was involved in doing many things which the foundation was supposed to do. This thin line became a challenge for us, as those who joined the company at later stages struggled between the 'development' and 'travel' roles.

Our product development phase is time-consuming as it's always issue-based. By the time it's developed into travel experiences that can be marketed and sold, we used to face a lot of challenges.

We have seen two extremes in the company when it comes to values. Those who want us to address far more development challenges, and those who want to dilute the triple bottom line responsibility to a simple CSR model, where they propose to share profits of the company to a social cause. The Challenge—and key—obviously is about striking a balance.

Can you describe to us the purpose of the 'A to Z of Responsible Tourism' campaign? Who was the intended audience for the campaign and how is it a powerful tool to promote responsible tourism?

It was an attempt to demystify Responsible Tourism and to help those in the fringes as well as in the mainstream travel business understand what benefits sustainability can bring to their business, lives, and global outlook. Boring academic jargon and 'Let's save the world' philosophies weren't helping to achieve that. We needed a fun way to celebrate Responsible Tourism.

Moreover, many DMCs [destination management companies] in India, as well as elsewhere, were reaching out to ask how they could be part of the Responsible Tourism movement and we had to explain it in a way they could relate to. Travellers obviously were also part of the target group. Interestingly, many universities started using the A to Z of Responsible Tourism as a communication tool to explain Responsible Tourism, and many travel companies used it as training material for their tour guides.

The campaign started as an initiative by my colleague Zainab Kakal, who got tired of explaining to her friends and family what she was doing at a responsible travel company called The Blue Yonder. It later evolved into one of the most exciting crowd-sourced initiatives targeted at the travel industry, as well as consumers.

During 26 days, we asked our online friends and followers about the values they could associate with responsible tourism, linked to each letter of the alphabet. We got many suggestions from all around the world, which we compiled into the cards that you can now see on www.theblueyonder.com/atoz.

The concept behind each letter and word was later edited by Jeremy Smith of Travindy, sketches were done by students from the National Institute of Design in India, and young designer Vinit Basa from Mumbai put it together as the colour postcards.

We were invited by India's first Biennale to be a partner and we launched the beta version there, followed by the big launch at ITB Berlin in 2013. We are glad that this year's World Travel Market London Responsible Tourism Awards Scheme has long-listed this amongst campaigns to promote Responsible Tourism.

Voluntourism, also known as volunteer tourism, has become popular as a way to combine holidays with volunteer work. Unfortunately, good intentions don't always lead to doing good. What is your view on this?

The world has changed a lot, and there are loads of resources available locally that can be channelled effectively. Decades of funding and volunteering haven't brought the promised sustainable solutions to the suffering. Yes, a country in crisis might need intervention from elsewhere, I agree. However, the volunteering we are talking about is either happening years afterwards or in destinations where there isn't any major crisis.

In my home state of Kerala, for example, local communities run a neighbourhood network called the Pain and Palliative Care Society. Along with its associated organisations, today this network led to more than 1,000 palliative clinics. Many of those are now run by the Government after officials saw the potential of this to reduce the tremendous stress on governmental institutions and the very limited private infrastructure in Palliative Care.

These clinics have catered to more than 100,000 terminally ill patients. They are run by more than 42,000 local volunteers. Micro-donations are raised locally by students and other volunteers and well-wishers. And all this was built up—without international volunteers—from a one-room clinic with two doctors, nurses and two volunteers 23 years ago.

So, for us the big question is: How can we all channel our energy towards cleaning our own neighbourhoods first—before we set off flying 3,000 miles to 'save the world'? I urge all well-meaning travel volunteers to think about this for a second before plunging into 'saving the world' and making another tour company in the source market rich at the expense of some poor community elsewhere!

Volunteering can be tremendous fun for both travellers and locals, if the attitude is about 'learning' from locals, rather than 'saving' the poor or rural populations. If travellers are looking for rewarding 'experiential' travel, then do so by choosing a company that will help you travel through destinations where sustainable development projects are promoted, pioneered and supported. Just don't call it volunteering. It's an immersive travel experience. Isn't it?

A new initiative was recently launched in the city of Kozhikode in Kerala called Compassionate Kozhikode, developed by the District Administration. What can other destinations who want to implement responsible tourism learn from this initiative?

Compassionate Kozhikode is a platform to celebrate the lives of such people driven by passion, hospitality and who can create a holistic destination out of

Kozhikode. It's literally an attempt to 'Create Better Places for People to Live & Visit.'

A plethora of initiatives has been launched in the last year by the District Administration, in partnership with civil society, government agencies and citizens, who volunteer to create an authentic and inspirational destination for them to live and invite people to visit and be part of their lives and experiences.

A lot of destinations launch tourism projects and make claims of how it's benefitting local people. Here it is working the other way around, where the hypothesis is that if efforts are taken to facilitate and create a destination that cares, such an ecosystem ensures experiential holidays.

Interestingly, the Hotels and Restaurants Association of Kozhikode has been recognized in the long list of World Travel Market's Responsible Tourism Awards for the 'Food With Dignity' initiative, run in partnership with the District Administration.

Why did you become a Board Member of the new Asian Ecotourism Network (AEN)?

AEN was long due. I used to follow The International Ecotourism Society and was actively involved in 3 conferences, from Norway to the USA. However, it was very visible that—apart from one or two core team members from the region—Asia wasn't a priority. I saw AEN as a good opportunity to bring together peers in ecotourism and responsible tourism and to bring business to the region.

Except for a handful of initiatives, networking among professionals in ecotourism is confined to conferences, advocacy and campaigns. AEN as a new platform can help members to find a business in this segment. Also, instead of following a holistic universal set of standards, this Asian Network gives us an opportunity to develop criteria that are culturally, socially and economically relevant to the Asian region.

Link to the interview: https://sustainability-leaders.com/interview-gopinath-parayil-the-blue-yonder-india/

Goverdhan Rathore on Tiger Conservation and Sustainable Tourism in India

43

India | As the son of Fateh Singh Rathore, one of India's best-known Tigermen, Dr. Goverdhan Singh Rathore was privileged to grow up in and around Ranthambore National Park. Having seen the uphill battle to save the park and its tigers, Goverdhan wanted to continue his father's legacy. After graduating from medical school in 1989, he started a mobile health care program visiting villages around the park. His visits provided insights into the local communities, their challenges and how to solve them while educating them about the tiger, its habitat and the need to save it. He is actively involved with NGOs like the Prakratik Society, Tiger Watch and Ranthambhore Foundation. Goverdhan started a Jungle Lodge called Khem Villas with his wife Usha. Built on a barren tract of land adjacent to the park, it has been re-wilded using local fauna to merge with the park habitat so wildlife can move freely through the lodge.

Areas of expertise: conservation, wildlife, ecotourism, responsible tourism, business, India, Asia, UNSDG 15 (Forests & Biodiversity), UNSDG 8 (Equal & Fair Economic Opportunities)

The following interview with Goverdhan was first published in May 2020 on Sustainability-Leaders.com.

© The Author(s), under exclusive license to Springer Nature Switzerland AG 2022
F. Kaefer, *Sustainability Leadership in Tourism*, Future of Business and Finance,
https://doi.org/10.1007/978-3-031-05314-6_43

Goverdhan, your late father Fateh Singh Rathore is a renowned figure in India for tiger conservation and the revival of Ranthambore National Park. What motivated you to continue in your father's footsteps and create the non-profit, Prakratik Society?

Growing up with my father was an adventure in the forest. In 1981, my father was brutally attacked when he tried to reason with a group of villagers illegally grazing their cattle inside Ranthambhore and he almost died. However, this changed the way park managers were looking at conservation, as a matter of enforcement rather than partnership with the stakeholders i.e., the people living around the park.

I was in Medical School in the 1980s and was very concerned about this, and was already thinking of how I could get involved to help this situation. I loved Ranthambhore and tigers. We would walk, swim, and stay in the small park guesthouses with no electricity. These were wonderful times and I could not think of living anywhere else.

Once I graduated from Medical School, I returned to Ranthambhore wanting to work with the local communities. Being a doctor, I thought that by visiting villages and providing preventive and primary health care, it would help me meet the local people and understand their needs better and what they thought about the park and its tigers.

We started a mobile health care program and we would visit 33 villages every week. This provided deep insight into the minds of the people and how they thought the park gave them nothing and that it was only for the tourists that came to see the tiger. I soon realised that this battle to save the tiger would not be won unless the local people became a part of the conservation effort.

Along with a group of friends, I founded the Prakratik Society in 1994, with the objective of trying to find alternate livelihoods for the local people to help reduce their dependence on the limited natural resources of the park. So, we started programs like animal husbandry and with artificial insemination, we improved the breeds of the local cattle. This way, fewer cattle could give more milk, and stall feeding could become viable, reducing the demand for free grazing in the park. Once stall feeding became the norm, cow dung became available in every home and we then started the biogas program, where we helped people build biogas digesters to create methane, which could then be used for cooking, reducing their dependence on wood from the park. Prakratik Society was a recipient of the Ashden Award for Sustainable Energy in 2004 in London.

Similarly, most houses were built using timber from the park, so we started a tree plantation scheme along the edges of the fields that would provide good quality timber after a few years. Farmers were given a cash incentive over 7 years for each surviving tree.

We worked with the local dairy cooperative run by the government to help market the increased milk yield from improved breeds of cows and buffaloes.

As we travelled through villages, we also realised that the educational standard in the region was abysmal, and without education, it would be difficult for people living at the subsistence level to understand why saving the forest and the tiger was important for their own future survival.

For the local community, it was a simple question that if their ancestors had taken natural resources from the park and grazed their cattle for thousands of years, why should there be a problem now. They could not see the impact of rapidly increasing population and therefore, demand as a problem.

In 2001, we started the Fateh Public School to help provide the highest quality education and offered scholarships to village children. Some of these children have gone on to become doctors and engineers from some of India's premier institutes.

Prakratik Society continues to work with local communities trying to find solutions to help them understand their roles in conservation and why it is important for their future generations.

According to the latest census, India has just below three thousand tigers left—a far cry from the number of tigers that once roamed this country. What do you think are the main challenges plaguing tiger conservation efforts today?

One of the greatest challenges facing tiger conservation is the fragmenting of habitats due to the rapidly increasing population, putting pressure on elected governments to prioritise the economy before ecology.

Park management is the next big issue. While India has some excellent park officials who have brought back species from the brink of extinction, the absence of a pure wildlife service means park management is totally dependent on who the Park Director is. You get a good director for the park, its biodiversity thrives and if you get a bad one, it deteriorates rapidly.

There is also a lack of political leadership that further compounds the problems, as very little independent research is carried out in the parks and published. How the parks are doing is totally dependent on what the park management shares with outsiders.

The reality may be completely opposite, as was seen in 2004–05, where the park management was claiming that there were tigers in three parks when there were none. Finally, independent evaluations declared that Panna and Sariska had no tigers, and Ranthambhore had lost more than half of its tiger population.

If we need to see tigers and their habitats survive long term, one of the most important things to do would be to create a separate Parks and Wildlife Division, so that park management is in the hands of people with experience. Strategies will need to be thought of, where corridors between parks can be created. Independent research and monitoring by scientists should be encouraged and their findings published on public platforms and peer-reviewed.

A certain amount of accountability has to be placed where park authorities are held accountable for their actions.

How difficult was it to convince villagers to relocate and co-operate in initiatives under Tiger Watch?

Getting local people and the poaching community together has been a slow process, taking nearly two decades, but it is paying huge dividends. However, I would like to elaborate a little more on Tiger Watch and what it does. From the past 2 decades, it has been at the forefront of tiger conservation—evolving and being innovative in its approach to conservation.

While Prakratik Society was set up purely to help bridge the gap between people and conservation, there was a big gap in independent assessment of what was happening inside the park, in terms of poaching and other illegal resource extraction.

In 1991, the first incident of tiger poaching was reported by an independent undercover operation conducted by my father, even though he was no longer in-charge of Ranthambhore. He had been coming to Ranthambhore from his posting at Sariska and because he had spent so many years in Ranthambhore, he had an inkling, like a sixth sense, that something was not right. As some of the lower staff were still people that had worked with him, he started talking to them and they too said something is not right.

His undercover work uncovered a dirty nexus of rampant poaching inside Ranthambhore. He got a lucky break and passed on the information about a poacher with a fresh tiger skin travelling on a train, to the local police chief who happened to be an old friend. They nabbed this man and recovered the skin and caught the poacher. What unfolded was a nationwide outcry, and even though there was evidence that more tigers were missing for lack of independent assessment, it was covered up and the official line was that only one tiger had been killed.

However, this brought to the notice of the people of India and the world, that not all was right with Project Tiger. In 1997, my father founded Tiger Watch, an NGO with the objective of doing research and monitoring, in and around Ranthambhore to be able to provide an independent insight into the functioning of the park.

Once again in early 2002, my father had an inkling that things were not right and that many tigers were missing. Fortunately, at that time, Tiger Watch had a MOU with the government to conduct research and monitoring inside the park. A young biologist, Dharmendra Khandal, had just joined my father at Tiger Watch. He was extremely knowledgeable on this subject.

During this time, Tiger Watch had started to identify and work with some local tribes like the Mogiya who are traditional hunter-gatherer tribes and carry the stigma of being criminal tribes. Using tracking skills honed over generations and using rudimentary self-made muzzleloaders, they would easily venture into the forest at night and kill a tiger. Within hours, they would skin it and take it away, burying the body underground with salt to return later to take the bones.

Tiger Watch started to create a directory of these tribes putting together a photographic record along with the location of common residence. They are nomadic, so sometimes difficult to pin them down. Identity cards were issued and we provided free treatment to them when they came to our hospital. Slowly they began to trust Tiger Watch and we started a free hostel for their children, so they could, for the first time, go to school. As this trust formed, they started sharing information regarding how easy it was to kill a tiger inside Ranthambhore and many had already been killed.

As Tiger Watch started to analyse data from the camera traps installed by the forest department, they soon realised that 22 tigers were missing inside Ranthambhore and even submitted a confidential report to the State Government. Sadly, this was ignored and brushed aside. Not someone to sit around, my father made this report public and it created a massive national and international uproar.

Finally, even the government report acknowledged that indeed 22 tigers were missing and the most plausible reason was that they were poached. As a direct result of this, the government took the matter seriously and good officers who knew Ranthambhore well were posted.

One of the positive outcomes of this was that the government and Tiger Watch started to partner in their work. Today, 40 children from the Mogiya tribes stay at the Tiger Watch-run hostel, their parents vowing never to kill a tiger.

Tiger Watch and the government have started a Village Wildlife Volunteer program that recruits young village folks living along the periphery of the park. They are paid a stipend and they monitor the periphery using smartphones and camera traps, passing on information in real-time through the internet regarding any poaching or any illegal activity in their area. They have now become the mainstay of information gathering in the villages, helping reduce poaching dramatically, so much so that the tiger population inside Ranthambhore has now more than doubled, with many tigers now occupying the peripheral forest areas, where there were none before.

The latest Economic Valuation of Tiger Reserves in India report mentions that those bring millions of dollars annually as economic benefits. Given the huge monetary value, how can India utilize this opportunity to develop more tiger reserves?

The potential to generate enormous revenue to further the cause of tiger conservation can only be realised if it is not always seen from a socialist lens. Wildlife tourism is a form of capitalism and the only social role it plays is to educate children and therefore, if any concession is to be given, it should be to school children.

However, the current dispensation of the park authorities is to see it as a social activity and not as a commercial activity. The result is that an incredible opportunity to generate revenue, to be able to provide incentives to villages in corridor areas to move, and therefore freeing up almost double the area for tigers, is lost.

But, just charging more money will not be enough. Efforts will need to be put to improve the experience for visitors coming to Ranthambhore. Currently, job creation through guaranteed work to guides and vehicle owners means visitors don't have the choice of having the best. Although some changes have been made, where an additional payment lets visitors choose, this is not how it should be; every visitor should be able to get the best service possible.

The whole scheme has taken the incentive away from guides and vehicles to better their craft, so many of the best guides have now become travel agents just booking safari for clients. This attitude has to change, a visitor to the park is contributing much-needed revenue and must be provided the best possible experience.

There should be innovative thinking i.e., higher rates in peak season and lower rates in low season, like summer, to spread the tourists out. There should be zonings like premier zones and less prime zones, and traffic and charges should be regulated accordingly.

At present, only 20% of the park is accessed by tourists. This is unwarranted because a huge area of the park is not accessed and for the most part, no one knows what is happening there. Tourists form a very important part of park patrolling and will report any illegal activity and even share pictures of tigers and other animals seen to help in establishing a better database.

How do you assess the situation in India in terms of sustainable tourism?

In my opinion, India may have missed the opportunity to create a policy for sustainable tourism. Most of the wilderness tourism has seen a huge increase in the last two decades, however, no policy has been created to make it sustainable. The only thing that has been implemented as a policy is how far hotels can be from the park.

Once a business gets the clearance to build a hotel by meeting this criterion of distance from the park, there is no building code, there is no correlation to carrying capacity of the park and the number of beds allowed to establish. It has been a completely flawed approach. The distance policy now means many projects have come up too far from the park gate, where guests have to travel half an hour or more before they reach the park gate, adding nearly an hour to their safari which is a waste of time.

Instead, an area near the park gate should have been declared a commercial zone with building bylaws like:

- Build only using local materials.
- One cannot build a hotel with more than 25 or 30 rooms or construct beyond one floor.
- Build on 5% of the total landholding and the remaining area has to be planted with local trees of the forest so it blends in with the adjoining forest.
- Enough gaps need to be created between hotels for animal migration, should the zone be in a corridor.
- Lighting should be minimum after a certain time at night
- No DJ or loud music allowed, and so on.

More importantly, the number of rooms to be allowed should have been capped, according to the carrying capacity of the park.

An absence of this has meant that unregulated development has been allowed to take place. Rooms with more than 100-bed capacities have come up. There is no mandatory green zone per hotel. The number of rooms compared to the carrying capacity of the park is way off the mark.

What this has meant is hotels are no longer selling a wilderness experience and more and more hotels are catering to conferences, marriages, and DJ parties, all contrary to sustainable tourism. Huge air-conditioned dining halls, conference rooms, reception areas guzzle power. Tremendous pressure on the groundwater reserves is drying up wells.

There is little that can be done now, other than overrunning the park with visitors to meet bed capacity or allowing a large number of hotels to close down. The only viable escape route is to create a large, enclosed tiger safari where the tourist overflow could go and see their tiger. This is the sad reality.

Your three bits of advice for sustainable tourism entrepreneurs or travel business owners on how to support wildlife conservation and at the same time ensure their own economic sustainability?

- Encourage small wilderness hotels that have a small inventory and provide a wilderness experience by encouraging nature walks and other experiences that enhance the guest's understanding of conservation and the abundance of biodiversity.
- Find new, smaller wilderness areas to start your business. Avoid the overcrowded parks as there is no future there for tourism. Many are already saturated beyond their capacity.
- Use alternate energy as much as possible.
- Conserve water by harvesting/recycling as much as possible.
- Reduce carbon footprint by using local produce and not getting exotic foods that need to be transported in cold chain and kept in deep freezers for many days.
- Get involved with the local community to help in the conservation process.

Link to the interview: https://sustainability-leaders.com/goverdhan-rathore-interview/

Graham Miller on Destination Sustainability Measuring and Challenges 44

UK | Professor Graham Miller is Pro-Vice-Chancellor at the University of Surrey with university-level responsibility for sustainability and employability. Graham is also Executive Dean of the Faculty of Arts and Social Sciences. He holds a Chair in Sustainability in Business with his research interests in creating a more sustainable tourism industry. Graham is chair of the Considerate Group,[1] a company formed to promote sustainable practice in the hotel industry by measuring energy and water usage data. He is the former co-editor of the Journal of Sustainable Tourism,[2] the leading academic journal dedicated to research that promotes sustainability of the tourism industry and sustainability through the tourism industry. From 2014 to 2019 Graham was the lead judge for the World Travel and Tourism Council's Tourism for Tomorrow Awards[3] and was also an appointed member of the World Economic Forum's Global Agenda Council for the Future of Travel and Tourism.

---

[1] https://considerategroup.com

[2] https://considerategroup.com

[3] https://wttc.org

F. Kaefer, *Sustainability Leadership in Tourism*, Future of Business and Finance,
https://doi.org/10.1007/978-3-031-05314-6_44

Areas of expertise: sustainable tourism, research, tourism business, UK, UNSDG 11 (Sustainable Human Settlements)

The following interview with Graham was first published in April 2017 on Sustainability-Leaders.com.

Graham, when did you discover your passion for tourism and sustainability?

I lived for nearly four years in Japan when the yen was really strong against most of the currencies of the world. I was paid in yen and so it was cheaper for me to travel outside of Japan than it was just to stay in Japan and live a normal life. So, I travelled lots throughout south-east Asia and started thinking about tourism as an industry and what the effect of it was on the places I visited.

The key moment for me was in visiting Langkawi in Malaysia. I travelled and travelled using planes, trains, boats and motorbikes to get to this place that seemed like it was the most remote place in the world. In the early 90s there was really very little development in Langkawi, but as I was leaving there was a sign being put up saying that the European Union was building road infrastructure and a major new resort was 'coming soon'.

I left Langkawi thinking about what impact this would have on the wonderfully relaxed and undeveloped experience I had just enjoyed. By the end of my four years in Japan, there were daily flights direct from Narita to Langkawi. What had seemed like an incredibly remote and special place had been connected to one of the world's busiest airports.

As a result, nobody will ever again be able to have the experience of discovery, adventure, and privilege that I had in Langkawi. There are huge implications in all of the above and with no right answer, and this is why I have remained fascinated intellectually, and challenged morally, with the development and sustainability of tourism.

What was your view of sustainability in tourism when you started your career?

When I first came to the topic, the dominant view of sustainability was very much one of keeping places special for the privileged few and then using sustainability as a way to charge more to discerning customers who didn't want the mass experience where tourism had been destroyed.

I was initially attracted to the idea that the consumer had the power to force change in the industry, but not through the elite consumer, but instead through the mass consumer.

My PhD was on how we could use indicators in order to give consumers the information they would need to demand more sustainable products.

Now in 2017, how has this view changed?

I am afraid I have largely given up on the view that the consumer is interested to change sustainability. I don't see this as being a force for change in any meaningful way. I still see organisations as being the key to driving sustainability, but I am now much more convinced that this can come through shareholder pressure for strong and resilient returns, the need to attract and retain the top employees and corporate reputation, which is linked in many ways with the values of the leaders.

The best examples of sustainability that I see have a leader who 'gets it' and drives this through the organisation with sheer force of will and determination.

Not all of these leaders are the person at the top, but in all instances, they are people with considerable talents and strength.

Where do you see the main challenges in terms of implementing sustainability initiatives in tourist destinations?

The increasing drive towards low tax environments means that public authorities have less discretionary income to spend on the public good. Instead, financial imperative dictates that priority is given to organisations that generate jobs and bring inward investment.

People continue to believe promises from politicians that the public realm can be maintained while driving down the taxes they pay, but this is just impossible. What we are seeing now is the political reality meeting the scientific inevitability.

How to effectively measure the sustainability performance of tourist destinations?

We developed the European Tourism Indicator System (ETIS), which is designed to be a way for tourism destinations to begin to assess their sustainability. It assumes no previous knowledge or experience in measuring sustainability. This is a journey and so the system is designed to help destinations with the process of identifying their needs and priorities, how to collect and analyse the data, and then how to report it back.

The challenge with ETIS is that the EU that funded the original work is now not able to continue to support the work, and so destinations are left to work things out for themselves.

I would love to see this initiative get adequate support and not depend on a committed group of people working on it in their own time.

Your thoughts on the current state of sustainable tourism in Europe?

There are great examples of sustainability everywhere, but my concern is whether there are enough examples of sustainability to keep pace with the general growth of tourism.

In simple terms, are the sustainability examples we see providing a fig leaf of respectability to justify the further development of a still unsustainable industry?

In your experience, what characterizes sustainability leaders in tourism? Which qualities or traits do award-winning professionals and change-makers have in common?

Strength, determination, and flexibility. Trying to change a system when there is huge inertia not to change is very difficult. Different arguments work with different people in different circumstances. Trying to drive change requires a leader to be aware of all the arguments, to be armed with evidence to support all, and be able to keep making these arguments until the opposition gives up!

Your thoughts on this year's WTTC Tourism for Tomorrow awards—the winners, process, and your role as a lead judge?

We have some wonderful examples of sustainability in this year's awards. As a judge, I put aside my concerns about whether there are enough examples and just focus on how truly amazing the examples I read about are. The people driving these projects are really inspirational and demonstrate that people can make a difference.

I am really pleased to see an evolution in sustainability from being something driven by passion, to something driven by strategy.

This may sound dull, but it is important because every organisation has a strategy, but not all organisations have people with passion. Making all organisations see the value of sustainability and plan for it with a commitment of resources will make a bigger difference. So, while the committed eco-lodge in country A is impressive, seeing an organisation develop a 20-year plan to protect a community, reforest an area or eliminate hunting is what really makes me more hopeful.

We also see more use of science and data in the awards now. More innovation and a greater variety in business models from charities and social enterprises to pure commercial organisations and government agencies.

In combination with the complete global coverage of the applicants, this demonstrates to me that whatever the problem being faced, wherever you are, whatever scale you are at, and whatever the type of organisation you are—there is a way you can be more sustainable.

Which sustainability verification and accreditation schemes do you consider most useful for destinations?

Whatever works. Destinations are so varied and complex that if destinations use anything that helps them to progress, then that is positive. As I said above, I don't really believe in the consumer perspective, so I don't think the accreditation needs to have consumer recognition. I see the advantage of all these schemes in how they help a destination to improve its internal processes.

Your favourite sustainable tourism success story?

The WTTC T4T has some great case studies. My personal favourite is probably Reality Tours and Travel. The founder of the company, Krishna Pujari, is a real hero for me.

As a professor, I educate students and provide research that has a long-term impact, though this is sometimes difficult to measure. As co-editor of the Journal of Sustainable Tourism, my work is helping to put out knowledge that informs academics who teach other students, so my work moves a step even further away from the impact.

Krishna's work at Reality has an immediate impact every day. If he does his job well, then tourists come to a part of the world that they would never think of coming, and someone in a slum in Delhi eats. That kind of direct and immediate impact has a huge appeal to me and provides a frequent reminder to me to ensure I can always see the route to impact my work.

In his interview,[4] Xavier Font differentiated between sustainable and responsible tourism as follows: "Responsibility is the process and attitude, sustainability is the goal. Nobody is sustainable, but I don't want to do business with someone who is irresponsible." Do you agree?

---

[4]https://sustainability-leaders.com/xavier-font-interview/

Xavier is a great colleague, so I am not going to argue with him! Sustainability has a scientific meaning, so I have always talked about the 'transition towards sustainability', recognising that nobody actually achieves sustainability.

However, we can identify components of sustainability and work to achieve those—such as safe water, elimination of poverty, no hunting, etc. I don't think organisations should be trying to do everything. They need to think about what the key challenges are where they work, and what can they do best to help with these. It may be, that in some places this is just paying their taxes and letting the relevant authorities deal with the situation. If they are not paying enough taxes though, then this is an issue.

My concern with the concept of 'responsibility' is that this is too flexible and open to interpretation—hence my preference for thinking about sustainability, albeit recognising this is a journey, not a single destination.

Link to the interview: https://sustainability-leaders.com/interview-graham-miller/

Rwanda | Greg Bakunzi is an experienced Rwandan tour organiser and guide for various international tour operators. He started operating Amahoro Tours back in 2002, as a sole tours company that promotes sustainable local development and improves the living conditions of people who reside near the protected areas, promoting local tourism that is both responsible and sustainable. Greg founded Red Rocks[1] in 2013, a Rwandan community-based organisation where the local community can gain from tourism activities. This sustainable source of income will bring underserved communities into the tourism supply chain and support

---

[1] http://redrocksinitiative.org/

© The Author(s), under exclusive license to Springer Nature Switzerland AG 2022
F. Kaefer, *Sustainability Leadership in Tourism*, Future of Business and Finance,
https://doi.org/10.1007/978-3-031-05314-6_45

community development projects through its non-profit organisation. The idea of the Red Rocks Initiative for Sustainable Development is to integrate tourism, conservation and sustainable community development, to enhance locals to be ambassadors of protecting their natural environment.

Areas of expertise: conservation, ecotourism, responsible tourism, tourism business, Africa, UNSDG 15 (Forests & Biodiversity), UNSDG 8 (Equal & Fair Economic Opportunities).

The following interview with Greg was first published in December 2020 on Sustainability-Leaders.com.

Greg, do you remember the first time you heard about conservation in connection with tourism? What got you interested in the topic?

When I started the gorilla business 18 years ago as a tour operator, my idea was to link tourism and conservation together. This benefits the local population residing on the slopes of the volcanoes, by involving them in conserving the rare mountain gorillas, since they benefit from the resulting tourism economy. I started a non-governmental organization which emphasized initiatives related to rural development projects—Red Rocks Initiative for Sustainable Development.

What distinguishes Red Rocks Rwanda from other tour operators in Africa?

Red Rocks as a social enterprise differs from other tour companies in that we try to incorporate tourism with cultural activities, which directly benefit the locals and include them in the tourism value chain. Locals get a say in the organization structure of my company, and we involve them in all our daily activities.

Which issues does a responsible travel business like Red Rocks Rwanda struggle most with?

At the moment, the most critical issue we are facing is to cope with the effects of COVID-19. No tourists are coming to visit us, and the local population is struggling to make ends meet. We try our best to assist them during these challenging times. At the moment, because the disease is still with us, trying to support the locals and making sure they are well and fed can be hectic. That's the only critical issue we are encountering right now.

The current coronavirus pandemic has hit the travel industry hard this year. Has your strong focus on responsible travel somehow helped you master the situation (has it made you more resilient)?

The current situation of the coronavirus pandemic drastically affects our business. We used to entertain tourists who in return participated in our cultural activities. And also, tourists used to purchase souvenirs from the arts and crafts made by the locals, which helped those to be in a better position to pay for their children's school fees, health insurance, food and other necessities needed to make a family happy.

At the moment we do what we can to assist the locals, but if the pandemic continues, we will not be able to help them sustain themselves. But all in all, we are in the process of advocating for domestic and regional tourism. Once we are able to open up our borders and allow regional travel, maybe we will be in a position to be more resilient.

Which entrepreneurial lessons or key sustainability insights from Red Rocks Rwanda can you share with other tourism professionals and businesses?

What I can assure my fellow tourism professionals is to use and/or develop local networks. It is important in these challenging times that tourism businesses come together to draw upon each other's strengths, experiences and ideas; to develop a plan for the future.

Marketing: To stay ahead, this is an ideal time to work on creating or re-examining your marketing strategy and re-evaluating and refreshing website content.

Maintenance and upkeep: This is an opportunity to consider a refresh; a touch of paint, a rearrangement of furniture or soft furnishings, a deep clean and completing any upgrade on accommodation units.

Which are the keys to a successful, responsible ecotourism business nowadays?

Make optimal use of environmental resources that constitute a key element in tourism development, maintaining essential ecological processes and helping to conserve natural heritage and biodiversity.

Respect the socio-cultural authenticity of host communities, conserve their built and living cultural heritage and traditional values, and contribute to inter-cultural understanding and tolerance. Ensure viable, long-term economic operations, providing socio-economic benefits to all stakeholders that are fairly distributed, including stable employment and income-earning opportunities and social services to host communities, and contributing to poverty alleviation.

Where do you see opportunities to be seized in Africa right now, in terms of making its tourism offerings more sustainable, or encouraging responsible travel?

We all need to advocate for responsible travel, by educating overseas tour operators of the need of supporting responsible tourism, which obviously means that the locals are to be part of the tourism value chain. And by involving them they become ambassadors for conservation processes, which are needed for the sustainability of the national parks in Africa.

Secondly, responsible tourism will also assist Africans to preserve their cultural heritage for future generations.

Link to the interview: https://sustainability-leaders.com/greg-bakunzi-interview/

Costa Rica | Hans Pfister is Co-founder, CEO and Marketing Director of the Cayuga Collection of Sustainable Luxury Hotels and Lodges in Costa Rica, Panama and Nicaragua. He strongly believes in sustainability and is very passionate about customer service in hospitality. Hans is a frequent speaker at conferences and is often invited to share his experiences at Ivy League Universities. He has served on the Advisory Board of the National Geographic Unique Lodges of the World and as the Regional Vice President for Mexico, Central and South America for the Cornell Hotel Society. Born and raised in southern Germany, Hans began his career with an apprenticeship at a luxury hotel in the Black Forest, followed by a degree from Cornell University's School of Hotel Administration in Ithaca, New York. He has more than 25 years of experience in hospitality in Europe, Asia and the Americas.

Areas of expertise: tourism business, hospitality, responsible tourism, Latin America, UNSDG 11 (Sustainable Human Settlements), UNSDG 12 (Resource Efficiency), UNSDG 8 (Equal & Fair Economic Opportunities)

The following interview with Hans was first published in November 2019 on Sustainability-Leaders.com.

© The Author(s), under exclusive license to Springer Nature Switzerland AG 2022   297
F. Kaefer, *Sustainability Leadership in Tourism*, Future of Business and Finance,
https://doi.org/10.1007/978-3-031-05314-6_46

Hans, do you remember what first got you interested in tourism and sustainable development?

Since I was little, I was always interested in hospitality and travel. I loved foreign countries and dreamed of travelling on the Trans Siberia Railway. I did an AFS high school exchange year in the US (Maine) in the 80s and was out and about on Interrail every summer in Europe. I loved learning languages and exploring different cultures. I am passionate about service and knew that I would want to work in hotels. My first job was running a little bar in my hometown in southern Germany.

Sustainability? I love pristine nature and being in the outdoors and growing up in Europe in the 70s and 80s, I was exposed to the green movement that started there. When I then came to Costa Rica and saw this incredible country, I was hooked.

Central America has been the focus of your venture Cayuga Sustainable Hospitality. There are currently ten sustainable luxury hotels in Costa Rica, Nicaragua and Panama under your wing. Do you have plans to add more hotels to your collection?

Yes, we are planning to add more hotels to the Collection, but we are very selective. It has to be the right fit. We have 10 different hotels and 10 different owners that consider their hotels their life projects, or 'babies'. So, we need to have great chemistry with those owners, and we need to make sure we share the same values.

We have walked away from great opportunities because the owners of the project did not want to go the extra mile to take care of the staff and were not genuinely interested in investing in the infrastructure to make their hotel more sustainable.

We work with hotels with no more than 50 rooms (we prefer around 25 to 30), that are focused on the luxury segment of the market and that have a strong commitment to sustainability. We love opening new markets and being the first in the region, like we did in Panama with Isla Palenque, or Aguas Claras in Costa Rica.

We are currently focusing on Central America but might be looking at South America and the Caribbean if something interesting comes up. Travel time to and from the hotels, language and knowing the local legal, tax and labour law systems could be barriers to entry to new countries.

We are getting requests for our services from faraway places like Indonesia, Kenya or Canada, but right now we are not set up to work there. Maybe in a few years, we can team up with a partner in a different part of the world and expand.

Which was the first hotel that joined the Cayuga Collection? How did you convince them to trust in your mission towards sustainable luxury travel?

Lapa Rios Lodge on the Osa Peninsula in Costa Rica was the first hotel. We started there 20 years ago. We were the result of a process that the founders of Lapa Rios started, called the 'Ecolodge Dilemma'. They had built a very successful lodge and conservation project, but wanted to get out of the day-to-day management and eventually sell the hotel. It took almost 20 years to find the right buyer and the hotel was finally sold in June 2019. The owners, now in their late 70s, are very happy that they had found an exit strategy.

Today, the criteria to be part of the Cayuga Collection are the following:

- We need to make sure we share the same values in terms of sustainability, conservation and especially how hotel owners treat their staff and the local community.
- We need to be clear that it is about the triple bottom line and not just short-term profits.
- We need to make sure that the location and concept of the hotel have a chance of being successful in the market (location is still very important in the hotel industry) and we want to inspire our guests.

There is not really a checklist. It's more about using our instinct and experience for making the right decision.

Cayuga Sustainable Hospitality is dedicated to managing and developing sustainable luxury hotels, resorts and lodges in Latin America. Why the focus on luxury travellers?

We manage small hotels. They can only be profitable and have a real impact if there is enough revenue. And the way to achieve this is at a high rate.

We also found that only luxury hotels can afford to hire a management company like Cayuga. It is also important to know that you cannot be everything to everybody, so we decided to focus on luxury.

One thing which we also learned over time is that wealthy guests often have influence over others. So, if we get those guests and inspire them about sustainability, we can have a multiplier impact. We have had many guests that are company owners, CEOs, celebrities, Nobel Prize winners, politicians, etc. and we think that we have made an impact with them.

Congratulations on Cayuga Collection winning multiple awards in recent years! What impact does such recognition have on your business?

It was a great honour to win the National Geographic World Legacy Award in 2017 and then a year later WTTC's Tourism For Tomorrow Award.

I don't think it had an impact on hotel sales, in terms of revenues. But it had a huge impact on our staff members. We took the award from hotel to hotel, and everybody took a picture with the award trophy and shared it with friends and family.

It also gave our work as a management company credibility with new clients. We are very proud to have won those two awards, especially knowing how many other great companies had applied and who the other finalists were.

With close to 17 years of experience with Cayuga, how has your understanding of sustainability changed since you first got involved?

There has been a strong shift in the tourism industry to understanding that sustainability is all about people. At the beginning it was more about smart use of resources and conserving the natural environment. And that is still an important focus. But we learned that without the 'buy in' of our staff and the local communities where we operate, no nature conservation is successful in the long term. So, creating employment opportunities, offering training and opening career paths is our main focus.

The other thing I learned is the focus on local. We only hire local staff and that includes management. We only serve local products in our restaurants, and we produce and buy local furniture, décor and equipment whenever possible.

Which are the main concerns linked to tourism sustainability at the moment in Central America?

We know that tourism is a great force for the good if done right. But it can destroy whole areas if done wrong. I hope that Central America does not repeat the mistakes of some Caribbean destinations and sells out to mass tourism. There is a chance to do things differently here. Costa Rica is a great example, and we try to inspire our neighbours in Panama and Nicaragua also to focus on quality tourism, not just the number of visitors.

Like many other places around the world, there is concern in Central America about the economic and political situation and the potential turmoil which this can bring to a destination.

And of course, we are also feeling the signs of climate change through extreme and unusual weather patterns, excessive rains, droughts, increase in temperatures, etc.

In your blog, you mention that, because of more and more international tourists, the vast jungles and amazing wildlife of Africa and tropical forests of Costa Rica are increasingly becoming safe from poaching and deforestation. Do you think that the onus has shifted to travellers and responsible businesses like yours in conserving natural resources, compared to governments?

Really successful conservation needs to be done in public and private sector alliances. Just the government or private companies by themselves won't make it.

Costa Rica is a great example of public-private partnerships in tourism over 30 years or more, and how sustainable development goals can be put in practice.

Sustainability is about inspiring each other. Sharing success stories. Looking for synergies and then showing the value of what has been achieved to everybody involved. The creation of national parks in Central America is not always a welcome event at first, but when you can show that it generates more wealth than cutting trees down or agriculture, you will soon get a buy in from everybody.

Prince Harry just started a new sustainable travel initiative called Travalyst. How unique is this initiative compared to other global sustainable travel initiatives?

I am not sure how unique it is as the plans are still very vague. What is unique is that it comes from him, and he and his wife have the opportunity to inspire as they are part of the royal family, but then again also the rebels. Harry is following the footsteps of his mother in that way and if there are concrete actions planned, this can be successful.

You have extended a cordial invitation to Prince Harry and his family to come and visit your hotels and lodges in Central America and to show them what you've learned about sustainable travel over the last few decades. What key learning would you share with him?

I think it is not about just one thing. It is about how it all comes together. We talked about our focus on the people and the concept of local. But what I would do with him is the same thing that we do with all of our guests that are interested. We offer a complimentary back of the house tour where we show our guest all we do in the hotel in terms of sustainability.

There are no 'staff only' areas in our hotel. We are a transparent organization. We take our guests to the kitchen, to the purchasing department, the laundry, the maintenance area, staff cafeteria and housing. We show them how we avoid using single-use plastic, how we separate waste, compost, grow vegetables or produce our own furniture from fallen wood.

There is so much to see and talk about that the tour often takes over two hours. I think with the Prince, we would have to plan for at least half a day. And we could finish making tortillas at the end.

You take pride in creating a warm and inspiring culture at Cayuga as a workplace. What role does social sustainability play nowadays in the context of sustainable tourism?

It is all about people. If our staff feel taken care of and that there are opportunities, they will take great care of our guests. We have long time employees that started as construction workers before the hotel was open and moved up to top management positions.

But it goes beyond the staff in the hotels. If the community feels that a hotel is bringing value to the people that live there, they support the operation and may even defend it. As our minister of tourism, Maria Amalia Revelo, recently said: You cannot have five-star hotels in one-star communities.

Anything else you'd like to mention?

I often get asked about sustainability certification for hotels. We participate in the local certification program, but I am not a big fan of certification in general. I think it can lend itself to greenwashing, which is a huge problem in our industry.

So, when I get asked how a traveller can find out if a hotel is really sustainable, I ask them to look for service-level employees in the hotel and ask them what sustainability means to them. How are they contributing to sustainability every day on their job? How is the hotel where they work making a difference?

If you get a blank stare from them, even if they have this fancy certification plaque at the entrance, you know something went wrong...

Link to the interview: https://sustainability-leaders.com/hans-pfister-interview/

Harold Goodwin on Taking Responsibility for Tourism

UK | Professor Harold Goodwin is the Founder Director of the International Centre for Responsible Tourism, which promotes the principles of the Cape Town Declaration that he drafted. He is the WTM's Responsible Tourism Advisor, putting together their flagship Responsible Tourism programme at WTM London. The programmes run at WTM Africa, WTM Latin America and Arabian Travel Market. Currently, he is developing a Platform for Change for the sector. He is the MD of Responsible Tourism Partnership and chairs the panel of judges for the Global Responsible Tourism Awards and the other awards in the family in Africa, India and Latin America. Harold works with industry, local communities, governments and conservationists and undertakes consultancy and evaluations for companies, NGOs, governments and international organisations.

Areas of expertise: responsible tourism, tourism research, UNSDG 12 (Resource Efficiency)

The following interview with Harold was first published in October 2015 on Sustainability-Leaders.com.

© The Author(s), under exclusive license to Springer Nature Switzerland AG 2022   303
F. Kaefer, *Sustainability Leadership in Tourism*, Future of Business and Finance,
https://doi.org/10.1007/978-3-031-05314-6_47

Harold, what was your view on sustainability when you started your academic career in tourism—was it a topic back then?

I started working on tourism in 1994 when I was research director at the Durrell Institute of Conservation and Ecology. It was a government-funded project investigating ecotourism, the relationship between tourism, conservation and sustainable development—so yes, sustainability was an important issue for me.

I was quite shocked by how little attention was being paid to sustainability by tourism academics and the industry. Ecotourism was all the rage but there was little understanding about sustainability among those promoting ecotourism and we were already two years on from Rio and seven from Brundtland.

Now in 2015, how has your view on sustainability in tourism changed?

By 1996 when I was asked to assist with the VSO campaign for ethical tourism, I had realised that since we couldn't define sustainability in any operational way it was best to avoid the term. There is a triple bottom line sustainability agenda, it is easy to draw up a long list of issues.

What is more important is to identify what matters in particular places and address those issues and set local targets. It makes sense to conserve water in lots of places, but it is not a priority everywhere.

The sustainability agenda is less dominated by environmental issues now than it used to be, but it is still a challenge—despite the MDGs [Millennium Development Goals] and now the SDGs [Sustainable Development Goals]—to get economic and social issues included on the sustainability agenda.

Why the focus on responsible tourism, and what does it mean?

Responsible tourism came to the fore in my work with the Association of Independent Tour Operators, the first business group to adopt Responsible Tourism, and in the work I did with the ANC government in South Africa.

Responsible tourism is about taking responsibility for using tourism for sustainable development. It is about what you do, what you take responsibility for achieving.

What motivated you to publish 'Taking Responsibility for Tourism' in 2011?

I have been teaching about tourism and encouraging people to make it better since 1998, I wanted to make the ideas accessible to more people and to offer an account of how responsible tourism emerged and resulted in the Cape Town Declaration.

I also wanted to write about some of the mistakes of sustainable tourism, carbon offsetting, ecotourism and certification and some of the controversial issues like volunteering. There will be a new updated edition in 2016.

As chair of the WTM Responsible Tourism awards, do you see real progress on the responsible tourism front?

The change since we started the World Responsible Tourism Awards in 2004 has been dramatic. Competition is much stiffer, and the standard is much higher too.

The Responsible Tourism programme at WTM London each November attracts 2000 participants and the programme now runs across the portfolio in Cape Town, Dubai and São Paulo.

Your key insights from 7 years as Professor of Responsible Tourism Management at Leeds Beckett University (2006–2013) and now at Manchester Metropolitan University?

There are many but two stand out. The field is rapidly changing, and course material needs constantly to be revised. Much of what professionals need to know to be effective comes from traditional disciplines in the social and natural sciences—tourism courses too rarely include this literature.

Tourism is a product of travel and the interaction of consumers and producers in destinations requires an interdisciplinary approach, particularly if you want to change it.

Knowing what is in the tourism books and journals is not enough and theory needs to be tested against experience.

Why did you start the International Centre for Responsible Tourism?

The ICRT resulted from the 1st International Conference on Responsible Tourism in Destinations in 2002 and the Cape Town Declaration. It has always been an independent network of individuals and national groups who come together to pursue the aims of the Cape Town Declaration. It has never had a legal form or bank account.

Which is the best way to teach responsible tourism?

In my view, responsible tourism should be taught as Continuing Professional Development to practitioners for people with several years of relevant experience and able to test theory against practice through assignments which draw on their work experience.

Which part of steering the tourism industry towards more responsibility and sustainability do you find most challenging?

One of the big challenges is to get people to see the difference between the abstract goal of sustainability and taking responsibility for dealing with the specific issues in particular places. It is so much easier to talk the platitudes of sustainability than to deal with the particulars.

The first UN conference on man and the environment was in 1972, we saw those pictures of spaceship earth, our finite world, in 1968. That is a working lifetime ago. We've made some progress, but the problems get larger faster than we can deal with them—we're losing the race.

How to measure responsible tourism on a destination level?

There is no way of measuring responsibility, you can report what people are doing and how successfully they are tackling the issues that matter locally.

Too much effort is being put into creating frameworks for measuring destination sustainability holistically—we need to focus on monitoring and reporting on the local issues that matter, the ones that need to be managed. They got that right in Calvia in the nineties.

Which issues do destinations struggle most with in terms of sustainability, and how can they overcome those?

The big challenge at the destination level is to create and maintain the partnership among and between businesses, communities and the different spheres and agencies of government—good political skills are required to identify and agree the issues, agree a solution and then to engage all the partners in making the changes.

Link to the interview: https://sustainability-leaders.com/interview-harold-goodwin/

USA | Hitesh Mehta is one of the world's leading authorities, practitioners and researchers on ecotourism, physical planning and architectural aspects of ecolodges. Hitesh has specialised in working with indigenous communities and ensuring that their settlements are self-sufficient and the fauna protected from the money earned through tourism. He has over 18 years of work experience improving human settlements, promoting ecological restoration and low-carbon development, and the socio-economic benefits gained to the local communities. In July 2006, National Geographic Adventure magazine identified Hitesh as one of five Sustainable Tourism Pioneers in the world mainly because of his master planning work to protect endangered habitats and alleviate human poverty. He is the author and photographer of the award-winning Authentic Ecolodges (launched by Harper Collins in November 2010). Hitesh is also the editor of the International Ecolodge Guidelines and co-author of the chapters on Site Planning and Architectural Design.

Areas of expertise: ecotourism, sustainable tourism, UNSDG 11 (Sustainable Human Settlements), UNSDG 12 (Resource Efficiency).

The following interview with Hitesh was first published in June 2017 on Sustainability-Leaders.com.

© The Author(s), under exclusive license to Springer Nature Switzerland AG 2022   307
F. Kaefer, *Sustainability Leadership in Tourism*, Future of Business and Finance,
https://doi.org/10.1007/978-3-031-05314-6_48

Hitesh, you have worked and consulted as architect, landscape architect and environmental planner in more than 62 countries. Do you remember what initially triggered your interest in holistic design and socially and environmentally friendly practices?

You could say that my interest in holistic and eco-friendly practices was latent— one that was dormant within the confines of my genes and only became alive once I had completed my traditional education training.

Whilst pursuing a Bachelor degree in Architecture at University of Nairobi (1979–85), I began to travel around Kenya with a group of like-minded adventure-oriented friends aka The Rough Gang. The six of us purchased a used Land Rover and before long, we had observed most of the National Parks and learned about the conservation challenges faced by park authorities.

It was during these safaris that I noticed the striking disconnect between the tourist lodge building architecture and the surrounding pristine landscapes. I developed a deep interest in relationships between lodge building architecture and surrounding landscapes. To become a more holistic designer, I then pursued a Masters in Landscape Architecture at Berkeley, which was followed by an academic career as Assistant Professor at the University of Nairobi.

Whilst teaching and working part-time, I became more and more convinced that there were better ways to integrate sustainability and tourism (protecting both endangered species' habitats and local communities). So I decided to take all three of my professional interests (architecture, landscape architecture and conservation) and combine them into one. Thus, was born my obsession with holistic design and socially/environmentally friendly practices of Ecolodges.

I could also make a case that holistic thinking has been in my genes for over 3000 years! The main bedrock element of my Jainism (Indian philosophy that began circa 1000 BC) upbringing is the doctrine of Ahimsa...which simply means 'non-violence to your fellow humans and non-violence towards non-human beings'.

In my family, we have most probably been vegetarians for over 40 generations! And I have now been a vegan in 52 countries for the past 11 years.

Jain values and principles are evident in all my projects—there is respect for animals, plants, local people and the soul of the place. The approach right from the outset is that of low-impact development. As such, my focus in landscape architecture moved to pristine and fragile natural areas where tourism was uncontrolled, had large social and environmental impacts and required a new planning paradigm to protect the sanctity of those places. I took this concept and started implementing it in my work, which I call vegan planning and design.

Search for authentic experiences has become a veritable trend in tourism. What role does design play in this regard? And what does an 'authentic' accommodation look like? How does it differ from conventional accommodation?

Ecolodges are as authentic as accommodations can get in the tourism industry. They are indeed the environmentally and socially friendly tourist accommodation components of Ecotourism and by their definition, can only be found in natural locations and not urban areas.

As such, the design of an authentic ecolodge encourages close interaction with the natural and cultural environment and has an atmosphere that is appropriate to the site's specific setting.

It is this metaphysical 'sense of place' that is one of the key ingredients which distinguish ecolodges from conventional accommodations: They provide a spiritual communion with nature and culture.

The role design plays in ecolodges is that it provides context and conveys a sense of 'isolation' and 'wilderness'—of being away from the negative visual impacts of modern civilization—glass, steel, concrete and aluminium facades, a common aesthetic of conventional accommodations.

After 10 years of research, interviews with architects and professors, eco-consultants, developers, operators, many indigenous communities, and feedback from stakeholders in the ecotourism industry, I developed the following criteria to determine visual authenticity in ecolodges:

- The lodge needs to fit into its specific physical and cultural context through careful attention to form, landscaping, and colour, as well as using vernacular architecture.
- The buildings and plantings need to have a minimal visual and physical impact on the natural surroundings and utilize traditional building techniques during construction.
- Both the architecture and landscape design need to use the locally available and environmentally friendly building and furnishings materials.

Judging by your experience, which aspects of building and operating an ecolodge tend to be the most challenging? And how can design help to overcome those challenges?

Since building and operating an ecolodge are two totally different phases, I will deal with them separately:

There are several challenges to building ecolodges:

First: Compared to the long history of conventional tourist accommodations, ecolodges are the 'new kid on the block'. Because of this, it is not easy to find an authentic and committed team of eco-consultants who have a good understanding of what it requires to master plan, design and build ecolodges.

Ecolodge planning and design in this global age of heightened cultural and environmental sensitivity needs to be participatory, holistic and sustainable in all aspects of the planning and design process.

Landscape architectural interventions are crucial for creating authentic experiences and this profession is lacking in most countries in the less developing world.

Second: Whenever there are profits to be made, there is going to be greenwashing. Developers are jumping on the bandwagon so that they can reap rich harvests by touting themselves to be 'eco' when they are NOT. I call these people the 'fish and chicken vegetarians' of the tourism world.

I have had many experiences where clients come to me stating that they want to develop an eco-project without really knowing what it means. Then, as the project goes along, they change their minds as they begin to understand the commitment it takes to build an ecolodge.

Third: Many ecolodge projects are still developed by husband and wife teams, or by single individuals, and therefore the money is always a constraint. The latest ecolodge project that I am building in Dominica, Eastern Caribbean is one such example. The project, which started in 2006, has stopped and started three times now because of insufficient funds. Good news is that it will be ready this coming winter!

There are several constraints to operating ecolodges:

- Because of the remote locations, servicing the ecolodge is always a challenge, and lack of government-funded infrastructure creates logistical problems.
- Freshwater resources and electricity from the grid are few and far between, so ecolodges need to be self-sufficient.
- Solid waste disposal is a serious challenge, especially in areas where there is wildlife.
- Sewage seep-off can cause groundwater pollution.

The secret to how the developers can overcome the above challenges and save money in ecolodges is through passive design.

If you use passive design techniques to reduce costs of avoiding retaining walls, water and energy resourcing, air conditioning systems etc. and construction techniques that use local labour and materials, then ecolodges end up being cheaper than conventional accommodation, as they use renewable sources of energy and natural air-convection cooling.

I believe strongly in using materials with the least energy embodiment. By this, I mean materials that have travelled the least distance and those that have little energy used to manufacture them. In most cases, I strive to use local natural materials. I carry out research during the site analysis stage to identify local artisans and craftspersons who work well with local materials. For the ecolodge in Dominica, no earth-moving machinery was used in the project.

Good design does not need to be expensive. It is a myth that ecolodges are more expensive than traditional resorts. If well planned, designed and constructed, ecolodges on the contrary are cheaper.

In 2006, National Geographic Adventure magazine featured you as one of five Sustainable Tourism Pioneers in the world. In your view, (how) has a sustainable tourism scene evolved since then?

The last 11 years have been a game-changer in the sustainable tourism world! It was in 2006 that Al Gore released his Oscar-winning documentary, An Inconvenient Truth. This was followed by other documentaries, like Leonardo di Caprio's 11th Hour and Louie Psihoyos's The Cove, which graphically brought to light the dangers of global warming, the threat to planet ecosystems, food famines, unsustainable homocentric approaches etc.

Many people in the tourism and travel industry now recognize that environmental and social stewardship is good for business, good for the planet, and is vital to the future of travel and tourism.

We have also reached a tipping point whereby the global traditional hospitality industry is more widely accepting sustainable tourism.

In the past 10 years, a lot of progress has been made in sustainable travel: with input from scientists and conservationists, the top 100 travel and tourism companies have written a joint report to lead the way forward to low climate-risk travel.

Several international airlines have tested biofuel-operated commercial flights. The United Nations Foundation initiated the now well-established Global Sustainable Tourism Council (GSTC), whose Global Sustainable Tourism Criteria are helping ensure that the tourism industry continues to drive conservation and poverty alleviation.

There are currently numerous sustainable tourism organizations, several conferences every year and even an International School of Sustainable Tourism in the Philippines, where I am a board member.

The future of sustainability in indeed good, with the United Nations designating this year as the International Year of Sustainable Tourism for Development.

And eco-friendly design has evolved in leaps and bounds. I would say that in the last decade, eco-design has taken a quantum leap because of climate change concerns, rising oil prices and, most importantly, heightened awareness amongst developers and designers.

There is now also a proliferation of green design certifications, like LEED, BREEAM, S.I.T.E.S., EARTHCHECK, etc. The EARTHCHECK Building Planning and Design Standard (BPDS) provides guidelines, tools and indicators used for assessment and certification and to assist designers, architects and developers in the planning and design phase of sustainable buildings and precincts.

Ecolodges are mushrooming all over the world and the only way is UP. I see them becoming increasingly popular because we have more and more people moving to urban areas, and so when they go on vacation, they need to visit natural areas for rest and relaxation.

The baby boomer generation not only have a lot of expendable income but also want to make a difference when they travel. This will ensure that sustainable tourism will establish itself over the next decade and beyond.

From an architecture and design point of view, which trends do you observe which might support or hinder a better, more responsible or sustainable tourism?
Trends that support more responsible tourism:

- Rise in community-owned and operated ecolodges, as well as projects that respect the local cultures and architecture.
- The influx of smaller but higher-end boutique hotels that are eco, luxurious and many of which have wellness centres/spas.
- A growing number of eco-refurbishments and expansions at existing lodges.
- Experience-oriented hotels and lodges. Tourists are searching for experiential travel and authentic connections with local peoples and nature.

- Developers and architects who are warming up to green design and technologies, such as fuel cell technology, more efficient solar and wind power, wave generated power, low impact light fittings that give out warm glows, and environmentally friendly materials for interior design that are made from recycled products.

Trends that hinder more responsible tourism:

- Adulation of so-called 'star architects', whose ego-full designs are mostly aesthetically driven but lacking in environmental or social consciousness.
- The sexy-looking modernism style, which uses bare concrete, steel, glass and aluminium—materials that are resource depleting and which are not contextual to the local setting.

Your firm HM Design practices a quadruple bottom line philosophy. In a nutshell, what does this mean? How does it lead to competitive advantage and client benefit? And to what extent could it be applied by tourism businesses, such as hotels or tour operators?

Every one of my office projects in the last 15 years has espoused what I have termed as the 'quadruple bottom line' approach to sustainability: Financial, Social, Environmental and Spiritual. After having worked in over 60 countries and in some of the most remote parts of the planet, I have felt that without the fourth aspect— spiritual, one cannot attain true sustainability.

Our 'continuity of the vernacular' eco-friendly projects is those that provide our clients' guests with a spiritual union with nature and culture. We design places that cater to travellers who want to learn about other places and local people, rather than just escape their familiar surroundings, but at the same time make a positive difference to the locations they visit. These unique experiences provide a competitive advantage to our clients.

Depending on the wishes of the clients, we employ metaphysical analysis in on-site workshops. Those help all consultants and clients to use their respective senses to understand and connect at a deeper level to the land. And then through both landscape and building planning and design, we ensure that our clients' hotel/lodge guests can employ all their six senses to experience the 'gestalt' of the site.

Tourism businesses, such as hotels or tour operators, can apply these metaphysical approaches to learn about the land. And in doing so, they are able to immerse their clients into the 'chi' (energy) of the land.

The UN has declared 2017 the International Year of Sustainable Tourism for Development. In your view, which are the main destination challenges right now regarding the sustainability of travel and tourism?

The rising middle classes in the top two MOST populous countries in the world—India and China—mean that there are more people willing to travel internationally. By the end of this year, a new record of annual international travellers will have been set: 1.2 billion.

Unfortunately, the infrastructure (roads, trails, solid waste management, sewage treatment, power supply, water sourcing etc.) is not growing in parallel to tourism arrivals, and this is placing extreme strain on established destinations.

I was in Cancun, Mayan and Tulum Riviera, and that whole destination is expanding to a point where mangroves are still being destroyed to cater to conventional tourist accommodation demands.

Another destination challenge regarding sustainable tourism is cruise tourism. As statistics have shown, it is anything but sustainable and is growing exponentially. Ships that accommodate over 6000 people are not rare. Since cruise ship visitors are not staying overnight, the host destination hardly makes $30 per person per day!

Economic leakage will continue to be a challenge in the years to come. In the conventional mass-tourism world, almost 80% of the profits leave the host community and ends up in the country where the hotel (chain) has its headquarters.

In contrast, over 90% of the money made in an ecotourism operation stays in the country, and more specifically in the local community. Ecotourism encourages traveller philanthropy, and this helps to sustain local livelihoods. Through ecotourism, local communities have benefited through the building of schools, health clinics, student scholarships etc.

Because of sea-level rise due to climate change, ocean-based destinations are the first ones to suffer. I foresee this as a serious destination challenge in the coming decade. The Maldives is already affected.

Which are your key lessons or insights gained during your 30 years of work as an architect, designer and planner?

- Learn to compromise but never let the 'tail' wag you. If you believe in something, don't let anyone or anything stop you. Never give up in what you believe. Keep on knocking at doors and one day, the door that will open will change your life!
- Education and awareness are the keys to making change. Spend time to educate clients and local communities and be patient.
- In our field, the best results are achieved when you have practised ego-less design with an eco-full approach! Employ a bottom-up approach to planning so as to empower communities and local consultants. Bring to the drawing table local people and consultants on day one of the planning and design processes.
- Never look into imposing your designs/ego onto a landscape. Use your six senses to immerse into the landscape and then lightly place your buildings, so they harmonize rather than dominate.

Imagine you could turn back time and start all over again. Knowing what you know now about business and sustainability, what would you do differently?

Great question! I would have started much earlier to practice green principles in my work. In my first 5 years as an architect and landscape architect, I ran a conventional design practice without any environmental or social consciousness. But then again, everything happens for a reason, and it happens when it is meant to happen. I could not have forced the situation. I needed to have gone through the traditional world to truly appreciate the eco-world!

My philosophy in life is to never regret anything because I look at myself as a confident person and capable of making rational, logical decisions at any given point in time.

Sustainability and tourism marketing professionals often don't quite speak the same language. How can design help to connect the two, to encourage and support sustainable tourism practices?

Design is a great connector. If you look at the conventional tourism destinations, the first image that comes to mind is of architectural design: Paris is Eifel Tower, Sydney—Opera House, London—Big Ben, San Francisco—Golden Gate Bridge, New York—Empire State Building or Statue of Liberty etc.

Through good planning and design, you can create memorable experiences that help people transcend into another level. Their memory and imagery of the place are influenced by its architectural and landscape design.

For example, in the Crosswaters Ecolodge and Spa, Southern China, for which I was the team planning and design leader, the arrival bamboo bridge has become both the marketing 'image' and brand for the project, as well as making the sustainability (bamboo construction) statement right at the entrance.

One of my great achievements as an ecoArchitect and ecoLandscape Architect is in the way my work, research and latest book have led to a beautiful integration of eco' and 'luxury', thereby creating a truly holistic experience in the tourism and travel industry.

In the 1980s and 1990s, the perception of an eco-lodge was an accommodation facility for low and middle-end travellers who would sleep on the floor and share common bathrooms.

I strongly believe that my presentations on authentic ecolodges over the past 10 years around the world (over 50 countries), as well as the sale of over 8000 books, have been successful in connecting sustainability and tourism marketing professionals.

My book *Authentic Ecolodges* has also focused on design interventions that make ecolodges aesthetically pleasing whilst also being sustainable.

Your 3 bits of advice to independent hotel owners eager to embrace sustainability?

- Do your homework and research about lessons learned by others. Visit a few properties and learn about the dos and don'ts. Also, read books on case studies on how other hotel owners have embraced sustainability and created a profitable business.
- Make sure you have a good group of consultants whose heart is in the right place and who also want to make a difference. Select architects, interior designers and landscape architects with proven records in sustainability.
- Do the 'cradle to cradle' research on building materials. There are loads of great information on the internet. Compare prices. Check out reviews.

Link to the interview: https://sustainability-leaders.com/interview-hitesh-mehta/

Ishita Khanna on Homestays and Sustainable Rural Development in India

India | Ishita Khanna is the Founder & Director of Ecosphere, an award-winning social enterprise in Spiti, Himachal Pradesh. Ecosphere aims to create a synergy between responsible travel and sustainable development. She is also the co-founder of MUSE (an NGO) that has been working in Spiti since 2002 on various community programs. Her experience working in grassroots and passion for the environment and mountains been the driving force for Ecosphere and MUSE which enable her to live her philosophy. She was voted as an MTV Youth Icon and has been featured in various magazines and articles as one of India's 50 leading social entrepreneurs. Ishita has been awarded the Real Heroes award by CNN-IBN and is an Ashoka Fellow.

Areas of expertise: community-based tourism, destination sustainability, ecotourism, responsible tourism, sustainable development, UNSDG 12 (Resource Efficiency), UNSDG 5 (Empowering Women), UNSDG 8 (Equal & Fair Economic Opportunities)

The following interview with Ishita was first published in June 2020 on Sustainability-Leaders.com.

F. Kaefer, *Sustainability Leadership in Tourism*, Future of Business and Finance, https://doi.org/10.1007/978-3-031-05314-6_49

Ishita, you started your career in sustainability and development at the Tata Institute of Social Sciences—How has your education and experience in social work impacted your approach to sustainable tourism?

While studying at the Tata Institute of Social Sciences, I undertook a thesis on the impact of tourism on the socio-cultural, economic, and ecological environment of a destination. The fieldwork associated with the thesis brought to light the negative impact which tourism can have on a destination. It helped me to understand how these negative impacts could be minimised or averted. This was the foundation for the work that I later started in Spiti.

Does your current view of sustainability and tourism differ compared to the one you had when you graduated?

My view of tourism and sustainability has definitely grown from the time I graduated. Academics often flow in the space of idealism, but ideal situations rarely exist on the ground. It has been a huge learning curve for me to live in one of the remotest and hardest parts of the world, working on a multitude of issues that come into play in shaping sustainability.

Ecosphere works on a broad spectrum of topics, such as:

- Helping communities build resilience to climate change
- Enabling year-round access to water and nutrition
- Carbon reduction and harnessing the abundantly available solar energy
- Reducing, recycling and upcycling waste
- Enhancing health care access, diagnosis, and treatment
- Enabling access to education for girls
- Helping to build community entrepreneurship, additional livelihoods
- Supporting and kindling the growth of compassion amongst communities

Considering that 15 years ago Spiti was not even found on any of the mountain tourism itineraries, Ecosphere has been instrumental in putting the valley on the map. How did you achieve this?

There are multiple factors that have put Spiti onto the global map and, most prominently, social media in the last 5 years.

The starkly beautiful locale of Spiti with breathtaking views of the mountains, interesting trekking trails, magnificent night sky, diverse flora and fauna, rich cultural heritage and fascinating folklore are some of the many reasons why Spiti has gained popularity around the world.

It is also aptly referred to as the 'Valley of Monasteries'. Some of the oldest Buddhist monasteries and temples dating back more than 1000 years, along with unique aspects of Tibetan Buddhist culture are well preserved and have flourished in the Spiti Valley.

How has Ecosphere been able to prevent any erosion of local heritage with the increased number of visitors?

From the inception of our tourism initiatives in the region, we realised the ease with which local heritage is destroyed to make way for modernism, often brought on not just by tourism but a multitude of factors. To address this, we decided to try and make tourism a tool to conserve it.

Local families were encouraged, and awareness created on the concept of homestays. This was done both to create an alternate and equitable source of income for local communities, but also as a means to conserve the local mud architectural building styles of the region, which were under threat by the mushrooming of concrete guesthouses. Local arts and crafts were similarly incorporated into the fold of tourism, enabling a source of income for the artisans and an incentive to continue practising their craft forms, thereby enabling their conservation.

Every destination faces the risk of overtourism. Spiti, too, is at cross-roads right now and measures will need to be taken soon by the community to create a balance between economics and ecology on which the tourism industry thrives.

What are the market opportunities Ecosphere can exploit to increase its competitiveness as an ecotourism destination?

We believe collaborations are the way forward: with like-minded organisations or individuals keen to make a greater impact, combined with greater overall market outreach.

'Authenticity' in destinations is nowadays a key expectation of modern travellers. In your experience, is the market now more engaged and interested in the sustainable livelihoods approach offered by Ecosphere?

We have noticed an increase in the experiential traveller, keen on meaningful experiences, and making a difference. Such travellers are keen on understanding and experiencing local life, or even volunteering while they travel.

For us, this is positive in every regard, as such travellers are more sensitive about how they travel and about their impact while they travel. They are more inclined to travel responsibly and are on the lookout on how their travel can make a difference.

As a social enterprise working to build sustainable local economies, we have been able to integrate various livelihood opportunities for local communities thanks to such mindful travellers. Moreover, through varied volunteering opportunities, we have been able to link travellers to the developmental needs of Spiti as well, matching their skills and interests with these needs.

Which issues does a responsible travel business like Ecosphere struggle most with?

Being a responsible travel business, we believe in paying our partners and employees fairly and ploughing back revenues towards various projects that work on addressing developmental needs and challenges in the region.

However, this gets difficult when regular tour operators try and copy the ideas and philosophy without putting them into action in their practices. So, while they may appear to be offering a similar product with a similar philosophy, in reality their on-ground practices are far from responsible or social.

To link nature conservation, culture preservation, income generation, and gender empowerment is a very ambitious set of goals and yet, Ecosphere has found a way to do exactly that. Which of the many projects during your work with Ecosphere have you personally found the most rewarding?

Our homestay program in Demul, a village on the highlands of Spiti Valley at 4,300 metres above sea level, has been the most rewarding for me—both personally and professionally. The homestays in Demul were started by Ecosphere in 2004 as a means to create alternate income for an otherwise agriculture-dependent community.

A 6-month-long cold winter with temperatures dipping to as low as −30 °C, leaves the village community with an extremely short working season to earn the year's livelihood. Agriculture, which is limited to one crop a year, is solely dependent on the winter snow for irrigation. This leaves the village community highly susceptible to slight changes in climate, which are increasingly becoming the norm.

Demul has approximately 55 households, of which over 48 operate as homestays. The spare room in these households was done up and converted into the homestay room. Hence, while in a Demul homestay, one is staying with the local family which makes the whole experience authentic.

Set across the beautiful trans-Himalayan meadows of Spiti, the Demul village homestays follow a unique rotation model to ensure even distribution of guests and the associated financial benefits across the village. The homestays are modeled on traditional patterns of village governance that favour cooperation over competition.

Every year, two village coordinators are appointed by the village, who allocate the guests to homestays on a rotational basis, ensuring equitable benefits between all homestays.

The Demul Homestays require minimal investment, providing an equal opportunity to both the rich and poor to get involved in tourism-related activities. The rotational system maintains equal benefits to the rich and poor. At the end of the year, the money is distributed equally amongst all the homestays and directly into the hands of the women. With over 1500 room nights annually, homestays in Demul are able to earn up to INR 40,000 (~USD 530) in a short span of 3 to 4 months, which is equivalent to an average 50% of their annual income.

Which aspects of developing a destination like Demul sustainably do you find especially important?

The advent of tourism often changes the traditional architectural integrity of a place, replacing them with concrete guesthouses often modeled on architectural styles in complete contrast to the local architecture, creating real eyesores.

One of the objectives behind the homestays in Spiti valley was also to maintain the architectural integrity of the villages, along with conserving the local architectural styles.

Along with the homestays, a host of activities and day trips have been developed, providing deeper insight for travellers into local culture and way of life, enriching their experience while encouraging travellers to stay longer. And on the other hand, enabling conservation of art forms and traditional knowledge.

With changing times, often traditional culture and crafts undergo a change and witness a slow death. In Demul however, to address the concern for dying crafts, activities that showcase and involve learning of traditional arts and crafts have been developed, such as yak rope spinning.

Making yak rope was an activity that the men in the village would do. However, the younger generation was not keen on taking it forward. Travellers can now learn the art of spinning yak wool into a rope, which encourages the younger generation of men at the village to learn the skill in order to teach it to travellers.

On similar lines, the younger generation is not learning the numerous traditional songs and dances. With the help of Ecosphere, the elders in the village have trained the youth, who are now actively showcasing this to travellers who come and stay in village homestays.

Demul village also houses one of Spiti's more well-known practitioners of traditional medicine, which visitors can experience in a day trip developed to enable both an immersive and learning experience for travellers and an incentive for the Amchi (local doctor) to continue passing this rich tradition to the younger generations.

The homestays have also provided a much-needed outlet for the women in the village to showcase and sell their traditional crafts. This is providing an additional source of income for women, from tourism.

Volunteering opportunities have also been developed which are not only helping the local community but are a great experience for travellers and a way to learn traditional systems of living and farming of Spiti. While staying with a family, travellers can choose to join them in the fields, herd their livestock, prepare meals, help with their children's homework, learn the local dialect or just immerse themselves in Spitian culture. Volunteers have also helped restore 300-year-old village stupas, fix drains, collect garbage and build an artificial glacier for the village.

While the homestays are just one program, we have found that they lead to a multitude of benefits for the local community, culture, and environment. Not just income but also a sense of pride amongst the community for their culture, heritage, way of life, traditional architecture, arts, and crafts enabling their conservation and further development.

We have also noticed that the community becomes more conscious of personal and village hygiene and cleanliness, helping villagers to take initiatives around garbage management and cleanliness drives.

With travellers keen on wildlife coming and staying in their homestays, the community has developed value and appreciation for the wildlife as well, which were earlier looked upon as predators or a nuisance. Cases of human-wildlife conflict were not uncommon earlier, which have now totally stopped. In fact, the community not long ago rescued and freed a snow leopard that had gotten stuck in a livestock pen, after killing numerous herds of livestock.

Which lessons or key insights from the entire sustainable development project of Ecosphere might serve other tourist destinations around the world?

We have found that it is crucial while working in destinations, especially when it involves local communities, that a holistic approach be adopted: one that reduces dependence and builds resilience. While developing tourism as a livelihood option in communities, it is therefore equally crucial that this is combined with developing other livelihood streams as well.

In Spiti, our endeavour from the start was to build tourism as an alternate livelihood option, while simultaneously strengthening and building other livelihood options as well, so that if there is a bad year for one, then the other can supplement.

Mostly till now, tourism was providing a much-needed alternate income stream for communities during tough agricultural years, and now in the times of COVID when all tourism is at a standstill, communities in Spiti, fortunately, have a safety net to fall back on.

Link to the interview: https://sustainability-leaders.com/ishita-khanna-interview/

USA | Jake Haupert is the Chief Vision Officer of The Transformational Travel Council. His work is rooted in the belief that transformational travel can incite change by breaking down divisions within ourselves, between us and among us. Through his work, he is cooperatively shifting the travel paradigm by encouraging a mindset, approach and model that measures success based on a more holistic, virtuous and prosperous outcome for all. The Transformational Travel Council and its global Allyship is solely committed to co-creating more personal fulfilment and societal thrivability through the development of rich and accessible methods, tools and practices guided by travel, diversity and equity, psychology, spirituality and neuroscience. He has founded multiple values + impact-driven travel organisations, Co-Creator of The Transformational Travel Journal, founding Circle Member of the World Experience Organization. He has been featured in The New York Times, Conde Nast, Skift and NPR among others.

Areas of expertise: entrepreneurship, sustainable development, responsible tourism, tourism business

The following interview with Jake was first published in June 2020 on Sustainability-Leaders.com.

© The Author(s), under exclusive license to Springer Nature Switzerland AG 2022
F. Kaefer, *Sustainability Leadership in Tourism*, Future of Business and Finance,
https://doi.org/10.1007/978-3-031-05314-6_50

Jake, you are an industry veteran with two decades of experience in the leisure travel industry. How has your journey been so far?

Ha, it has been a blur, strange to even be referred to as a veteran, but I guess that is the case, it certainly feels like I earned it. However, I am always a student, always a little childlike in my approach to life, stretching, learning, growing—which I think helps me stay where I prefer to be, on the bleeding edge.

That said, looking back, I've definitely been around the block once or twice, perhaps evidenced by the grey coming in my beard and a few more wrinkles of joy around my eyes. I feel blessed, I am one of the few that identified a calling early in my career, in my early 20's to be exact. Since then, my journey has been incredibly rewarding, rich with relationship and real success, but also riddled with classic entrepreneurial strife and struggle.

Looking back, it is clear that the bumps and bruises, the sleepless nights, the obstacles along the way are what guided me to the work I am doing today in transformational travel. In a way, I now have way more perspective and find myself way more comfortable with being uncomfortable, simultaneously being intentional and surrendering at the same time.

Just trying to make subtle adjustments here and there to stay in the current and flow down the river of life.

What inspired you to create The Transformational Travel Council (TTC)?

Good question, one that I have contemplated on many levels. I've always been good at the idea thing, also pretty good at bringing ideas to life. I seem to have good 'vision,' certainly, a very big heart, am purpose-driven, and I harbour great hope for humanity.

However, as I traveled, I grew, and as my awareness expanded, I found myself with this dual perspective, one that was deeply drawn to and appreciative of the beautiful things in life. But I was also painfully aware of the growing disconnect and separation between the beauty of humanity and the society we've built. It seemed that as a civilisation, we've not been heading in the right direction, and I have a burning desire to help steer us back on course.

I've always felt that travel is our most powerful catalyst to positive change, but after many years in the space, I can see that the dysfunction within our society is mirrored in the travel space, and travel becoming digitalized, commoditized, sterilized and centered around serving the privileged few. Travellers—often because they were simply conditioned to—are traveling for entertainment, with a sense of expectation and entitlement, that results in an increasingly toxic travel dynamic, one based on consumption and extraction.

This is obviously to the detriment of local communities and the environment, but it is also to the detriment of personal advancement, consciousness, and cross-cultural understanding.

In my mind, travel is broken, so I decided to dig into the roots of why we travel, understand why, and how it evolved, discover what is working and what is not.

Perhaps through that deep and rigorous study, I started to reimagine and rebuild travel, fully aware that others in the space have had similar thoughts, perspectives, and visions. We've convened and are activating change, #BetheChange if you will.

Now for the short answer to that question—This work is simply my calling and every experience I've had led me to it.

I fully believe I am meant to convene a global community and co-create a more conscious, purposeful travel that maximises the power of travel to positively transform how we live our lives, how we live with others, and how we live on our planet.

How does TTC help travel businesses or destinations to promote or facilitate transformational travel?

Our Ally Program is a convergence of change-makers and doers from all sectors of travel, masterminding, mutually caring, and advocating for better travel. We believe that by uniting, sharing, learning, and activating, phenomenally rapid change is possible.

Our emerging ecosystem of Allies around the world is convening to source our individual and collective inner capacities, knowledge, and wisdom to manifest change in travel that embodies universal truths such as dignity, compassion, fairness, and courage. Activated, one traveller at a time, by guiding and empowering them in their quest for deeper meaning, connection, growth, and change.

Based on the rigorous, multidisciplinary study we've done on travel from the perspective of anthropology, neurobiology, and psycho-spirituality, we've created a methodology, framework, and guiding practices that can be integrated into travel businesses and destinations. We strategically support and strengthen ethical, emotional, equitable, and ecological travel standards to benefit travellers, hosts, communities, and the environment.

The work we're doing at the TTC is split into 4 Houses:

- Introspection—'Connection to Self'
- Bridging—'Connection to Others'
- Expansion—'Connection to the Infinite'
- Integration—'New Ways of Being and Engaging with the World'

The latter is the key, transformational travel leads to positive change, one traveller at a time, by making new travel and lifestyle choices natural and infectious.

You define transformational travel (or TT) as any travel experience that empowers people to make meaningful, lasting changes in their life. While such journeys are customer-centric and beneficial to travellers, how does this form of travel contribute towards achieving sustainable development goals?

That was actually our definition early on. As our learning evolved, we felt compelled to change it last year.

Instead of focusing on a singular, passive experience or moment in time, we felt it was more about engaging actively and purposely.

Just like a yoga practice, a meditation practice, or exercise regimen, we believe transformational travel is a results-driven, holistic travel practice, rooted in mindset, awareness, and outcomes. Transformational travel is 'intentionally traveling to stretch, learn, and grow into new ways of being and engaging with the world.'

Modern, transformational travel is the missing link in the conscious and sustainable travel movements—it is about how to create deep, personal stretching, and growth that raises consciousness individually and collectively.

We often refer to the old saying, 'you can take a horse to water but you can't make it drink.' The same holds true with sustainable tourism: just because we inform travellers of dire circumstances and impacts, preach of better practices, this doesn't mean these people are becoming better stewards or making more responsible choices. To do that, we need to get them more deeply connected to nature, to others—and connections must be made from the inside-out.

We believe this is where we awaken to the interconnectedness of our world and all sentient beings, understand that together we stand, apart we fall. When we're connected to the world, to nature, to each other, we move from being mindless to being mindful of our choices and impact, travelling, and living more consciously.

We all know that people care for and protect the things they love, so with that wisdom in mind, we need to rebuild the travel industry, society, from the inside-out, heart-centered and soul balanced.

When travel is catalyzed as a source for good, and tourism has accessible frameworks, tools, and practices to create deeper meaning and connection, together we'll be co-creating a more empathetic, grateful, steward-centric human being, one that makes better, more mindful, conscious, and regenerative decisions, not just for one, but for all.

The Sustainable Development Goals are far more accessible and achievable in a transformation economy when growth mindsets are embraced.

As tourism continues to lean into these concepts and modalities, strives toward these goals, transformational travel will be identified as a real way to accelerate the process of personal transformation and amplify the benefits, and make real progress in reaching the SDGs.

And with many luxury travellers just now starting to travel more sustainably, it's vital that we continue to challenge the status quo, move from being simply sustainable to being regenerative. A regenerative system behaves like any living system, with healing occurring from within, adapting, evolving, and improving conditions, not simply sustaining, but regenerating.

Will microadventures and domestic travel be the norm in the near future?

Our TTC Director of Research, Jasmine Goodnow, just finished an in-depth study on microadventures and according to her research, time and distance away from home are not as important as one's frame of mind when traveling. Disconnecting from email, phone, and other distractions is important, 'Depth over Distance!'

Savvy and resilient travel companies, accommodation providers, and destinations will be able to shift their brand, message, value proposition, and experiences in a way that cultivates this mindset and attract travellers seeking 'depth over distance' regardless of where they are traveling from.

Aren't travel experiences more about how it makes you feel and how it touches you? More often than not, it really doesn't matter where it is, it is not to say the place doesn't play an important role, it does, but it is really about how deeply the traveller connects with travel companions, the experience, the hosts, and the destination and how all of that harmonises to support and empower the traveller in their pursuit of deep, rich, immersive, and potentially life-changing travel.

By inviting travellers to embrace a growth mindset in travel, focusing on desired outcomes, setting intentions, being present, reflecting, meaning-making, and taking action, travel can be transformative, no matter how far away from your front door it is.

Following the coronavirus pandemic, what trends will we observe once travel restrictions are eased?

As travel restrictions ease, we will see a traveller emerge that is a little more aware and thoughtful than before. People will be more inclined and receptive to embrace practices like intention setting, gratitude, being more present and reflective, and considerably more open to the unfamiliar, unforeseen, and the unknown.

I don't know if it will be the end of leisure travel, per se, to travel for the sole purpose of fun and relaxation. But if people are going to leave home and hop on a plane, they're going to be more contemplative of their why, motivations, desired outcomes, and more conscious of their choices and their impact.

Difficult times at scale have a way of cracking humanity open, inviting introspection, providing space and time to take stock of what's important, reevaluating, and reimagining. During times of the coronavirus lock-down, we've gone back to our roots of joy, simplicity, happiness, and fulfilment, even reconnected with values, beliefs that support family, community, and connection to self, others, nature, and the universe.

At the TTC, we see the travel market on a spectrum of awakening. As confidence, curiosity, and openness develop, tourists become travellers, then explorers, and finally seekers, who are conscious and pursuing change and growth every day and everywhere.

These challenging times undoubtedly have provoked some introspection. Perhaps we are a little more conscious and aware than we were before. And if we aren't, the travel industry that serves us certainly will be and that is a very good thing.

Imagine you could turn back time and start all over again. Knowing what you know now about business, travel, and sustainability, what would you do differently?

Honestly, I wouldn't change much. I feel the path has served me well, hardships included. But if I had to turn back time and start all over again, I would have embraced 'Caentsu Daja' sooner. In the Ecuadorian Amazon, the Cofan tribe practice 'Caenstu Daja,' which means—let it happen, let it flow.

On a scouting trip for Explorer X, The Sound Tracker® Gordon Hempton of Quiet Parks International, and I had an epiphany that we were always the 'make it happen' type of guys, but that changed in the jungles of the Amazon. Before we flew out, we both got 'Caentsu Daja' tattoos and life has been a little sweeter, a little more peaceful, a little more spacious ever since. Living in the jungle, off of the land, immersed into the wild, with the Cofan, taught us an important lesson that week and I am eternally grateful for it.

Do you have any favourite business or destination which you think is leading the way in how it approaches sustainability—or being especially innovative in how it is dealing with the current pandemic?

ROAM Beyond is certainly one of those businesses that have sustainability at their core and have also adapted well. We provide a travel solution that is arguably more sustainable than living in a traditional home. Each glamping trailer or mobile dwellings, contains robust solar power systems, compostable toilets, and have the ability to 'leave no trace' in remote locations.

ROAM Beyond also created the Haven Experience in response to the current pandemic's impact on travel. It provides a solution to those that are comfortable traveling in 'pods' or 'quaranteams'. Like-minded friends and family that are comfortable with sharing space who don't interact with others and minimising contact with on-site staff. Even having chefs sending ingredients for meals in a cooler and even live streaming outdoor cooking instructions.

Anything else you'd like to mention?

In regard to destinations, many of them are wisely using this time to reimagine their brand, message, infrastructure, and sustainability practices, some even leaning beyond, like Joe Pine and James Gilmore's Experience Economy and dabbling in the nascent 'Transformation Economy.'

The Transformational Travel Council is extending an invitation to tourism destinations to become an early adopter of our Transformational Travel Destinations Program. This is an invitation to destinations that wish to lead the efforts in transitioning to a transformational economy at a destination level and implementing and promoting sustainable destination good practices through the adoption of the TTC guidelines and tenets for Transformational Destinations.

The early adopters will be widely recognised and their findings will help shape the final framework for Transformational Destinations. Guyana, Australia, Willamette Valley, Greenland, and others have expressed initial interest.

Link to the interview: https://sustainability-leaders.com/jake-haupert-interview/

Jalsa Urubshurow on Pioneering Responsible Tourism in Mongolia

USA | In 1992, Jalsa Urubshurow founded Nomadic Expeditions,[1] a pioneer in adventure travel to Mongolia. Providing immersive conservation and cultural journeys for adventurous travellers since its founding, Nomadic Expeditions is hailed for its unwavering commitment to sustainability. Mr. Urubshurow also co-founded the Golden Eagle Festival to preserve this centuries-old tradition from near extinction. In 2002, he opened Three Camel Lodge, a remote wilderness lodge that serves as a centre of education and conservation in the Gobi Desert of Mongolia. From its all-Mongolian staff to its numerous projects to protect the rich natural and cultural heritage of Mongolia, the lodge continues to serve as an emblem of sustainable hospitality under his leadership. Mr. Urubshurow is the recipient of the Order of Polar Star—the highest award the government of Mongolia can bestow upon a foreign national.

---

[1] https://www.nomadicexpeditions.com

Areas of expertise: conservation, destination sustainability, ecotourism, responsible tourism, UNSDG 12 (Resource Efficiency), UNSDG 15 (Forests & Biodiversity)

The following interview with Jalsa was first published in September 2020 on Sustainability-Leaders.com.

Jalsa, you built Mongolia's first community-based eco-lodge, Three Camel Lodge. What led you to become a leader in sustainable tourism practices in Mongolia?

I was asked in 1990, by the first Prime Minister of the newly formed Democracy of Mongolia to promote travel and tourism from North America to Mongolia. At that time there were fewer than 100 tourist visitors per year from America. I have always been convinced that Mongolia could be a world-class travel destination, but in those early years, it was clear that most of the available experiences in Mongolia were utilitarian, the carryover of the Soviet model of tourism.

I created Three Camel Lodge to further my mission of promoting Mongolia as an adventure and ecotravel destination while elevating the visitor experience standards.

My vision has always been to help Mongolia benefit in positive ways from tourism to support economic development. To borrow a famous line from the movie Field of Dreams, 'if you build it, they will come'. I knew that if we had a luxury ecolodge, it would help put Mongolia on the map of sophisticated travellers.

Of course, in my passion to develop tourism in Mongolia, the need for sustainability was always foremost in my mind. I built Three Camel Lodge to reflect the sustainable design and cultural authenticity, using local materials and hiring local Mongolians to construct the lodge to ensure that the lodge's operations wouldn't disrupt the delicate ecosystem in the Gobi Desert.

For me, sustainable tourism was never about marketing; I believe it was, and still remains, the right thing to do for any hotel or lodge. I'm pleased that our sustainability practices, which we constantly adapt and continue to innovate, have set a model for tourism in Mongolia, and also, we are proud to have been recognized for our sustainable tourism leadership throughout the world.

Your Golden Eagle Festival attracts scores of international travellers to Mongolia to watch this spectacle. Briefly, what makes this event special and how has the festival evolved since you started it in 1999?

The Golden Eagle Festival may be one of my proudest achievements, but not because it's now known worldwide or because it attracts travellers from around the globe. I am proud of this event because it's proof that a few people can truly make a big difference.

It was the late '90s when I first encountered this remote part of Mongolia and learned about this culturally rich and unique tradition of falconry with golden eagles. I was awestruck by it, but I knew that with just a few families (and mostly older men), this ancient falconry tradition in Mongolia would soon disappear. I wanted future generations, including women, to honour this great tradition, so I co-founded the Golden Eagle Festival as a way for locals to preserve and celebrate their tradition together.

What started out as mainly for local communities to keep their tradition alive, has grown into an international event and as a member of the committee for the festival, I continue to advocate for more sustainability practices and to ensure that the local economic benefits from tourism benefit to the people.

One of my proudest moments came in 2015 when 13-year-old Aisholpan, a young woman took home the top falconry prize. The world took notice and the BBC filmed a documentary (The Eagle Huntress) that gained international acclaim.

However, while it may be better known now, fewer than 1000 international travellers attend annually, given the festival's remote location. In that sense, it still remains an off the beaten path travel experience.

I'm delighted to share that now over 300 families practice this tradition and young men and women are leading the way. The festival has gone a long way towards sustaining a unique tradition for future generations, yet its evolution requires us to monitor and study its impact on the golden eagle population and treatment of animals.

Which are the main challenges right now, potentially threatening the success of the festival and your accommodation business?

Our biggest challenge is ensuring that the preservation of this rich cultural tradition is balanced with the preservation of wildlife and nature. To that end, Nomadic Expeditions has funded a new research initiative partnering with the Wildlife Science and Conservation Center of Mongolia (WSCC) and The Peregrine Fund to support golden eagle conservation and cultural heritage preservation. This new project will establish a baseline study to ensure the health and safety of these majestic birds of prey. We never want this celebration to negatively impact the golden eagle population, and this initiative will work to further their conservation and protection.

In terms of our luxury eco-lodge and from an accommodation standpoint, our biggest problem has always been our relatively short tourism season in Mongolia (May through October) given the harsh winters in the country. What we can promise is a truly incredible and sustainable travel experience during our short season, both at our Three Camel Lodge and on our customized tours throughout the country.

Innovation and imagination are keys to success in both business and conservation and you are known for being good at both. Can you shed some light on your recent Mongolian Bankhar Dog project at the Three Camel Lodge and how it helps you to promote sustainable development and responsible travel?

Our recent partnership with the Mongolian Bankhar Dog Project is a perfect example of how we are always looking for ways to connect with the community. We constructed a new kennel facility for breeding and training Bankhar dogs (some refer to as Mongolian mastiff) at Three Camel Lodge. This protective breed was used by nomads for centuries and was slowly disappearing due to cross-breeding in recent generations. However, we're not just bringing back a breed; we're developing a wildlife conservation partnership because these herding dogs will be donated to nomadic families to help to protect their livestock.

Livestock is essential to the nomadic way of life, yet they often fall victim to a number of predators, including wolves and even snow leopards in the more remote, mountainous parts of the country.

Survival and partly tradition have perpetuated the belief that if a predator kills your animal, you must seek it out and kill it. Enter the Bankhar dog, that protects the livestock from the predators and as a result, there has been a decrease in retaliation killing. Of course, the puppies and dogs are endearing to guests at Three Camel Lodge.

Mongolia has extended a ban on all international travel as a response to the ongoing outbreak of COVID-19, due to which 70% of tourists have cancelled their travel plans. How are you coping with this situation—as a business but also as a community?

We have zero international travellers to Mongolia this year, and while that is hard to deal with, we've taken the time to focus on the positive. We are proud that we haven't laid off a single employee in Mongolia yet. From our staff and our facilities to our trips themselves, we have been focused on rebuilding, innovating, and improving.

At Three Camel Lodge, we're viewing low occupancy as the perfect time to finally build the spa we've been talking about adding. We've trained our staff on COVID-19 health and safety protocols and how to adjust to a new normal when travel resumes. We have been using the downtime to explore and dream again, and have spent these months traveling Mongolia on numerous scouting trips. It turns out that you can never stop exploring your own backyard, because we continue to find places that we are excited about including in our new itineraries, and in many cases, we've gone back to the drawing board and redesigned entire trips.

We're into conservation through tourism business and we all deeply miss those inspiring connections we make with our travellers who also believe in what we do, but this extra time has been valuable for taking a deep look at who Nomadic Expeditions really is and what we want to offer our clients.

We decided to refund 100% to all our travellers who had booked and cancelled a trip in 2020. It's a bold step, as it meant we took a hit to our bottom line, but we've always believed that people are more important than profits. While most of our competitors offered credits on a future trip, we made a conscious decision to ensure that our travellers experience zero financial loss at this very difficult time. It's been greatly appreciated by our clients.

At the same time, we've continued with our ongoing sustainable tourism efforts, since we believe that this is needed now more than ever across the entire travel industry. We've launched new projects (like the conservation partnerships we mentioned above) and we've even let this time stoke our creativity, as we will be publishing a children's book on the natural and cultural treasures of the Gobi in Mongolia to benefit the local community as well as our visitors. The book will be published in Mongolian and in English.

Mongolia ranks low in the 2019 Travel and Tourism Competitiveness Index. With a mission to ramp up tourism to diversify the Mongolian economy, there is a

risk of unsustainable growth. To your mind, what concrete steps should the government take to make tourism development sustainable?

The Mongolian government has been notoriously slow when it comes to supporting tourism development—the airport in Ulaanbaatar took years to open. Big projects like that are important but I also think that the government needs to partner with the so-called 'little guys', the boutique tour operators who have been successfully providing experiences while protecting and preserving the environment. We have on-the-ground experience and know what works. Most importantly, we will continue to encourage the Mongolian Government to embrace destination stewardship based upon sustainable tourism best practices.

How do you envision the future of tourism in Mongolia, at a time when the world is bracing up to significant environmental changes, due to climate change?

I strongly believe that the very nature of travel will change in a post-COVID world. Fewer travellers will want to travel long distances for short trips. Instead, they'll want to truly immerse themselves in a destination, staying longer.

I think it's a win-win for a place like Mongolia. It will attract those travellers who may have thought it was 'too far' before since they'll be less interested in jetting off for a quick getaway and more interested in a longer, more immersive experience.

In the end, it will mean fewer travellers but longer stays, and I think that combination is not only better for the environment but better for travel in general.

Reflecting on your experience as a tourism innovator and entrepreneur, how has your view on sustainability and tourism changed over the years? Which key lessons or insights could you share with us?

We are seeing increasingly examples that more sustainable companies are also more profitable companies. And that is an important message. You can have a profit with purpose.

The lesson I would share is that you have to think about the future and not necessarily what you need at the moment. It may take more time, and more money to be a sustainable tourism company, but in the end, it will serve you and the planet well.

Anything else you'd like to mention?

I would like to see more travellers understand the connection between sustainable tourism and a great holiday. Whether it's travel media, tourism professionals, or governments leading the charge, we need to help travellers better understand why it's critical to choose the right travel partner and by that, I mean the sustainable travel partner.

Link to the interview: https://sustainability-leaders.com/jalsa-urubshurow-interview/

Saint Lucia | With a Geology degree and RYA Yachtmaster certification, James originally came to the Caribbean as a yacht captain, divemaster and Earth Science instructor for Seamester Programs in 1997. His Master's degree in Responsible Tourism Management (2010) brings together his interest in the environment, community and tourism. He helps provide a holistic framework to assist communities to develop sustainable tourism strategies of appropriate scale and suitability for fragile ecosystems whilst ensuring livelihoods from market-ready and in-demand products for discerning tourists. His own company, Jus' Sail Ltd. in St Lucia, is the recipient of the Caribbean Tourism Organisation Sustainable Tourism Award (Community Benefit) and was runner up in the WTTC Tourism for Tomorrow Awards in the same category. James has undertaken numerous responsible tourism consultancies across the Organisation of Eastern Caribbean States and currently managing the Union Island Climate Change Adaptation Project for International Conservation Charity Fauna & Flora International.

Areas of expertise: responsible tourism, tourism business, Caribbean, UNSDG 11 (Sustainable Human Settlements), UNSDG 8 (Equal & Fair Economic Opportunities)

© The Author(s), under exclusive license to Springer Nature Switzerland AG 2022   333
F. Kaefer, *Sustainability Leadership in Tourism*, Future of Business and Finance,
https://doi.org/10.1007/978-3-031-05314-6_52

The following interview with James was first published in November 2017 on Sustainability-Leaders.com.

James, do you remember what made you turn your passion for environment and local cultures into a master's degree in responsible tourism? What triggered your interest?

My interest in responsible tourism was born from my passion for environmental issues and the sustainable development of SIDS in the Eastern Caribbean. From 1997–2001 I worked as an instructor and Captain for ActionQuest and SeaMester Programs, yacht based educational voyages for teens and college students. I was fortunate enough to sail the length of the Lesser Antilles from BVI to Grenada several times.

During those voyages, you saw the juxtaposition of tourism development, fragile ecosystems (often denuded) and poverty within local communities. There was a clear disconnect and the idea of pursuing a form of tourism that could be a driver for sustainable livelihoods and local economic development that also enhanced, not destroyed, the environment was compelling.

From your experience, which aspects of balancing sustainability, guest satisfaction and generating tangible benefits for communities are the most difficult? How have you dealt with those challenges?

Profitability of the business is the biggest challenge when you pursue the above. I have not found it hard to balance sustainability with guest satisfaction or generating tangible benefits for communities, because quite often they can be sewn together nicely. However, that takes an awful lot of work and energy.

When you care about many aspects outside the direct running of your business to deliver on the bottom line, you find your attention and resources are not always focused on profitability as your primary goal or motivation. So, balancing the profitability of the business with the points you cite above is our biggest challenge. How to deal with this challenge… Remind yourself every day that without a profitable business you cannot stay in business and therefore cannot continue to deliver to your guests and local stakeholders!

Your motto at Jus' Sail, "Better places to live and better places to visit, in that order" is a great way to define sustainable tourism and a simple way to promote it. What else would you add to it?

That motto is the tagline for Responsible Tourism, I learned it from Professor Harold Goodwin when first starting the MSc in Responsible Tourism at ICRT in 2008. I have yet to find a better way to express what RT means to me… I try not to add to it because I believe it's an all-encapsulating statement.

Yes, obviously there are details you can get into, such as ensuring that a memorable experience for a tourist is also a beneficial one for the host so that both parties come away happy. But to convey the ethos of RT, it has never been bettered!

What was the most important thing you learned during your postgraduate studies—and during your time 'on board'?

I think the most important thing I learned during my studies was that no one has all the answers and that there is no magic silver bullet. Every situation must be dealt with individually and woe betide the 'expert' who arrives in a destination and believes they know the answers better than the local stakeholders know the issues.

If it was easy to deliver sustainability, it would already have happened. Arrogance from an outsider who has 'the knowledge and the answers' is not helpful when working in a destination, and will not help to build trust or long-term results.

Too many projects are consigned to history once the project funding dries up and projects end—thereby demonstrating their unsustainability.

I would also suggest that I have learned more after graduating, since starting my own business and striving to make it work in the real world. Academic study is no substitute for working in the field.

Life has a way of showing you that your best plans can come to nought if you are not flexible, adaptable, determined and humble. It's extremely hard to make a small tourism business survive, let alone thrive, and the reality of running a business can sometimes overwhelm all the best intentions for focusing upon sustainability issues.

How do your sustainability practices differ from those of other companies in the region? And where do you think you could still improve?

I believe our primary point of difference is that our company has an explicit focus upon developing the local youth through our training program for unemployed youth, from which they can build the skills and confidence to pursue a rewarding career in the marine or wider tourism industry. Our focus upon youth development has generated goodwill and interest from our clients, the private sector and public sector bodies.

The type of person who wants to enjoy an authentic sailing experience aboard a traditional local trading sloop also happens to be the kind of person who wants to know that the company they choose to spend their money with is actively engaged in helping to build the resilience of the community and the destination to which they have travelled.

There is always room for improvement, this is what keeps you awake at night. Gaining greater levels of engagement with the accommodation providers and tour operators is probably where we could do better.

You work a lot with the local youth. Are there any particular challenges linked to engaging adolescents with sustainability?

Not really. Young people tend to be energetic and passionate. If you inspire them and encourage their line of thinking towards evolving their consciousness and putting emphasis upon being a solid global citizen, they will often run with it and end up being more passionate than you are. The youth just need to be given the care and attention and opportunity to thrive. In the Caribbean, opportunities for that are sadly sorely lacking.

In your experience, which are the key factors responsible for tourism businesses and entrepreneurs need to keep in mind when implementing commu-nity programmes as part of their sustainability commitment?

Take your time to work at a pace that the local community and stakeholders are comfortable with. Too many projects are based upon the timelines that are suitable for the donor or funder. This does not take into consideration the reality of the lives of those persons that are supposed to be the beneficiaries.

Invest time in extensive dialogue with local beneficiaries and stakeholders to ensure that the programme is really going to deliver for them and that it will be 'owned' by them. Imposition from above or from outside will, without sufficient engagement by local persons, rarely deliver in the long-term.

Be prepared to commit financial resources to get the locals involved. Many community members live hand to mouth subsistence existence. This leaves them little opportunity to undertake activities that do not directly deliver immediate economic results. Or put another way, put bread on the table at the end of each day.

It is unreasonable to expect local community members to always commit their time in kind to a project without any financial gain. They simply cannot afford to take that risk, no matter how worthwhile the project may be in the long-term. Long term thinking is a luxury they don't have.

So, setting aside budget to cover a stipend, at least, is essential. But obviously, this must be looked at on a case-by-case basis.

As a finalist of the WTTC Tourism for Tomorrow awards, how has this recognition impacted you personally, and the business?

Despite the warm fuzzy feeling at the time, I have to say I do not think there has been any tangible benefit to the business or to myself personally. We are proud of the recognition that we received, but it's old news already and there is no long-term impact that we have managed to gain from it. That is not to say we are disappointed or do not value the recognition ourselves. We feel good about it.

Advocacy for policy change is one of the objectives pursued by Jus' Sail. In your view, which policies are the most in need right now of being updated, linked to tourism and sustainability? And what is preventing such changes from happening?

At a national level, in the Caribbean context, policy to pursue a level playing field for local companies to receive the same concessions that are often offered to foreign investors would be advantageous for the sustainability of the local economy.

A move away from the development of large resorts on beachfront land is long overdue. Such developments ignore the realities of a changing climate and require the removal of habitat which helps to protect and sustain coastlines.

In addition, building back from the remaining beaches and allowing greater freedom of access to beaches for others, be they locals or other visitors, will ensure more balanced access to beach resources. This in turn can help develop a more balanced economy.

Greater protection of fragile habitats and cultural heritage is essential if destinations are to retain their uniqueness, which attracts visitors in the first place. Developments should only be allowed at a scale that is appropriate for the location. Megaprojects are not in my view the way to go on small islands.

Which are the main challenges for local communities regarding access to international funding for sustainable tourism projects—and value chains? And how can these be overcome?

The disconnect between the requirements of the funders—administration, financial management and project management skills—and the skill-sets available locally, within a community.

Projects should be reverse engineered and designed to fit the needs and abilities of the beneficiary, not expect the beneficiary to adapt to the complex and bureaucratic methods of international donor agencies.

Looking forward, which developments do you plan or foresee for Jus' Sail and the St. Lucia destination in the next years, regarding sustainability performance?

For Jus' Sail a greater focus upon the measurement of our impacts, but this requires resources beyond our reach for now. Nationally, I hope that more companies will get involved in the Saint Lucia Hotel and Tourism Association and its efforts to push the sustainability agenda. We are stronger together and greater collaboration is always beneficial to my mind.

Link to the interview: https://sustainability-leaders.com/jus-sail-sustainability-strategy/

USA | Jamie Sweeting is the President of Planeterra Foundation and Vice President for Social Enterprise and Responsible Travel for G Adventures, leading the company's social enterprise, stewardship and community development work. He is responsible for establishing the company's long-term social enterprise strategy and working to ensure responsible corporate sustainability performance. Previously, Jamie served as Vice President for Environmental Stewardship and Global Chief Environmental Officer for Royal Caribbean Cruises Ltd. He guided their work to conserve and protect the environment. Sweeting has over 25 years of experience in tourism, conservation, development and business management. Before working with Royal Caribbean Cruises, he served as a senior business advisor and senior director of the Travel and Leisure program for Conservation International's Center for Environmental Leadership in Business. He received a Master of Tourism Administration from George Washington University in Washington, D.C., and a B.A. honours in Leisure and Business Management from Manchester University in England.

© The Author(s), under exclusive license to Springer Nature Switzerland AG 2022
F. Kaefer, *Sustainability Leadership in Tourism*, Future of Business and Finance,
https://doi.org/10.1007/978-3-031-05314-6_53

Areas of expertise: responsible tourism, tourism business, UNSDG 10 (Reduce Inequality), UNSDG 11 (Sustainable Human Settlements), UNSDG 8 (Equal & Fair Economic Opportunities).

The following interview with Jamie was first published in January 2018 on Sustainability-Leaders.com.

Jamie, throughout your extensive professional career in sustainable tourism you have provided support on a wide range of topics to companies, non-profits, governments and destinations. Do you remember the first time you heard about 'sustainability'? What led you to dedicate your life to the development of sustainable tourism?

I got into SCUBA diving when I was a teenager back in the 1980s, and as a passionate diver, I fell in love with the marine world. From there it seemed pretty clear to me that the common-sense approach was to protect the places I wanted to visit and enjoy.

Has your view on the topic changed since you first got involved?

Yes and no. I still very much believe that tourism can, and should, be a major contributor to conservation and community development. However, I now know that this isn't necessarily that important to the majority of the world's travellers.

I spent many years struggling with coming to terms with this fact. 'Why don't they care?' I'd incredulously ask myself... 'How can they not care?' Eventually, I realized that, while it's easier to work in scenarios where the traveller does actively care, it's still very important to work hard on these issues even if they don't.

Why? Because if tourism activities are fundamentally unsustainable, eventually destinations will reach a tipping point whereby the negative effects will degrade the travellers' experience. In turn, demand will drop and it's exceptionally difficult to regain market share once it's been lost.

After 50+ years of mass tourism, we're unfortunately on the precipice of the mountain, about to witness the decline of many tourism destinations that have failed to manage themselves sustainably and are soon going to pay a hefty price.

You are the President of the Planeterra Foundation, which focuses on connecting and supporting women, youth and indigenous communities. In your view, which are the main barriers with regards to succeeding as an entrepreneur and developing successful tourism 'products'?

Tourism is a global industry. This means that tourism entrepreneurs developing small businesses are not only competing within their local market, but the destination and country they operate in are competing with others to get travellers to choose their destination.

Another barrier is a lack of knowledge amongst consumers with regards to understanding what is responsible travel, in connection with local people and the environment—often well-meaning travellers fail to seek out responsible travel choices simply because they don't know what is good and what is bad.

The wonderful thing about Planeterra's model is that we partner with travel companies to see where they already have a business. We determine whether developing a social enterprise and integrating it into their product mix will make their tours more attractive and provide a better experience for their guests. We've literally flipped the 'build it and they will come' model on its head, by building projects where people are already coming to visit.

Various of our previous interviewees, including Carole Favre and Megan Epler Wood, pointed to lack of market access as one key issue preventing local communities from fully participating in—and benefiting from—the tourism supply chain. How does Planeterra help them overcome this hurdle?

As I mentioned before, we start with the market first. Building a tourism market from scratch is extraordinarily difficult unless you have some exceptional assets in your favour. Chances are there are already travellers going to (or near) your location if you have some of those assets.

If you are trying to be a pioneer, getting travellers to come to your remote community, you have to recognize that it may well be a long, hard slog. Unfortunately, there are many more examples of failure than there are successes.

That said, we have a number of Planeterra projects that are in areas where there are already travellers, but those community enterprises weren't benefiting from that traffic. We were able to work with those enterprises to build their capacity to work with groups, rather than waiting on passing individual travellers.

In these cases, it was a matter of identifying the opportunity for both the travel company and the community enterprise and making the market linkage that has made these ventures successful. Planeterra's long-term successful partnership with the leading adventure operator G Adventures[1] has allowed us to rapidly increase tourist visitations to such initiatives.

Last year we partnered with a Women's Cooperative in Belize, which was receiving approximately 100 travellers per year. Now they are receiving over 1000 travellers per year as a result of being integrated into G Adventures Belize trips.

In your view, are international events and organizations like UNWTO, GSTC and the 2017 UN Year of Sustainable Tourism adequate ways for overcoming the challenges of sustainability in tourism globally?

These are important organizations and events. However, until we truly mobilize the power of the private sector and governments, I think the goal of sustainable travel will remain elusive.

I believe that governments need to do more to safeguard their natural and cultural assets, and to reward companies that act responsibly and support community development and conservation.

At the same time, the global tourism industry is suffering from a massive case of the tragedy of the commons. None of the major travel brands have truly committed to internalizing the costs of maintaining the proverbial goose that lays the golden egg. For now, they are getting away with it, as are the vast majority of the literally millions of tourism enterprises around the world—but for how long?

In the coming decades we are sadly most likely going to witness some major cultural and environmental catastrophes; the global rapid decline of coral reefs, for

---

[1] https://www.gadventures.com

example. Not only will this deprive us of some of the world's most spectacular environments and species, but also the critical ecosystem services they provide.

It is time for the travel industry to start paying much more attention to cultural and environmental issues, both from a business and an ethical perspective. These massive impacts are now no longer lifetimes away. They are happening now.

As VP of Social Enterprise and Sustainability of G Adventures and avid traveller yourself, which destinations have inspired you the most this year, for their innovative sustainability initiatives?

Peru continues to impress me. I think some of this may have been fate—their international tourism industry has been built around selling nature and culture and not sun, sea and sand. As such, the vast majority of the industry is locally-owned and run.

As some of the main destinations of the country have developed over the past decades, so has the quality of the product at all service levels. More recently there has been a growth in the diversity of places and products that can be visited, as well as a blooming food tourism market.

The government is also actively engaged in working with the private sector to address major challenges regarding visitor management at prime sites such as Machu Picchu.

Throughout your professional career, have there been any specific organizations or persons that have served you as role models or source of inspiration?

Too many to list here. I have been very blessed that I have had many, many people support me in my career. I am eternally grateful to each and every one of them. I'm part of the first generation of people who actually actively embarked on a career in sustainable travel. I studied tourism as an undergraduate at Manchester University in the UK (and even did my thesis on ecotourism) and then went on to do a Masters in Tourism Administration at The George Washington University.

The pioneers in this space all came from different and diverse backgrounds— tourism planners, landscape architects, environmentalists, filmmakers, marketers, designers, anthropologists, entrepreneurs, chefs, etc. etc.

Beyond my pure passion for travel and experiencing different cultures and places, I believe it is this diversity of different people and professions that has made my career in this area so rewarding.

Overtourism has become a critical sustainability issue for popular destinations. In your view, what should destination managers, but also tour operators like cruises, do to prevent it?

There is nothing particularly new about the phenomenon of overtourism, except the very apt word that's been recently popularized by Skift.

There were complaints about tourist crowds in the Lake District in England as far back as the 1850s. When I lived in York, England in the 1980s, disgruntled residents ran a campaign against the intrusion of too much tourism on their daily lives by printing and disseminating T-Shirts that said 'F*** Off Tourists!'

The answer to these challenges is better to run tourism. It's really not rocket science—better coordination, scheduling and management can often dissipate many of these problems.

Concerns about overtourism are often not just about overcrowding, but also indicate an imbalance of who is really benefiting from the tourism in a given destination.

In the travel industry we are often guilty of looking at places solely as destinations, rather than as the collective home of the people living there. When local people are seeing adequate benefits from tourism, and travel experiences are managed effectively and efficiently, we can witness the win-win of tourists having a great visit and the host community seeing tangible and measureable benefits, and therefore becoming even more motivated to make tourism a sustainable, inclusive endeavour.

Specific to the cruise industry, I'll share a little-known fact: cruise lines are not allowed under legal anti-trust provisions to meet and coordinate schedules, even if it is to minimize overcrowding. National and local governments that control the ports of call can do this, but the cruise lines cannot.

So, you can blame the cruise lines for overcrowding in places like Dubrovnik. But apart from avoiding that port (which, one could argue, would not be meeting their fiduciary obligations to their shareholders, given the significant customer demand to visit the historic city) there is little they can do to solve this situation without government intervention.

Another issue, which you talked about in your keynote at the T+SG summit, is that only 5 out of every 100 tourist dollars spent stay in the destination visited. Do you have an idea of how to change this and make the tourism system fairer?

Buy local! At G Adventures, 91% of our on-tour hotels, restaurants and tour providers are locally owned. Of those, 9 in 10 purchases more than half of their supplies from local producers, markets and farms. That means there is very little leakage of the tourism dollars G Adventures spends in-country, to put together our trips.

And there is a significant multiplier effect of those dollars spreading through the local economy.

The key is to follow where the money goes and ask who is benefiting from the tourism economy.

Which major trends do you observe in tourism right now which might support or threaten the ability of tourism businesses or destinations to become more sustainable?

There is little doubt that more and more people are craving the opportunity to have meaningful interactions with nature and local people in the places they go and visit; so much so that we are heading for this positive phenomenon going mainstream rather than being niche.

I'm certainly not suggesting that everyone wants this kind of experience or that they want it all day, every day on their holiday. But if we continue to see the traditional cruise passenger and sun, sea, sand tourist seeking out one of these experiences when they take a trip, then we rapidly move from this being a niche offer to a segment of the travel industry catering to tens of millions of people.

What excites me about this is the potential to increase the quality of life for hundreds of thousands of people currently living in poverty, by providing wonderful tourism experiences to people who want more from their holiday.

Finally, what's next for Planeterra? Are there any upcoming projects that you feel particularly excited about?

We're busy trying to bring new social enterprises to market that benefit marginalized women, disadvantaged youth and rural communities, particularly indigenous communities. We're always on the lookout for new projects and certainly encourage people with ideas and opportunities to reach out to us.

We've also been working with G Adventures and other partners to develop and publish some new responsible tourism guidelines. Last year we worked in collaboration with George Washington University to publish good practise guidelines for working with indigenous communities.

Currently, we're partnered with the ChildSafe Movement[2] and Friends International[3] to develop similar guidance for travel companies on child welfare.

We hope other companies and organizations will take up these guidelines and integrate them into their day-to-day business practices.

Link to the interview: https://sustainability-leaders.com/jamie-sweeting-interview/

---

[2] https://thinkchildsafe.org
[3] https://friends-international.org

Denmark | Janne Liburd, DPhil, PhD, is Professor and Director of the Centre for Tourism, Innovation and Culture at the University of Southern Denmark. She has developed the field of sustainable tourism co-design. Maximising the impact of co-design is not simply a question of introducing new methods but involves building capability, trust and ownership with—not for—others such that the innovation is sustainable in the long term. Her professorial dissertation (2013)—Towards the Collaborative University—contributes significantly to the debate about the being of the university, by ambitiously uncovering and reclaiming the very aims and freedom of the university through collaboration. By ministerial appointment, Janne is the Chair of the UNESCO World Heritage Wadden Sea National Park (2015–2022).

Areas of expertise: overtourism, tourism research, destination sustainability, Denmark, Europe, UNSDG 11 (Sustainable Human Settlements), UNSDG 14 (Oceans & Marine Life)

The following interview with Janne was first published in October 2017 on Sustainability-Leaders.com.

Janne, do you remember what inspired you to focus your academic career on sustainable tourism? When was the first time you heard about 'sustainability'?

© The Author(s), under exclusive license to Springer Nature Switzerland AG 2022   345
F. Kaefer, *Sustainability Leadership in Tourism*, Future of Business and Finance,
https://doi.org/10.1007/978-3-031-05314-6_54

After I finished high school in Denmark, I took two years off to work and travel around the world as a backpacker. Having ample of time and a limited budget, I conversed with people along the way and was often fortunate to be invited to stay with locals, whether in Canada, Fiji, New Zealand, Australia, Indonesia or Thailand. I was curious to get a glimpse of how they lived, their relations to their place of residence, nature and other people.

This led me to anthropology, where I first encountered tourism as a means of development. Inspired by French philosopher Michel Foucault, part of my PhD research was centred on the development discourse. I was fascinated by the ability of the development discourse to reinvent itself, especially in the form of sustainable development.

I never had any classes on tourism whilst studying anthropology. But the 'impact studies' of the 1980s and 1990s championed by Valene Smith were of great inspiration. I remember wondering how anthropologists (who for the most part never wrote about encountering tourists during their rite of passage fieldwork) were able to judge what tourism 'did' to culture, i.e., either had negative or positive impacts and how that could be applied to sustainable development.

As a world-renown researcher and teacher of tourism and sustainability, has your view on the topic changed since you first got involved?

During my PhD research and thesis on 'Sustainable Tourism Development in the Eastern Caribbean' (1999), I often felt lost in attempting to link theory to practice. Mass tourism in the Eastern Caribbean was dominated by tourism enclaves and the cruise industry, both of which were characterised by an ethnic division of labour. Global sustainability ideals of equality, natural and cultural integrity were overshadowed by a profit-maximizing industry. And regulators seemed to be elected based on a mandate to make life easier for foreign investors (and themselves).

Today, I am proud to say that I am educating philosophic tourism practitioners, many of whom come to share my commitment to transforming tourism from a world-taking to a world-making phenomenon, which touches on so many aspects of human values.

My research is no longer limited to observing and wanting to understand how and what others do with tourism. I advocate sustainable tourism development and am delighted to work with stewards who care beyond selfish interests.

You have long been involved in the BEST Tourism Education Network. Which are the major trends and innovations in tourism education?

I should like to say that tourism higher education has moved from telling students what to know and do, assessed in multiple-choice tests, towards educating with students, and bringing forth their individual potentials, taking them into unsettling territory, and creating space for transformations of the self, and to engage in future world-making.

Unfortunately, I think we're still far from the latter aims of tourism higher education. Most tourism programmes are more concerned about disciplining means (represented by pressures to maintain league table rankings, professional accreditations, research excellence frameworks, predefined curricula, learning outcomes).

Are there any topics that you find underrepresented in Tourism degrees, and which should be covered more in-depth?

The management school approach to tourism higher education leaves much to wish for in terms of influences from the Humanities and the Arts. Those would be essential for understanding the importance of culture and communication, appreciation of other norms and forms of knowledge creation, and understanding the importance of collaboration with others—students, teachers, tourism professionals, policy-makers, community groups, and the larger society while studying.

As Chair of the Board of the Wadden Sea National Park in Denmark, how do you approach and encourage destination sustainability?

The sustainable development of the Wadden Sea National Park, which is also UNESCO World Heritage, is embedded in everything we do! We strive for transparent action and reporting in all of the interrelated pillars of sustainability: People, Planet and Profit. Nature conservation and use it at the centre of steering sustainable development, as the national park is the home of millions of migrating birds, fish and human activity, which also includes residential areas and commercial use.

The nomination of the Wadden Sea as a World Heritage Site in 2014 followed the 2009 nomination of the neighbouring German and Dutch parts of the Wadden Sea. It rests on more than 30 years of political commitment of protection of the largest, uninterrupted stretch of tidal flats in the world. Trilateral collaboration and everyday ownership of the national park by stewards who care drive sustainability in the Wadden Sea.

*Concern is growing among sustainable tourism practitioners about the unintended, potentially negative effects of being listed as a UNESCO World Heritage Site (***i.e.,*** overtourism, environmental degradation, heritage loss, ***etc.***); do you think this concern is warranted?*

The concerns of unregulated tourism to World Heritage Sites are genuine and negative examples plentiful, which is why UNESCO now requests that new World Heritage Sites produce a plan for sustainable tourism. We have a trilateral sustainable tourism development strategy for the Wadden Sea, which has been adopted into operational plans by the four municipalities of the Danish Wadden Sea.

The Wadden Sea is visited by approximately 10 million tourists per year, and the carrying capacity is still far from being exceeded. This does not mean, however, that certain places have become hot spots and bear the majority of visitors, and thereby earnings from tourism. A range of visitor management strategies is currently in the making.

Governance of the WH Site is also contingent on residential pride in the National Park. Residential pride is increasing, which is indicative of ownership and that people care about nature conservation (also documented in my latest research). This alignment of personal values with those of the Outstanding Universal Values, which is how UNESCO describes sites of global importance, is key to enabling sustainable development in practice.

Denmark recently changed its rather strict regulation of coastal areas to facilitate the development of 'sustainable' tourism facilities. In your view, does

the new regulation do justice to the holistic understanding of 'sustainable tourism'
as used in academia?

Denmark's coastlines have been protected from tourism development and construction for more than 80 years. In 2014, the Danish politicians opened up for softer regulation of the coastlines and invited proposals for tourism development projects within the hitherto protected coastal zone.

The call explicitly requested nominations for sustainable tourism projects, but our comparison between academic sustainability discourse and the approved projects suggests that tourism actors do not address sustainable tourism development as a holistic concept.

Moreover, our research has documented how long-term perspectives are largely absent, whereas economic benefits are emphasized.

Denmark—and especially its capital, Copenhagen—is internationally highly regarded for its commitment to sustainable development, particularly in the fields of energy and environment. In your view, which are the sustainability priorities for Destination Denmark in the next years?

Unfortunately, and as documented in the research above on the coastal tourism developments, weak political leadership characterizes the envisaged transfer towards sustainable tourism development.

Denmark is still internationally leading in green energy. We pride ourselves of global commitments to CO_2 reduction, but increasingly, practice is lacking. The new 2025 Strategy for Tourism in Denmark fails to demonstrate a clear commitment to sustainable development, as the concept sits uneasily with a neoliberal government and industry alike.

As Professor at the University of Southern Denmark, you also teach in the European Master in Tourism Management, a collaboration between three countries which was just awarded by the European Commission for its exemplary achievements. How have you personally—and the university—benefited from being part of this program?

I can honestly say that I feel the benefits and joy every single day. Whereas program administration has been quite demanding, especially in order to meet regulations in the three, very different consortium countries of Denmark, Slovenia and Catalonia (Spain), the benefits of working with excellent students, visiting scholars, industry partners and partner universities on all continents are simply amazing.

The university and EMTM student cohort further benefit from enrollment of students who have heard about tourism studies at the University of Southern Denmark, often from the global promotion of the EMTM. Some chose us for their MA in International Tourism, which runs parallel to the EMTM programme, but without mandatory mobility between the three consortium partner countries.

Your career advice to students of Tourism university programs interested in dedicating their career to sustainability?

Follow your heart and mind. Just do it!

Link to the interview: https://sustainability-leaders.com/interview-janne-liburd/

Thailand | Jens Thraenhart has over 25 years of international travel, tourism and hospitality experience. In 1999, he founded Chameleon Strategies, an Affiliate Member of the UNWTO, advising tourism organisations and travel businesses in leveraging digital engagement to drive sustainability. He co-founded a leading China travel marketing agency Dragon Trail and headed the marketing strategy and digital departments at Fairmont Hotels & Resorts, Canadian Tourism Commission, and Dusit International. He was appointed as the Executive Director of the Mekong Tourism Coordinating Office (MTCO) by the tourism ministries of Cambodia, Lao PDR, Myanmar, PR China, Thailand and Vietnam in 2014. He is currently serving as the Vice-Chair of the Affiliate Members Board of the UNWTO, was the Chair of PATA China and has been on the board of industry organisations like PATA, HSMAI and IFITT.

Areas of expertise: destination sustainability, responsible tourism, sustainable development, UNSDG 10 (Reduce Inequality), UNSDG 8 (Equal & Fair Economic Opportunities)

The following interview with Jens was first published in February 2020 on Sustainability-Leaders.com.

349
F. Kaefer, *Sustainability Leadership in Tourism*, Future of Business and Finance, https://doi.org/10.1007/978-3-031-05314-6_55

Jens, you founded the award-winning digital marketing firms Chameleon Strategies and Dragon Trail and created the much-applauded Mekong Tourism initiative, among many other accolades. What motivated you to take up the Executive Director role to head the Mekong Tourism Coordinating Office (MTCO)?

I was actually the Special Advisor to the previous Executive Director of the Mekong Tourism Coordinating Office since 2010, while I was Co-Founder/ President of Dragon Trail and Chair of PATA (Pacific Asia Travel Association) China, living in Beijing.

The countries of the Greater Mekong Subregion (Cambodia, Lao PDR, Myanmar, Thailand, Viet Nam, and the provinces of Guangxi and Yunnan in PR China) have always fascinated me. It is such a beautiful and stunning region, rich in culture, heritage, and environmental assets—to be able to assist in promoting and developing tourism in these countries was a tremendous honour, to get the trust from the six governments. On the other hand, it was an exciting challenge to create a platform for stakeholder engagement and collaboration.

I also saw this role as a unique opportunity to build a new model when it comes to sustainable destination management.

Being a dual citizen of Germany and Canada, what excites you about Asia that you moved to the other side of the world, to currently live and work in Bangkok?

When I was around 16 years old, growing up in Germany, my father told me that the future will be in Asia and that Chinese will become the most important language. This was in the mid to late '80s, during a time when China was still closed, and the world viewed China very differently than today. My father's words always stuck in my head.

When I was invited to a conference in Singapore in the early 2000s, right afterward I booked myself a ticket to Shanghai, not knowing anybody or anything. After that, I had occasional trips to Asia, especially China, while working for Fairmont Hotels & Resorts (we developed a mini-Chinese website back in 2004), and for the Canadian Tourism Commission.

Then in 2008 I moved to Beijing, originally as President of a small boutique hotel company, but stayed for over 5 years, co-founding and developing Dragon Trail and China Travel Trends.

As Chair of PATA China, we launched the China Sustainable Travel Forum & Awards in 2010. It was a fascinating time to be in China during the 2008 Olympic Games in Beijing and the 2010 World Expo in Shanghai; a time that truly was influenced by growth, change, and confidence.

In 2014, I decided it was time to leave Beijing for Bangkok, as I believed that Southeast Asia was to be the next frontier for economic growth. In 2014, the six-member countries of the Greater Mekong Subregion (GMS) appointed me to head the Mekong Tourism Coordinating Office (MTCO). Now China is the number one tourist source market for all of the GMS countries.

As the Executive Director of MTCO, your objective is to promote the Mekong region as a single tourism destination and to foster responsible tourism development. What challenges do you face with respect to the coordination and implementation of strategies while dealing with the various tourism ministries?

The member countries of the Greater Mekong Subregion have been incredibly committed to driving regional tourism collaboration. This year, we will celebrate our 23rd Mekong Tourism Forum in the UNESCO World Heritage site of Bagan in Myanmar.

Obviously, a lot has changed in the region, as well as in global tourism, in the past 20 years. While some countries were closed to tourism, others were in early development phases. Today all countries focus on tourism as a key economic pillar.

The rising Asian middle class has shifted key source markets from the west to the east, resulting in more regional tourists, demanding different products and communication channels. But despite of all that, the key challenge that we have faced, and are trying to overcome, is to drive engagement.

In order to overcome this obstacle, the Mekong Tourism Coordinating Office has created various innovative initiatives to encourage public and private sector engagement, such as:

- The social commerce marketing and capacity building platform—'Mekong Moments'
- The regional social media campaign Mekong Mini Movie Festival
- The responsible travel guide Experience Mekong Collection
- The Mekong Trends delivering tourism insights
- The Mekong Innovative Startups in Tourism (MIST) program, mentoring startups in the travel tech and social impact categories
- The institutional website MekongTourism.org, housing an e-library
- The Mekong Tourism Contributor Program
- The annual industry conference Mekong Tourism Forum

Regional tourism collaborations between multiple countries can be a real challenge. How do you create value for the travel and tourism industry in the Mekong region?

The strategy of the Mekong Tourism Coordinating Office (MTCO) over the past five years has been built entirely around contribution, collaboration, and stakeholder engagement. If engagement can be achieved, the collaborative nature of marketing the destination is immensely powerful—so long as it's facilitated in a way that each stakeholder in the visitor economy contributes, benefiting their own business and ultimately collaboratively driving value to the overall exposure of the destination.

This collaborative model could, in theory, enable any low-budget destination to effectively compete with any cash-rich destination, building virtual consortia of travel businesses in the destination.

We have developed a collaborative marketing and capacity building ecosystem to drive sustainable tourism and inclusive growth in the region. This strategy has also transformed the traditional destination management model to an inclusive crowd-marketing model. The result: a sustainable brand powered by authentic experiences which are created and delivered by small, responsible travel businesses. Those are shared by real travellers, inspiring the world to visit the region as conscious and purposeful travellers.

The Experience Mekong Collection brings together small, medium and social enterprises committed to responsible tourism. What are the criteria a business has to fulfill to be a part of this group?

Consumer trends have fuelled the desire of travellers to engage in authentically local and sustainable experiences. This is positive for inclusive tourism, and the economic value that tourism brings spreads out to secondary destinations and into communities where micro, small, and medium businesses have the opportunity to benefit.

These small responsible travel businesses are curated to make up the Experience Mekong Collection, serving as a trusted consumer label for social enterprises in the travel sector. The collection is aligned to the 17 United Nations Sustainable Development Goals (SDGs) and functions as a capacity-building platform for other businesses to learn and share best practices.

The Experience Mekong Collection now comprises close to 350 small, responsible travel businesses. Those offer sustainable travel experiences in the categories of stay, taste, do, shop, cruise, and tour.

Anybody is able to nominate businesses to be added to the Experience Mekong Collection and the Mekong Tourism Advisory Group (MeTAG) makes sure these businesses, in fact, do create a social impact on their communities and provide an authentically local experience.

Place 'consumers', including travellers and residents, share their experiences with these businesses on social media, building a content cloud of responsible travel experiences in the Mekong Region.

This sustainable model has these 'promoters' organically building a stakeholder-driven destination brand around authentic and sustainable experiences, preserving the heritage and culture of the destination, supported by the local community.

Have you seen any results that these programs have been successful?

It is always challenging to truly measure the impact of such initiatives. However, we do have some encouraging indicators which suggest that we are on the right track. Regarding the Experience Mekong Collection, we were able to measure that an increasing number of travellers are using it to plan their itineraries when traveling in the region. This motivates us to now build a trusted brand and to further promote these businesses.

We strongly believe that the more successful small, responsible travel businesses and social enterprises thrive in the destination, the more inclusive and sustainable travel and tourism in the region will become.

When it comes to our Mekong Innovative Startups in Tourism (MIST) program, quite a few of our startups have received investments and found partners through the program.

In addition to the Travel Technology category, we have now also launched a dedicated category for Travel Social Enterprises. Helping these innovative and passionate entrepreneurs to get exposure and mentorship is especially important for ventures in countries in the Mekong Region that normally don't get as much attention.

Data collected and analysed by research firm Forward Keys shows that, between May 2016 to 2017 and May 2018 to 2019, multi-country trips in the Mekong Region from source markets USA, Germany, UK, and Singapore grew by 14%. For source markets located in the rest of the world, growth was slower, at 10%.

The same highlighted countries also make up for the highest visitor numbers to MekongMoments.com, suggesting that this campaign has a clear return on investment in attracting 4% more travelers, compared to organic growth.

In addition, we have seen a strong brand exposure through the Mekong Mini Movie Festival Campaign,[1] which reached over 22 million people last year and where we can see a large number of return participants from all over the world.

You believe in the power of visual storytelling to inspire people to visit new destinations and promote responsible travel.

I believe that visual storytelling and film can play an important role in promoting destinations and responsible travel. Travelers now have access to low-cost local SIM cards and Wi-Fi and can share their experiences 'in the moment'.

Visual content from photos and videos shared on social media about travel experiences is a powerful way to inspire and influence potential future travellers.

Our social commerce platform MekongMoments.com is powered by Enwoke, an award-winning innovative destination engagement system. The initial idea was to aggregate experiences in the form of visual content shared on social media and tagged with #MekongMoments and connect the content to the actual businesses, to increase exposure and sales. Once the businesses see the value created, they would encourage their guests to share their experiences by creating mini social media contests. The content cloud that is created is inspiring prospective travellers all over the world.

As the content shared has a focus on responsible travel businesses, including experiences from the Experience Mekong Collection, the perception of the Greater Mekong Subregion as an experiential and responsible travel destination is strengthened due to the influence of social media networks, such as Instagram, as travel decision-making channels.

Powered by the same technology and integrated into MekongMoments.com, we launched the successful Mekong Mini Movie Festival[2] campaign where people share their experiences on social media in video format of 60 seconds or less, and tagging their posts with #MekongMinis.

In the first year alone, we aggregated over 500 tagged videos, which generated over 1.2 million video views and reached 22.6 million people worldwide. The second-year was even more successful, with a significant increase in the quality of videos shared.

---

[1] https://youtu.be/jmgDd8hFkAY

[2] https://www.mekongmoments.com/minis

Via a partnership with WWF, the mascot of the Mekong Mini Movie Festival is the Mekong (or Irrawaddy) Dolphin, to raise awareness of endangered wildlife and the importance of sustainable tourism in the region.

The reason we called it 'festival' is the opportunity for the tourism ministries, as well as other travel companies that are active in the region, to screen the mini videos at travel trade shows, events, and conferences.

They even embrace the initiative by developing their own branded Mekong Mini Movie Festival campaign as part of the bigger regional initiative. That way, we not only promote the Mekong Region via powerful, user-generated experiences but also promote sustainable tourism via our mascot and in partnership with WWF.

Finally, we created our first Mekong Tourism destination film by curating the content from many of the shared #MekongMinis videos. You can view the video here and below.

Our second Mekong Minis Award winners were announced during the 2nd Asia Destination Film Forum in Bangkok, bringing together tourism boards and filmmakers.

How do local communities benefit from your work?

While all our programs are highly integrated, the Experience Mekong Collection is probably the most obvious initiative indirectly benefiting local populations.

Due to the fact that many of these small responsible businesses are in secondary destinations, and creating a social impact, however having limited marketing resources, the Experience Mekong Collection can not only disperse tourists to secondary destinations, but can also do good as these businesses give back to the local communities.

Another innovative practice for which we were recognized by the UNWTO, in the UNWTO Report on Inclusive Tourism, is how we engage the local tourism businesses when organizing our annual conference, the Mekong Tourism Forum, which rotates around secondary destinations among our six member countries.

While normally tourism conferences are hosted in a big hotel or conference center, we focus to make the Mekong Tourism Forum as inclusive as possible, to engage stakeholders in the destination.

At the 2017 Mekong Tourism Forum in Luang Prabang, Lao PDR, we made the destination the venue and hosted 16 sessions at 16 relevant small businesses. For example, a session on river tourism was hosted on a river cruise, a session on food tourism at a local restaurant, on nature-based tourism in the botanical garden, and on community-based tourism on a rice field which was part of a local village.

Lunches were catered by various small local restaurants all over town, spreading the economic benefit of the event to almost 50 small businesses, taking delegates out of the normal conference into the destination to engage and connect with local residents.

This MICE concept could help smaller destinations to compete for events by developing a consortium of businesses to collaborate and host delegates.

Peter Richards, International Trade Centre responsible tourism specialist, summed up the challenges and results: "While complex from a logistics standpoint, the organizers' efforts paid off to create one of the most inclusive and experiential tourism conferences yet, by making the destination the venue. This is a new benchmark for inclusive conferences."

Which of your sustainability initiatives could be easily replicated in other destinations?

All our programs can easily be replicated and adapted to other destinations. We are always happy to share our programs and to help implement strategies.

We have been approached by quite a few destinations, from cities, countries, as well as regions. One good example is the Greater Tumen Region of the four-member countries of Mongolia, Russia, South Korea, and China.

We have started working with the Ministry of Environment and Tourism of Mongolia to develop their digital tourism platform, containing a Mongolia travel inspiration website, an integrated in-destination Mobile app (soon to be launched), and the #FeelMongolia campaign—all powered by the same Destination Engagement System technology Enwoke—as well as a curation of responsible travel experiences (similar to the Experience Mekong Collection).

A well-integrated tourism app is a very strong tool for destinations to involve businesses, local communities, travellers, and other stakeholders.

While developing the Mekong Travel app, we have been working with other destinations to create a tool that benefits all parties involved, and which also helps to steer visitor flow and information distribution, based on visitor preferences.

What challenges should DMOs in Asia and elsewhere foresee and prepare for over the next few years, as the number of tourists continues to rise?

Tourism has done a lot of good for the planet. And still does. We need more tourism in the future to help address the challenges we face. Travel is a right and is also important, not only for economic reasons but also for mutual understanding and peace.

But tourism also creates challenges. Tourism numbers will continue to rise, and destinations will continue to struggle to manage the growth and leverage tourism for economic benefits.

Growth should not be the objective. If it is, it leads to imbalance. We have seen too many instances where a few benefits and many more are stuck with the burden of increased visitor numbers.

The benefit of tourism is typically defined as economic, with the assumption that more tourists create more benefits. But this is not always true, as tourism has a hidden cost.

More important than growth is to find the right visitor, at the right time, in the right numbers. Only this way can we protect the heritage, the people and the environment that give a destination its unique vibe and character.

Our theme for this year's 23rd Mekong Tourism Forum in Bagan, Myanmar, hosted by the Ministry of Hotels and Tourism of Myanmar, is focused on 'Achieving Balanced Tourism'.

In order to achieve sustainability, there needs to be a balance between the various stakeholder groups as well as the environment, and heritage of the destination.

While we are witnessing over-crowding in many destinations, I don't believe in the underlining rhetoric of overtourism. Imbalance in many instances is a result of greed, which can then create over-crowding, unhappy residents, pollution, danger to lose intangible heritage and endangered species, and many other issues that appear as a snowball effect.

A trip should leave a positive impact on visitors and residents while contributing to a sustainable planet. Achieving balanced tourism requires the collaboration of all stakeholder groups, to ultimately heal the diseases deep inside of the destination or the place.

Just like a healthy person is a happy person, a healthy destination is a happy destination in the long-term if it generates economic benefits for all.

I hope that destinations in the future will shift their mindset from overtourism to balanced tourism. Short-term fixes and gains will not drive sustainability in the long-term.

Link to the interview: https://sustainability-leaders.com/jens-thraenhart-interview/

USA | Jeremy Sampson is the CEO of the Travel Foundation, a leading NGO in the travel and tourism sector. He has been instrumental in spearheading the Future of Tourism Coalition and was named its Chair in October 2020. With experience in tourism, conservation and sustainability arenas, Jeremy has lived on three continents and worked extensively in more than 30 countries. He spent 5 years as VP of Partnerships at Sustainable Travel International and 2 years as the President of international tour operator GreenSpot Travel. Jeremy served as Adjunct Professor at the George Washington University International Institute of Tourism Studies and currently serves on the Destinations Working Group of the Global Sustainable Tourism Council. Before joining The Travel Foundation, Jeremy worked at the IUCN Centre for Mediterranean Cooperation in Malaga and helped launch the MEET Network, a regional ecotourism network and destination management association for protected areas across the Med.

Areas of expertise: destination sustainability, responsible tourism, sustainable development, UNSDG 11 (Sustainable Human Settlements)

The following interview with Jeremy was first published in March 2020 on Sustainability-Leaders.com.

Jeremy, with close to a decade of experience in the field of education, what inspired you to switch to sustainable tourism and run a charity like the Travel Foundation (TF)?

When I joined Sustainable Travel International after some years working in the education sector, it was a natural transition for me. I have always been passionate about travel and curious about the world, and I've primarily sought out mission-driven social impact work throughout my career. Actually, my very first job out of college was in marketing at Ambassador Tours, one of San Francisco's oldest travel agencies, so it was great to come back to the industry after some time away.

When I started on my current path, I surely couldn't have explained to you exactly what 'sustainable tourism' was. But after a short time, I realised that tourism has the potential to leave powerfully positive impacts on people and places. I am an optimist by nature, but I also know first-hand that this vision is far from a reality in many places.

I feel very fortunate to call sustainable tourism my passion, my hobby, and my career—all rolled into one. It was a no-brainer for me to apply for the CEO job at the Travel Foundation, where I have the opportunity to work with an incredibly dynamic team and contribute in meaningful ways towards the evolution of our sustainable tourism movement.

TF has worked on some very interesting projects over the last years, like the 'Lionfish on the menu' in St. Lucia or 'Taste of Fethiye' in Turkey. How do you choose such topics, and how do you mobilize funds and personnel to train the locals in so many countries?

The Travel Foundation was originally established in 2003 by some of the big outbound operators in the UK, such as TUI, Thomas Cook, and Virgin Holidays. The idea was to mobilise funds and other resources from those companies to ensure tourism had a positive local impact in destinations popular with UK holidaymakers.

For many years, our projects at the Travel Foundation have been designed and implemented in collaboration with businesses and destinations, primarily focused on tackling key issues and opportunities that impact the tourism product and supply chain.

However, a few years back, we decided that to have the impact we wanted, we should engage with a much broader set of international stakeholders, and work to change policies and practices at a more systemic level.

Our recent work on the Invisible Burden[1] report, released last year in collaboration with Cornell University and Epler Wood International, enabled us to take a step forward in our thinking.

The report calls for a real shift in the way tourism is managed and explores the root issues that prevent destinations from delivering on this mandate—issues like a lack of information on the costs of tourism and accounting for the risks to shared assets, a need for entirely new skills within the public and private sector, and a need to innovate the financing of solutions.

---

[1] https://www.thetravelfoundation.org.uk/invisible-burden/

This requires detailed analysis to report on the true cost of tourism in destinations and design solutions. Our work now and going forward will be about tackling this systemic change head-on, so the industry can better deal with the growing number of big issues we're facing.

Of course, we still care about projects that address conservation and improve livelihoods, and these are a critical part of our DNA. But we are increasingly working to ensure that destinations and businesses around the world (not only in the UK!) use better information and knowledge to prioritise the biggest risks and opportunities and have the tools they need to collaborate around a shared agenda in order to innovate new solutions.

And we plan to put all this great learning back into the sector through a portfolio of Knowledge Products, designed to inspire and inform change on a global scale.

In your recent interview with Ecoclub, you expressed that the measurement of tourism costs and benefits remains inadequate at all levels. What KPIs does the TF follow to measure the impacts that your projects bring?

We have a rigorous approach to monitoring and evaluation, which we are very proud of. The metrics vary from project to project, but they will be aligned to our organisation's strategic aim: to build the capacity of the sector to develop and manage tourism in a way that maximises its value for people who live and work in destinations.

We are nearly done with our new 3-year strategy, which has encouraged us to look closely at our value proposition and clearly identify the value we bring to our industry partners.

We are going to be measuring the public and private sector behaviour changes that will help to bring about a new paradigm in destination management and improve livelihoods for people in destinations.

We obviously need to move beyond the traditional metrics used, to better understand tourism costs and benefits.

We'd like to see our partners demonstrate a more holistic understanding of the impacts of tourism—the assets that need to be managed and protected, and their associated costs. And then we'd like to see our partners change the way they operate, take responsibility for the impacts they control, and invest in actions that contribute meaningfully to both the problems and the opportunities for local people.

A recent article in The Telegraph refers to 2020 as the year of the tourist tax, detailing how countries reeling under overtourism can and should impose taxes. Do you think such taxes are justified? Are they the only means to generate funds to maintain a destination's infrastructure and to manage it well?

Taxation is certainly an important part of the equation, but I don't think it should be viewed as the only financing mechanism to address challenges. One of the problems with taxation is that the allocation of revenues often lacks transparency and is frequently misappropriated, either for political reasons or due to lack of information.

Beyond taxation, we should be looking at other, more innovative funding solutions that the tourism sector has not traditionally leveraged, like conservation financing, impact investment, and green bonds.

Destination trust funds is another salient idea addressed in The Invisible Burden report.

You know, I often think money really needn't be the problem. There's no shortage of money flowing this way and that in tourism. Perhaps with better accounting for and understanding of risks, we can start making sure there's enough funding to innovate and problem-solve those big-ticket challenges.

A recent Skift report suggests that 53% of global travellers are willing to pay more for products that demonstrate environmental responsibility. When there seems to exist a clear market demand, why are tourism business so slow in catching up and seizing the competitive advantage?

I think, on the whole, businesses really are starting to wake up to the new reality, though ever so slowly. Big news items like overtourism and plastics have, ironically, helped in this regard because these stories in the media impact consumer perceptions. There are actually many good business practices in tourism, but these are often limited to the assets which businesses are directly responsible for.

The biggest concern is the lack of shared responsibility for the assets that the destinations share with their tourism providers, often for free, like beaches, green spaces, and infrastructure. There is a lack of accountability and funding for maintaining and protecting those assets.

Meanwhile, we talk to companies all the time who are exploring how to evolve products towards emerging market trends. And tourism supply will necessarily need to improve, in order to deliver on new experiences. But right now, there remains limited high-quality local supply to service that demand, so there's still a lot of work to do to embed skills locally and train up the sector in how to develop, manage, and promote a new tourism economy. This remains a big area of focus for our organisation.

Sustainability being at the core of TF, which initiatives are you particularly passionate about right now?

We've just announced the next steps in our partnership with EplerWood International and Cornell University (our partners on the Invisible Burden report), this time working closely with PATA, the Pacific Asia Travel Association too. We're focusing on addressing the capacity gaps highlighted in the Invisible Burden report. In other words, the critical skills that destinations will need to manage tourism going forward.

We'll be creating practical training tools and resources, covering things like holistic accounting methods that measure the invisible burden of tourism; data management skills, better reporting systems, and innovative financing mechanisms that enable tourism destinations to cover the costs of new solutions

You can't just shift from destination marketing to being a destination 'management' organisation with the wave of a wand. This requires an entirely new set of skills.

Actually, there are a lot of exciting ideas and opportunities in the pipeline right now, and I feel it is an incredible time to be working in the sustainable tourism field.

Your key insights from your time at the IUCN?

I spent most of my time helping to manage fantastic large-scale transnational EU projects, like MEET [Mediterranean Experience of Eco Tourism] and DestiMED.[2] During that time, I had the opportunity to see first-hand the power of cooperation at scale and the success that is possible when the public and private sectors come together around shared objectives and needs.

Living in Malaga, Spain, I also saw and experienced the pressures of a rapidly increasing tourism sector and the very real impacts this is having on residents and the community at large.

DestiMED (and its spin-off project DestiMED PLUS, which is now underway) are great examples of how concrete efforts to change the tourism model in the Mediterranean can work, by focusing on the core areas of shared governance, practical use of data, and innovative product development.

Overtourism and the effects of climate change are now frequently discussed. Apart from those, which issues in tourism do you think are the most important on the sustainable tourism agenda right now?

It's a mistake to look at the things in the news, like overtourism and plastic pollution, without acknowledging that these issues are symptoms of bigger problems—as described in The Invisible Burden report. Unless we fix the big problems in tourism fundamentally—the way it's developed, managed and delivered—then we're simply putting band-aids on the wound.

There is a trend towards destination management, rather than traditional destination marketing, but there are a lot of major gaps to plug, and our work at the Travel Foundation is now primarily focused on accelerating this transition into a new tourism economy.

Link to the interview: https://sustainability-leaders.com/jeremy-sampson-interview/

---

[2] https://destimed.interreg-med.eu

Joanna Van Gruisen on Regenerative Tourism in India

India | Joanna Van Gruisen was born in the UK but has lived in the Indian subcontinent for over 40 years. She is a wildlife photographer, writer and conservationist. She came from a community work background in the UK, but after coming to Asia worked in the wildlife field, initially making wildlife documentary films and later doing still photography and writing. Since 2010 she has been involved full time with tourism as she and her biologist partner opened and manage a small mud-cottage 'Sarai' in central India—the Sarai at Toria. They also run a small trust, 'Baavan—bagh aap aur van', working to extend biodiversity conservation beyond the government's system of protected areas in an inclusive, not exclusive, manner. They seek to use tourism as the economic driver to create a conservation-based economy that brings development and agency to the local community.

Areas of expertise: ecotourism, hospitality, responsible tourism, tourism business, UNSDG 12 (Resource Efficiency), UNSDG 15 (Forests & Biodiversity)

The following interview with Joanna was first published in November 2020 on Sustainability-Leaders.com.

© The Author(s), under exclusive license to Springer Nature Switzerland AG 2022
F. Kaefer, *Sustainability Leadership in Tourism*, Future of Business and Finance,
https://doi.org/10.1007/978-3-031-05314-6_57

Joanna, with Sarai at Toria you have, together with your husband, created a unique eco-lodge in India which has received impressive reviews on Tripadvisor. What motivated you and Raghu to start an ecotourism business in this area?

That has a bit of history and is rather a long story! I will try to be brief. Neither Raghu nor I come from a tourism background, nor indeed had we any experience in running a business: I came to India initially (from the UK, via Nepal and Sri Lanka) to shoot wildlife documentaries (later morphing into still photographer and writer on wildlife and conservation) and Raghu is a scientist, a conservation biologist.

Our connection with this area arose through his 10-year study on the ecology of tigers in the Panna Tiger Reserve. We had had to leave the area in 2005 when we spoke out about the poaching that, unattended, led to the extinction of tigers there a few years later. We had always intended that the research project would continue long-term and have a conservation component based on the scientific information of the study.

Once some of the dust had settled, although there was no possibility for Raghu to undertake research within a Protected Area, we wanted to extend conservation in the region to areas beyond the park boundaries. We believed this could be best done while bringing appropriate development to some of the remote communities of the region. We felt that being dependent on grants for such work would create an unstable existence, so we decided to open a small tourist lodge in the hope that this would give us an economic base from which to work.

We also felt that a positive synergy could operate between the tourist concern and the conservation and development activities we hoped to pursue. The iconic 1000-year-old temples of Khajuraho, recognised as a UNESCO World Heritage site, are close by and as there was a lack of culturally appropriate accommodation in the area for visitors, we felt there was a niche for us.

How do you approach sustainability at the Sarai at Toria?

We both come from wildlife conservation backgrounds and are aware of climate change and the adverse impact of human activity on the planet. Our bottom line and foundation for the Sarai were to keep our carbon footprint as low as possible; we wanted to create accommodation that was as earth-friendly as possible, in build and running. We, therefore, looked to the vernacular architecture of the region and quickly settled on mud as our main building material, since this was clearly the best medium for handling the extremes of hot and cold that this area enjoys—from 45–50 °C in summer to 3–4 °C in winter—without us having to resort to energy-hungry cooling and heating appliances.

The aim was to provide a healthy built-up environment in which our guests could discover that comfort and elegance ('luxury') could be achieved without it costing the earth (environmentally). We also had in mind that doing so could raise the status for mud in the local community and so hopefully lessen the effects of the disdain which is generally meted out to those still living in traditional dwellings.

I guess our approach to sustainability was from a personal level, an extension of the way we lived. We aimed to keep things simple and avoid excess and unnecessary waste. So, for example, since we never used bottled water ourselves, we saw no reason why our guests would need it! More seasoned tourism professionals threw up their hands in horror telling us this would never work for a hotel; but we put our water filter where it would be visible and in 10 years, we have never provided a single plastic bottle of water and only ever had one party who complained.

In such a rural area, one has to be self-sufficient in many spheres that others may take for granted—since where we are there is no electricity, no running water, no sewerage lines, no rubbish collection. Perfect situation to encourage sustainability! Besides we are in such beautiful countryside and on the banks of one of the cleanest rivers of India, so I think one would need to be very callous or totally blind to its merit if one did not take care to preserve and not destroy this.

We began 100% solar with a 10 kW power plant—this made us very conscious of the wattage of lights and appliances, searching out the most efficient and low energy ones. We naturally avoided single-use plastics and as far as possible use natural items in our furniture, furnishings, consumables and cleaning items. Knowing about the benefits of effective microorganisms means we do not have to worry about adding toxic chemicals to the land or waterways.

I guess overall our approach was to find a way others could share our enjoyment and appreciation of all the natural, cultural and social wonders of the area without this spoiling any of them.

Social sustainability being one of the keys to achieving sustainable tourism: how do you involve the local community at the Sarai at Toria?

The local community are the Sarai really, it is their warmth and natural hospitality that makes the experience what it is. All our staff hail from the villages around the Sarai at Toria—all that is except one cook who had already been working with Raghu for over 15 years when we began the hotel. Several came as labourers to build the cottages and stayed on, training with us in housekeeping, F&B, etc. Our driver, boatman, guides and naturalist who accompany our guests on safaris, walks, excursions, etc. are all from the neighbouring villages. Our guests get authentic and expert local information and this provides a social as well as physical sense to the visitors' experience.

Of course, we showcase and encourage our guests to appreciate pottery and other local crafts and this helps provide revenue to the artists. As far as possible, we buy locally and use local amenities so the tourism income and benefit is spread as widely as possible.

The current coronavirus pandemic has hit the travel industry hard this year. How does it affect your business?

Yes, indeed and for us personally we are yet to experience the worst of it. The Sarai at Toria runs seasonally, usually from 1st October to 15th April. This is due to the fact that after mid-April the weather is just too hot to enjoy all the outdoor activities, and in the monsoon, we are often cut off by a swollen river.

So, until now, October, we have in effect only lost one month's business and most of our guests and travel agents who had to cancel have been good enough to postpone and leave their credit with us. We had just completed 10 years of operation so loans were mainly paid and we had been able to build up a small corpus fund. So, all this combined has enabled us to retain all staff which is our main priority. The bite may happen now; I fear this season will see very low occupancy, especially as in earlier years around 75% of our guests were international travellers. We are yet to see how much movement there will be at the domestic level.

However, we also hope that life becomes more normalised so that we can again take up our NGO activities in the region. If this happens it may be possible to divert some of the hotel staff to work there and thereby share salary expenses. Several aspects of the way we operate are advantageous: growing some of our own fruit and vegetables has helped, especially during the initial strict lockdown; staff being able to go home easily (and return), sourcing consumables locally, not being dependant on distant items etc.

Which aspects of running the business sustainably do you find the most challenging? And how do you overcome those challenges?

One of our biggest challenges has been in trying to take the tourists and industry along with us. In the early years especially, it was only a small handful of tour operators who understood what we were doing. Others would just look at aspects like no air conditioners and no swimming pool and take their bookings elsewhere. This is still an issue but is improving as attitudes change and both the clients and the operators become more conscious. I think the buzz of sustainability in the industry throughout this lockdown period could prove very beneficial for a new travel ethos.

When we started, we had hoped that the Sarai might serve as a stepping stone for our employees to move up in the industry career ladder. But in fact, most of our staff have remained with us throughout. We had sponsored several local youths to go on hospitality training courses which guaranteed a job at the end of this. However, all have since returned to their home areas.

Sadly, there are still many prejudices across India and being an outsider in a different part of the country is not easy especially for someone from a rural village with little outside exposure. Their experience was discouraging to others.

As I mentioned earlier, we operate seasonally and are only open for six and a half months of the year. One could have extended this into the summer season by installing air conditioners and swimming pools and other indoor activities, but this was not the kind of tourism we were looking to encourage, since it is not really appropriate for these agricultural community areas. But financially this restricted season can be challenging.

Balancing our footprint with benefit in our local area is relatively simple as we can see the advantage tourism can bring for communities in remote rural areas and for the wildlife there. In India wildlife tourism has tended to get a bad press with accusations of resort owners making big profits and providing no benefit to the areas in which they are situated.

Through our trust, Baavan, we did a study with TOFTigers[1] to look at the finances of tourism around 4 tiger reserves in our State of Madhya Pradesh. Even we were surprised at how much is fed into the local community. We also found that 85% of visitors are budget travellers and average occupancy overall is only around 30%, resulting in lodges not being hugely profit-making enterprises. Even though

---

[1] https://www.toftigers.org

tourism has developed without plan or intention, it still contributes quite considerably to social development and to conservation in the immediate environment where it is situated.

However, the majority of Sarai's guests come from overseas, so we also felt the need to consider the carbon footprint of their long-distance travel. Being wildlifers, running a responsible tourism concern is necessary but does not seem sufficient. We like to look beyond and are hoping to use the Sarai as a stepping stone to creating an opportunity for a true Conservation Tourism enterprise.

Through our trust, we have set out on a path and are working in some villages beyond the park and buffer area. Our objective is to use tourism as a tool for creating a new area of biodiversity conservation, one where the local community are the main stakeholders and beneficiaries.

At present, India has only a handful of parks where there is a good chance to see a tiger. Entry is limited and demand exceeds the supply. Also, the focus is mainly on the charismatic carnivore species and in most places, only one mode of entry—by jeep—is permitted. Developing areas outside the park system would allow a more varied and enriched visitor experience too.

With 10 years of experience with the Sarai, how has your understanding of sustainability changed since you first got involved?

When we first started, our idea was to contribute as little as possible to the warming of, and damage to, the planet while running our operation. But a few years ago, we realised that this is not enough, we needed to become more proactive and actually be part of an attempt to reverse the damage. This was when we first came across, and began to use, the term regenerative.

Our aim is always to try to spread our views to our guests and also to the tourism industry as far as we react with it. This is where our small trust Baavan comes into play and we are taking small steps to spread conservation beyond India's Government Protected Areas and aiming to complement their exclusive model with an inclusive one that will benefit the local human population along with increasing the area's biodiversity.

We see tourism as a potential driving force for this. Broadening support for conservation tourism, we have also been able to successfully encourage urban-based local agents to materially support such endeavours.

This COVID period has given all of us time to reflect and discuss ways to move forward with greater purpose. Several of us responsible tourism lodge owners are now coming together to pool resources and create an independent enterprise that can take up key conservation projects. We hope to also attract all other wings of the tourism industry—the tour operators in India and abroad and the travellers. We want to bring everyone together on the premise that 'alone we are a drop but together we can make an ocean'!

Your three bits of advice for accommodation owners or managers on how to care for the local community and wildlife, and at the same time ensure their own economic sustainability?

Use common sense, science (that is, facts) and empathy. Rather a general comment but I think it is important to understand that each situation is unique. What may be advisable or beneficial in one area may be quite the reverse in another.

To me 'care for' has a bit of a parent-child connotation, so I would rather say we need to respect the local community and wildlife—it is not so much that we need 'to care for' the local community nor the wildlife, but we do need to treat both respectfully. Respect allows there to be an equitable and positive relationship between all parties. Similarly, it is not an individual's job to 'care for' the planet but it is our duty to respect all in it and to always interact with thoughtfulness and consideration. Indeed, the planet cares for us way more than we do it!

Balance is key, as in so many things. Too many tourists and the destination and community can be spoilt. The landscape, local community and visitor need to co-exist in equilibrium so the first is sustained and the other two may benefit from their interaction with each other.

If you had to pick one—which would you name as your most rewarding experience so far, linked to your work at The Sarai?

It is rewarding when guests understand and appreciate what we are doing; this is especially so when those I might refer to as the 'marble bathroom, aircon, 5-star' tourists arrive with some trepidation but leave happy and with some appreciation that such items are not essential to a luxurious stay.

Also seeing the joy on the village school children's faces and hearing their enthusiasm for the natural world during the nature camps we organise locally, gives reward and hope for the future.

Anything else you'd like to mention?

In India we are part of a dynamic marketing cooperation of like-minded hotels— RARE India, started by Shoba Mohan. This has been invaluable as a mechanism and forum for sharing ideas and receiving inspiration and education.

The Sustainability Leaders Project would appear to offer the same on a much broader scale. It is thrilling to read of so much happening all over the world, and to feel part of a bigger movement is hugely stimulating and cheering. I'm really happy and honoured to be part of such an energetic, positive and forward-thinking community.

Link to the interview: https://sustainability-leaders.com/joanna-van-gruisen-interview/

USA | Journalist, editor, speaker and consultant, Jonathan Tourtellot runs the non-profit Destination Stewardship Center (DSC), successor to National Geographic's former Center for Sustainable Destinations, a program he founded and directed. He is active in the Future of Tourism Coalition, where the DSC is one of six founding partners. During his three decades as a senior editor with National Geographic, he wrote about travel, geography, the environment and science. In 2020 he founded the online quarterly Destination Stewardship Report in cooperation with the Global Sustainable Tourism Council, with articles contributed from around the world on how to handle destination management more effectively and sustainably. Recently, he worked on the creation of destination stewardship councils, citizen inclusion in tourism planning, sense of place as a key element of sustainability, and overtourism—notably as a contributing author in the book Overtourism: Lessons for a Better Future (Island Press, 2021).

Areas of expertise: destination sustainability, ecotourism, geotourism, overtourism, North America, UNSDG 11 (Sustainable Human Settlements)

F. Kaefer, *Sustainability Leadership in Tourism*, Future of Business and Finance,
https://doi.org/10.1007/978-3-031-05314-6_58

The following interview with Jonathan was first published in May 2017 on Sustainability-Leaders.com.

Jonathan, you have been involved in sustainable tourism and destination stewardship for many years. Do you remember what brought you to the topic in the first place? And your view/thoughts on destination sustainability back then?

In the 1990s I happened to return to two places for which I had fond memories from visits a decade earlier: the coast of Crete and northern Iceland. I was stunned. Resorts and villas had sprung up all along Crete's Aegean coast, catering to international resort tourism and blunting the very Creteness of the place. Around Lake Myvatn in Iceland the changes were more subtle: A hot spring closed here pathways roped off there. Growth in tourism was taming the wild appeal of the area. The two experiences constituted an epiphany for me. Tourism was transforming the world—even in once-remote northern Iceland! But it seemed the world was not paying much attention.

I was a writer/editor at National Geographic then, where I had always been most focused on both travel and Earth-system sciences—ecology, climate, human geography. That led me to learn about ecotourism, which in turn led me to tourism impacts generally.

Once I wandered into the valley of tourism policy, I never came out again; tourism interacts with almost everything! Never boring. I've been trying to call more attention to it ever since.

Now in 2017, how has your view on tourism and sustainability changed?

I love travel. I still believe that tourism can do a lot of good, not only from revenues that incentivize conservation and fight poverty, but also in terms of public education and public diplomacy.

A world with tourism tends to be more peaceful than one without. You're less likely to drop bombs on a place where you've had a nice holiday.

Originally, I thought the push for tourism sustainability would have to be consumer-led, driven by eco-friendly traveler demand. Instead, it was led more by businesses and NGOs, supported by modest consumer interest. The industry has begun to clean itself up, and that's great. Unfortunately, though, the industry has not done much to impose limits on itself.

The volume and style of tourism growth concerns me. The sustainability movement has not really tackled these issues head on.

From the beginning, 'sustainability' has been seen as primarily environmental. Without question, Earth's environmental health is critical, but that concern has tended to preempt the other aspects of sustaining the quality and character of a destination—culture, social health, aesthetics. Ironically, even conservation of natural habitats can get shoved aside. I know of at least one resort that proudly touts its LEED-certified buildings while leaving not one centimeter of its shoreline in a natural state.

Now, with the power of the sharing economy on tap, I'm coming back to a belief that the general public, or at least the proactive portion of it, has a role to play, by providing useful feedback to destinations.

Why did you establish the Destination Stewardship Center, and what is it about?

It actually began in 2001 as National Geographic's 'Sustainable Tourism Initiative' and evolved into the Center for Sustainable Destinations, a part of what was then called National Geographic Mission Programs.

In the course of that work, we initiated the annual Destination Stewardship surveys that ran as the cover story in National Geographic Traveler for 7 years. We initiated the Geotourism MapGuide program, which enlisted local participation so as to raise perceived value of destination distinctiveness.

With National Geographic Traveler and the Travel Industry Association of America (now renamed U.S.T.A.) we introduced in 2002 the concept of the 'geotourism approach', defined as tourism that sustains or enhances the geographical character of a place—its environment, culture, aesthetics, heritage, and the well-being of its residents.

Inspired by the principles of good ecotourism, my wife, Sally Bensusen, and I had coined that term in 1997 as an umbrella approach that would include not only natural habitats, but all the endemic characteristics of a destination.

Despite some confusion with the narrow geological use of the same word (especially in the context of geoparks), the 'National Geographic definition' has been adopted in various parts of the world, including an endorsement by the tourism ministers of the Organization of American States in 2013. Today we're working with geotourism advocates in Mexico, Albania, Alaska, and Indonesia, among many others.

As part of a multiyear restructuring, National Geographic terminated funding for the Center for Sustainable Destinations in 2010, so I reestablished it as the independent, nonprofit Destination Stewardship Center—CSD became DSC—still with the same mission: 'To help protect the world's distinctive places by supporting wisely managed tourism and enlightened destination stewardship.'

So far, that support is largely informational—resource directories, news aggregation, presentations and workshops for destinations and conferences, and a collaborative blog platform open to anyone with an appropriate contribution to make.

We are now launching a pilot video project to tell and disseminate stewardship success stories. The DSC website is itself a kind of demonstration project, hinting at what needs to happen on a much broader basis, and we're open to any future arrangement—informational services, partnership, sponsorship, membership, even merger—that will advance the mission.

Having worked with the National Geographic Traveler magazine from 1991 until 2015, which are your key insights and lessons from that period?

The world is in the middle of the greatest media revolution since Gutenberg. Information technology has disrupted most conventional media business models, and National Geographic's media have not been spared, resulting in their eventual merger with Fox late in 2015. Even before then, magazine and institutional economics required pleasing readers and advertisers first; sustainability themes had to fit in around the edges, or not at all.

In fact, it has been frustratingly difficult to interest conventional media in destination stewardship, partly because so many editors have preconceptions about where to put tourism topics. Travel articles are supposed to be upbeat, entertaining, inspirational, with a touch of consumer advice how to get a good deal or airplane seat. No negative impacts, please. The only other travel option is the business news category, which only covers tourism industry profits, losses, and such things as which hotel chain added the most rooms that year.

Stories on the overtourism phenomenon are among those finally pushing destination stewardship into the general-news category.

As for success and disappointment, my greatest were the same: I succeeded in mustering National Geographic financial support for a sustainable-destination program for 9 years, but I failed to get it institutionally locked in for keeps, as I had hoped. Not surprising, perhaps, given the sea change that was underway. At least the work we did then continues in other ways today.

Overtourism has become a critical sustainability issue for many popular destinations now, especially cities like Barcelona. In your view, what can destination managers do to address this?

I call it the numbers problem. It remains stunningly unrecognized by both governments and the tourism industry.

International tourist traffic has increased forty-fold since the days of the first commercial jetliners, but all those people must fit into spaces and places that remain the same size they were in 1958. And still tourism ministers everywhere keep calling for yet more arrivals. Large cruise ships exacerbate the problem by dumping thousands of low-benefit day trippers into historic ports and cities. World Heritage sites and other famous attractions are especially susceptible.

Bragging rights have always been an incentive for travel, but in many places it lately seems to have become the sole incentive, with busloads of selfie-stick-wielding tourists crowding in, taking their proof-of-place photos, and zooming out. They experience nothing, learn nothing, and provide minimal economic benefit while taking up lots of space.

Whether it's an entire city like Barcelona or a few square meters of standing room in front of the Mona Lisa, we cannot fit ever-increasing numbers of visitors into finite spaces without consequence. Destination managers must publicly confront arithmetic reality. We need policies that incentivize longer stays for serious travelers and disincentivize hit-and-run, high-traffic, selfie-stick tourism. Measure success not by arrival headcounts, but by benefits per tourist.

The U.N. have declared 2017 the International Year of Sustainable Tourism for Development. In your view, which are the main destination challenges right now regarding the sustainability of travel and tourism?

So far, the impact of the U.N. year has seemed underwhelming. Perhaps something usable will come out of it by year's end.

Measures of success are the biggest challenge. The relentless drive for ever more arrivals trumps any real progress in sustainable practices for destinations. Too many hotel-dominated DMOs measure success by heads in beds, not by benefit for the destination as a whole. (And they want those heads be in HOTEL beds, not those of

Airbnb hosts!) Token green gestures by the industry do not address the larger issues of caring for the destination. Linen re-use in itself does not achieve sustainability.

Aside from purely manufactured attractions like theme parks, the ultimate tourism product is the place itself. That reality remains strangely unrecognized by government and industry (but not by tourists!). There are encouraging exceptions, though. I was party to one hotel chain's 'responsible business day' competition. One of the best entries showed hotel staff helping to restore local shrines and parks. The slogan: 'It's our product. Of course, we want to protect it!' That attitude needs to become pervasive.

From 2004–2010 you ran the annual Destination Stewardship surveys in National Geographic Traveler magazine. Judging by your findings from those surveys, and your current research at the Destination Stewardship Center, how important is a destination's (actual and perceived) sustainability performance for its reputation and competitiveness?

Not important enough, not yet. The real goal of those surveys was to stimulate public discussion, and in some cases it worked. Whenever local advocates cited some destination's National Geographic score as a reason to adopt a more sustainable policy, that was a victory, just because the interaction between tourism and stewardship had never been discussed before. That alone did not ensure any changes, of course.

Until a sustainability stewardship rating is as well-recognized as a Michelin star, developing destinations will have trouble incentivizing long-term sustainable policies in the face of short-term economic needs.

We have had numerous inquiries about resuming those surveys, which involved two rounds of soliciting experts' opinions from all over the world. The surveys were trailblazing at the time—and expensive to conduct. Based on lessons learned, I personally would like to see them restart with a combined methodology—part expert opinion, part hard data from such sources as Albert Salman's destinations database, and part proactive public opinion. That would require obtaining public opinion in a form more structured than the random remarks you can see on crowd-sourced websites.

Ratings that always average out to 4.3 stars won't help. I'm intrigued with the rise of Airbnb and its kin, whose memberships include both hosts and guests—residents and tourists. Together, these people have the potential to send powerful messages to destination leadership.

Imagine you could turn back time and start all over again. Knowing what you know now about the media business, travel and sustainability, what would you do differently?

In one way, not much. For several years, the then editor-in-chief of National Geographic Traveler, Keith Bellows, gave me space to write about destination issues. We were able to operate the Center for Sustainable Destinations at National Geographic. These occurred at about the only point in that institution's history when the conditions were right, and the right people were in the right place and cautiously willing to give it a try. The National Geographic brand affiliation and our alignment of missions was impossible to beat.

But I knew continued institutional support was uncertain, so I concentrated on trying to cement a permanent role for the Center inside the National Geographic Society. In retrospect, I would have instead given highest priority to our proposed, self-sustaining global geotourism network, with multiple sources of support.

Today we still need that global knowledge network to inform and promote better destination stewardship. The Sustainability Leaders Project is a good step in that direction.

Your thoughts on the current state of tourism sustainability in North America, USA especially?

Even pre-Trump, the U.S. has been unusual in that 'sustainable' is regarded with active hostility in conservative sections of the country, as is 'U.N.' and 'environmentalism'. The concept does have traction in the Pacific and northeastern coastal states, but nationally, tourism sustainability isn't even on the radar. Canada by contrast—despite dependence on extractive industries—still seems to do a better job in destination stewardship than the U.S., as demonstrated in our National Geographic survey results.

Trump's America First and anti-environmental stances have now added a giant and very unpleasant distraction, complete with a disagreeable impact on international tourism into the U.S. People who would be working on sustainability are instead concentrating their efforts on fighting Trump's policies. For now, it's three steps backward, and none forward.

Your 3 bits of advice to destination managers eager to embrace sustainability and to turn it into a reputational and competitive advantage?

- Discover, protect, and responsibly develop your unique cultural and natural assets. Authenticity and distinctiveness cannot be undersold by cheaper competitors.
- To do that, form a destination stewardship council, with public, private, and civil-society membership, tasked with protecting nature and culture, enhancing quality, and marketing responsibly to appropriate tourism segments.
- As noted above, established destinations should incentivize high-quality tourism and disincentivize high-quantity tourism. Count success not by increased arrivals, but by benefits per tourist.

Link to the interview: https://sustainability-leaders.com/interview-jonathan-tourtellot/

USA | Dr. Jonathon Day, an Associate Professor at Purdue School of Hospitality and Tourism Management, has over 25 years of experience in tourism management. Jonathon is the co-author of The Tourism System eighth edition, author of the Introduction to Sustainable Tourism and Responsible Travel and over 30 peer-reviewed articles in journals including Tourism Analysis, Journal of Travel and Tourism Marketing, Annals of Tourism Research, International Journal of Contemporary Hospitality Management. He currently chairs the Travel Care Code (travelcarecode.org), a network of academic and marketing organizations promoting responsible travel. Dr. Day's research and engagement activities include projects in Nepal, Colombia, Uzbekistan, the United States and Australia. Jonathon's work is focused on sustainable tourism and responsible travel, and how tourism can be used to not only enrich travellers but support destination communities.

Areas of expertise: destination sustainability, tourism research, UNSDG 11 (Sustainable Human Settlements), UNSDG 8 (Equal & Fair Economic Opportunities)

The following interview with Jonathon was first published in November 2017 on Sustainability-Leaders.com.

© The Author(s), under exclusive license to Springer Nature Switzerland AG 2022

F. Kaefer, *Sustainability Leadership in Tourism*, Future of Business and Finance, https://doi.org/10.1007/978-3-031-05314-6_59

Jonathon, do you remember what inspired you to focus your academic career on sustainable tourism? When was the first time you heard about 'sustainability'?

It has always been with me—although it took me a while to get to the full triple-bottom-line. Growing up in Australia I was blessed to be able to enjoy nature in many ways. From my home in Brisbane, I was able to take weekend trips to the rainforest or the beach—sometimes both the same weekend. I grew up understanding the special natural wonders of our country and the importance of protecting the environment.

In my work at Tourism Queensland, I came to appreciate the challenges of small tourism organizations marketing—particularly internationally. Helping business, especially smaller businesses to enter the North American market, was one of my favourite things about the work.

It was later that social justice issues came on my radar. Bringing the potential of tourism to positively impact these issues—social, environmental and economic—came as a natural progression for me.

As a world-renown researcher and teacher of tourism and sustainability, has your view on the topic changed since you first got involved?

I really found my calling when I joined the tourism industry. I consider myself very lucky to have spent time in tourism. That said—these days I have a more nuanced view of tourism than I had when I first joined the industry.

Like any tool, tourism can be used positively or negatively—I am now more aware of the failures in the system than I was.

Nevertheless—I think that we have a great body of knowledge about what needs to be done to achieve positive outcomes. And I remain optimistic that the system is improving. But. . . there is still a lot of work to be done.

Based on your experience, what are the main challenges for sustainability in destinations?

There are a couple of challenges that are on my mind these days. I think there are some big picture issues that we need to deal with to achieve the best possible outcomes.

Lack of systems view of tourism

First, I think that it is important that we take a 'systems view' of tourism. We often talk about destinations like they are corporations—where there is someone in charge, and everyone must follow that leader. Destinations don't work like that. They are not hierarchical—they are networks. The best destinations recognize this and have teams of motivated people working toward shared goals. Shared leadership, agile strategy and collaboration: those are the ideas that really resonate when getting things done in groups and systems.

Not enough focus on destinations communities

The second challenge is a system issue too. We need to recognize that the tourism system involves many more actors than just tourism companies. We need to stop taking tourism for granted. Bigger isn't always better. We need to truly behave in a way that proves we recognize that tourism takes place with the consent of the destination community.

Tourism officials sometimes say they are working to improve the quality of life of their residents, but it is often from a perspective of "I know what's good for you." Few destinations track resident sentiment.

This summer, the story of overtourism has surfaced and the discussion seems to have taken many by surprise! That surprise shouldn't happen—we need to be tracking our stakeholder's issues and responding to them. We—tourism—need to be engaged with our communities.

Traveller behaviour

I am also very interested in how we can effectively improve traveller behaviour. I am excited that more organizations are designed for sustainability. It is one thing to say people 'should' do this or 'should do that' (and I believe that they should!) But just saying they should change, without enabling the change, won't lead to the best outcomes.

People are complex. Even when they really want to do the right thing, they don't. 'Nudge' policies, that make it easy to do the right thing can be very useful in achieving better outcomes.

Where do you see the priorities for destinations now in terms of sustainability? And which topics not yet broadly discussed should destinations prepare for?

Destinations need to realise that sustainability is mission-critical. DMOs, particularly here in the United States are going through a significant change.

As the marketplace has changed, DMOs have recognized they need to take an active role in the delivery of the destination experience. It is not sufficient to produce a great promotional campaign if the product doesn't live up to the messaging.

It's not surprising really—your perception of the Apple brand is based on the user experience more than any ads you've seen. Why should it be different for destinations?

As a result, DMOs are adopting product and destination development strategies. Although they have started to adopt product development—there has been little said about incorporating sustainable tourism principles into their development plans.

There are some great exceptions. Both Oregon and Lake Placid work to align destination development with community needs and aspirations. As I mentioned— big isn't always better.

There is a great opportunity for DMOs to adopt sustainability as they become more involved in place building.

DMOs also need to recognize their role in the destination system. They can be important catalysts and facilitators. The best DMOs—and the best leaders—help their teams become better. It requires a specific set of skills to be a great collaborator and we need to build destination members capacity to do it. DMOs can help lead this process—and make the whole team stronger.

In your view, are international events and organizations like UNWTO, GSTC and the 2017 UN Year of Sustainable Tourism helping to overcome the challenges of sustainability in tourism?

Absolutely. There is a marketing saying that "you'll get sick of your campaign long before your market does". If there is ever a moment that we feel we have 'told our story', we need to remember that saying. There are still a lot of people that need to get the message about adopting sustainable tourism principles. Each day these organizations are promoting the benefits of sustainable tourism—it is very important work.

The United Nations General Assembly declared 2017 as the International Year of Sustainable Tourism for Development. It is an important platform and a great way to add a new dimension to the sustainable tourism conversation. Shining a spotlight on sustainable tourism for development stimulates new thinking and prompts action that will endure long after 2017 is in the history books.

Where do you think the future of sustainable tourism research is leading? Are there any topics you personally think that academic research in sustainable tourism should be more engaged with?

Sustainable tourism research covers a broad range of topics. There is some great work being done on the marketing of sustainability by Xavier Font and his colleagues.

I think that some of the big picture work that is being done in tourism, the transition from the oil economy, and climate change by Susanne Becken and the team at Griffith University is very important.

I am happy to see that sustainable tourism is moving from being a type of tourism (like ecotourism or agritourism) and being described as a set of principles.

I think the sustainable tourism movement has much in common with the quality movement of the 1980s. Sustainability needs to be completely embedded in tourism—not a small part of it.

There is as much research that can be done to support effective sustainability in cities as there is in ecotourism.

I am becoming increasingly interested in the 'how' as opposed to the 'what'. We are getting a pretty good handle on what needs to be done—I think important insights will come from looking at how we get it done. Whether that is looking at individual behaviours or group dynamics—understanding how we can effectively encourage behaviours that lead to sustainable outcomes is important work.

Having worked for leading Tourism Boards, such as Tourism Queensland, can you explain in a nutshell how destination branding can support the sustainable development of a destination? Are there any aspects that you consider are crucial for its success?

Destination branding and sustainability are both strategic activities that build and maintain important community assets. They are mutually supportive activities.

Simply put, destination branding is building the long-term reputation of the destination. That reputation, that destination brand, is built in the user experience of the destination. If the destination is not paying attention to the triple bottom line—

making sure the environment is protected, the culture is vibrant and people and businesses thriving—it will not maintain a good reputation or a strong brand.

At the same time, the best brands reinforce and celebrate the actions of people and companies undertaking sustainable tourism. These brands celebrate the environment or culture and heritage. In doing so, they generate strong justification to maintain and support them.

Throughout your distinguished career, have there been any specific organizations or persons that have served you as role models or source of inspiration?

I am a great believer in the work of the Global Sustainable Tourism Council. Randy Durband and the team at GSTC understand that even small organizations can make a big impact. GSTC thinks big and makes a difference.

Earthcheck is an organization that is making a global impact. They really walk the talk. They have taken strong research and applied it in ways that help people achieve their objectives.

I am always really inspired by social entrepreneurs; people that see a problem and develop entrepreneurial ways to address it. I have also been inspired by the work of Gavin Bate and Adventure Alternative[1] and its sister charity, Moving Mountains[2] in Nepal. Gavin thinks about positive, long-term change for not just months or years, but generations.

On a smaller scale, Samrat Katwel of the Hands On institute is doing great work using community-based tourism to overcome prejudice in Untouchable communities in Nepal.

3 pieces of advice for emerging researchers interested in tourism sustainability?

Impact is more than how many journal articles you have written. Work to ensure that your research reaches people that need it, in ways they can use it.

Think big—but go deep. Remain curious—and enjoy.

Anything else you'd like to mention?

I am really excited about the work we are doing with travelcarecode.org. I am working with a group of academics and marketing professionals on this project, which is designed to promote responsible travel. The program includes both consumer-focussed marketing messages and support materials to assist tourism businesses and DMOs to promote responsible travel through their own marketing efforts. The project has the added benefit of providing students with real world cause-marketing experience.

Link to the interview: https://sustainability-leaders.com/interview-jonathon-day/

---

[1] https://www.adventurealternative.com
[2] https://www.movingmountainstrust.org

The Netherlands | After several roles in the travel, tourism and hospitality industry, Jos has been the CEO at NBTC since late 2007. In this capacity, he and his team at NBTC have driven the sustainable development of destination Netherlands through the initiation, co-creation and activation of Perspective 2030[1] (2018), a new vision and ambition for destination Netherlands. It provides new paradigm shifts for short-term recovery strategies as well as mid and long-term development, branding and marketing of the country's visitor economy. Focus on the net impact of visitors' ecological, social and economic activity and a shift in its focus from promoting the country to 'managing' it. Among other things, NBTC supports cities, regions and provinces—through public and private stakeholders—with the adoption and implementation of Perspective 2030s strategic fundamentals, orchestrating a coherent process towards the sustainable development of destination Netherlands, respecting local and regional differences throughout the country.

---

[1] www.perspectief2030.nl

Areas of expertise: destination management, destination sustainability, sustainable development, Netherlands, Europe, UNSDG 11 (Sustainable Human Settlements)

The following interview with Jos was first published in October 2020 on Sustainability-Leaders.com.

Jos, what motivated you to take up the role of Managing Director at NBTC Holland Marketing, the organisation in charge of promoting the Netherlands as a destination?

After having handled various roles within the travel, tourism, and hospitality industries, I got the opportunity to head the National Tourism Board in late 2007. After being employed by NBTC in another role some years before (Director UK & Ireland, London, 1999-2003) the love for the organisation, its purpose, position, and potential never died. Hence, when I got in the position to take the helm, it was a no brainer. It turned out to be a tough start, given the financial and economic crisis that occurred in December 2007.

In hindsight, I identify roughly three distinctive stages in my capacity of CEO: breaking, building, and transforming. After having reorganised and rebuilt the organisation followed by a strategic re-development, we are now in the middle of a systemic transformation; as an industry as well as the organisation.

Currently, the transformation process has, of course, been heavily affected by the COVID-19 crisis, forcing us to focus on short-term support, re-thinking and re-designing our activities for 2020 and 2021. These dynamics in a public-private setting to develop a valuable and enjoyable sector still makes me tick and keeps me going since that decision back in late 2007.

In your view, what does Holland stand for today? And what makes it a 'green' destination?

As you know, The Netherlands is mostly man-made. Reclaimed from the water. In a relatively small country with a relatively high population density, we constantly need to think about the optimum way to utilise our (natural) resources. It turned The Netherlands into an inventive nation, constantly searching, designing, and applying solutions for a sustainable future in an environment where space and resources are limited.

I'd say that our open attitude, inventiveness, and entrepreneurship create a great breeding ground for innovation and practical solutions for everyday needs and future challenges alike. As a destination, this should translate to a more balanced approach: taking the common interest and needs of visitors, citizens, and entrepreneurs as a starting point and focusing on the economic as well as ecological and social impacts.

Sustainability reaches well beyond the ecological angles. It also incorporates the social sustainability of our visitor economy. Our conviction is that only then we can ensure a livable, loveable, and valuable country.

With regard to the ecological angle: the process that is taking place will drive a decrease in our ecological footprint of destination Netherlands. For instance, via a focus on source markets that can reach the Netherlands by electric car, benefitting from the excellent e-infrastructure, or by train. Within the destination, the e-bike and our world-class infrastructure provide ample opportunities to discover our country sustainably.

Driven by the growing interest and demand for 'green' products and services, entrepreneurs are genuinely 'greening up their act' fast.

Given the importance of sustainability in the bids for meetings and conventions we clearly see that our M&C industry is greening up very fast and is among the global front runners, hopefully presenting them also with a head start when it comes to sustainable recovery post-COVID.

NBTC's Perspective 2030 has ambitious goals with regards to handling the anticipated 50% rise in tourists by 2030. Is this scenario still realistic, post-COVID-19? (How) have you had to adjust your destination management strategy?

COVID-19 will affect the estimated numbers, the growth trajectory and speed. One of the scenarios particularly in the case of a working vaccine or remedy could still be a 50% rise. It may take a little longer, but in that scenario, most of the world population will still be willing, able, and committed to travel.

As a vision, ambition, and a strategy, we feel confident that Perspective 2030 will be as relevant as before COVID-19 since it aims to deliver sustainable development of destination Netherlands, based on the common interests of citizens, visitors and businesses and focus on a balanced approach to economic, ecological and social impacts. Something that will be relevant even without (extreme) visitation growth.

We will use scenario planning to assess what the possible impact of various conditions will be to adjust our planning as and when needed. Tactics, however, do change. Our short-term approach and activities are influenced heavily by the current crisis.

The pandemic has set many destinations back, with no hints of a full recovery anytime soon. To what extent will the current situation impact the commitment of cities, regions, or countries to sustainable tourism?

It will be very interesting to see how governmental interventions and market dynamics will interact. Market dynamics are very strong and these market reflexes, both from businesses as well as consumers, are deeply rooted and old habits die hard, as we all know. In other words, the market could easily fall back to its pre-COVID-19 habits as and when a vaccine or remedy has been introduced, driving volume to previous heights based on the intrinsic motivation to travel and myriad of (discounted) propositions to feed that desire.

However, if destinations would like to see COVID-19 act as the much-debated reset, personally, I think they need to start thinking and acting as a change agent fast by putting in place relevant interventions to steer the market in the desired directions (and those differ per destination, even within a country). Of course, you'd also like to see businesses and consumers acting as change agents, arguably being much stronger and more inspiring.

However, I am fairly sceptical about the autonomous and organic change or even reset, considering the absolute need for liquidity (and thus visitors) to survive as a business and people's strong desire to travel. So, change can be facilitated and actively guided by governments to redirect future visitor streams and the conditions to realise that—in our specific case much like our ability and reputation to redirect water streams and shield ourselves from flooding.

Brian Mullis, in his interview with The Place Brand Observer,[2] *highlighted that Guyana is trying to attract "travellers who seek out authentic nature, culture and adventure experiences as they tend to stay longer and spend more during their vacations, travel with a lighter environmental footprint." What type of visitor are you targeting, in your efforts to achieve destination sustainability?*

Well, considering our broader definition of sustainability, striving for balanced economic, ecological, and social impacts, we'd like to attract those travellers who add value to the destination, mainly by being respectful of and interested in our core values; who value sustainable travel and are willing to spend money in our local communities as well.

Guests who travel beyond our hotspots have a higher propensity to spread out, travel independently, and absorb well in their temporary host environment: those who travel with intrinsic motivation for a place or theme and respect their hosts and their communities. So, looking at more angles than just the economic value to define 'valuable visitors' or 'quality guests'. Clearly, this is still work in progress and, to make matters more complicated, there's huge differentiation per city or region, given their different 'fabric' and challenges.

One of the strategic cornerstones of Perspective 2030 is to spread future visitors throughout the country, though the effectiveness of distributing visitors seems to be limited, judging by the experiences of Amsterdam Marketing. How can The Netherlands ensure that the locals in less travelled destinations are prepared to receive the influx of visitors when they are already aware of the negative implications of mass tourism observed in Amsterdam?

First and foremost, we do not try to seduce people to visit places where they are not wanted. This strategic pillar of Perspective 2030 relies heavily on the right definition and intended purpose of 'distribution'. It is not trying to convince visitors to Amsterdam to travel elsewhere. Why would they? They made a deliberate choice to visit our capital—and they should, it's beautiful.

What we do mean in the framework of Perspective 2030 is to identify those destinations that see an active role for the visitor economy to contribute to their local/ regional challenges. Their ambition and challenges in combination with a clear and credible positioning act as a compass for destination development and marketing, seducing the desired guests at the desired times. This process or cycle should always be based on the place's own identity and DNA, to attract and cater for future guests who fit the profile.

In other words, based on the principles of Perspective 2030 we try to drive a nation-wide bottom-up process, developing attractive destinations based on their own unique identities and subsequently attracting 'matching guests', who recognise and are driven by these qualities; this also applies to the impact on the acquisition of

---

[2]https://placebrandobserver.com/how-guyana-promotes-sustainable-tourism-through-destination-marketing/

businesses and future citizens. By doing so, we aim to steer and distribute future guest flows into desired directions.

The low hanging fruit is to deliver themed propositions and partnerships, as we do through co-created storylines, attracting specific guests to specific places. A decentralised or deconstructed destination if you like, consisting of several cities or regions sharing the same theme and catering for the same desired guests. If Amsterdam is part of such a theme, it acts as a gateway location to the other partners within the themed partnership, thus stimulating visitation to these other places and decreasing some pressure in Amsterdam itself.

Destinations like Amsterdam need to address more the pre-COVID 'overtourism' in their city centers. A clear vision plus a coherent action programme, including demotivating and limitation measures for some and development and seduction measures for others, need to be put in place.

Successful destination marketing used to be mostly about innovative, engaging campaigns and selling a destination's offerings to potential visitors. Yet, in times of overtourism, it is more and more about brand stewardship and active networking internally. In your view, are DMOs prepared for this paradigm shift—ready to take on the role of facilitator and brand/destination manager, rather than 'just' promoter?

It's a work in progress. In The Netherlands, most DMOs are well aware of the need as well as the opportunities for this transformation. In many destinations, they are starting to act on it. At NBTC, we restructured our organisation along the lines of 4 working domains that—for us at least—define destination management:

- Intel & insights
- Strategy & branding
- Destination development
- Marketing

So, marketing (or promotion if you like) is still very much part of our scope, however, it's now part of a broader, coherent, and interdependent set of expertise that allows us to deliver on the challenges and opportunities of destination management. It requires new knowledge, expertise, and capabilities that we sourced externally.

It also meant decreasing our resources for marketing and building and strengthening other domains, including the introduction of a new domain to us: destination development. If we ask other stakeholders to make their contribution to Perspective 2030, we may as well start by practising what we preach ourselves. The restructured organisation in line with Perspective 2030 enables us to do so. The jury is still out when it comes to its effects—we have only just started on this exciting journey.

How important is a destination's sustainability performance nowadays for its appeal to visitors and its competitiveness?

Well, for a relatively small yet growing group it's a must, and for most others a nice to have. We are convinced that sustainability will become a hygiene factor, so as a destination we should start thinking and acting based on 'sustainability by design or default' throughout the entire customer journey. Quite a challenge is such a fragmented travel and tourism landscape. A nice example is Hotel Jakarta in Amsterdam, almost completely CO_2-neutral yet hip, cool, and luxurious.

One of the key issues with sustainability in tourism is to overcome the gap between our attitude as citizens versus our behaviour as consumers. In 2019, the term 'flight shame' popped up in the public debate, expressing the fact that an increasing group of people is becoming aware of and now opposed to the downsides of (cheap) flights/flying. Yet, in that same year, Dutch people flew more than they had ever done before. The sector has its work cut out and we believe that it's a matter of time when attitude and behaviour will get aligned: we better be ready for it.

Which trends do you observe in destination management that are likely to influence the work of tourism leaders and sustainability stewards in the years ahead?

- Tourism as a mean to various ends, rather than just an end in itself
- Shared interests as a driver instead of a single sector or stakeholder's interests
- Marketing becoming less relevant
- Cross-sectoral collaborations, both in government as well as companies
- Liveability and sustainability becoming as important as 'the economy'
- Tourism becoming more complicated

Link to the interview: https://sustainability-leaders.com/jos-vranken-interview/

Bhutan | Candid and proactive, Dr. Karma Tshering has had an array of enriching experiences in his career. He has interned in some prominent US national parks like Yellowstone National, evaluating tourist campsites in Masai Mara and Serengeti, and conducting tiger surveys in Bhutan and Nepal. He served with the national parks and conservation sector in Bhutan for over 20 years. As a firm believer in incentive-based conservation, he played a key role in institutionalizing ecotourism and nature recreation programs within forest management. Recognizing the need to foster partnerships and collaborations between tourism stakeholders and businesses, he founded the Bhutan Sustainable Tourism Society[1] to strengthen sustainable tourism development in Bhutan. He presently works as a freelance consultant at Eco-Call Consultancy Service,[2] pursuing his passion for promoting sustainability. He obtained his masters from the University of Edinburgh and his PhD from the University of Sydney.

Areas of expertise: ecotourism, sustainable development, destinations, Asia, UNSDG 15 (Forests & Biodiversity)

The following interview with Karma was first published in June 2016 on Sustainability-Leaders.com.

Karma, you started your career in sustainable tourism as a National Coordinator for the Tiger Conservation Project—can you describe this project and what you learned from this role?

Bhutan, though a developing country, has given top priority to the conservation of the natural environment. The Tiger Conservation Project was initiated in 1995 with the support of WWF as the country's first species-specific project. As a young conservationist, I felt privileged to be given the opportunity to be the National Coordinator for the project. Being passionate and curious to learn, this role provided the perfect platform for me to be out in the field to gain experience and insights in conservation.

To protect the tiger meant we first needed to know where exactly there were tigers in our country since a scientific census had never been conducted. Due to the rugged terrain of the country, many of the survey techniques used in the foothills and grasslands of India and Nepal did not seem easily applicable in our country.

We had to rely on interviewing local communities as our major source of gathering data. Along with the technical assistance of a renowned tiger expert, within two years, the project made commendable progress. GIS mapping showing the presence and absence of tigers within the country was produced for the first time. Baseline surveys were established and a tiger conservation strategy for Bhutan was developed.

The major findings of the surveys showed that the tigers occupied habitats beyond the protected areas. This prompted the government to take a landscape approach to conservation through the declaration of several biological corridors that connected the protected areas of Bhutan. While all these were significant achievements for conservation in Bhutan, I realized through interacting with local communities that saving the tiger required a greater focus on people.

The Tiger Project affirmed my belief that people are at the center of conservation, and hence their support and cooperation is critical to development. I have been a firm advocate of a people-centered approach for conservation ever since.

You completed your master's in resource management and obtained your Doctorate looking at community-based tourism as a positive force for the protected areas of Bhutan. What were the main insights for you personally?

---

[1] www.bhutantourismsociety.com

[2] www.ecocalling.com

Bhutan has declared an extensive network of protected areas, covering almost half the size of the nation's land (among the highest proportions in any country). The responsibility for its management within the government was entrusted to the Nature Conservation Division, the office I was working for.

At the time, protected area management was relatively new in my country. I consider myself lucky to have been given the opportunity to become, almost from its inception, a member of a group of protected area professionals who grew with that vision.

Unlike many other countries, Bhutan's conservation policy allows people to reside within the protected areas. This circumstance demanded an integrated approach to conservation and development.

Since most of the conservation programs were donor-funded, they lacked sustainability. The people's interest in conservation activities was short-term. Therefore, identifying projects that ensured long-term commitment and benefits to both people and conservation was essential. That's when my attention was drawn to tourism.

Through my research I was able to demonstrate that sustainable tourism development is a viable option to stimulate the support of the public for biodiversity conservation and cultural preservation, and that such development is necessary for the sustainability of the protected areas in Bhutan.

It was evident that if planned and implemented in consultation with the local people and other relevant partners, tourism has the potential to offer a symbiotic relationship in promoting socio-economic development, cultural preservation and biodiversity conservation. Tourism is not a single sector responsibility—as it used to be perceived by people in my country—but a multi-dimensional concept which requires constant communications, collaboration, and partnerships.

Conservation of the environment is one of the four pillars of Bhutan's Gross National Happiness philosophy. To what extent does tourism in the country currently act as a positive force for conservation and happiness?

The four pillars to the vision of happiness, in addition to the conservation of the natural environment, are the preservation of the cultural heritage, equitable socio-economic development, and good governance. While the country seeks to focus development based on each of the happiness pillars, in my view we cannot focus on each pillar in isolation but will need an integrated approach that combines all four pillars.

It is important to identify programs that can successfully combine all the pillars, resulting in an overall positive impact. Tourism has the prospect to provide this critical link.

Especially for my country with limited potential for industrialization, but with a unique culture and an intact natural environment, the capacity of tourism as a major force for its development is apparent. There are several examples around the world where tourism is becoming the driving force for conservation. Likewise, here in Bhutan, tourism has made a substantial contribution to conservation and happiness.

The stringent conservation policies weigh heavily on the local people living in and around the forests and parks with increased wildlife predation impacting their subsistence livelihood. This problem has been alleviated through seed funding established for livestock compensation established through tourism, other interventions and promoting of community-based tourism to provide supplementary income.

Sustainable tourism development is able to generate positive benefits for conservation of the cultural and natural heritage while offering socio-economic benefits to the local people. This symbiosis contributes to happiness.

Can you explain the idea behind 'incentive-based conservation' as a practical approach to safeguarding nature?

I am sure many of us would agree that human motivation to a large extent is incentive-driven. Incentives can come in a couple of forms: either direct or indirect. Likewise, an incentive-based approach to conservation to me is the most logical and practical to fulfilling conservation needs.

It is only natural that people become more encouraged to participate in conservation if they see some benefits for themselves. Economic benefits are no doubt attractive, but there are also far-reaching benefits for people's appreciation towards conservation through nature recreational activities.

I had the opportunity to demonstrate this in Bhutan by starting nature recreational programs. I realized that although we had an extensive network of parks and protected areas in Bhutan, it lacked public interest in its appreciation and consequently its support.

Incentivizing through responsible enjoyment to me is a sustainable approach. I played a key role in advocating the establishment of a Division within the Department of Forests and Parks that was specifically mandated to promote nature recreation and ecotourism programs within the forests and parks of Bhutan.

The Division, since its establishment in 2011, has created several nature recreational areas and programs for the public. This has led to an increased understanding and appreciation of natural areas, resulting in increased public support for conservation.

People are a vital link to nature conservation and strengthening this link is fundamental to achieving the conservation objectives. A conservation policy that engages people's participation through an incentive approach has the prospects of delivering positive benefits for people and nature.

In a world that is threatened with climate change, Bhutan is one of the few countries that have achieved carbon neutrality. What lessons from Bhutan might serve other tourist destinations around the world?

The state of the natural environment is one of the most important attributes for developing sustainable tourism. Bhutan is fortunate that the stringent conservation policies within a large subsistence farming community have led to an intact natural environment consisting of over 72% of the country under forest cover.

While quick monetary benefits from forests have lured many countries to indiscriminately log large areas, Bhutan on the contrary, under the visionary leadership of the 4th King was not tempted towards these short-term gains. Instead, the forests were conserved and nurtured like the goose that lays the golden eggs for the future economy of Bhutan. The Bhutanese people now enjoy that future. The pristine state of the natural environment has generated substantial revenue from clean energy production through hydropower and enhanced opportunities for tourism.

Bhutan has demonstrated to the outside world that supporting conservation is not only about fulfilling the ecological need, but an economic investment endowed with long-term benefits. Hopefully, many countries are encouraged and will learn from Bhutan and pledge their commitment to follow this path.

Travellers are increasingly seeking destinations that are more natural. The authentic cultural and natural landscape of Bhutan has branded Bhutan as one of the top tourist destinations.

Anything else you'd like to mention?

With the unique opportunity to be one of the pioneers for ecotourism development in Bhutan, I am currently laying the foundations to establish an ecotourism society for Bhutan. I welcome any ideas and suggestions. I am happy to make any contributions to the promotion of sustainable tourism development.

Link to the interview: https://sustainability-leaders.com/interview-karma-tshering-bhutan/

Kaspar Howald on Connecting Agriculture, Local Trade and Tourism

Switzerland | Kaspar Howald studied classical philology and philosophy at the University of Zurich. After completing his doctorate, he moved to Rome, where he first worked at the Istituto Svizzero di Roma, the Swiss Cultural and Scientific Institute. After a 1-year working stay on behalf of Pro Helvetia in Egypt, Kaspar returned to Rome, managed the cultural programmes of the Goethe-Institut, the German cultural mediator abroad, for 3 years. Since 2014, he has been the director of Valposchiavo Turismo. With the project '100% Valposchiavo', he is committed to the valourisation of local products in the tourist offer and tourism marketing of Valposchiavo.

Areas of expertise: destination sustainability, mountain tourism, Alpine destinations, Switzerland, UNSDG 11 (Sustainable Human Settlements), UNSDG 12 (Resource Efficiency)

The following interview with Kaspar was first published in November 2019 on Sustainability-Leaders.com.

Kaspar, in 2016 your community received the Swiss Milestone Award of Excellence in Tourism, for your initiative '100% Valposchiavo'. Do you remember how it all started? How and why did you get involved?

F. Kaefer, *Sustainability Leadership in Tourism*, Future of Business and Finance, https://doi.org/10.1007/978-3-031-05314-6_62

The story of 100% Valposchiavo started already in 2012 when the local farmers were confronted with a huge infrastructural project by the local energy provider. One of the consequences of this project would have been the loss of some acres of farmland in one of the most fertile zones of the Valposchiavo. The farmers had to ask themselves how to use the remaining farmland more effectively and how to become more competitive on the market, to compensate for this loss.

In response to this challenge, the local farmer associations started working on a project for regional development of which one aim was to reinforce existing and to build new locally closed value chains to increase the sale of products based on local raw materials. The local DMO was from the beginning partner of this project.

I got involved in the process when I came to Poschiavo as director of Valposchiavo Turismo, at the beginning of 2014. From the start, I was convinced that '100% Valposchiavo' could be a very strong USP for a destination like Valposchiavo.

The healthy, innovative and diversified agriculture (more than 90% of the farmland are cultivated by farms certified by organic farming label bio suisse), the strong food processing sector and the rich and lively culinary tradition make the Valposchiavo an interesting destination for conscious tourism.

What is 100% Valposchiavo about—what are its objectives?

100% Valposchiavo is a collaborative project between agriculture, local trade and tourism to build and strengthen locally closed value chains—from the production through transformation to consumption.

The main objectives are to increase the local added value generated through tourism and to distribute this added value to other economic sectors of the valley.

In this way, tourism can contribute to the sustainable economic development of the region and can play an important role to preserve (and increase) local workplaces, as well as to maintain know-how and tradition.

Looking back over the 5 years of existence of 100% Valposchiavo—would you say that the purpose of the initiative has been reached—mission accomplished?

100% Valposchiavo is a very ambitious project and even if we have been working on it for five years now and can see positive results, the mission is by far not accomplished yet. For instance, the collaboration between producers and gastronomy has to become much closer. The chefs have to adapt their menus much more to the different seasonal offers and they have to become more resourceful in the use of local raw materials (meat, in particular, is a very difficult topic).

Also, we would like to offer our guests more possibilities to have a real experience of different examples of our locally closed value chains: from farm to table. There is still a lot of work to be done.

During our visit to Valposchiavo, we had the opportunity to experience firsthand how passionate restaurant and hotel managers are about 100% Valposchiavo. To your mind, why has it been able to become such a success?

I think it's hard to find an alpine valley where you have such a large offer of locally produced food. Then there was already existing cooperation between agriculture and gastronomy because there were family ties or friendships between the actors of these two sectors.

Most important, however, I think is pride and self-confidence. People from Valposchiavo are very attached to their territory. They are convinced that Valposchiavo is one of the most beautiful alpine valleys, that here you eat the best food made from the best ingredients. This self-confidence is an important part of authentic hospitality.

Urs Wohler in his interview[1] emphasized that "anyone who wants long-term success must work together regionally"—How do you encourage collaboration among stakeholders in Valposchiavo?

People from Valposchiavo—perhaps because of its geographically peripheral position—have always been used to work together and to find local solutions to different challenges. This collaborative spirit helps a lot and it allowed us to build up the project with a strict bottom up-approach. So, we tried to involve all relevant stakeholders from the beginning.

It helped a lot that all the main interest groups (the farmers' associations from Poschiavo and Brusio, the local trade association and the gastronomy sector) were convinced by the potential of 100% Valposchiavo right from the start.

Another important factor is quick wins, like prices and media coverage, which are good stimuli to keep on working. In the long run, however, you need economic success. As long as the stakeholders are sure that they get some benefit out of the project, they will stick to it.

All too often tourism businesses and destinations develop well-meaning sustainability strategies but ultimately fail at implementation. To your mind, what does it take to succeed? Which pitfalls to avoid?

The most important thing is to build your strategy on existing foundations. Your possible stakeholders have to embrace the strategy and consider it a part of themselves. So, a top-down approach surely won't work.

Then there must also be some fun in it. The most gratifying effect of 100% Valposchiavo for me is to see the enthusiasm with which our stakeholders invent new products—pizza, gin, eight types of different ravioli, ketchup and much more—and create new collaborations. This shows that they really share the 100% Valposchiavo spirit.

The iconic red Rhaetian Railway being such a well-known symbol for Switzerland and particularly the Graubünden region, what role does the railway play for Valposchiavo—and the sustainable development of the destination?

The Rhaetian Railway is our most important transport connection to our two main markets, Switzerland and Northern Italy. Unfortunately, still, a great majority of our guests arrive by private transport. But the Rhaetian Railway has a huge symbolic value for our valley.

Indeed, the Valposchiavo is far away from everything, but you can get there with one of the most beautiful alpine railways, enlisted since 2008 as a UNESCO World heritage. This distinction played an important role for the self-awareness of the people from Valposchiavo and somehow enforced a decidedly sustainable approach

---

[1] https://sustainability-leaders.com/urs-wohler-interview/

to the further development of the region: If UNESCO considers our valley as a World Heritage, it is our job to preserve its natural beauty and its qualities.

Which are the main challenges of creating and managing a destination brand such as 100% Valposchiavo, with a strong focus on sustainability?

The main challenge is to keep the partners going and to keep up the enthusiasm. Now, after five years of intensive work on the project—not only by the DMO but also by our partners—sometimes I hear some partner who tells me: It's time to invent something new. So, I have to explain that it's true that we were getting used to 100% Valposchiavo and it is not so thrilling anymore as it was in the beginning. But still, the biggest part of our potential guests does not know the project and that's why we have to go on with it.

In tourism, we are used to giving too much importance to the concept of 'novelty'. "What is your news for next season?" is a common question between the operators, but I think—at least for small destinations—marketing using "novelties" is not sustainable. We lack the big investments to finance novelties (at least the infrastructural ones) and most of all we lack the marketing power to let our potential guests know about our novelties.

That's why small destinations (and perhaps also bigger ones) should stick to marketing by topics, for example, Valposchiavo with the topic of local products. The market for this topic is big enough (and probably growing in the future). We have to be persistent and self-confident to play this tune for the coming years.

Is cooperation beyond tourism players—e.g., with farmers—becoming more important for destinations in alpine regions?

In alpine regions like Valposchiavo, where tourism is not the main economic sector, I'm sure that cooperation between players from different sectors will become more and more important.

The main goal of tourism in alpine regions should be to contribute to a sustainable development and a viable future for the region. To reach this goal, tourism has to integrate itself in the local economic setting. Standing alone, tourism won't be able to generate enough added value to keep alive peripheral alpine valleys—and alpine valleys who cease to live do not attract tourism.

What's next for 100% Valposchiavo—which projects are you currently planning, which ideas are you thinking about?

We will have to work a lot more on the formation of our partners. The farmers have to know more about the needs of the hoteliers, and vice versa, to coordinate the balance between supply and demand and to improve the quality of our touristic offer.

Furthermore, we would like to put into practice our concept of the so-called 'agriturismo difuso', adapted from the 'albergo difuso'. The idea is that our farms are too small to offer accommodations for guests like classical agrotourism.

So, we try to encourage collaborations between farmers and the accommodation sector (hotels, pensions, holiday homes …) in the sense that the accommodation sector will accommodate the guests and the farmers will offer 'agricultural experiences' during the day, which allow our guests to fully grasp the concept of 100% Valposchiavo.

Our main touristic season is summertime, which coincides with the most interesting season in agriculture. I'm sure that this idea has great potential.

Link to the interview: https://sustainability-leaders.com/destination-valposchiavo-interview-kaspar-howald/

Kelly Bricker on Ecotourism Challenges and Opportunities

USA | Kelly Bricker, PhD is Professor, Director of the Hainan University-Arizona State University International Tourism College, a practitioner in ecotourism, and a consultant. She completed her PhD with Penn State University in sustainable tourism and protected area management, has experience in ecotourism, visitor and protected area management, and the impacts of tourism. Kelly has authored books on sustainability, highlighting case studies of tourism's role in environmental and societal issues, such as *Sustainable Tourism & the Millennium Development Goals: Effecting Positive Change*; on adventure education in *Adventure Programming and Travel for the twenty-first century*, and on graduate education in *De-Mystifying Theories in Tourism Research*. She serves on the board of the Global Sustainable Tourism Council as Vice Chair and on the Executive Committee of the Tourism and Protected Area Specialist Group of the IUCN.

Areas of expertise: sustainable tourism, ecotourism, tourism research, UNSDG 11 (Sustainable Human Settlements)

The following interview with Kelly was first published in April 2017 on Sustainability-Leaders.com.

© The Author(s), under exclusive license to Springer Nature Switzerland AG 2022   399
F. Kaefer, *Sustainability Leadership in Tourism*, Future of Business and Finance,
https://doi.org/10.1007/978-3-031-05314-6_63

Kelly, do you remember the first time you heard about ecotourism? What got you interested in tourism and sustainability?

While I had been involved in nature-based tourism through marine programs in the British Virgin Islands and the Florida Keys, I really first learned of ecotourism through TIES, back when TIES was The Ecotourism Society (TES).

My husband and I were working for Sobek Expeditions (now Mtn Travel Sobek) overseas and one of the employees for Sobek at the time was involved in TES as an Advisory Board Member, her name was Leslie Jarvie. Leslie, unfortunately, passed at a very young age, but certainly was a dedicated advocate and my link to TES, and invited me to a board meeting in Washington DC during the very early days. After attending the meeting, I joined and started to receive the then, paper newsletter.

As we had travelled over the years for Sobek and then Sobek Travel Group, the World Heritage, we began to see patterns in the places we visited. In some places, tourism appeared to help a destination, the environment and local communities. In other places, tourism did the exact opposite—areas where you saw tourism degrade and devastate areas and destinations.

Ecotourism, based on what I was learning from TIES, was a strategy and tool for conservation and improving the quality of life for local people.

After working in the adventure travel realm for nearly 10 years, I headed back to school to study what sustainable tourism and ecotourism were all about, and to apply the first-hand experiences we had in the field.

So, in 1994 I went back for a PhD and began my academic journey in unpacking these concepts and observations. The academic side provided opportunities to learn about how to study these impacts, and how to objectively evaluate.

My minor in geography also introduced me to how other disciplines viewed tourism—not always so positive. This fueled my interest in finding ways to utilize the economic power of tourism to effect positive change.

Now in 2017, how has your view changed regarding ecotourism?

I think my understanding of what makes ecotourism work has broadened and is constantly evolving. There are so many intricacies, complexities, and contexts that one size does not fit all. Many things have to align to make it all work. However, my view of ecotourism as a tool for conservation and associated community benefits is perhaps stronger than in earlier days.

With all that is happening with the world today, our disconnect from nature, our abuse of the planet, and all the associated challenges with climate change, poverty, and health issues, I believe ecotourism continues to have enormous potential to influence the world in a positive way. It is a mechanism for health and well-being, from the individual to a global level.

We know nature is good for humans, however, humans must also be respectful of the very nature in which they are a part, for this complex exchange to work.

As a professional dedicated to tourism sustainability for many years, which have been your main insights?

There is still a lot of work yet to do! I am deeply encouraged by recent efforts to bring practical application of sustainability and ecotourism concepts and principles to the industry. I support initiatives by the Global Sustainable Tourism Council,

TIES Conferences and workshops, and applaud the work of so many dedicated organizations and individuals.

As colleagues and I have suggested in the book *Sustainable Tourism & The Millennium Development Goals: Effecting Positive Change*, it is critical that we tie the work we all do to the societal and environmental challenges we face. For example, we now have plenty of evidence that demonstrates the importance of biodiversity conservation as the cornerstone to human health and well-being. This becomes an essential foundational aspect of all tourism development and subsequent management programs and projects.

Biodiversity is critical to economic, social, and cultural sustainability, and ultimately quality of life and well-being.

I am encouraged by the recognition of the important links between healthy ecosystems and healthy people. This assists us all in the call for attention to sustainability concepts and the way in which to initiate and manage tourism.

I have also come to realize that ecotourism cannot operate in isolation of other types of tourism and other industries. What happens 'next door' to ecotourism has huge implications for the success of ecotourism.

Ecotourism must work with its tourism 'neighbours' and other stakeholders to achieve sustainability at a destination level. TIES has worked on this idea a bit with the expansion of the conference to the Ecotourism and Sustainable Tourism Conference idea.

Instead of narrowing my focus, I seem to be seeing even more connections and how sustainable tourism can influence society. This is at times overwhelming, yet useful.

For example, a colleague and I have written about industrial farming in the USA, and the connections to sustainability. This is an area largely ignored by our society, even though the impacts are severe. This has really driven home the need to spend more time on how we are sourcing all products within the tourism industry if we are to be truly sustainable.

I am intrigued by new concepts being brought to light, such as the idea of planetary boundaries and the idea of resiliency in society and ecosystems. All have applicability to how we manage tourism sustainably, for the sake of the planet and all that inhabit it.

Where do you see the main challenges at the moment in terms of tourism sustainability, and destinations specifically?

The primary challenges have to do with reducing extreme poverty, sustainable consumption and production patterns, climate issues, and resource depletion.

As mentioned above, sustainable farming and food production have so many implications for sustainable tourism. It is obvious that we need a healthy ecosystem for ecotourism to occur, but in reality, we need that for everything—we need it to survive. So, these challenges apply across the board.

I think the sustainable development goals summarize challenges across the board and what we need to do to improve the environment and society as a whole. Tourism is inextricably linked to all aspects of society, therefore connected to these goals as well.

How do we construct and develop society to address increased populations and diminished resources? How can tourism assist in alleviating extreme poverty? These and many more questions apply significantly to the travel industry and the enormous power it has to make a positive difference.

Your key research insights as academic investigating tourism sustainability?

Sustainable tourism, especially ecotourism, serves as a tool for conservation and sustainable community development—with several caveats. Despite the attention given to sustainable tourism development, it has proven difficult to define and operationalize. However, progress is being made through the efforts of the GSTC, and many others.

Critical decisions about tourism development and management are made at local, national, regional, and international levels. Unfortunately, despite the apparent vertical integration of each level, decision-making is not always made in and amongst these levels. There are competing interests, contradictory developments, and climatic issues which all influence strategies at each level.

From the public to private, to voluntary sectors, and tourists, there is a need for increased collaboration and partnerships.

Sustainable tourism, of course inclusive of ecotourism and many other types of tourism, is about sustaining both the industry as a whole, and the attributes associated with various products (social, cultural, and environmental), on which it is all based.

Regardless of the difficulties in operationalizing various forms of sustainable tourism, I am of the belief (and have written with colleagues about this) that sustainable tourism encourages long-term perspectives, fosters notions of equity, promotes inter-sectoral linkages, conservation of natural and cultural resources, and cooperation amongst many stakeholders.

There is enormous, continued potential in sustainable tourism to effect positive outcomes for society and the environments.

We need to support an appreciation for nature, ecosystems, cultural resources early in a person's development. Youth are our future, and we need to invest in them and their knowledge today. My parents invested in my understanding of the natural world through camping trips, animal husbandry, and exposure to diverse communities.

This exposure as a young person had a lasting impression, as I am certain it has for many of the tourism professionals interviewed by the Sustainability Leaders Project.

Your most memorable moment during your time as President of the Intl. Ecotourism Society (TIES)?

There are many, and these come to mind in themes.

Certainly, the early days brought excitement, meeting so many people dedicated to ecotourism and learning of their efforts to support a healthy planet.

ESTC conferences are amazing. The people who attend, the keynotes, and learning about efforts on the ground leave one with feelings of hope and promise for a sustainable future.

Ecotourism Professionals: many people have made enormous contributions to the principles of ecotourism around the world. Board members, past and present, dedicated tourism operators, and destination management visionaries have created significant memories in my life.

Which aspects of leading an international organisation like TIES do you personally find the most challenging?

Leading an international non-profit is achieved through the cooperation of many dedicated folks. TIES is expected to be many things to many people, and finding ways to support expectations, the full mission of TIES and at the same time do it with primary volunteers, is challenging.

After every conference, I do see the relevance of this organization, yet finding sustainable ways to support the organization economically with the number of staff it would ideally need, challenges the organization every day. There are so many things we would like to do, yet everything has a cost and finding the balance between what can be done with our resources and what should be done is always a challenge as well.

TIES is currently focused on building networks and advocates for ecotourism and providing resources to support education and training, through our online education programs, conferences, research and workshops.

In a world where our connection to nature is more critical now than ever before, ecotourism can be the connection and impetus for conserving nature. It can help to get people outdoors, and benefit communities for increased quality of life and well-being.

There is no doubt that the organization will continue to promote ecotourism as a tool for conservation and a mechanism to effect positive change. I do think the ways in which people will learn and utilize the resources offered by TIES will change.

Partnerships for ecotourism will become even more critical, and it is our hope that the organization and global network of TIES can help lift ecotourism to even more prominence and importance in the future. I think the organization needs to build its constituency with scientists, practitioners, conservationists, and wider sustainable tourism industry, and feel this has started.

TIES has experienced many changes over the years, from Executive Directors to Board Members and Advisory Board Members. I have most likely been at the helm way too long and this will need to change. TIES must rebuild its leadership in a way that reflects the reality of non-profits today.

This process began by creating a virtual office, reducing the costs associated with maintaining a location; continue to build resources and information that assists all aspects of ecotourism, and rebuild the board and structure of administration of TIES.

Some time ago, TIES Advisory Board members announced their departure. This event caused people to question TIES and its administration on many levels. This of course is not how you want things to go, and was very difficult for me personally, as well as many others.

While it would have been perhaps time to quit with the departure of the Advisory and many other challenges, I chose to stay and try to reset the organization. This is happening, yet there are many, many, years of challenges to overcome. This is not easy and will take time, yet TIES is slowly making headway.

I believe in what TIES stands for and like many organizations, believe these struggles are a part of life. There is so much work to do and I believe with new energy and creative dedication, the organization will find the best way to be relevant.

Link to the interview: https://sustainability-leaders.com/interview-kelly-bricker/

Norway | Linda Veråsdal is the Founder of Ethical Travel Portal, a platform that provides inspirational stories from the grassroots and in-depth responsible holidays to different countries in the world. Linda is a responsible tourism activist and consultant. She has been working on product development, training and coaching in various countries and communities, including lecturing on sustainable tourism at the university level. Her latest initiative is The Gambia Cotton Trail, which connects the whole supply chain from seed to finished product, and is a journey from east to west in The Gambia. She splits her time in Norway and The Gambia, where she also is involved with Footsteps Eco Lodge.

Areas of expertise: destination sustainability, responsible tourism, sustainability challenges, tourism business, UNSDG 8 (Equal & Fair Economic Opportunities)

The following interview with Linda was first published in July 2020 on Sustainability-Leaders.com.

Linda, as a passionate ethical traveller, do you remember the first time you heard about sustainable tourism?

Not exactly, but there are 2 – 3 occasions that I know of that had an impact on me! When I grew up, our holidays were always in Norway. Not too far away from where

F. Kaefer, *Sustainability Leadership in Tourism*, Future of Business and Finance,
https://doi.org/10.1007/978-3-031-05314-6_64

we lived. My parents have always been interested in our heritage & culture, which became a part of our holidays. From old building techniques to traditional arts and crafts, offbeat with local communities, eating slow food, and so on. The interest must have started from there.

Post that, the explorer in me woke up and it was time to step outside of the Norwegian borders. During one of my first longer overseas travels, I encountered responsible vs. irresponsible experiences during a voluntourism trip to beautiful Guatemala. It was completely out of control.

Sustainable tourism as a subject I heard for the first time around 2001. I was studying Visitor Management at the University of Stavanger and a guest lecturer gave us the basics of sustainable tourism.

Adding those early experiences together, I knew what my field in tourism would be.

What distinguishes an ethical traveller from a 'normal' one?

I believe it is a combination of what the traveller wants to experience and chooses how that experience is being delivered to them. The curiosity to get to know another place, its nature, learn about traditions, culture, arts & crafts from the local people themselves.

Instead of rushing, it is about travelling slow, experiencing, learning, and moving into a more transformative way of travel. All that, and at the same time making sure their holiday is having a positive impact and contributing to the local community they are visiting.

How has your view on the potential of tourism as a tool for a more sustainable travel industry changed since establishing the Ethical Travel Portal?

Being in the middle of the COVID-19 crisis, the need to use tools for sustainability and creating a more ethical travel industry is more important than ever. The world is changing, travellers' needs are changing and we—the travel industry—need to change and adapt as well. It will not be business as usual, and as we build tourism back, it has to be done better than how we left it pre-COVID-19.

Sustainability is a part of that, not only in the experiences we want to offer but also in preparedness for another crisis. We all have to reset and look at what we can do to be better, for our communities and their people, our planet, and also, for our travellers.

What is the Ethical Travel Portal all about?

Ethical Travel Portal is a hub for selling responsible tourism packages, an online ethical magazine of inspirational travel stories, and a consultancy.

For the tour operating part, we offer trips to various countries. All local partners and experiences have been vetted for their sustainability before they become part of our portal. We are also working on the ground in Norway, The Gambia, and Nepal where members of our team are present.

In our ethical travel magazine, we write about people and places, good practice, and ethical issues. It is also open for external writers to contribute as well. Currently, we are working with Tina Hudnik and Maja Čampelj from G-Guides in Slovenia on presenting the monthly winner of the Green Microphone Award.

The consultancy part: together with my business partner Raj Gyawali and a handful of selected specialists, we use our expertise in responsible product development and marketing to work on the ground with different projects. We connect those projects to our marketplace at ETP.

We often see the development of great sustainable local products, but unfortunately, some of them never reach the market. The Ethical Travel Portal as a platform seeks to close that gap.

Would you say that the tourism industry as a whole has improved in terms of its sustainability in recent years?

I would like to say yes, but I also think it is going too slow! However, I see how the community is growing with more and more passionate people and companies pushing forward for sustainability. It is definitely happening! The bigger leaps are yet to be achieved. With COVID-19, as discussed above, this is our time to build back better!

The Gambia is among your favourite destinations and also your second home. What attracts you to the country?

I first came to The Gambia in 2006, as part of my Masters in Responsible Tourism Management, doing a course on Responsible Destinations led by Professor Harold Goodwin and Adama Bah. The week combined classroom lectures and field excursions, a perfect combination to understand how theory is working in practice. It most certainly helped me understand how responsible tourism theory can be adapted locally.

That trip was the first of many returns to The Gambia. In January 2008, ETP organised our very first learning experience, which went to The Gambia! Now, I live here several months per year, with my partner David White, who owns Footsteps Eco Lodge.

The Gambia is a country that captures many people, reflected by the high percentage of repeater tourists. It is very relaxed, and easy to get to know the amazing people living here.

Adama Bah in his interview[1] expressed his frustration that tourism in The Gambia has traditionally been controlled mainly by foreign powers and investors. In your experience, is this still the case?

To a large degree, that is still the case. But foreign ownership is not all bad. It depends on the responsibility the owner personally has towards the business and local surroundings. Foreign or not, to run a sustainable business the owner has to make sure it is well anchored locally, including employing local people on fair salaries and year-round contracts, sourcing food produced in the country where possible, promoting locally made souvenirs, activities, local guides, and drivers, respect the culture and local custom, environment, make sure the money stays in the country—register legally, pay taxes, re-invest into the country, and so on.

---

[1] https://sustainability-leaders.com/adama-bah-interview/

Footsteps Eco Lodge[2] (owned by my partner, so I am quite biased here) is foreign owned, but we practise what I mentioned above, and more. Running for 20 years now, with all staff on full-year contracts and continues to have 50% of its original staff, which is a good sign of sustainability.

Another initiative is led by Adama Bah and Lucy McCombes, the driving force behind (together with a team in The Gambia and UK) the Ninka Nanka Trail: a responsible community experience, including community-based tourism in rural communities along and on The River Gambia, bringing together a wide range of stakeholders and communities.

Fair Play is a community initiative well-anchored locally in Janjanbureh upriver that offers private chartered riverboat for multiple days, kayaking, birdwatching and fishing, and community activities.

The Gambia Cotton Trail[3] is an initiative Daouda Niang and myself are working on. We work with local organisations and across industries from farmers, spinners, weavers, tailors to local artists. Some of the produced fabric is exported to ethical brands, and some sold within the country per metre or as ready-made souvenir products for responsible shoppers to bring back home. Travellers can also join the trail across the country with cotton, local handicraft, culture, and tradition as a theme!

All the mentioned initiatives involve spreading the travellers from the coast to the upcountry. It attracts adventurous and ethical travellers, well needed to spread the benefit of tourism throughout The Gambia.

What are the main challenges in The Gambia right now, preventing businesses or destinations from becoming more sustainable?

The season here is very short, from November to April, with peaks within the season. Business owners need to start planning for the whole year and not as a project starting in November and ending with the last flight that departs at the end of April.

For example, during the off-season, a lot of people are laid off without any salaries—or retainers, until they might be lucky to get a new contract in November again.

In the past few years, we have had to face Ebola (though no cases in The Gambia), change of government issues, end of Thomas Cook, and now COVID-19. The result has either been a very reduced season or one suddenly cut short. We need a holistic crisis management plan in place, before the next set back.

Another challenge and threat are the rate at which natural resources are being destroyed. The Gambia is an eldorado for bird watchers and nature lovers. It has miles of trails that are not yet developed for walking and hiking tourism.

But we're seeing the coastline being destroyed by sand minders, fish factories polluting the environment, and deforestation happening at a high speed. I am afraid if this is not controlled in the near future, locals will lose great resources. And, the

---

[2]https://footstepsinthegambia.com
[3]https://www.gambiacottontrail.com

reasons due to which many travellers visit The Gambia for, time after time after time, will disappear.

What does it take to become a responsible tour operator?

In short, drive and passion for what you are doing and believing that what you do, is the right way forward! Build great partnerships, be transparent, and don't be afraid of sharing good (and bad) practices. Listen, learn, and adjust when needed!

Tourism professionals sometimes avoid engaging with 'sustainability', since it is not something usually part of their KPIs. Reflecting on your own experience, which advice can you share with tour operators in terms of how to deal with 'selling' sustainability to their stakeholders and clients?

If you work on the story behind what you are selling, a lot is already done, in particular with clients. Let sustainability be a natural part of the conversation—in a language everyone understands.

I don't necessarily believe that sustainability needs to be listed by bullet points. It is important to tell what you are doing, why, and how you want to move forward together with your stakeholders!

But sustainability measurements should be part of a key performance indicator. Maybe we will see that now, post COVID-19.

Link to the interview: https://sustainability-leaders.com/linda-veraasdal-interview/

Lisa Choegyal on the Pandemic and the Sustainability of Tourism in Nepal

Nepal | Lisa Choegyal is an international tourism marketing and product development consultant specialising in regenerative, pro-poor, sustainable and responsible tourism based in Kathmandu for nearly 50 years. With a background in the private sector, she worked for 25 years as marketing director with Tiger Tops and the Tiger Mountain group, specialising in wildlife, trekking, adventure and village tourism in South Asia. She is currently a director of the award-winning Tiger Mountain Pokhara Lodge. Since 1992 Lisa has undertaken consultancy roles in Nepal and throughout the Asia Pacific region, often with TRC Tourism. Clients include governments, tourism boards, development agencies, NGOs, local communities and private sector operators. Lisa is a writer and editor, regularly publishing books and articles on tourism and conservation, and experienced in production liaison for documentary and feature films. She serves on several pro-bono organisations and Boards, and since 2010 is New Zealand Honorary Consul to Nepal.

Areas of expertise: destination sustainability, ecotourism, responsible tourism, tourism business, Asia, UNSDG 8 (Equal & Fair Economic Opportunities).

The following interview with Lisa was first published in May 2020 on Sustainability-Leaders.com.

© The Author(s), under exclusive license to Springer Nature Switzerland AG 2022   411
F. Kaefer, *Sustainability Leadership in Tourism*, Future of Business and Finance,
https://doi.org/10.1007/978-3-031-05314-6_65

Lisa, you are a leading sustainable tourism consultant in the Asia Pacific region. Do you remember what got you first interested in tourism and sustainability?

I've always been interested in conservation. After studying art history in Paris, my first job was selling antiques in New York. But when I arrived in Nepal it seemed entirely natural to adapt art and heritage conservation to nature and wildlife.

Nepal tourism was in its infancy in the 1970s and was receptive to innovation and change. The Royal family were keen on hunting and led the impetus for wildlife conservation. FAO was forging national parks and protected areas all over Asia, and Nepal was at the forefront. Nepal IUCN and UNESCO World Heritage Site listings were the first in South Asia.

In 1974, I was lucky to stumble upon Tiger Tops,[1] which was actively practising ecotourism and sustainability before the words were even invented! The lodge had started operations in 1965, whereas Chitwan National Park was only gazetted in 1970 and established in 1973. It was exciting times. And Nepal led all Asian destinations in forging the links between wildlife and heritage tourism with their conservation by involving local communities.

How have your views on sustainable tourism changed over the years?

In this current environment, I think we all have to agree that the only hope is sustainable tourism. In the Himalayan context, this is inextricably linked with climate change. Our response to COVID-19 has taught us some harsh lessons. Anyone who does not heed them is unlikely to survive.

Despite the preaching of academia and idealists, the first rule of sustainable tourism is to be financially viable and to make money. Otherwise, you will not be in a position to help anyone—neither the local people, climate change, heritage, nature or wildlife. The current coronavirus pandemic is really putting that to the test, and convinces me that serious, practical solutions are the only answer.

Being based in Nepal, which would you consider the destination's strongest tourism sustainability assets?

Nepal boasts of amazing assets in terms of scenery, wildlife, people, living culture, and historic monuments.

The mountain people of Nepal possess an innate sense of hospitality, which is one of our most enduring attractions.

Also, over the years the success of development programmes with partners such as ADB, DFID, UNDP, World Bank Group and many other bilateral and multilateral donors have left Nepal with a strong legacy of sustainable tourism. Thanks to their support, we have a cadre of highly skilled grassroots specialists who understand the concepts and can help translate them into tangible benefits for the country and its people.

What are Nepal's main challenges?

Nepal's sense of inferiority, surrounded by giant neighbours, means it often does not give itself enough credit or have confidence in its actions. Nepal does not need to

---

[1] https://tigertops.com

feel threatened by foreign investors. Tourism is by definition an international business with the potential to bring meaningful FDI to strengthen the country.

There are many examples where Nepal has been a leader—especially in Asia—in hands-on sustainable tourism initiatives, but it seldom gives itself the due credit.

Examples include the private-public partnership model of the Nepal Tourism Board, using trekking and the Great Himalaya Trail to spread tourism benefits east-west along the Nepal Himalaya, as well as much-admired heritage tourism models, such as Dwarika's Hotel.[2] And the national parks in Nepal's Tarai lowlands provide some of the very best wildlife viewings in Asia, able to show visitors iconic species.

The coronavirus pandemic has turned travel upside down. How are you coping?

I personally am relatively lucky in that our Tiger Mountain Pokhara Lodge has no outstanding debt and is likely to weather the storm under the MD, Marcus Cotton.

But we have had to postpone the Sustainable Summits 2020 Nepal conference that we had won for Nepal and which I am involved in organising. It is sad, as we had strong links as part of the Hindu Kush Himalaya voice for CoP-26 (which is also postponed) with a galaxy of star speakers confirmed, including the former New Zealand Prime Minister, Rt Hon Helen Clark, Ben Fogle, Reinhold Messner, and Sir Chris Bonington.

In personal terms, I usually work from home, and we live on the edge of Kathmandu Valley, where lock-down is not onerous.

But Nepal's travel industry is devastated.

It is only now that many will realise how far-reaching are the tourism benefits within Nepal, how important the indirect as well as direct impacts of tourism, and how harsh the fallout is going to be for all but the more professional or deep-pocketed operators.

Apart from the many challenges that COVID-19 poses, do you think this crisis also presents an opportunity?

The coronavirus pandemic presents wonderful opportunities to clean up the industry and to reward the most innovative and responsible operators. We will have a renewed appreciation of the importance of the domestic market, which in Nepal many remember kept us going through the 10-year insurgency that ended in 2006.

We have never regained the full spectrum of markets since then, having (as a broad generalisation) lost the high-end and become established as a low-cost budget destination, driven by growth from the cost-conscious segments from our neighbours India and China.

The recent story of Nepal tourism is that we have had 25% growth but a 23% reduction in daily expenditure by visitors, despite work by the Nepal Tourism Board and Government of Nepal initiatives, such as Visit Nepal Year 2020 (now abandoned, of course).

---

[2] https://www.dwarikas.com

The current crisis allows us to attract international visitors prepared to pay for our amazing assets. Why are we selling ourselves so cheap? This is a chance for Nepal to decide what sort of visitors the people of Nepal need and want, that will bring them the maximum benefits throughout the country.

To make the best of the COVID-19 challenge, I believe that deep within our hearts this is a chance for us global travellers to readjust our unsustainable sense of entitlement and expectations from travel, to appreciate and value rather than to trample and consume.

Imagine it being May 2021. How do you think the pandemic will have changed travel and tourism, from a sustainability point of view?

In a year, many markets will still be resetting and re-evaluating the fallout from the devastating effects of COVID-19. The essential meaning of the word 'sustainability' will be truly tested. Many national and international companies will not survive this crisis, and the hope is that the ones that do, will be stronger for it.

In Nepal, the current oversupply of hotel rooms in Kathmandu and Pokhara, for example, may become adjusted more healthily by market forces.

Domestic tourism will become massive worldwide. I would like to think that courageous, responsible tourism operators in Nepal and elsewhere will be rewarded for developing innovative and sustainable products that are market-driven and will stand the test of time.

If you had to choose just one recent project of yours, for the impact it has had on you and the involved community: which would it be?

I'd have to say Tiger Mountain Pokhara Lodge[3] (of which I am an owner). Since opening in 1998 (by Sir Edmund Hillary) we have survived both internal upheavals and external forces to survive as a model of high-value responsible tourism, setting standards in Nepal and working to benefit our adjacent communities. But then you could say I am biased!

Link to the interview: https://sustainability-leaders.com/lisa-choegyal-interview/

---

[3] https://tigermountainpokhara.com

UK | Lucy McCombes is a responsible tourism and community development professional with an academic background in Social Anthropology of Development (MA) and Responsible Tourism Management (MSc). Offering an applied and social perspective on responsible tourism development, her work has involved social impact assessment, community participation and consultation, participatory and multi-stakeholder planning, monitoring and evaluation, socio-cultural analysis, and qualitative research. Lucy's main research interests are around the responsible management of the social impacts of tourism and tourist-host interaction within community-based tourism. Since 2009, Lucy has been working for Leeds Beckett University as a Research Fellow and Senior Lecturer teaching on the MSc Responsible Tourism Management (in Leeds, The Gambia, Mauritius and Kenya), as well as working on a range of tourism-related external consultancy and research projects. Outside of work, she is co-founder of Ninki Nanka Encounters Foundation—a Gambian charity supporting community-based tourism along the River Gambia.

Areas of expertise: responsible tourism, destinations, tourism research, Africa, UNSDG 8 (Equal & Fair Economic Opportunities).

© The Author(s), under exclusive license to Springer Nature Switzerland AG 2022
F. Kaefer, *Sustainability Leadership in Tourism*, Future of Business and Finance,
https://doi.org/10.1007/978-3-031-05314-6_66

The following interview with Lucy was first published in June 2020 on Sustainability-Leaders.com.

Lucy, you have been teaching the MSc Responsible Tourism Management course at Leeds Beckett University for over 10 years. Do you remember what first got you interested in the topic, and in teaching?

Family holidays to places like India and Mexico sparked my early interest to travel and work in overseas development. This shaped my decision to take a few years out volunteering to get some work experience in Ghana, Jamaica, and India, which is when I first started thinking about the impacts of tourism and its potential for poverty reduction.

I went on to study a MA in Social Anthropology of Development at SOAS (University of London) and then spent several years of my career working in community development in Mali and Plymouth (UK).

I really enjoyed working as a practitioner, trying to engage communities in an incredibly wide range of development projects, but ended up feeling like a 'jack of all trades but master of none'.

It was in this headspace that I found out about the MSc Responsible Tourism Management course when I read an article by Professor Harold Goodwin in the Geographical Magazine and decided to do the course myself to help me specialise in development within the tourism sector.

I studied between 2006-08 when the course was based at the Medway Campus of the University of Greenwich, with a couple of modules being taught in The Gambia alongside Gambians practising responsible tourism. This I found really inspiring and I have ended up studying, volunteering, or working in The Gambia ever since!

In fact, I did my first ever teaching in The Gambia a couple of years after I had completed my MSc, when I was asked to cover teaching for one of the modules and found it surprisingly enjoyable!

After this I joined a fabulous team at Leeds Beckett University, where I have been teaching on the MSc, researching and doing consultancy around responsible tourism management for almost 11 years now.

Briefly, who is this course for and how does it prepare the next generation of sustainable tourism champions?

As my own story shows, our MSc Responsible Tourism Management course is aimed at early and mid-career professionals who are keen to apply their existing experience to the tourism industry, or to those who want to apply responsible tourism management and practices to their existing work in tourism. It covers responsible business practices that improve the quality of life for communities and conserve the environment and local cultures.

Our approach to teaching applied field trips and working closely with industry partners (UNWTO TedQual,[1] certified) encourages our students to apply their learning and assignments to their own interests and current work in tourism.

---

[1] https://www.unwto.org/UNWTO-ted-qual

Our modules support our students to assess and monitor the complex impacts of tourism and implement responsible tourism strategies within marketing, operations and product development. It is delivered face-to-face as a full-time course at Leeds or via distance learning supported by tutorials, our university's Virtual Learning Environment (VLE) and regular contact points to meet in person.

Engagement with our inspiring network of alumni and responsible tourism practitioners around the world is a fantastic asset and is something that has really motivated and helped me personally. This network includes five cohorts of students hosted by partners in The Gambia, Mauritius and Kenya who have studied/are studying our MSc through funding from the Commonwealth Scholarship Commission.

Where do you think the future of sustainable tourism research is leading? Are there any topics you think should be addressed more by academics?

I think the future of sustainable tourism research has received an immense shakeup with the rapidly evolving impact of the COVID-19 pandemic. More than ever, academics will need to justify their research in terms of the impact it has on society and engage in research that is applied and industry-driven to support the sector through the current crisis and to recover.

I think sustainable tourism research will need to focus more on identifying bold (and perhaps unpopular) alternatives to 'business as usual' that support the transformation of business practices and economies if the goals of sustainability are ever to be achieved. This is a tall order in the future context of a global economic recession. But there is some hope that this is possible.

This is the first time in history that there has been such widespread acceptance that tourism and travel need to be different and better for the planet and its inhabitants, which means the sustainability movement has the attention of more people than ever before.

To your mind, which are the main topics and concerns linked to tourism sustainability at the moment, especially for destinations?

The main topics and concerns that come to mind are:

- The potential of the circular economy for revolutionising the industry and how destinations are managed;
- Growing consumer demand for transformational and experiential travel shaping new tourism products;
- Responsible management of the environmental impacts of tourism.
- conservation of endangered species;
- The meaningful engagement of communities and other stakeholders in planning and pioneering a way forward for tourism at a global and local level.

Having been involved in the development of the Ninki Nanka Trail in The Gambia—touted as a responsible community-based tourism experience: how does this project benefit the local community and wider destination?

The Ninki Nanka Trail is a Gambian initiative which was recently launched by the Gambia Tourism Board, that aims to pioneer a responsible approach to community-based tourism in rural communities along The River Gambia, bringing together a wide range of stakeholders and communities. It is envisaged that the trail will benefit the local community and the wider destination in a number of ways, such as creating additional livelihoods through tourism for rural communities, demonstrating and pioneering responsible tourism practices that can be rolled out in managing other destinations in The Gambia.

It helps in raising awareness of the local cultural diversity, natural and cultural heritage, providing a more sustainable alternative to the country's sun, sand and sea offer, and creating more mutually beneficial encounters between tourists and the locals.

In practice, a responsible approach to planning and product development has been adopted for the trail which has engaged local communities, tour operators and other stakeholders right from the start.

As part of the process, those involved at various stages have carried out

- impact assessments
- market research
- participatory interpretation
- planning to capture different stories along the trail
- responsible tour operator and guide training
- development of a code of conduct
- development of a product catalogue for local tour operators who meet responsible tourism criteria
- pilot trips with our Leeds Beckett students and other adventurous guests
- industry FAM trips
- community capacity building
- product development training
- collaborative branding and
- boat safety training

To keep the momentum going and to support a responsible return to tourism in The Gambia post-COVID-19, myself, Adama Bah and a great team in the UK and The Gambia have just co-founded Ninki Nanka Encounters. This is a Gambian charitable foundation that aims to benefit Gambian communities by supporting the development of the trail and a wide range of associated activities aimed at creating mutually beneficial tourist-host encounters.

What are some of the challenges that the team had to overcome when building this trail?

The reality of 'just' developing a trail along the River Gambia for tourists has been far from easy. The Gambia is an old hand at delivering holidays in the tourist development area and resorts along the coast for a predominantly package market. But community-based tourism is a comparatively new industry requiring different systems, infrastructure, and skills in a very different rural context.

There are inevitably challenges and concerns with trying to scale-up community-based tourism as a vehicle for poverty reduction in a fragile environment. The development of the trail has required working towards addressing issues such as:

- The need for community capacity building to deliver new tourism products and services;
- Lack of environmental management systems;
- Gaps in quality tourism accommodation along with some parts of the trail;
- Funding of tourism infrastructure;
- The tricky situation of ownership and management of the trail in practice, since it covers such a large area and includes such a wide and rich range of communities, World Heritage Sites, national parks, private and public owned land and assets, and The River Gambia itself.

However, these challenges have been embraced as part of the long-term plans and hopefully, the Ninki Nanka Trail will be able to play a special part in the future of tourism in The Gambia.

Link to the interview: https://sustainability-leaders.com/lucy-mccombes-interview/

Spain | Luigi Cabrini has been the Chairman of the Board of Directors of the Global Sustainable Tourism Council since 2014. He has led several initiatives on sustainable tourism at a global level, focusing his activities on tourism and climate change, biodiversity, observatories for sustainable tourism, the green economy, heritage and global partnerships. He was the Director of the UNWTO Sustainable Tourism Programme, Director for Europe and Secretary of the General Assembly and Executive Council. From 2014 to 2018, he was an Advisor to the Secretary-General of the UNWTO. Before joining the UNWTO, he was engaged for 20 years in United Nations programmes for development and for protection of refugees, in Guatemala, Mexico, Somalia and Pakistan. He was UNHCR Representative in Poland and Spain.

Areas of expertise: sustainable development, tourism, UNSDG 11 (Sustainable Human Settlements).

The following interview with Luigi was first published in May 2015 on Sustainability-Leaders.com.

F. Kaefer, *Sustainability Leadership in Tourism*, Future of Business and Finance, https://doi.org/10.1007/978-3-031-05314-6_67

Luigi, before you joined the United Nations in 1982, you worked as a freelance journalist in Rome, Italy—was there any specific moment or experience that triggered this career change?

At that time, I was writing articles in support of the 'No Nukes' (anti-nuclear) movement for the Italian edition of Rolling Stones. Joining the UN (thanks to an interview held with a UN recruitment delegation) gave me the possibility of actually doing things that could lead to positive changes, not only write about them.

My first assignment was in Guatemala, where I was coordinating projects for the United Nations Development Program. These ranged from turtle nest protection to improvement of maize crops. Unfortunately, those were very sad years for the country, ravaged by civil war and ruled by military dictators.

When did you discover your passion for sustainability and tourism?

I joined UNWTO after 20 years working in development and refugee protection with the United Nations High Commissioner for Refugees. It was quite a change! It was however becoming increasingly clear that tourism can also be a powerful tool for development and poverty alleviation. In many countries that I have visited, tourism provides a way forward.

I see tourism as the sunny side of countries and peoples, willing to show their best and to welcome anybody interested in their nature, culture, music, gastronomy. It is the opposite of the dark side, which produces wars, conflicts and displacement.

Which moments will you always remember from your time as head of UNWTO's sustainable tourism program?

I had the chance of getting to know amazing people and unique places. I was surprised by the contrasts in Asia: very modern cities with the latest technological outfits, together with idyllic landscapes recalling life as it was centuries ago.

Visiting remote villages in Nicaragua, reachable almost only by boat, where the Garifuna people were developing community-based tourism was an experience I will never forget.

How has your view on sustainability and tourism changed during your time at UNWTO, and now as chairman of the GSTC?

I have definitively learned a lot these last years. I have always been positive about tourism's potential to improve people's lives and to be a driver for enhancing the value of nature and culture, which implies, of course, their conservation, but also their sharing and interpretation.

There is of course a lot of badly managed tourism as well, but the pressure to be sustainable is quite strong today.

I have witnessed an evolution of sustainability from an elitist concept to a way of doing business, which is adopted by an increasing share of the industry.

Why did you decide to chair the GSTC?

I have been associated with the GSTC since its foundation. UNWTO has always been a strong supporter and as Head of Sustainability, I was the person dealing with it. Being elected Chairman of the GSTC has given me the possibility to continue to be engaged with sustainable tourism and with an organization that is actively contributing to making it happen.

GSTC is a 'virtual' organization with staff, board directors and members scattered all over the word. I am based in Madrid, where UNWTO has made available an office and is supporting some of the activities.

My institutional role is to represent publicly the GSTC and I participate in several events, making the case for our organization. However, I am also involved in several operational matters as a member of the Executive Committee.

In your view, how will the sustainability agenda in travel and tourism evolve in the next years? Which trends do you observe?

The number of tourism companies embracing sustainable tourism is continuously increasing, not least because natural and cultural resources are the capital of the tourism industry, so there is a strong interest in preserving them.

I also see mounting pressure toward tourism operators to do things better, mostly from younger generations who are more sensitive to the vulnerability of our planet.

Which achievements at UNWTO are you most proud of?

The network of UNWTO Observatories for Sustainable Tourism has been a good initiative that motivated destinations to establish a permanent mechanism to monitor their vulnerabilities and to take corrective actions when required.

I am also satisfied with the work done to include tourism as one of the sectors that can steer the change toward a Green Economy. Together with UNEP [United Nations Environment Programme], we produced good evidence of this potential.

Which is the most difficult part of working for a UN agency, such as UNWTO?

While UNWTO refrains from getting involved in political issues and instead encourages governments to build on their common interests, sometimes underlying conflicts surface and these need to be addressed with a combination of common sense and diplomatic skills.

As sustainability adviser to the UNWTO Secretary-General, where do you see the main challenges for sustainability in travel and tourism?

Almost all countries today include 'sustainability' in their tourism strategies. Tools are available, such as the Global Sustainable Tourism Criteria, but we are still far from the mainstreaming of sustainability parameters in the industry as a whole.

UNWTO has made it clear that with the predicted growth of tourism (1.8 billion international tourists in 2030), embracing sustainability is not an option, but a necessity.

Which is the best way to measure the success of sustainability initiatives in travel and tourism?

I believe that the GSTC has made an important contribution by offering a global system that is credible and recognized by more and more stakeholders, The Council's Criteria provide an objective measure of whether a business or destination meets the standards to be considered sustainable.

Your favourite travel or sustainability book right now?

Overbooked by Elizabeth Becker[1]—a well-documented and well-written book, to which I contributed my two cents.

Link to the interview: https://sustainability-leaders.com/interview-luigi-cabrini/

---

[1] https://sustainability-leaders.com/elizabeth-becker-interview/

Belgium | After a master's degree in marketing and communication, Maja Vanmierlo (formerly Campelj) did her postgraduate studies in Responsible Tourism Management at Leeds Beckett University in the UK, where she explored ways to implement sustainability criteria and UN Sustainable Development Goals in tourist experiences. With her work, she empowers tour guides to be cultural connectors, sustainability communicators, unforgettable experience creators, educators and ambassadors of sustainable tourism for all. She shares her passion and knowledge at G-Guides where she helps destinations, tour operators and experience creators to design and sell sustainable, responsible, and enjoyable experiences that support local communities, accelerate local development and contribute to the environment. Maja believes that the art of travelling is the art of communication. Communication between different cultures and diverse people. Her life mission is to help people create meaningful and life-lasting relations among different cultures and to interpret the unknown to the travellers of the world.

Areas of expertise: responsible tourism, tourism business, tour guiding, Europe, UNSDG 8 (Equal & Fair Economic Opportunities).

F. Kaefer, *Sustainability Leadership in Tourism*, Future of Business and Finance, https://doi.org/10.1007/978-3-031-05314-6_68

The following interview with Maja was first published in May 2019 on Sustainability-Leaders.com.

Maja, having been involved in tourism for over a decade now, do you remember what first got you interested? And has your view on tourism changed over the years?

I always wanted to see as many parts of this world as possible and meet different people and cultures. But this was not enough for me. I wanted to show these places and connect people from different cultures. And I could best bring this idea to life by being a tour guide and tour director.

My view on tourism hasn't changed that much over these years. For me it was always about immersing myself in other cultures.

Being in contact with tourists and travelers every day since more than 15 years, I see that people are no longer in search of sightseeing, but rather in search of experiences.

So, I would say that what first brought me to tourism is what is now motivating more and more tourists to travel.

You are the co-founder of G Guides—in a nutshell, what is it about?

G-Guides is a private School for tourist guides and a private research institute, where we strive to bring to life academic and governmental as well as international efforts to move towards more sustainable development and responsible tourism.

We are achieving this first through our research contributions, and second through education and trainings for different stakeholders in tourism, especially tour guides and tour directors. Our third pillar are events and conference contributions, where we try to raise awareness about the importance and role that tour guides have in the whole journey towards more sustainable development and responsible tourism.

You would be surprised how underestimated and overlooked this profession is by other stakeholders in tourism, due to lack of knowledge about the profession. Consequently, the communication power that tour guides possess is not used effectively. This is a big opportunity missed by destinations to communicate sustainable tourism efforts and to bring tourists onboard.

What motivated you to establish the G-Guides School for tour guides?

Like most purpose-driven companies, the main motivation was and is to change something that is unacceptable.

Firstly, I never understood why there is no mandatory training in Slovenia to obtain the national license for tour guides. When I started to explore this issue more in-depth, I learned that training and education all around the world don't prepare tourist guides for the role(s) they will have in the tourism industry.

The most disturbing, painful and unacceptable fact for me was that most of the training still promote an idea of tour guides being mostly about knowing historical facts and such, not taking into account that we can easily find most data online now.

I was (and still am) determined to change this with G-Guides. This was proven right as we were chosen by the EU Commission for the promotion of the profession in all EU countries, and our work was recognized by the UNWTO as a contribution to sustainable tourism and the SDG [sustainable development goals].

Ljubljana has done some impressive work over the past years in making the city more environmentally friendly and livable. What role do tour guides play in the responsible tourism context?

Tour guides play a bigger role in the responsible tourism context than most of the stakeholders in tourism understand. They are sustainability communicators, promoters of responsible tourism, accelerators of local economic development, as well as cultural brokers.

But they need to get this knowledge during their training. You cannot expect guides to provide five-star service, to communicate the values of the destination and sustainability and to properly communicate across cultures if they are not trained for this.

Tourist guides who have had proper training and are thus prepared for the role they will have as a main link between the visitors and destination can be the most valuable part of the whole marketing mix of the destination.

No successful brand would invest resources into development, marketing and branding, but then not care about the point of sale, where the product meets the (potential) customer. And yet, this is exactly what is happening in tourism at a destination level.

With the Green Microphone Award, you seek to reward and recognize tour guides who are going the extra mile for promoting responsible tourism through their work. How has your experience been so far, and how does the award process work?

Green Microphone—Voice of Responsible Tourism[1] is the award we give annually to the guide who has done the most to bring responsible tourism into practice. This was a brilliant idea of G-Guides co-founder Tina Hudnik, who could not stand the fact that there is an award for responsibility and sustainability for just about every segment of the tourism industry—except for tourist guides. So she thought: hey, why don't we start giving one. So, we decided to start with Green Microphone.

2018 was the first year of the award and we are positively surprised and honoured to have had tour guides from almost all continents participate—and to gain media support from the Ethical Travel Portal.

Tour guides can be nominated by somebody or can submit their own application online. Also, our final ceremony and the last part of the competition is 100% online. We want to minimize the carbon footprint of our event.

Your thoughts on the current state of tourism sustainability in Slovenia?

Many people in Slovenia have done an amazing job in the field of tourism in the last few years. I am very proud every time I see Slovenia on the list of top ten green, clean, safe, sustainable and must-visit destinations. At the same time, I am also very afraid.

There is one crucial mistake which many brands and destinations have done before: over-promise and under-deliver. It is indeed very flattering to be on all those lists, but this alone does not bring sustainable tourism into life. Today's tourists are

---

[1] https://www.green-microphone.info

very much aware of what is and what is not responsible for tourism and sustainable development, so we should by no means underestimate the knowledge and awareness of our guests. It can backfire easily.

Because sustainability seems to be a must-use word now in Slovenian tourism, there is a big potential danger of greenwashing.

How do Airbnb Experiences and other portals affect the traditional tour guiding business—and which other trends or challenges are you facing right now?

I am very happy you are asking this question. From my point of view, there was never a better time for being a tour guide than now. With the right set of knowledge, skills and basic use of modern technology you can have a thriving, purpose-driven, lifestyle business, where you can contribute to intercultural exchange and accelerate responsible tourism every day.

However, there are two main challenges in our profession at the moment. First, how to prove the positive impacts and benefits of tourist guides in terms of sustainable tourism and local economic development—and this is what I am addressing right now as part of my master studies at Leeds Beckett University in the UK.

The second challenge: how to persuade traditional tourist guides that they need to gain a whole new set of skills for the new century, in order to improve the reputation of our profession and to provide five-star services for visitors.

What does it take to become a good tour guide?

There is no single answer to this question because there is no single type of tourist. And this is what I am trying to tell first to tour guides and then also to the visitors. Today, the market is so segmented, that different target groups need different approaches.

That said, I would point out two skills that every single guide needs. Communication competence and cultural intelligence. This is far from just being able to tell a good story. A good guide will be able to communicate the same story to a wide variety of audiences, according to their knowledge, interests and cultural background. And this is not easy. And very few guides actually have these skills. Just think of telling the same joke to different age groups, in different languages to an audience with different cultural background. It doesn't work.

Tourism professionals sometimes avoid engaging with sustainability, since it is not something usually part of their KPIs. Reflecting on your own experience, which advice can you share with tour guides or tour operators in terms of how to deal with sustainability?

True, but I think it is because they do not have sufficient knowledge about what this is and how should they implement it to the core of their business. A worse case would be when they lack the knowledge but feel that they need to engage somehow and then engage in the wrong way.

My advice is very simple: talk to somebody who has the knowledge. Despite sustainability being a well-known term in academia, it is still relatively new to tourism businesses in practice.

Guides and tour operators can do so much to make the responsible tourism mainstream, but they need to put sustainability in the focus of their product design. And this needs to go far beyond avoiding plastic bottles and straws, which are the absolute buzzword now in tourism.

On the other hand, many businesses are doing a great job, but do not know how to communicate what they are doing to their audience.

Link to the interview: https://sustainability-leaders.com/maja-campelj-interview/

Egypt | Dr. Manal Kelig is the Executive Director—ATTA MENA region and the Co-Founder of GWE companies for Sustainable Tourism Development. Manal has a wide range of experience as a historian, entrepreneur, and sustainable tourism specialist, with expertise in tourism business development, industry collaboration, and customer orientation. Manal focuses on expanding the reach and impact of her work through developing and facilitating networking and professional growth opportunities with various travel organizations, media contacts, tourism boards and their representatives at all levels. She has over 20 years of work experience with sustainable tourism projects across countries with different socio-economic profiles. Throughout her career, she developed strong strategic planning and management experience that considers the multidimensional nature of tourism as well as the complexity of the stakeholder networks it involves. Manal's diversified and extensive skills led her to consult for many leading development organizations on sustainable tourism development and cultural heritage preservation.

Areas of expertise: destination sustainability, responsible tourism, sustainable development, Africa, UNSDG 5 (Empowering Women), UNSDG 8 (Equal & Fair Economic Opportunities).

© The Author(s), under exclusive license to Springer Nature Switzerland AG 2022   431
F. Kaefer, *Sustainability Leadership in Tourism*, Future of Business and Finance,
https://doi.org/10.1007/978-3-031-05314-6_69

The following interview with Manal was first published in April 2020 on Sustainability-Leaders.com.

Manal, what led you to dedicate your life to the development of sustainable tourism experiences in the Middle East?

When I went for a BA degree in Guidance, I was attracted to the study of Ancient & Modern History and World Archaeology subjects that formed the core of this degree. As I started working in the field, the available work opportunities were with mainstream tourism trips.

I always had an affinity for creating positive change and this got me to build stronger ties with the local communities. I instantly felt that tourism and community development seemed at odds. The growing need and demand for culturally and environmentally responsible tourism was very evident, and I explored means of how tourism can have a positive impact on nature and people.

From there I was drawn to sustainable tourism, as I saw how it could have a positive impact by affecting people in their daily lives and also create experiences for the travelers that can translate into positive action back at their homes.

Muslim-friendly travel is one of the topics you have been focusing on over the last years. In a nutshell, what would this entail?

In the past 40 years, Muslim-friendly travel, or Halal travel, has not been limited to Faith Tourism to the Holy Mecca in Saudi Arabia. Muslims travel on business, family holidays, honeymoons, and back-packing, just to name a few. The new Muslim traveler belongs to young demographics from the educated middle & upper class, who want the latest, most beautiful, and best brands.

Muslim-friendly travel products need to provide adequate services based on the observance of a range of Halal requirements, so Muslim travellers can experience vacation and recreation. The application of these religious principles to tourism is not to restrict the scope of Islamic tourism, but mainly to ensure that Muslim travellers will enjoy products and services of good quality within their spiritual and moral terms.

To date, the Muslim traveler has been commonly misunderstood, highly stereotyped and his/her needs oversimplified. There are economic opportunities and a need to genuinely understand the Muslim consumer.

Apart from being a tour operator, advisor, and the Executive Director MENA region for the Adventure Travel Trade Association, you are also a 'Philanthropreneur', as you call it. How has one role led to another?

I always had an affinity for creating positive change and my work in tourism was the main enabler for my role as a Philanthroprenuer.

As I recognized the linkages between tourism, environmental protection, economic development, social equity, innovation, and a variety of other areas, it became a natural process to take on a greater role in providing solutions while still managing my business.

Working with the local communities allowed me to better understand their specific social and environmental issues. The main key is to discuss these issues with the locals and to work together on finding innovative solutions to tackle them.

I learned to study how the various factors feeding into a specific issue interplay with each other and to identify their underlying causes, before acting. Some of the issues that were tackled included access to education, recycling and closing the gender gap.

Are there any barriers that you think prevent communities in Egypt or the wider MENA region from participating more significantly in the responsible travel industry?

In the MENA region, tourism is an industry that is often misunderstood due to its complexity. This lack of understanding of tourism as an economic sector with associated benefits in job creation and improved economic activity creates a lack of both financial and human resources dedicated to sustainable tourism development.

On many occasions, there has been poor strategic planning that led to missing out on including the SMEs and representatives from the local communities in the planning processes.

Also, the stakeholders lack the motivation and community participation to invest in sustainable tourism models.

How do you think tourism, as an industry, needs to change or evolve to accelerate the empowerment of women? What concrete steps can tourism organizations take towards this goal?

All types of tourism generate employment for women. Despite the advantages and advancements for women in some regions, significant inequality persists in other destinations, where women are concentrated in low-status jobs, are underrepresented, and not receiving enough education and/or training.

If tourism policymakers work closer with the private sector to promote gender equality and include women's empowerment as a fundamental component in the planning and implementation processes, tourism can then be harnessed as a vehicle for promoting gender equality, increase women's participation in the workforce, in entrepreneurship and in leadership—more than in other sectors of the economy.

There are many local champions—men and women—challenging the status quo, and they are taking actions to close the gender gap and to influence the development and growth of travel and tourism in their country. Now is the time for different tourism organizations to connect with these change-makers and to support their initiatives.

The current coronavirus pandemic has put travel upside down. Apart from the significant challenges this poses, do you think it also presents us with opportunities, especially regarding tourism sustainability?

The tourism industry is no stranger to 'crises'. We all have experienced all sorts of terrorism, Sars, economic recession, and others. However:

The COVID-19 has hit us so hard, and it will take time to comprehend its full impact. On a positive note, governments worldwide may finally begin to notice the tourism industry more due to its 'absence' and hopefully come forward to support its community with tangible initiatives geared towards freelancers, micro & small businesses and their staff to survive this crisis on the short term. And on the long term, help them with the needed resources to start their work on sustainable tourism development.

Your three bits of advice to the many passionate travel entrepreneurs out there, on how to manage this difficult time?

Well, 'business as usual' has changed. During the past months, many of us lead with empathy and support our travellers beyond the standard services, to ensure their safety and getting them home.

Now we are hit with a new reality that might need us to reinvent our entire business system. Still, we need to lead again with empathy as we look after our teams and reassure our entire supply chain that we will work together to reactivate our business when the time is right.

Let us try to use this time to gain new skills in business management, strategic thinking, and to understand the new tourism trends.

Link to the interview: https://sustainability-leaders.com/manal-kelig-interview/

India | In 2004, Manisha and her husband Himanshu, along with a few friends co-founded Village Ways to help sustain the villages of the Binsar Wildlife Sanctuary in the Himalayan state of Uttarakhand in India, which were under serious threat of outmigration due to lack of livelihood opportunities. Village Ways has established 26 community-owned, small guesthouses in India, Nepal and a community tourism project in Simiens—Ethiopia. Village Ways is a unique concept of tourism, collectively owned and run by the community, where benefits are shared with the entire village. Manisha is also the founding member of the International Centre of Responsible Tourism India Network. Village Ways has won several awards like the British Guild of Travel Writer's award, Guardian Travel Awards, Times Green Spaces Award, WTM Responsible Tourism Awards (2009, 2013, 2017, 2020), TOFT

© The Author(s), under exclusive license to Springer Nature Switzerland AG 2022
F. Kaefer, *Sustainability Leadership in Tourism*, Future of Business and Finance,
https://doi.org/10.1007/978-3-031-05314-6_70

wildlife Community Award, Responsible Employer Award by PHD chambers of commerce and Industry, Government of India etc.

Areas of expertise: community-based tourism, destination sustainability, hospitality, responsible tourism, sustainable development, India, Asia, UNSDG 5 (Empowering Women), UNSDG 8 (Equal & Fair Economic Opportunities).

The following interview with Manisha was first published in December 2019 on Sustainability-Leaders.com.

Manisha, with a degree in English Literature, Fashion and Textile Designing—what motivated you to create a tourism initiative like Village Ways that benefits rural communities?

The main reason for us to get together as a team and to start a community tourism initiative was to assist village communities living inside a protected area in the Binsar Wildlife Sanctuary of the Northern Indian state of Uttarakhand. These villages were under serious threat of out-migration, due to a lack of livelihood opportunities.

Tourism brought the much-needed income stream into these villages, led to a revival of local culture and traditions and a sense of pride in the communities about their surroundings. This was the starting point and paved the way for working in other villages and regions.

Village Tourism Enterprises helps set up small guesthouses in remote rural regions of India, Nepal, and Ethiopia. How do you market these guesthouses to attract travelers?

We market these enterprises directly through our own travel company, Village Ways Travels, which is based in India, and also through other like-minded partners who understand the ethos of our trips.

We create attractive and interesting itineraries, usually combining various guesthouses into one great holiday. In the Himalayas, we specialize in wonderful walking trips, where guests hike from village to village, accompanied by local guides. In the south, the focus is on relaxed exploration and discovery of local life.

Our guests don't just visit these villages as visitors: they become a part of the whole concept. The guests always say they are privileged to be able to interact with the local communities, understand their way of life while enjoying some of the most beautiful and serene locations.

Apart from the economic gains that the locals make with their involvement in rural tourism, what are some of the social benefits to these communities?

Our partner communities form strong village committees. Those are often the first time these villages have had a body to represent their interests. Village committees liaise with neighboring communities, which increases social reach and brings friendships and shared learnings.

We also ensure that everyone involved in tourism has the benefit of language classes and other training, to increase self-confidence and self-esteem. Community members become proud of their villages, and we have seen a decrease in littering, and a new respect for the landscapes and nature.

Mobilizing funds is a common challenge in community tourism initiatives. How do you collaborate with investors who wish to support responsible tourism?

With over 10 years of experience now, we have a good track record in delivering sustainable tourism projects. We started the project with self-raised funding and then created an enterprise concept where profits were reinvested in new projects, which were developed in India.

When we started expanding to other regions and countries, Village Ways Partnership was set up to collaborate with funding bodies who support the local communities of rural areas to develop responsible tourism practices.

Since then, we have worked in partnership with government agencies, such as forestry departments and state tourism authorities, as well as non-governmental agencies, such as wildlife charities. This aspect of our work is incredibly important for being able to deliver well-funded and sustainable guesthouses and to assist more villages that are in need of an additional source of livelihood.

How does Village Ways ensure equal representation between both genders when it comes to earning through tourism?

We are passionate advocates of gender equality and indeed of all kinds of equality. We work in some traditional, socially conservative areas, and we work closely and sympathetically to encourage equality.

We find that making sure women are represented at the committee level is key, as this empowers women and gives them a role in the decision-making process. We have also encouraged some of the first female guides to be trained in the Himalayan regions of India.

Has your view on tourism and sustainability changed since founding Village Ways?

Sustainability is the essence of what we offer. The whole reason for setting up Village Ways in the first place was to help sustain village life in areas that were being depopulated. We wanted to find a way to provide a further income stream and job opportunities that could run alongside existing livelihoods.

To some extent, I see sustainability as the only way to conduct a tourism business such as ours. Our entire ethos is about sustainable tourism, really from the bottom to the top of the business.

Sustainability does not relate to economy alone, but is also about creating positive impact through tourism regarding social, traditional and environmental aspects.

What challenges do you foresee for the future of sustainable travel, with respect to climate change?

We encourage all our communities to adapt to climate change. This can mean taking measures to conserve water, which is becoming an increasingly precious resource in the mountain regions, and also to adopt farming methods to changing climate patterns.

We also feel that it is important that sustainable tourism makes a strong case for itself in regard to international air travel.

Our partner communities would suffer economically and socially if visitor numbers fall, so we have to argue that, if you do travel by air, then make sure it is to a sustainable and responsible destination, where the trips bring benefits to hosts.

At the same time, we are now doing more marketing to domestic guests as we recognize that long-distance travel may become less frequent, as consumers make sustainable choices.

With more than a decade dedicated to community tourism, which would you consider the key milestones or achievements of your professional journey so far?

Overall, this has been an incredible professional journey for me and one which has changed my perspective on life. The motivation is being able to change the lives of people in a positive manner through this special kind of tourism.

Since we have advocated the involvement of women in our enterprises, the success of our female guides in many areas, a village committee in Kerala which is run by a full women team and the representation of women in village committees and their close involvement in decision making, have made us all very proud of our efforts.

In Madhya Pradesh, we are working in a village where there is no literacy and communities were not confident of being able to host and interact with visitors. With training and constant support to this village, we are now receiving very encouraging feedback from our guests. This includes one which I particularly like to quote, about a guide who has been appreciated by our guests for his confidence in explaining the story of Ramayana (epic) to them, and they thought he was now very capable to take guests independently on day walks and village tours.

In Binsar Sanctuary, a guide once said in a committee meeting that, "Earlier we used to throw stones at birds and now we open a book to identify which one it is." We see this as a big achievement.

There are many such stories from our villages, which are a huge source of motivation for me and our team.

What tips do you have for those who want to engage with sustainable tourism business practices but don't know where to start? What are some of the common pitfalls to avoid?

We feel that the most important factor is that guests must have a compelling reason to visit. The holiday experience must be rewarding and enjoyable as a whole, and it is vital to have a full picture of the kind of trips guests wish to make.

Too many sustainable destinations do not get the numbers to be sustainable long-term. Linked to this is the need to have a marketing plan to generate interest, and to be as flexible as possible in terms of the market, perhaps linking with tour operators and agents, as well as marketing directly to guests.

How do you measure your success in terms of sustainability performance?

We have worked under the guidance of Dr. Harold Goodwin on impact studies. In 2013, we started an economic survey system through which we are able to collect data from all villages, to assess the economic benefits reaching up to the household level. The first report has been published. We are still working on this system to improve it, so that it can be implemented more broadly.

Link to the interview: https://sustainability-leaders.com/manisha-pande-interview/

Nepal | Marcus Cotton first visited Nepal in 1983, where he worked for Nepal's leading conservation and development charity, the King Mahendra Trust for Nature Conservation (now the National Trust). Later, he joined Tiger Tops at the invitation of the owner, Jim Edwards, to oversee the group's environmental and responsible tourism activities and enhance administrative and financial management. Currently, he is working at Tiger Mountain Pokhara Lodge[1] since 2001, where he continues to be the co-owner and Managing Director, a mentor to the team running this iconic property that sets the standards for regenerative tourism in the middle hills of Nepal.

---

[1] https://tigermountainpokhara.com

F. Kaefer, *Sustainability Leadership in Tourism*, Future of Business and Finance,
https://doi.org/10.1007/978-3-031-05314-6_71

The Tiger Mountain Pokhara Lodge is one of the handfuls of properties in Nepal independently audited for its sustainability. A keen observer of Nepal's socio-political evolution, his keen interest is that tourism be a force for good in Nepal's economy, especially in the aftermath of the coronavirus pandemic.

Areas of expertise: ecotourism, hospitality, responsible tourism, sustainable development, tourism business, ecolodge, mountain tourism, Asia, UNSDG 12 (Resource Efficiency), UNSDG 8 (Equal & Fair Economic Opportunities).

The following interview with Marcus was first published in March 2020 on Sustainability-Leaders.com.

Marcus, Tiger Mountain Pokhara Lodge in Nepal is widely praised as a real treat for environmentally conscious, discerning travellers, and a worthy example for other accommodation providers to follow. What makes the lodge so unique and successful in offering sustainable, authentic experiences?

I think we have several advantages. First, our location; with such an awe-inspiring Himalayan panorama, all (our guests and staff, we all gather to take the same pictures!) are uplifted on a clear morning by the sight of the snow peaks. Like Switzerland's iconic Matterhorn, we have Machhapuchhare (Fishtail) framed right in our main doorway.

Coupled to that are our small size (only 18 rooms) and our almost zero staff turnover (all except of one have been with us for over 10 years and most of them from the start, 22 years ago). After all, it is that amazing team of dedicated staff that actually realize and implement our sustainability and authenticity. Add to that a wonderful heritage from Tiger Tops (Tiger) and Mountain Travel Nepal (Mountain) pioneers since the mid-60s in conservation tourism and our 'founding fathers.'

Authenticity is, like ecotourism, such a right word but has become so over-used as to be devalued. This is a shame as there is an authenticity—we sum it up in the concept that, even if we had no guests, we would do things just the same—this is who we are. Heritage and continuity are vital, as the processes of being sustainable and being original or authentic take time to evolve and mature.

Yes, superficial things like zero-plastic can be implemented rapidly, but the deeper psychology and ethos can only evolve over time. I think also, we dare to be a bit different, and do not follow standard tourism/hotel patterns, structures, etc.

Why do you put so much focus on sustainability at Tiger Mountain Lodge?

Because it is right!

For me, and all the team, sustainability is non-negotiable. If one is in a business for the long term, then sustainability is an inherent part of your strategy and corporate DNA. Well, if it is not, you are not going to be sustainable.

I like the various aide-mémoires around, be it People/ Planet/Profit or The Long Run's 4 Cs: Community, Conservation, Commerce and Culture—these provide focus and a framework to guide decisions and visions.

The inspiration came from my youth in England, seeing a village flora prepared by my mother in the Second World War when evacuated to stay with friends in rural Buckinghamshire and one summer holiday I tried to replicate this at our home only a few miles away—and could only find half the flower species. Some fifteen years later, repeating the exercise floral diversity had fallen a further 50%. This shock, coupled with the drive for agriculture on an extractive and for-the-subsidies approach, struck me as inherently wrong and a dangerously short-term approach. Time and events appear to have proved me right and the world has woken to the global environmental threats we face.

Leap forward to the late 1990s when I had the immense privilege to live and work inside the then-named Royal Chitwan National Park, heading up the incomparable staff team at Tiger Tops. There I saw, from the inside, the amazing diversity, richness and unutterable beauty of the subcontinent's wet sub-tropical jungle.

Being able to support the Department of National Parks and Nepal Army in their protection of this global asset, while, concurrently, providing livelihoods and social support for communities surrounding the park was an amazing and deeply fulfilling challenge. As was meeting so many 'big names' in conservation and related fields and sharing experience, ideas and concepts. Having the wonderful Dr. Charles (Chuck) McDougal as a mentor, guide and incomparable tiger conservationist iconoclast living at the lodge was the icing on the cake. No one could have a wiser, kinder or more mischievously funny guru.

Moving to Tiger Mountain in 2001 provided me with another challenge—after all, being sustainable inside a premier national park is relatively easy, the regulations are all on your side.

Doing true sustainability outside a park and in a community with diverse and different aspirations was a whole new focus for me and also for the lodge staff. Many of whom are from the local village. It required a re-focusing that took some time and a slowing up to accept that things taken for granted are not always the same for different communities. Written differently—don't go like a bull at a gate!

Which aspects of running the ecolodge sustainably do you find the most challenging?

A hard question to answer, perhaps because the staff at Tiger Mountain are so enthusiastic and supportive that it is quite easy!

For tangible issues, it can be difficult in the supply of suitable products—for example, there are no effective eco-detergents for hotel laundry in Nepal. I tried to import a leading European brand, but the hurdles proved insurmountable, before considering the cargo miles.

Swiss winemakers support a vineyard in Nepal, but they still have a way to go before we can confidently offer their product!

It is the intangible elements that are the challenge and can be complex. . . building the psyche of the staff to think sustainable, to be sustainable and to enthuse about sustainability was perhaps the biggest challenge. . .that subtle move from head to both head and heart, as it were.

How do you overcome those challenges?

For the supply issues, we just have to work around them—surface cleaner made from vinegar infused with local orange peels being one example: kills the bugs, smells lovely and natural, uses local raw materials, provides local commerce, easy to produce in our kitchen too!

For the intangible, I think it is persistence, reminding, learning together and growing together as a team. Inevitably there is an element of 20:20 vision with hindsight, so I can look back and see where we were; where we are, and how we came to be here—well, that's more complex as I know we are here, but I am not entirely sure how it happened! It did, it works, and we must have overcome the odd hurdle en route: leaped over, dodged around or flattened the hurdle? Probably a bit of all.

Back to my answer to your first question—having a great staff team is key. Also, having supportive guests who can see that one does not have to wear a hair-shirt to be sustainable.

To your mind, which are the main benefits of putting sustainability at the core of a hotel's business strategy?

A brutal focus on money to start! At Tiger Tops we were able to double operating (or gross) profit just by being more sustainable in our stores and supplies: moving from expensive imported canned goods to local fresh products, profit and pleasure for our guests. Having a python voluntarily in the kitchen proved an interesting and effective way to control pests too!

Sustainability tends to incur some capital costs—for example to offset the use of LPG in our staff kitchens meant constructing a bio-gas digester. So, we had to work hard to earn the profits to allow us to invest. This will pay dividends environmentally and financially in the years ahead.

Similarly, we need to invest in special cylinders to be able to buy commercial CNG from a local supplier, Gandaki Urja, that has just started production; CNG does not liquefy at relatively low pressures like LPG. Once we can afford this, we will offset or replace our guest kitchen LPG with locally harvested CNG.

Emotionally, knowing you are doing what you believe deeply to be right, drives confidence throughout the organisation and this has its own rewards across all aspects of the business.

Sustainability or 'responsible conservation tourism' as I like to define it, has received an enormous global boost from the likes of Sir David Attenborough and Greta Thunberg, as thought and activist leaders who have tapped effectively into mankind's deep unease with the state of the earth.

Thus, being sustainable is something our guests (and thus our paymasters) are increasingly seeing as the sine qua non, a must-have of a business. This is no longer a niche element of tourism and as today's challenges play out, it will move even closer to centre-stage.

How do you monitor sustainability at Tiger Mountain?

We both verify (audit against our own policy) and certify (audit against someone else's standard). However, we do have difficulties with certification as it always tends to be a one size fits-all approach and takes little or no account of local diversities and ground realities.

Thus, we prefer independent verification which we do through Yardstick UK. This is a small enterprise founded by Jenefer Bobbin who did her postgraduate studies in Responsible Tourism at Leeds Metropolitan University under Professor Harold Goodwin. She came to work alongside us to develop a verification scheme primarily to counter greenwashing. We have continued with her to evolve this to a wider model—the process is ongoing—and we hope shortly to be able to integrate our sustainability actions with the relevant Sustainable Development Goals for the tourism business. With this we will lead on, I hope, to working towards carbon neutrality.

For certification we currently use Travelife, an EU-based GSTC affiliate, achieving Gold standard in 2017 and we currently are awaiting the result of our recent audit, late last year.

The challenge for certification agencies is to come up with a scheme that takes account of locale, locality, and humanity and does not try to shoe-horn everyone into a single box. I believe this is possible but requires a greater focus on the auditor's discretion and local understanding.

I feel monitoring and knowing the facts of your performance is essential—for that all-important reality-check (pilots are taught early on to rely on their instruments, not their instincts), to justify and back up any claims made; to share with all the staff and stakeholders to show what you are doing, what you intend, where you want to go and how you hope to get there.

It is also an excellent means of promoting and marketing the property and business, and a way to raise awareness and consciousness about sustainability in both tourism as an industry and as individuals.

A fellow lodge operator once asked me about monitoring and what was the purpose. My answer was, whatever you want it to be. You can use it just for self-satisfaction and validation, for marketing, for developing community support fundraising, all sorts of things.

The coronavirus pandemic has been hitting tourism and travel business hard. How does it affect your business?

The current pandemic has an inevitable and significant impact on our business, being in the heart of the tourism sector. However, long experience in Nepal shows me you always have something going on! Political upheavals in 1990, 2006; Maoist Insurgency from 1996-2006; Earthquake 2015, etc.!!

A sustainable and responsible business has to manage these issues. I see everyone screaming for government bailouts (how is the government to pay for medicines, protective gear, etc., if we all have tax holidays?). What about shareholder bailouts?

To me, the shareholders are responsible for their businesses and must retain reserves and provide capital to cover these issues, like now the coronavirus pandemic. Their genesis and specifics are largely irrelevant—we know there will be something and therefore the business must be geared to covering whatever adverse issue arises.

We all have an absolute duty to our staff to support them in their careers—when the going is good we hear much of this from business leaders…who go strangely quiet when headwinds arise! It is not good enough to put out mealy-mouthed announcements that 'protection and safety of our staff are paramount' while issuing forced redundancy notices.

For Tiger Mountain, we have 'been here before' and we will work our way through the current crisis. An absence of guests does not mean a shutdown of the ongoing show—there are many opportunities to handle matters we are often too busy to address—wider training options, increased community liaison and idea-sharing, initiation of new conservation projects, etc.

Apart from the many challenges that COVID-19 poses, do you think this crisis also presents an opportunity?

Absolutely, this epidemic has provided a perfect pause to review issues such as overtourism. It gives us time to explore better ways of doing things, gentler ways of diversifying tourism and a chance to promote sustainability in its widest remit to the industry.

It is a golden moment to listen—to our staff, our communities, our stakeholders and to see how we can do better, how we can enhance our linkages, increase our beneficial impacts and mitigate negative ones.

Commercially, we can explore different, alternative and diverse source markets.

Imagine you are at a conference and someone approaches you who is just about to start with a lodge similar to Tiger Mountain. What advice would you share, in terms of how to get started, and pitfalls to avoid?

I will share a story about a gentleman that visited Tiger Mountain to talk about building a lodge on his land some 60 minutes away. Douglas and I had a long chat about tourism and wandered around the lodge grounds.

My concluding advice was, above all else, one must get the environmental aspects right from the very start.

There are things I'd have loved to incorporate into the lodge design that was not and cannot easily be retrofitted. This, I am thrilled to say, he has done and way above anything I had imagined; he has set the bar high for small lodges in Nepal. Beyond that, it is essential to take time—as long as it takes—to listen to the local community, understand their expectations, fears, desires, and aspirations.

Far too many times tourism rides roughshod over local stakeholders or pays lip-service only to stakeholder concerns. This is perhaps tourism's Achilles-heel and an aspect that needs far more focus.

Finally, be yourself, be your project, don't mimic others. I am still trying to understand what that ridiculous coloured runner 2/3 way down the bed in most hotels is for! To me, it serves only to show that the owners/managers lack vision and are copying… 'because that's what hotels do.'

Link to the interview: https://sustainability-leaders.com/marcus-cotton-interview/

Marcus Curcija on the Importance of Measuring Sustainable Development Success

Australia | As Co-founder & Managing Director at Insight into Impact, Dr. Marcus Curcija makes stakeholder engagement simple. By streamlining the qualitative inquiry and analysis process, Marcus helps organisations with ESG reporting, meeting the Sustainable Development Goals, and achieving greater outcomes by communicating stakeholder input at scale. His signature process, 'Perspective Analytics' improves organisational decision-making and helps businesses achieve greater success both internally and externally. His methodology provides a means to measure and monitor social performance, thus establishing greater accountability and transparency to mitigate impact washing and ensure funds achieve their intended purpose. For the past 15 years, Marcus has gained experience implementing sustainable practices and achieving community benefits all across the world. Marcus has also ventured into software development with Goalie, which enables end-users to utilise his patented process at scale. To date, his methodology has been adopted in 36 countries across industries and sectors.

© The Author(s), under exclusive license to Springer Nature Switzerland AG 2022 445
F. Kaefer, *Sustainability Leadership in Tourism*, Future of Business and Finance,
https://doi.org/10.1007/978-3-031-05314-6_72

Areas of expertise: tourism business, sustainability management, Asia, UNSDG 12 (Resource Efficiency), UNSDG 8 (Equal & Fair Economic Opportunities).

The following interview with Marcus was first published in September 2020 on Sustainability-Leaders.com.

Marcus, as a consultant and software developer you combine your passion for community development with the need to measure social impact—something which projects aimed at promoting sustainable development sometimes lag. What inspired you to pursue this career path?

My passion for community development sprouted from my roots growing up in the Pennsylvania rust belt. As the unemployment rate rose, the population continued to sink and the opportunities for prosperity seemed to be eradicated from the region. I continued to ask myself if anyone was paying attention to the increasing percentage of individuals and families that were doing it tough.

Driven by wanting to understand how community dynamics can impact economic growth, for my PhD I focused on conflict management during community-based tourism initiatives. From this, I was able to better understand how the perceptions of stakeholder groups could influence a project's success. I evaluated examples from across the globe before landing consultancy and managerial contracts.

I found the lack of accountability between organisations, their beneficiaries, their reporting, and their funding agencies to be alarming. A lot of organisations were not measuring their impact or gaining an appreciation for whether or not their services were even needed, and companies were not getting input from the community.

I came across many organisations that used storytelling to validate their existence. Storytelling is a great marketing tool; however, when trying to understand the impact, it does not convey the whole picture. Typically, the story only covers the most beneficial aspects that occurred for one particular stakeholder group or another.

I wanted to find a solution that encapsulated all stakeholder groups' perceived impacts, particularly the project's intended beneficiaries. For far too long their (intended beneficiaries) opinions have been overlooked, so I sought to find a way to place community input into the heart of decision-making.

I realised that the biggest opportunity for improving decision-making was in turning qualitative data into quantitative insights. So, I sought a more complete means to obtain, synthesise, and present rich, robust data.

Fortunately, this means happened to co-exist with a new, exciting way to measure social impact. I was also able to use the outputs from my process to aid storytelling and more importantly validate claims of positive impact.

How important is data visualisation in influencing organisations to implement sustainability in their operations and economic strategy?

Let's be honest, in a world where most of us are time-poor and accustomed to instant gratification, the vast majority of individuals aren't keen to read reports. Plus, the decision-makers simply do not have the time to really understand the context or the dynamics that exist between stakeholder groups.

So, at Third i Management we found data visualisation to be a key component when influencing others to be more sustainable, socially, and environmentally conscious. We've found data visualisation much more efficient than reports because of three factors:

- understand the complexity that often exists within a dataset in a simplified way
- significantly less time is required to become familiar with all of the content
- it eliminates many of the barriers that may exist with the reader, such as education level, professional experience, and cultural differences

Reading about interrelationships can be confusing for our audience, which ranges from top-tier C-level executives to community members who scarcely speak English, if at all. What makes data visualisation so advantageous for Third i is its ability to clearly and concisely present 'interrelationships.' Complex scenarios are explained through 'interrelationships.'

So, using data visualisation, we are able to establish greater communication between stakeholder groups. Once these stakeholder groups can communicate, the possibility of collaboration becomes a reality.

While working in community development, I noted that often different stakeholder groups wanted the same thing but were at a stalemate simply because they were unable to communicate their perspectives in a manner that made sense to the others. Unfortunately, as the conflict is occurring in a community setting, empathy goes out the window. Hence, I identified that if we could aid that communication through data visualisation, then empathy, mutual understanding, and collaboration would not be far off.

Individuals around the world have lost their livelihoods due to the coronavirus pandemic. How will this affect our ability to make progress in reaching sustainable development goals?

From what I've witnessed, the global consciousness seems to be awakening faster now than before the pandemic. That said, I say this from a place of privilege. So, when talking globally about what will it take to reach the SDGs, unfortunately, many, many people are doing it really tough, and their livelihoods have been severely affected. However, I don't think the pandemic has impacted our ability to reach them.

Let's not be naive, the SDGs are and have always been GOALS. What was always needed and remains needed is for governments and the vast wealth and influence of the private sector to really step up. Unfortunately, here we are, less than 10 years away and very, very few countries have national policies to address the goals. News over the past few months does seem to indicate that more are working towards these changes now though.

There needs to be a prioritisation implemented by governments across the world that, if you are adhering to sustainable principles and practices, you are not taxed more for wanting better.

That is what I see as the biggest hurdle: purchasing power needs to favour the conscious consumer. They should not be charged extra for remaining a niche market.

I have been supporting the GOALS because I believe in them, I believe in people, I believe in coming together for the greater good and to fix this mess we are in. I believe in stepping up and doing what needs to be done to convince those that repute science and the necessary lifestyle changes. This is not a question about right or wrong, it's a movement to make our own lives and everyone else on this planet healthier, happier, and communities more sustainable.

How keen are organisations and governments right now to invest in social impact measurement of sustainability initiatives, following the reduced economic activity through most of 2020?

In many ways, I think they are still trying to figure it all out. There are many organisations or individuals out there who say they can 'measure' social impact; however, many of them are measuring outputs. At Third i Management, we measure and monitor outcomes.

Among the biggest challenges to social impact measurement is the lack of a standardised approach that can be implemented at scale.

Another challenge that we frequently face is that organisations and governments are hesitant for us to identify how negative impact occurs, and what can be done better. I have to constantly remind them that we need to identify both what is being done well and not so well, to assess the context and provide solutions that will achieve greater outcomes.

Once they realise that measuring impact will help them more than hurt them, they are very keen. Only through measurement and monitoring are we able to experience sustained success.

Which of the UN sustainable development goals do you consider the most important to focus on right now?

I believe that a joint focus on Goal 11 (Make cities and human settlements inclusive, safe, resilient, and sustainable) and Goal 12 (Ensure sustainable consumption and production patterns) seem to address many of the focal points of the other Goals. Meaning, as you read down through the 21 Targets associated with those 2 Goals, everything from housing to waste management to inclusion & equality to air & water quality to recycling and the environment are covered.

Target 11.4 (Strengthen efforts to protect and safeguard the world's cultural and natural heritage) is the one I am most passionate about. After all, there is an argument to be made that if mainstream culture learned from indigenous customs and land management, then many of the challenges that seem to be escalating throughout society and with land use would be mitigated over time.

From a professional standpoint, SDG Target 8.9 (By 2030, devise and implement policies to promote sustainable tourism that creates jobs and promotes local culture and products) and SDG Target 12.b (Develop and implement tools to monitor sustainable development impacts for sustainable tourism that creates jobs and promotes local culture and products) have been the focus of a Sustainable Tourism Campaign that we recently initiated at Third i Management.

We work a lot in the community-based tourism space and the pandemic has hit many people hard. Due to this halt in global travel, we are working with local community tourism groups, their industry, and government stakeholders to identify pre-pandemic challenges and impacts. We help them find common ground so they can start discussing and working towards a better tourism offering, ready to roll out once the pandemic is over.

Knowing that people are looking ahead, have hope, and still think about working together has helped bring some of our communities together and form a stronger bond in these tough times.

Economic sustainability is on everyone's mind right now, do you anticipate growing interest in impact investing?

Yes, I do. Right now, people are more concerned about wanting to purchase from conscious companies. Well, the same goes for the funds, companies, and organisations that want to invest in opportunities that improve people's lives or the environment, while also receiving a financial return.

What is holding impact investing back is the lack of a standardised approach to measurement. Investors are seeking greater accountability throughout the industry because they want fund managers and intermediaries to implement measurement and monitoring standards. A great deal of current and potential investors are seeking to better understand the impact that is occurring in the community as a result of their investment.

However, in the Annual Impact Investor Survey 2020 published by the GIIN (Global Impact Investing Network) a few months ago, half of the impact investors surveyed state that measurement remains a 'significant challenge' and nearly 40% still do not independently verify their impact, which means they have no idea if their investment is having any benefit on communities or the planet.

In our conversation before the interview, you mentioned the need to mitigate 'impact washing' and to aid 'attribution'—what do you mean by this, and how can it be done?

Great question, I'm glad you brought that up. Finding a solution to mitigate 'impact washing' was another one of our guiding lights when conceptualising Third i Management and the process we use.

'Impact washing' is when a company, organisation, or fund makes unsubstantiated claims about the positive impact they are making. They may do this through their reporting to obtain funding dollars or through the promotion of their existence (such as the speculative storytelling I mentioned earlier) and/or activities. Think about it as 'sustainability fraud.'

In the GIIN Annual Survey 2020 that I previously mentioned, 66% of the impact investment industry's top professionals consider 'impact washing' to be the biggest challenge to the industry.

In an act of fraud, companies are stating that they are doing something good for the world when it could not be further from the truth. They also greatly over-exasperate their claims, their impact to bring in money. Hence, by making these claims, they are attempting to gain market share.

However, I'm happy to say that many consumers and investors are growing more sophisticated—seeking companies to measure and provide proof of their impact. That is where we (Third i Management) come in.

Attribution is what more sophisticated impact investors seek from their investment. It is a means to give value to a portfolio, which consists of gaining insight into the quantifiable benefits of that fund in the lives of its intended beneficiaries. No longer is taking someone's word for it good enough. These investors seek attribution through measurement.

Our form of measurement takes into account the perspectives of the beneficiaries, as well as other stakeholders associated with the project, to determine outcome-based measurement scores, which are used to demonstrate how the impact of a project is being perceived.

Which aspects of being an impact consultant do you find the most rewarding? And which the most challenging?

I enjoy working with social enterprises and community groups. It is very rewarding to hear about the positive impacts that have ensued following the delivery of our insights and solutions. Nothing is more personally and professionally rewarding than to give back to the community.

Concerning challenges, when it comes to social impact, many people simply do not understand what it means or why it is important to evaluate. Quite often they think they are doing enough because they are doing what has historically worked for their position or organisation. However, that is why so many sustainability-based projects have failed or not reached the success that was intended. Thus, sometimes trying to convince others about the importance of measurement can be challenging, but absolutely rewarding once they embrace/accept the change.

The same goes for informing people about the importance of the Sustainable Development Goals. It's truly a victorious feeling when you help an organisation realise that they can align with the Goals to grow their market share, build a better internal culture, and bring benefits to the world.

Anything else you'd like to mention?

For far too long, decision-making has been a top-down exercise. That's shameful. It is of absolute importance that the perspectives of those being impacted by a project are considered and used to formulate the decision intended to benefit them.

Nobody has the right to tell a group how they should live or what they need. Although it happens inadvertently, by not listening to the collective voice of the community, too many companies are forcing their opinions and placing their needs in front of those who live there.

I've heard too many times how hard it is to get input from some community groups. That's absolute rubbish. I have never faced any challenges when obtaining the input that I need for an assessment. It comes down to this: Be honest, build trust by being authentic and empathetic, value their viewpoint and their voice. Listen without judgement or expectation.

Only through taking into account the community's perspective, by understanding their wants and needs can true success, and subsequently sustainability, be achieved. It's important to not forget that we are all on this journey together.

Link to the interview: https://sustainability-leaders.com/marcus-curcija-interview/

Brazil | Dr. Mariana Madureira is a social entrepreneur, co-founder and director at Raízes Desenvolvimento Sustentável, a Bcorp engaged in vulnerable communities and women empowerment. Member of the 36×36 Network of women for an Economy in the Service of Life, member of the advisory board of Projeto Bagagem (NGO of community-based tourism in Brazil) and The Impact Hub BH. Mariana has been working on community-based and sustainable tourism projects, entrepreneurship, strategic management of associations and networks, and coordinating several projects that involve multidisciplinary teams and collaboration. PhD in Psychosociology of Communities and Social Ecology, Master in Cultural Heritage, a specialist in Urban Environmental Planning and Social Business Management, with a graduate degree in Tourism.

Areas of expertise: sustainable development, community-based tourism, tourism business, Brazil, South America, UNSDG 8 (Equal & Fair Economic Opportunities).

The following interview with Mariana was first published in August 2017 on Sustainability-Leaders.com.

Mariana, your home country of Brazil is widely admired for its breathtaking landscapes and rich cultural traditions, but also pitied for its social inequality and

F. Kaefer, *Sustainability Leadership in Tourism*, Future of Business and Finance, https://doi.org/10.1007/978-3-031-05314-6_73

large-scale environmental degradation. Do you remember when you first thought about the sustainability of tourism? What got you interested in the topic?

I've started thinking about tourism as a force for good when I travelled to the USA for an exchange program during high school. I had the chance to live with a family totally different from my own and to experience a different view of the world. When I came back, I noticed I was a much more open and tolerant person than most of my acquaintances. I realized then how important travelling is for self-consciousness, and the potential tourism has for teaching us about people, history and the environment, as well as connecting us as human beings. So, at the age of 16, I decided to study tourism.

At university, I had the opportunity to think about the wide impacts of tourism on destination communities and the environment. I became particularly interested in the topics of gentrification, heritage and authenticity—on which I focused my master thesis.

Reinforcing the importance of local people and their culture, and valuing simplicity, seems to be a great tool to fight standardization and culture loss. If we add up the income generated through tourism as a stimulus for keeping people in their original places, community-based tourism [CBT] can really help sustainable development.

With Raízes we developed a five-year project in the Jequitinhonha Valley[1] that we are very proud of and which we consider a successful case of CBT.

I'm also a volunteer for Projeto Bagagem,[2] an NGO that supports Turisol (Brazilian Network of Solidarity and Community Based Tourism) and promotes CBT in Brazil.

And I've decided to keep studying to better understand the potential of tourism for change: I'm currently a PhD researcher in psychosociology of communities and social ecology.

You are involved with Raízes Desenvolvimento Sustentável since 2006. Can you briefly explain what this is about and how (as a social enterprise) your approach is perhaps different from that of more conventional tourism advisors?

Being a social business is a double challenge: we must, at the same time, deliver to the needs of the market (as a company) and effectively support a cause (as NGO).

Our approach with Raízes is different from most conventional tourism advisors. Although there are some very committed ones, the majority tend to consider 'result' a report very well done with the attendance list, photos, explanations of the methodology and justification for the action delivered.

Of course, reports are important and we do work with them. But for us, result is the actual change in the reality of the communities we work with and the effectiveness of the actions in the mid to long-term. This means that we often spend our own resources on monitoring community projects that are 'considered done' by the clients.

---

[1] https://raizesds.com.br/en/projeto/sustainable-culture-at-jequitinhonha-valley2/
[2] https://projetobagagem.wixsite.com/projeto-bagagem

Which are the most common issues communities in tourist destinations in Brazil face linked to tourism and sustainability?

Unfortunately, we still have many issues to overcome to properly develop tourism in Brazil, especially sustainable and community-based tourism. We can see issues related to all tourism actors: business owners, employees, local inhabitants, NGOs and even tourists. I'll focus on some of the governmental issues here, as they seem to be the most pressing nowadays.

Brazil still lacks many public policies, such as one that supports small business and community-owned properties in tourism, and prevents international chains to take home a substantial part of the income generated in Brazil.

Although Brazil has some very well written policies (that could even serve as examples for other countries), many of them end up failing at the time of implementation—be it because of technical incompetence, insufficient inspection funds or corruption issues.

Our indigenous policies are a good example. In 2015 Funai (the government agency for indigenous matters) launched a policy for tourism in indigenous lands. The implementation, however, is now delayed by the resignation of Funai's president (for obscure reasons), and frequent land invasions due to lack of proper protection.

With its focus on fossil fuels, mining and soy (among other agroindustry products), the Brazilian government has never paid much attention to tourism. Our tourism ministry was created only in 2003. It had a good start, producing valuable short-term plans, but implementation is lagging due to corruption charges against some of the ministry's executives in 2011. Since then, the Ministry of Tourism hasn't done anything really relevant.

And right now, we are facing an economic, political and moral crisis across Brazil. This slows down tourism activities and developments. We know that many changes are necessary, and we hope that we can use the momentum created through this crisis to tackle them.

In 2016 you were involved in the Green Passport initiative. What was this project about? And your key insights/lessons learned from being involved?

The Green Passport is a very relevant campaign created by the United Nation Environment Program (UNEP) to raise awareness on production and consumption in tourism.

It's about time people realize that tourism is not a 'clean industry'. To the contrary, it is damaging our natural environment in many ways, especially through greenhouse gas emissions and excessive waste production.

Some of our choices can make our trips much more environmentally friendly, but we need to acquire a deep understanding in order to make the right choices and to convince others to consume differently—and less.

For the Olympic Games in Rio, the UNEP joined forces with the Olympic Committee for the Green Passport initiatives. I was very happy that they invited Raízes to support the initiative. It was an opportunity to take advantage of the Games' enormous audience to talk about sustainability in tourism. We tried to make it as fun and interactive as possible.

Brazilians were our main target since we realized that their consciousness about the impact of travel was low—probably because the sustainability theme is still being associated with big companies and industries or considered an academic term that doesn't refer to our daily routine. Transmitting the sustainability message through simple actions was our strategy to help Brazilians recognize the role their own actions play regarding tourism sustainability.

As a female social entrepreneur in Brazil, which three aspects of starting, running and growing a social enterprise do you find the most difficult?

Starting: creating a business model for a social enterprise is a big challenge. You have to match the needs of the community, people or cause you to want to benefit and the interest of the clients that might want to pay for it.

Running a business in Brazil requires us to be flexible and to multitask. Hiring is expensive and, in the beginning, entrepreneurs usually deal with all the areas of the company.

Growing a social enterprise can be easier if you develop a scalable model. There are some funds especially designed for social business scalability in Brazil: Artemisia, Vox Capital, Sitawi, Quintessa and others.

Which trends do you observe in tourism right now, which might impact (positively or negatively) the sustainability of destinations in Brazil—and the well-being of communities?

Globally I see the 2017 International Year of Sustainable Tourism declared by the UNWTO an opportunity to think about and discuss tourism and its (un)-sustainability. Unfortunately, I don't think that these discussions will lead to real change or transformation, because the necessary changes for more sustainable tourism might be very inconvenient for some industries, and even for consumers/tourists.

In Brazil, community tourism is a trend. Brazilian tourists are just starting to be interested in community destinations sold to them as 'experience tourism'. Overseas visitors have always been keen to explore different cultures and to visit indigenous tribes and traditional communities such as quilombolas, sertanejos and ribeirinhos—or favelas.

A growing interest in community-based tourism experiences is a great potential for economic development, but also a risk to their way of life and the environment in which they live. We need to be very conscious of this and careful.

Reflecting on the lessons that you have learned so far through your professional work, what 3 bits of advice would you give to women in Brazil keen to start their own responsible tourism business?

Be very conscious of your purpose (why do you want to start a business?), your abilities and resources (what and how). Make sure they match the expectations or needs of the territory/market you are planning to work in. Finding the right balance between what you like to do, the needs of the planet and what people are prepared to pay for is a big challenge. Chase it.

Don't think too small. We, women, tend to leave the big achievement for men and content ourselves with crumbs. It's about time we show presence in the higher positions. Not because we need to compete with men, but because the world lacks feminine energy. Women in leadership tend to embrace causes, care for the people and the planet much more than men.

Be persistent. It takes time and a lot of patience to achieve your goals. Be financially prepared because it might cost you some money too. I assure you'll learn a lot by running your own business, but sometimes you'll literally have to buy this knowledge.

Link to the interview: https://sustainability-leaders.com/interview-mariana-madureira/

UK ǀ Marta Mills is a sustainable and regenerative tourism specialist and educator, experienced communicator and researcher with a particular interest in tourism in mountains and Protected Areas. Regenerative Tourism Specialist at the Tourism CoLab,[1] and Researcher and Sustainability Consultant at the Mount Everest Foundation.[2] Over 15 years of experience working on sustainable development programmes in the Caucasus, Western Balkans and the EU, advising and educating on improving responsible tourism practices. In love with Georgia and the Caucasus mountains, where she has returned to 20 times since 2001, and one of the first enthusiasts that started building the Transcaucasian Trail.

Areas of expertise: destination sustainability, sustainable development, UNSDG 1 (Eradicate Poverty), UNSDG 15 (Forests & Biodiversity), UNSDG 8 (Equal & Fair Economic Opportunities).

---

[1] https://www.thetourismcolab.com.au

[2] https://www.mef.org.uk

The following interview with Marta was first published in July 2018 on Sustainability-Leaders.com.

Marta, you describe yourself as "a well-travelled backpacker, hiker, yoga teacher and a fan of the Caucasus". Having experienced so many places, what made you fall in love with this specific region?

I had been to over 50 countries when my friend and I decided to hike in the high mountains of Svaneti in Georgia, back in 2001. We had to travel overland across Poland, Ukraine and southern Russia for about a week which turned out to be quite stressful—the constant demands for bribes from the Russian authorities and the police were exhausting. When we finally made it to Georgia across the Black Sea, tired and not knowing what to expect, the Georgian border guards welcomed us like kings, offered wine and good food and organized a lift to the mountains. And it carried on like this: wherever we went, the local people kept offering us great food, wine and free accommodation, and we immediately fell in love with the food, wine, the Georgian hospitality and the incredible nature.

With an unusually high number of endemic plant and animal species, the Caucasus is one of the only 35 biodiversity hotspots in the world. The wild nature, the rich, complicated, diverse and fascinating cultural heritage and the geographical position between Europe and Asia make the region unique and enthralling. I have been to Georgia 14 times and 4 times to Armenia with 2 more trips booked already for this Autumn, and I still find something new that amazes me every time I go.

By the way, my main motivation to train as a yoga teacher came from a desire to settle somewhere deep in the Caucasus mountains and teach yoga, which I will do one day when I have a bit more time!

Your expertise as a sustainable tourism consultant spans across many areas, including sustainable destination strategies, local economic development plans and marketing and communications strategies. In your experience, are clients aware of how interconnected 'tourism' is with other areas and disciplines, such as economic development?

The international donor organizations I have worked for in the Caucasus are definitely aware of this. I have recently assessed GIZ Georgia (German Development Agency) to see how tourism and other industries (the wine sector, for example) can be linked to increasing the competitiveness of the Small and Medium Enterprises, and how this would impact on the economic development of the regions.

The local governments see tourism as the main driver for local economic development, particularly in the more remote mountainous areas where tourism is becoming a key source of income. However, way too often they have no vision/strategy on how to 'utilize' that growth in tourism for poverty reduction, how to ensure that it supports the local population by creating fair job opportunities now and in the years to come. They have no plans for long-term tourism, particularly for sustainable tourism where environmental, social and cultural impacts should also be taken into consideration.

Which are the main topics or concerns linked to tourism sustainability at the moment in the Caucasus region, especially Georgia and Armenia?

I do worry about the unsustainable and rapid growth of tourism in Georgia.

I was speaking at the Sustainable Mountain Tourism Forum in Georgia last November in 2017 and had less than 10 min to present the challenges to sustainable tourism development in the Caucasus. There are quite a few—the lack of awareness of climate change and its effects on mountain tourism; poor waste management; no consideration for the environmental and social impacts; the lack of leadership to drive sustainability—and it was impossible to fit them all in a short presentation.

So, I used these 10 minutes to challenge the audience—a range of tourism businesses, government representatives, donors, tourism NGOs—for showing little awareness, little understanding and no interest in developing sustainable tourism. I said that "the biggest challenge is to convince YOU—local tourism stakeholders— that the only option you have is sustainable tourism, and you need to work together to achieve that!".

They need to show more commitment to the principles of sustainable tourism as the economic angle still prevails over social and environmental issues. Have systems in place to monitor and measure impacts. Do more to preserve the local heritage and biodiversity. Focus on the quality of tourists rather than increasing the quantity. And cooperate, communicate and engage all relevant stakeholders in all stages of tourism development, from planning to implementation.

These are the key topics that need to be addressed to make tourism in the Caucasus more sustainable. If not managed responsibly, the key assets—nature and culture—will be lost.

To your mind, which are the global trends in sustainable tourism?

One of the biggest trends is growth in experiential travel, specialized and educational holidays ('learning-while-travelling'). An increasing number of travellers look for unique, exclusive, personalised and authentic experiences that will also benefit local people and enable interactions with them. They want to buy local products, stay with the local families, contribute to local social and environmental projects, actively participate in local festivals.

Growing public awareness not only of environmental issues but also of human rights and working conditions, as well as the welfare of animals in tourism, is also encouraging. This will create more and more customer pressure on travel companies and destinations to support the local people and local economy, engage in social enterprise projects and allow the customers to contribute and give back through their travel.

I also believe that safety will remain as one of the key factors in choosing a destination, regardless of the attitude to sustainability. As this useful report by the Dutch government states, geopolitical instability does not deter European travellers, but it influences their choice of destination. This is particularly important in the conflict-prone places when even a small mention of possible instability will affect visitor numbers—and something that the Caucasus tourism stakeholders should never forget about.

Together with the World Bank you are currently setting up sustainable destination management organizations in Georgia, as well as helping establish strategies for regional and national tourism marketing in the country. Can you

tell us more about this work, especially how you integrate destination marketing with—management?

We are setting up DMOs in two regions, Kakheti and Imereti, the first DMOs in Georgia. Ideally, they will comprise of the members of both public and private sectors, all of them contributing resources to the DMO but also benefitting from it. The work has been very interesting because a DMO concept is new to Georgia and we are working with all stakeholders to work out the most appropriate structure for each region, the type of governance that will actually work in the local context. We explain the benefits, opportunities but also the challenges that lay ahead of each DMO.

At the same time, we have developed new marketing and branding strategies for Georgia, as well as separate ones for Kakheti and Imereti. More integrated and region-wide destination marketing is one of the key priorities for the new DMOs, and ideally, the future marketing initiatives and priorities will be decided jointly by all DMO members.

I have delivered training on sustainable tourism and sustainability criteria for destinations to the new and prospective DMO members and staff and ensured that the actions in the DMO's Action Plans follow the principles of responsible tourism. Time will show whether they are actually followed, but I have provided the initial guidance and hope to support them as they develop over the coming months.

As an expert in mountain tourism, which are the main challenges in developing and managing e.g., alpine destinations sustainably?

I have already mentioned the key challenges linked to the lack of leadership in implementing sustainable tourism and poor cooperation between all stakeholders. In the Caucasus, I sometimes feel that the need and importance to protect and conserve the biodiversity is not a high priority for the local decision-makers who look at short-term economic gains. They take the mountains and nature for granted, without seeing the negative impact of unsustainable use of natural resources. In Svaneti, for example, 35 hydropower dams are planned to be built that will flood forests and communal lands in adjacent areas.

Additionally, there is the usual issue of human capacity: many of the local people who work in tourism in mountainous rural areas are untrained and unprepared for receiving the higher-spending and most desirable EU tourists to the standard those expect. The quality of accommodation, customer service, waste management, infra-structure and product offering need a lot of improvement. Developing tourism products, particularly in and around Protected Areas, needs to be done responsibly with biodiversity protection in mind.

Many local people in the most touristy—but also quite remote places, such as Svaneti, abandon their traditional jobs in agriculture and 'go into tourism', without any prior experience and basic knowledge about the industry, tourists' expectations etc. That causes all sorts of issues, for example, problems with local food supplies or conflicts between tourists and hosts.

These are the very local and Caucasus-specific issues, but more globally, manag-ing the effects of climate change with melting snow, retreating glaciers and unreli-able weather patterns are huge challenges.

Do you have best practice examples?

As for best practice, look at central Asia, Tajikistan and Kyrgyzstan. Many small destinations there promote sustainable tourism as a way of supporting local people by creating fair job opportunities, including the most disadvantaged groups. The Pamir Mountains of Tajikistan was recognised as one of the World's Top 10 Sustainable Destinations at ITB Berlin in March 2018 for their work with the local communities and on preserving the historical heritage and natural resources of the Pamirs.

The Kyrgyz Community Based Tourism Association 'Hospitality Kyrgyzstan' is an umbrella organisation uniting several destination communities (CBT groups) that trains local people in remote mountainous communities on tourism product development and helps with marketing. Its work has created new opportunities in the villages, where there was no prior tourism infrastructure.

In the Alps, many French resorts—Châtel, Chamonix, Les Orres and others—are worth mentioning for having strategies and measures in place for effective energy and waste management, educational and environmental awareness programmes in schools and local communities, and for promoting sustainable public transport options. Arêches-Beaufort was the first municipality to subscribe to the Mountain Resorts Sustainable Development Charter and promote innovations in more sustainable snowmaking and energy sourcing.

Successful destination marketing used to be mostly about innovative, engaging campaigns and selling a destination's offerings to potential visitors. Yet, in times of overtourism, it is more and more about brand stewardship and active networking internally. In your view, are DMOs prepared for this paradigm shift—ready to take on a role of facilitator and brand/destination manager, rather than 'just' promoter?

What can hold DMOs back is seeing themselves as Destination Marketing Organizations, rather than Destination Management Organizations that is much more than marketing. We have put a lot of work with 'our' new DMOs in Georgia to make them really understand that sustainable DMOs are not only about marketing. Their role is to link up, coordinate and manage all the elements that make up a destination (attractions, amenities, access, marketing, human resources, image/reputation and pricing).

The action plan for our DMOs in Georgia Marketing includes marketing and event management, but also training and education, product development, research and information/data management. We want to actively involve the DMOs, as well as the local population, in taking responsibility in protecting their own unique nature and culture and developing them responsibly with a long-term vision.

Effective DMOs steer the development of tourism for the whole region to improve the quality of service, increase the knowledge of their region nationally and internationally and improve its attractiveness for the visitors and investors. As long as this is done with all interested stakeholders with a focus on environmental and cultural preservation, and for the benefit both of the tourists and the local population, the DMOs are the true brand managers for their destination. And only in this way they can call themselves sustainable.

How important is a destination's sustainability performance nowadays for its competitiveness?

I am a huge advocate for sustainability so for me, this is immensely important for destination's image, reputation and attractiveness. I also believe that it will become the 'competitive advantage' of destinations. It will take time but it will happen, as more and more people realize that we will have to change travel patterns and do more to preserve nature and the 'authenticity' in destinations.

Much has been said about Slovenia being a sustainability leader, and rightly so. I love the fact that they have made sustainability their selling point, and they are so proud of what they do. Now and then I toy with the idea of doing an internship with the Slovenian Tourist Board, that would be amazing.

Ultimately, in my opinion, the sustainability performance for any destination, DMO or organization depends on its leaders—people who really 'get sustainability' and drive it through the organization and are determined to implement it.

Destination marketers tend to avoid engaging with 'sustainability' beyond using it for promotion aimed at specific niche markets since it is not usually part of their KPIs. Reflecting on your own experience, which advice can you share with destination marketers in terms of how to deal with sustainability?

Don't use sustainability as a marketing tool if you don't really do it. Greenwashing is very harmful to the tourism industry, for the organization itself, for tourists and the concept of sustainability as it creates mistrust and an excuse not to believe in the need and effectiveness of behaving sustainably.

But if you really are sustainable, communicate it using a simple language showing the tangible benefits to the customer (tourist, resident, investor, business owner). Complicated statistics charts, lots of data, academic jargon and language about 'saving the planet' are great switch-off points to the customers.

Remember that—as Prof Wolfgang Strasdas said in his interview—if sustainability entails higher costs, for example by paying fair wages, or if it requires more far-reaching behavioural changes, then limits of acceptance are quickly reached, both by the tourism providers and by the tourists themselves. Education and awareness-raising about the reasons and benefits of being a more responsible tourist or resident are therefore crucial—DMOs should support any initiatives that provide regular and practical training. The local population need to understand that their unique natural and cultural assets are their unique selling points, and consequently if they lose it, they will lose tourists.

And I will come back to leadership one more time—to be able to 'deal with sustainability', destination marketers need to get buy-in for sustainability at the highest levels of the organization and show a real determination in pursuing their goals, as the opposition is inevitable.

Anything else you'd like to mention?

I know that I may have been quite harsh in this interview on tourism decision-makers and other stakeholders in Georgia. I talked a lot about the challenges. But there are also many positive developments. And there are enough caring and ethnically-motivated people who want to preserve the region's rich and unspoiled natural and cultural heritage.

And although I am watching the rapidly growing numbers of tourists to Georgia with concern, and am trying to convince the tourism planners to favour quality over quantity and focus more on the environmental protection, I know how important it is for the local populations in the mountains to receive tourists. It is an absolutely fascinating place on Earth that is worth exploring, and if anyone is prepared to step off the beaten track a bit, they can still have the experience of the incredible hospitality, bizarre unforgettable adventures and the most pristine nature I had back in 2001.

Link to the interview: https://sustainability-leaders.com/marta-mills-interview/

USA | Megan Epler Wood has led academic, business, civil society and philanthropic organizations managing sustainable tourism since 1990. Her firm EplerWood International designs net positive regional tourism development projects with past support from the World Bank, IFC, IDB, GIZ and USAID. Her current research and online courses with Harvard and Cornell universities have tested a range of new methods to help protect destinations using digital systems with standardized social and environmental metrics. Her 2017 book, *Sustainable Tourism on a Finite Planet* and the 2019 report *Destinations at Risk; The Invisible Burden of Tourism* offer insights for all students in this profession. In 2020–2021 and beyond, Epler Wood and the team are focused on digitally training future professionals and financing destination sustainability for the long-term, post-COVID-19.

Areas of expertise: ecotourism, tourism research, sustainable development, tourism business, destination development, UNSDG 15 (Forests & Biodiversity), UNSDG 8 (Equal & Fair Economic Opportunities).

The following interview with Megan was first published in March 2015 on Sustainability-Leaders.com.

Megan, when did you discover your passion for sustainable tourism?

In 1986-7. I received a Fulbright Scholarship together with my husband to film for 6 months in a remote private reserve in the Andean cloud forest with extraordinarily high biodiversity, in Southern Colombia. We made a documentary for a television special in Colombia broadcast on Earth Day and distributed it to non-profits throughout the region as a result of the funding we received.

Our task was to communicate why and how local biodiversity could be preserved. The reserve, La Planada, where we filmed had been funded by WWF-US, via donations in the U.S, but there were many limits to funds for maintaining essential reserves like this throughout the region.

During our six months immersed in this remote wildlife refuge, we were able to reflect on the challenge of preserving this extraordinary biological treasure. Threats of logging and extensive slash and burn farms were all around us.

I travelled with the native Awa through their well conserved adjacent territory and saw the deep connection between traditional indigenous livelihoods and the stewardship of the land.

Ultimately, I went to Ecuador, where there were many early community-run ecotourism projects already starting to blossom. I realized that local people, especially indigenous people, were already the stewards of these amazing ecosystems, and yet they were struggling with funds and little understanding of what steps to take to protect their lands from encroachment.

I became highly motivated to find the best source of revenue to help local people to conserve their land, and decided that tourism was the obvious choice. It hit me like a lightning bolt, and I have never gone back.

Why did you start the International Ecotourism Society (TIES),[1] in 1990?

I returned from Colombia in 1988 and began to talk with major conservation organizations about making a documentary on 'ecotourism'. I pitched it as the very first global investigation of how tourism could contribute to conservation of natural resources and local well-being.

At the time, the National Audubon Society had a series called World of Audubon, funded by Ted Turner. I got a contract to do a global documentary on ecotourism. We filmed in Kenya, Belize and Glacier National Park in Montana, in the U.S.

The Coors beer company had been the sponsor and there was a protest against their environmental practices. They dropped out and I temporarily lost funds for the documentary. I had already assembled hundreds of contacts in my advance research for producing the film.

It was actually in frustration that I decided to start TIES!

---

[1] https://ecotourism.org

I began calling all of my contacts to suggest there should be an organization dedicated solely to the question of tourism as a tool for conservation and sustainable development. Once the filming resumed, I travelled the world promoting the idea and we picked up many supporters and founded the organization.

From your 12 years as president of TIES, which initiative or accomplishment are you particularly proud of?

Creating a global membership of over 1000, developing the first textbooks in the field, and fostering dynamic locally based conferences in Asia, Africa and Latin America, which were highly inclusive of indigenous people and local communities.

The question of how local people could remain stewards of their vast biodiverse land, retain their culture, and benefit from ecotourism was our focus and we made those dialogs extremely deep and profound in Kenya, Ecuador and Malaysia.

I am also very proud of the early sustainable business models we fostered, in particular the whole concept of ecolodges. We were the first organization in the world to hold an ecolodge conference and develop guidelines for them.

How has your view of sustainable tourism changed over time and now as Director of the International Sustainable Tourism Initiative at the Harvard Center for Health and the Global Environment?

I have always been inspired by the question of how ecotourism can contribute to local well-being. This became my passion, and has been the core of my work in the past 15 years.

While I began as a conservationist and a biologist, I have evolved into a sustainable business and economic development expert.

What I learned from years working in the field in developing countries is that ecotourism has to be a successful form of commerce first and foremost before it can be a conservation tool.

Ecotourism is a pipe dream unless it can deliver profits, not massive profits, but profits to small and micro enterprises that need to be connected to either global or regional supply chains.

I have stepped up my capacity to foster enterprise and have had to learn about sustainable tourism supply chains on a global scale to deliver on this part of my work.

This led to my work at Planeterra and ultimately now at Harvard and also at the Cornell Center for Sustainable Global Enterprise,[2] where I am a Senior Fellow and soon to become an associate on their team.

Why this new initiative at the Center for Health and the Global Environment at the Harvard T. H. Chan School of Public Health?

I have been teaching Environmental Management of International Tourism Development at Harvard Extension School since 2010. It is an open-enrolment, on-line class, which attracts students from all over the world to discuss how global tourism operates as a business, its supply chains, and specific technical challenges for managing its footprint.

---

[2]https://www.johnson.cornell.edu/center-for-sustainable-global-enterprise/

We have been able to attract extraordinary speakers, who come to us on-line from throughout the world from mainstream corporations such as Wyndham and Hilton, TUI Travel, Royal Caribbean and Carnival Cruise, and also, we have looked at the management of airports, with lectures from both Chicago and San Francisco, and destinations such as Belize, Mexico and South Africa.

The students have undertaken invaluable research on environmental management of specific sectors of the industry, looking at questions of managing tourism's impacts on air, water, biodiversity, and how to ensure there is management of waste, waste water, and proper land-use.

This expanded my base of information in an exponential way. It inspired me to write a book, which is in progress. And it inspired me to go to the leadership of my department at Harvard and ask them if there could be a research program that looks at the global impacts of tourism, convenes major tourism leaders, and develops more robust data analysis on tourism impacts, while offering innovative shared solutions on the management of the growth of tourism.

My timing was good, and the program was approved by Dr. Jack Spengler who directs the Center for Health and the Global Environment at the Harvard T.H. Chan School of Public Health. It is called the International Sustainable Tourism Initiative.

The Center is an extraordinary institution, which has provided global leadership on questions of how the loss of biodiversity will affect public health, how climate change will impact local populations, and how nature and biodiversity deliver extraordinarily important benefits to human health. It is a fascinating place to be, that fosters new ideas and asks us all to think of solutions on a very large scale.

What is the best way to measure sustainability in travel and tourism?

A set of benchmarks must be established with social, environmental and economic indicators. They must be measured over time and have a long-term data trajectory. Independent analysis is essential.

What our field needs is more field studies to measure environmental and social impacts on the ground consistently over time. This can be facilitated using new user friendly geodesign tools, which allow for benchmarking and regular study check-ins through geo tagging, and can result in visualized projections of how tourism will impact social and environmental systems over time.

With climate change ramping up, the need to establish benchmarking and visualized projections is urgent. I am calling for regional projects to undertake geodesign measurement systems in advance of large-scale development that are paid for with public private funds.

This will give local decision-makers the data they need to understand how ecosystems and social systems are faring during the next phase of rapid tourism growth.

With tourism reaching 9% of global GDP, and 6 billion air travellers expected to travel annually in the next decade, the time is now to define new systems for the travel and tourism industry to grow while also benefitting human health and the environment.

Your main insights as consultant at Epler Wood International, and as teacher at Harvard's Sustainability and Environmental Management Program?

As a consultant, I have largely worked to help set a structure for regional and national ecotourism projects that can preserve ecosystems and support local well-being. I went to the Angkor Watt region of Cambodia in 2005, where 95% of all food stuffs were imported into Cambodia from Vietnam and Thailand, according to international hotel groups.

All businesses agreed that buying more local products and services would be an important contribution to the economic development of communities in Cambodia. But the difficulty of linking the agricultural supply chain to local hotels was much higher than I expected.

Local rice producers were not growing rice for hotels, but rather for themselves. I reached out to every relevant agency, and there was total agreement that we need to connect the growing hotel food and beverage supply chain to tourism, but there was no funding to make that happen.

I have never forgotten that experience, and continued to seek funds for linking food supply chains to hotels. I am still working on it with a project at Harvard that we hope to foster.

Linkages to food supply chain for tourism could do more for poor households than any other action taken to support the economic development impacts of tourism. These opportunities appear to be waiting to be captured.

Judging by your extensive experience, which are the main challenges public and private organizations tackle on the sustainable tourism front?

Existing NGOs and research around the world have established the importance of the tourism economy as a tool for both sustainable development and conservation. Work to develop criteria for sustainability has been a primary focus for years.

The next step is to deliver on the promise of sustainability from tourism. This will require:

- independent measurements and benchmarking,
- policy analysis and review of how to move from intentions to regulations and legislation with funding to support the cost of managing a sustainable destination,
- a global effort to ensure tourism supports protected areas,
- public private solutions to ensure tourism development does not overwhelm local systems for water, waste, and waste water which at present are not in place to support the global growth of tourism.

Sustainable tourism being such a diverse field, which area do you think is the most crucial to focus on right now?

Sustainable infrastructure, based on all of my student work at Harvard.

We will have a crisis on our hands if there is no national and donor recognition that sustainable waste water treatment, energy, and solid waste management is required to accommodate the next generation of tourism development.

The provision of clean water for hotels is often in conflict with local use, and this situation promises to become graver with global climate change and population growth.

Tourism planning can take all of these issues into account and incorporate solutions for tourism and local economies, a tremendous opportunity.

Where are the main opportunities?

The fact that the tourism economy is growing and is profitable is an enormous opportunity. If the tourism economy was declining and corporations were not profitable, there would be no chance to get better systems in place.

The stakes are high now. Recognition of the growth of tourism in emerging economies and the essential nature of tourism for economic development is only beginning to dawn on top policy makers.

Local people can have access to a better livelihood through tourism, we have already proved this with our work with ecotourism, but strategic policies to make this happen on a grand scale must be on the table and soon.

In which region of the world do you see most interest and momentum for sustainability in tourism?

Europe, where there are stronger government policies and EU cooperation on managing tourism throughout the region.

What role does entrepreneurship play regarding a tourism business or destination's abilities to successfully implement sustainability?

Entrepreneurship is at the heart of how sustainable tourism can benefit local people. Women are incredibly important potential beneficiaries.

Women in tourism generally earn about 80% of a male's wage. Notwithstanding these disadvantages to women, there has been a broad increase in the participation of women in the tourism industry worldwide. Women want these jobs, and they want to enter the work force.

Despite the many barriers they face, tourism offers a better chance than many industries to move ahead. In Sub-Saharan Africa, women are 10% more likely to be the 'boss' in food and beverage operations than in other industries.

Helping men and women to develop businesses that connect to regional and global supply chains is the most effective means of ensuring local well-being.

We are seeing a lot of movement and mergers within sustainable and ecotourism—do you think this is a sign for sustainability moving mainstream?

My deepest concern is actually the fracturing of our field, the many definitions, and approaches that are so similar. We need as much cooperation as possible.

Which part of (1) the tourism value chain and (2) operational unit in a hotel is the most challenging in terms of sustainability?

Food! It may not be as challenging as all that, but it is so significant. Not enough work has been done on how to improve global supply chains to offer hotel food and beverage departments more sources of locally grown, healthy food. We hope to pursue this much more at Harvard.

For the Planeterra Foundation you worked on market access for small, community tourism enterprises in Latin America, and how to connect them with tour operators like G Adventures. Why is this an issue—what exactly are the challenges?

There has been a lack of support from the global donor community for connecting successful tourism operators to local community enterprises. While I was at

Planeterra, we were able to help the IDB come to the realization that connecting a working tour operator to local enterprise is a key to success for local enterprise. They have gone on to be highly successful with this approach.

Your (career) advice to professional and academic newcomers to sustainable tourism?

Come to an understanding of business and business supply chains first! I strongly recommend taking classes that review tourism business models and approaches to profitability, margins, and all aspects of business operations within the sector of tourism you are interested in. Then study how to manage the environmental, social and economic development side second.

Your favorite 2015 tourism, travel or sustainability book?

Well, it is actually not a book but a paper. A bit wonky, but I was thrilled to just find UNCTAD's paper on sustainable tourism from 2013. *Sustainable tourism: Contribution to economic growth and sustainable development.* I really like it and think it is correct, based on all of the research I have done.

Link to the interview: https://sustainability-leaders.com/interview-megan-epler-wood/

Austria | One of the Europe-wide frontrunners in sustainability and CSR—Michaela Reitterer is setting overall trends in green tourism since 2001 with her first zero-energy balanced hotel. Her utmost concern for innovation, courage and willingness for conscientious sustainability at all levels rightly earned Michaela the nickname-award of Green Queen. At her award-winning Boutiquehotel Stadthalle in Vienna, she has created a tranquil urban gardening oasis with a magnificent lavender roof and a colony of honey bees; she is also setting standards by showcasing all 17 of the UN's Sustainable Development Goals. As the long-time (former) president of the Austrian Federal Hotel association ÖHV, she has earned significant recognition for representing Austria's top hospitality industry and as a distinguished expert for sustainability as a successful business case.

© The Author(s), under exclusive license to Springer Nature Switzerland AG 2022   473
F. Kaefer, *Sustainability Leadership in Tourism*, Future of Business and Finance,
https://doi.org/10.1007/978-3-031-05314-6_76

Areas of expertise: green hotel, tourism business, Austria, hospitality, UNSDG 12 (Resource Efficiency).

The following interview with Michaela was first published in March 2018 on Sustainability-Leaders.com.

Michaela, you operate the most sustainable hotel in Vienna. You have received numerous awards, including the Austrian Climate Protection Prize and the Environmental Prize of the City of Vienna. What made you decide to operate your hotel sustainably?

To be entirely honest, there was no way around for me to build a house that generates as much energy as possible. Energy costs are a real factor in the hotel industry in Austria. But it has also to do with the expectations of our guests, our enjoyment of innovation—and wanting to make use of the latest developments, which is especially exciting.

What have you done to make the hotel sustainable?

In our 'Passiv Haus' part of the hotel we produce as much energy as we need during the year. The breakfast and all drinks and cakes are of certified, organic quality.

We motivate our guests to save energy and to participate in the project Zero Waste.

We are also a very family friendly business and try to fulfill all requests, to contribute to a great balance between work and family.

At Hotel Stadthalle there is equal pay for the same job, no matter if man or woman. We train many young people conscientiously and take part in a lot of charitable projects.

We grow our own herbs and harvest our lavender from the lavender roof, which we use and sell. We have 6 bee colonies that provide the annual amount of honey needed for the hotel, and we offer all guests who are traveling to us by train, bicycle or electric car a 10% discount on their room rate.

What do your guests say about this concept and which guest type books your accommodation?

Our guests live consciously at home and do not want to miss out when on holiday. As we are close to the Stadthalle, Vienna's largest event location, a lot of our guests come because of their concert visit and then realize our attitude to sustainability. Here we often create this wow effect and inspire them. We have old and young, hip and traditional, fans of organic food or friends of a green oasis in the middle of the city: simply a wide range of guest profiles, who are united in their enthusiasm of our house, and above all my great team.

To what extent does the focus on sustainability affect your costs and accommodation rates? Did you find that higher prices deter guests?

In Austria we have the saying that what costs little is worth little. We are not much more expensive than other hotels, but we never try to attract guests through price dumping or last-minute actions. That's not sustainable and is short-sighted, which doesn't suit us.

What do competitors say about your commitment?

As far as I know, they are pleased and find it great. But I also have to say that I would like to have more hotels in Vienna and Austria that are committed to sustainability. Then I would not have to pull the marketing cart alone!

Are you largely alone with your involvement or is there a trend towards more sustainable action in Austrian hotels?

There are some awesome colleagues in the Austrian and German hospitality business who are really great in sustainability. They are all member of the Sleep Green Hotels. Many have subscribed to offering regional produce and work with local farmers. I think this is especially important and it is also a USP of the Austrian hotel industry—the quality in the hospitality, and the food and drinks.

As President of the Austrian Hotel Association, you have direct contact with colleagues and competitors. Do you see this as an advantage to make sustainable action in the industry more attractive?

Yes, of course, and it is also very important to me. But it is also important not to preach to and judge colleagues, pointing fingers at them. Rather, I try to motivate and convince them by walking the talk.

But you also have to be able to afford and want sustainable business management, and here it will take a while for everyone to realize that ultimately it will be the cheaper form of entrepreneurship.

What would be the necessary condition for the entire tourism industry to switch to sustainability? Do you think this will happen in the foreseeable future?

Here you have to be careful with the word, because sustainability means something different for everyone.

I believe in the renaissance of summer retreats and that it will again be considered fashionable to spend (short) holidays in Austria. This is a trend that we are increasingly aware of, and even for our most important target markets, a summer holiday in the mountains, on crystal-clear lakes and in dreamy little villages is becoming increasingly popular.

Back to nature is important to all of us. The hotel industry has the potential to not only offer this to the guests, but perhaps also to inspire them.

Link to the interview: https://sustainability-leaders.com/michaela-reitterer-interview/

UK | Mike McHugo founded Discover Ltd. in 1978, one of the leading educational tour companies in Morocco. In 1990, Discover bought a ruined mansion in the High Atlas Mountain village of Imlil, opening as Kasbah du Toubkal 5 years later and becoming one of the most award-winning hotels in the country. From the offset, the Kasbah has been at the forefront of sustainable development in the Imlil Valley. Much of it is funded by the 5% levy added to the bills of all guests, an initiative unique to the Kasbah du Toubkal. With funds from this levy, in 2006, Mike founded Education For All to provide accommodation close to secondary schools for girls

F. Kaefer, *Sustainability Leadership in Tourism*, Future of Business and Finance,
https://doi.org/10.1007/978-3-031-05314-6_77

from rural families in remote villages of the High Atlas to continue their education. In February 2019, he was awarded the MBE for improving gender equality, presented by Prince Harry, Duke of Sussex.

Areas of expertise: ecotourism, mountain tourism, responsible tourism, sustainable development, Africa, UNSDG 10 (Reduce Inequality), UNSDG 8 (Equal & Fair Economic Opportunities).

The following interview with Mike was first published in June 2019 on Sustainability-Leaders.com.

Mike, what inspired you to set up Kasbah du Toubkal, back in 1989?

I started an adventure travel company called Hobo Travel in 1978 and came to Imlil for several days of trekking. Between 1978 and 1989 I had built a friendship with a local mountain guide called Omar Ait Bahmed (now known as Hajj Maurice). 1989 found me and my brother Chris in Imlil with our mother, as our father had recently died. Hajj Maurice's house looked up to the ruined Kasbah du Toubkal. Chris had recently read in the Financial Times that the King of Morocco, Hassan II, had made inward investment easier and he said why don't you find out who owns it. That was the start of 5 years of paperwork to acquire the ruin of Kasbah du Toubkal.

We said from the outset that we wanted the Kasbah du Toubkal to be run in a sustainable way and be of benefit to both the visitor and the local population.

Has your view on sustainable development and tourism changed since then?

Not really—I still think that sustainability is really about common sense and running the operation taking account of the local population and looking after the environment, but also being pragmatic.

I was surprised how we won several sustainability awards in the early days— when to me we were just doing what was pretty obvious and trying to be good neighbors. To me it's all about do what you can, with what you have, where you are.

When we started actually operating in 1995, there were few services or communal facilities in Imlil, so some of the first things we did through the money from the 5% levy we collect from guests were pretty obvious—a rubbish clearing system, an ambulance, and a communal hammam.

In your experience, which have been the main challenges that Kasbah Du Toubkal has had to overcome to be where it is today?

We were fortunate that the Internet was just coming, so for a small company that understood to a certain extent about marketing internationally, we could market more widely than would be possible otherwise.

We were, of course, early adopters and ahead of the start of the explosion of riads   the traditional guesthouses you now find everywhere in nearby Marrakesh and elsewhere. And real growth of tourism in Morocco came with the new king, Mohammed VI, in 1999. So we were able to get a reasonable amount of press coverage. We also had some good photography which has been key.

One of the issues we had—until we began to be commercially successful—was for Hajj Maurice to understand what the international tourist wanted, and they want authenticity as well as good service.

Funding is often a challenge for community-based tourism accommodation or experiences. Has this been an issue in your case?

I actually think that because we did not have deep pockets at the beginning, we understood through operation what worked and what we needed. We built Kasbah du Toubkal organically and were fortunate to have a profitable business in France, so we ploughed the profits from France into Morocco. Obviously, if we did not have that one, we would have needed deeper pockets. But having deep pockets and doing things too quickly can also be problematic.

One of the first messages visitors come across at Kasbah du Toubkal is "We are all guests of the local inhabitants of this area". What would it take for local communities in Morocco to be able to participate more significantly, or benefit more from tourism?

I think if all the tourist facilities had a 5% levy, much more could be done. Probably the biggest problem is that people who work in tourism see each other as competition, but we can be complimentary.

I would like the Imlil valley to have its own online booking system, rather than having to use which everyone else does. Following their discount-focused system is a race to the bottom. We have been fortunate enough not to have to offer and discount on Booking.com and have our own online booking system.

Part of the great success in implementing responsible and sustainable tourism at Kasbah du Toubkal is its close partnership and collaboration with the local Berber community. Can you tell us more about the role of the Imlil Village Association and how it has contributed to the mutual success of the community and the Kasbah?

Before we even started, we wanted the Kasbah operation to be of benefit to the whole community—not just to people who work with us or those we buy services off. We wanted our guests to be liked in the area. So, we came up with the idea of a 5% levy for the village association, which we formed when we received a donation from the filming of Kundun. The first project we did was something that we thought would be completely uncontentious: providing an ambulance. With the Kasbah's success, the amount collected through the levy has grown and has allowed us to fund more and more projects.

When we started the 5% levy there was considerable pushback, particularly from tour operators asking us why we were doing this, but also from many individuals. Over time, with the growth of sustainable tourism and people being more environ-mentally and socially aware, we have found that our guests' attitude to the 5% levy is very positive.

Together with others in the tourist industry I also founded Education For All in 2007, which provides girls from the outlying villages the chance of secondary education.

The local community has seen how we have behaved, and I believe the majority (if not everybody) have respect for this. A verse in the Koran says "Allah shall know them by their deeds".

In your view, which are the main tourism sustainability challenges that Morocco faces today?

Probably similar to those in many developing countries and where tourism is growing rapidly. Mass tourism owned by outside shareholders in their drive for more profits is not typically compatible with responsible tourism. Small-scale, owner-managed businesses are much more likely to be sustainable by nature. Fortunately, with sustainable tourism becoming more mainstream, hopefully, there is a greater chance of large operations also being run following sustainable principles.

Which projects or achievements with Kasbah du Toubkal are you most proud of so far?

The 5% levy and its potential for doing good is, I believe, a good model and it ought to be copied more widely. Maybe the tourist authorities could encourage this. On a personal note, Education For All has to be probably the work I am currently most proud of.

Looking ahead, which sustainability trends or topics do you think will have the most impact on the work of tourism professionals in 2019?

I think the issue of air travel will increasingly be a topic that is discussed and is not very compatible on an environmental basis or concerning long-term sustainability.

Link to the interview: https://sustainability-leaders.com/kasbah-du-toubkal-morocco-mike-mchugo-interview/

South Korea | Dr. Mihee Kang serves at the GSTC as Director for the Asia Pacific region and is an authorized trainer and destination assessor. She is also Managing Director of K TOURISM, a consultancy providing solutions for tourism destinations and businesses. She has taught and researched ecotourism and sustainable tourism at Seoul National University and other universities since her PhD degree in 1999 at Seoul National University. As the first PhD in ecotourism in South Korea and a UNESCO MAB Young Scientists Awards 2000 winner, she has been involved in sustainable tourism policy development and planning and sustainability assessments of various types of destinations in many countries over 25 years. Mihee has served several key positions in both domestic and international organizations including Jeju

World Heritage Committee, Korea Forest Education Committee, and Asian Ecotourism Network. She has actively participated in numerous research projects and ODA projects in Korea and abroad.

Areas of expertise: ecotourism, sustainable development, tourism research, Asia, UNSDG 15 (Forests & Biodiversity), UNSDG 8 (Equal & Fair Economic Opportunities).

The following interview with Mihee was first published in July 2016 on Sustainability-Leaders.com.

Mihee, do you remember the first time you heard about sustainability in relation to tourism?

Yes, I first learned about sustainable tourism during my first year of graduation school in 1993. 'Environmentally Sound and Sustained Development' (ESSD) was a buzzword in Korea when I started to study ecotourism as a good strategy for sustainable tourism development. Most people would be surprised to learn that forests cover 64% of the total land area in South Korea.

As an ecotourism consultant, which of your recent projects did you find particularly challenging?

Most of my projects are related to ecotourism certification and ecotourism development. The biggest challenge is that most initiatives are managed by the central government, and local stakeholders depend so much on the government's financial support. I am concerned that these initiatives cannot survive without this financial support. Consequently, there is a lack of business approach.

Another challenge is that there is little understanding of ecotourism and sustainable tourism criteria. Stakeholders have tried hard to conserve nature and benefit local communities, but they need to understand the requirements or detailed strategies to achieve sustainability of their activities.

But thankfully, the public's understanding of ecotourism has increased, and there are now more local-based ecotourism enterprises.

What motivated you to co-found the Playforest Cooperative?

The Playforest Cooperative was established to show that conservation can benefit local communities through tourism. Play forest has also registered as a tour agency legally to be able to sell local tour products.

Many local communities and organizations are dedicated to conserving their forests, but they hardly get any profit from their conservation efforts. Some of them develop their own tour programs but face challenges accessing the tourism market.

I co-founded Playforest together with some colleagues to guide those communities and to demonstrate how to develop and sell sustainable forest-based tours. 30% of our profits will go to conservation efforts. I hope it will teach people how sustainable tourism businesses work.

Compared to traditional tourism development, why is ecotourism especially important for Southeast Asian Nations?

Southeast Asian countries have rich nature and culture but the people there are relatively poor. Some countries have lots of tourists but I am not sure how much of the benefits go to local people to improve their lives.

To conserve nature and culture, tourism should be developed at a small scale by local people with active and meaningful local participation.

Many travel and tourism operators focus on short-term economic development, where adopting ecotourism can be a challenge.

You are a research professor at Seoul National University; in your view, is sustainable tourism research in Southeast Asia receiving sufficient attention?

I don't think so. There are quite a number of projects, but they are not for academic research but rather for developing short term policies or strategies. I personally had no research funding from our government and most of my projects were related to policymaking or on-site ecotourism program development.

As a board member of the Asian Ecotourism Network (AEN)—what do you hope to achieve? Which challenges are the most urgent for AEN to address for more sustainable tourism in Asia?

I hope the stakeholders in Asia feel that it is easier to raise their voices and have a better network based on similar culture. AEN can function as a bridge among different stakeholders. There are many Asians who feel language barriers, especially with English, and AEN board members from all Asian countries can help them to deliver their voice to the international society.

The most challenging and urgent issue for AEN is to get enough funding for carrying out our goals, such as providing eLearning tools, training opportunities, and market data. And, we need to involve more Asian ecotourism leaders and organizations and establish strategic road maps for achieving our goals.

To support more sustainable tourism in Asia, AEN needs to provide Asian criteria that are suitable for the Asian natural and cultural environment. We are going to cooperate with the Global Sustainable Tourism Council to train Asian stakeholders about sustainable tourism and ecotourism, using the GSTC criteria.

Link to the interview: https://sustainability-leaders.com/interview-mihee-kang/

Spain | Monica Vallejo has been running Hostal Grau since 1993, a unique green sustainable hotel concept in Barcelona. Thanks to the recent Leed Gold Certification, Hostal Grau is now a referent in Barcelona for sustainable hotels. It uses non-toxic materials and certified wood in the hostel construction and purifies the air with a new OH radical natural cleansing system. Proud of being the first hotel in Spain to have a room with electromagnetic wave reduction. Monica believes that a sustainable hotel is more than just respecting the planet. It is also about the social responsibility to help others.

Areas of expertise: green hotel, hospitality, hotel business, Spain, UNSDG 12 (Resource Efficiency).

F. Kaefer, *Sustainability Leadership in Tourism*, Future of Business and Finance, https://doi.org/10.1007/978-3-031-05314-6_79

The following interview with Monica was first published in March 2015 on Sustainability-Leaders.com.

Monica, how do you approach sustainability at Hostal Grau in Barcelona?

Our Hostal Grau has gone through some major renovations since 2013, an opportunity to turn the historic building inside out and to rebuild its interior using state-of-the-art natural, ecological and sustainable elements. During the renovation we took extra care that materials used were PVC (polyvinyl chloride) free and the painting natural, non-toxic. Rockwool makes for some great soundproofing between the rooms. Moreover, all wood used was from sustainable sources (FSC certification), and all removed furniture was given to charity or properly recycled.

Throughout the renovations process, we worked with the LEED program for advice on best practice, and are now on the way to become the first LEED-certified hotel in Barcelona.

At Hostal Grau you will sleep on handmade beds, complete with cocomat and naturalmat mattresses. Our bedrooms have double glazed windows, which blocks out just enough street noise for a good night's sleep in the heart of Barcelona.

In our daily operations, the housekeeping staff use organic cleaning products, and guests are highly recommended to consume Km Zero products from the ground floor cafe Centric Bar. We also encourage the use of bicycles while visiting Barcelona, which are available from a nearby bike rental shop.

Why the LEED Green Building Certification Program?

LEED is for buildings, and we wanted an internationally recognized standard that applies the same principles and criteria to all kind of buildings and uses, regardless of its main activity. Moreover, LEED honors the Cradle to Cradle vision to some extent, and goes well beyond energy efficiency.

Would you recommend LEED certification to other small hotels?

Yes, of course, it has been our guide and our project management tool for the project. LEED and the LEED advisory team was our beacon along the process, otherwise we couldn't have made it, it's not our core business.

If you could start renovations again, what would you do different?

Quite some things, starting with a new team organization from the very beginning, doing much better planning, and improving the execution time frame by making a more professional assessment of the construction partner.

How do you measure sustainability performance at the hotel?

We are metering energy and water use, we deal with air quality and material flow from the design phase (e.g., choosing non emitting materials, building green walls, green procurement policy, etc.), and we apply both LEED and C2C metrics for this work in progress project.

We think that quality hotels should embrace regenerative sustainability, so we're planning to promote some carbon management strategies as well (e.g., offsetting projects). Material is key, and hotels need clear guidelines not to get lost in the process, and not to frustrate our well-educated clients either.

Your views on the current state of sustainable tourism in Barcelona?

Barcelona needs a regenerative sustainability agenda to become a truly smart city. We like the way circular economy addresses that, and we would like the city in general, and the touristic industry in particular, to lead this venue.

Link to the interview: https://sustainability-leaders.com/interview-monica-grau/

Jordan | Muna Haddad is the founder of BARAKA, a consulting company established in 2011 specializing in sustainable tourism development. She led eco-tourism master planning and community-based tourism development in Jordan, Morocco, Tunisia, Lebanon, Palestine, France, and Italy. Muna founded Baraka Destinations to demonstrate the financial viability of innovative tourism models that contribute to poverty alleviation through job creation, cultural preservation and nature conservation. She is the Founding President of the Jordan Trail Association, an NGO tasked with developing and managing a 650 km hiking trail that helps in creating tourism jobs in poverty pockets and underdeveloped areas. She speaks at universities and international conferences on topics related to transformative travel, adventure tourism, community-based tourism, destination marketing and has recently been addressing defensive marketing and the impacts of the political crisis on the region. In 2021, she joined Jordan Prime Minister's Delivery Unit as the Tourism Priority Lead focused on a sustainable tourism recovery post-pandemic.

Areas of expertise: community-based tourism, destination sustainability, eco-tourism, destination development, tourism business, Middle East, UNSDG 12 (Resource Efficiency), UNSDG 8 (Equal & Fair Economic Opportunities).

© The Author(s), under exclusive license to Springer Nature Switzerland AG 2022   489
F. Kaefer, *Sustainability Leadership in Tourism*, Future of Business and Finance,
https://doi.org/10.1007/978-3-031-05314-6_80

The following interview with Muna was first published in February 2020 on Sustainability-Leaders.com.

Muna, having worked with the tourism industry for more than a decade; when did you first discover your passion for sustainable tourism?

I stumbled upon the travel industry after university. My first job was working in marketing at the Jordan Tourism Board. I remember the moment I fell in love with the travel sector so clearly. I was hosting a group of journalists, standing in front of them representing my country, culture, and history.

I got to see the power of the tourism sector in shattering misconceptions, opening people's hearts and minds to the world and its people, and fighting bigotry. Travel makes us better people. I wanted to be a part of that for the rest of my life. I got to know my country and its people at a much deeper level and fell in love with that too.

Then, three years in, I quit my job in search of a way to use tourism as a force for good, and to make sure local communities have a fair shot at benefiting from tourism financially, but also by using it as a platform for their stories.

What are your thoughts on the current state of ecotourism in Jordan?

I think the growth of adventure tourism and ecotourism in Jordan has attracted the right kind of clientele for the tourism sector. Those are travellers keen to dig a little deeper and get to know the people of Jordan better. They are the best ambassadors for the country, they fight to protect its ecosystem in the global community and represent its people everywhere they go. It's a beautiful thing.

At the same time, it is a worrying state: as Jordan has become a more popular destination, places like the desert of Wadi Rum are being threatened. As demand grows, site management becomes critical.

There is still a long way to go in terms of government commitment to nature conservation and reaching a balance between tourism and conservation. I also believe there is room for improving policies to become more inclusive of local communities engaging formally in the tourism sector.

Value, not volume is a topic we discuss often. I am keen on that—however without making travel exclusive.

Olive oil harvesting in Umm Qais is one of the many community-based tourism initiatives developed by your organization, Baraka Destinations. How can a seasonal business like this one generate revenue during lean periods, and also during off-peak travel seasons?

Seasonal activities are much more than bringing in guests and revenue during off-peak season. In the type of rural slow tourism that we are creating, we observe the land in all its seasons, and our menus change based on what is being cultivated from the land around us.

What community-based tourism projects are you currently working on?

We are now preparing to announce the third destination we'll start working in. We have worked hard in partnership with the people of Umm Qais and have been able to make a real difference in the tourism sector in the area.

What I am most proud of is the linkages we have created within the community in localizing the supply chain, this way ensuring that more than 73% of tourism income remains in the village, compared to the global average, 19%.

We are also looking at eventually replicating the model in other countries and influencing donor funding to support the impact-driven business.

It is time for community-based tourism to be viewed as a solid business rather than a charity, and I feel we were able to prove that through our work with Umm Qais. This is not just about money, but being able to put the local community at the center of the story of the place, which is a key to success in terms of ownership, resilience, and stamina.

In your experience, what is the average incubation period for a community-owned tourism business to turn profitable?

Three to five years.

There are a lot of variables relating to the budget available for marketing and sales, as well as capacity building and training.

One of the most valuable lessons learned for us was that we needed to regularly have a training budget, not just at the setup phase. As more tourists come, the businesses and entrepreneurs need to evolve.

Another important attribute to our model is reliability and longevity. We are there to stay, we are not a three-year project. So, the nature of our relationship with our partners is different and has a long-term vision, which is clear to all involved. Our vision is aligned, we all want to put their village on the tourism map, to tell their stories and share their experiences.

What are some important tips that you can share with other entrepreneurs in the responsible travel industry?

Over the years, meeting various people in the sector, I feel there is a fear of working with the locals. Understandably so, there are a lot of examples of where relationship have gone sour.

My attitude about it comes from perhaps the most valuable lesson I learned from a Bedouin Sheikh from Wadi Feynan. I met him a decade ago as I was just starting my work in community-based tourism. I was trying to be as invisible as possible and fitting into his world by dressing, speaking and acting in the way I thought would mimic his world. I walked into Abu Khalil's goat-hair tent and introduced myself briefly putting on a very heavy Bedouin accent, which clearly didn't belong to me. He saw right through me, and with a very kind look in his eyes, put his hand on my shoulder and said to me, "dear, there is no need for you to pretend to be something you are not. In this world, there is room for you and room for me."

Since that day, I am completely myself, honest and transparent with anyone I meet. Because, there is room for them, and room for me.

That should be the foundation of any relationship we have with the people we work with. Wherever they are from, whatever life experience they have had.

What were the criteria for choosing the 12 social enterprises which are now part of Jordan Tourism's Meaningful Travel Map of Jordan?

We created a meaningful travel map for the Jordan Tourism Board and for Tourism Cares. It was created as a tool for tour operators and has managed to attract more than 10k tourists to these places so far.

The purpose of it was to challenge the charity mentality and to support local communities by bringing business to them. The traffic has pulled some organizations out of debt, increased employment opportunities for others, and certainly got the stories out of the unsung corners of the country.

The selection of the current enterprises on the map was meant to take into account the impact, be it social or environmental. We looked at the owners, who are benefiting from the business, a variety of experiences and geographic representation.

Of course, it was also important to look at market readiness and worth noting that we insisted to have enterprises that were not exactly at that level and have tour operators invest in their upgrade.

For example, Iraq Al-Amir Women's Co-operative has a superb experience that supports the women in that village, and The Travel Corporation stepped in to support upgrading the gift shop so that the groups that have committed to visiting would spend more on-site and benefit these ladies by buying their souvenirs from there.

Amman, Dead Sea, Petra, and Wadi Rum are famous among international visitors and happen to be the most visited places in Jordan. Do you think ecotourism experiences in places like Dana Biosphere Reserve, Azraq Wetland Reserve or Shaumari Wildlife Reserve will experience a similar demand from tourists?

The nature of these sites can accommodate the masses (luckily), but at the same time, the ecology is a lot more delicate. I think Dana has certainly seen incredible growth over the last decade in tourism arrivals and the Royal Society for the Conservation of Nature and the local community have really had to work on finding a balance. As protected areas, they are held accountable at higher standards for visitor management.

To your mind, are there enough efforts by the government in handling overtourism issues in places like Amman, Dead Sea, Petra, and Wadi Rum?

I think the government is very conscious of overtourism on a national as well as a local level. Petra, being the most critical due to its UNESCO listing, has been taking active steps in visitor management. A lot of us who are keen on the ecosystem preservation of Wadi Rum are concerned about the lack of control of campsites being set up.

It is the right of the locals of Wadi Rum to benefit from their natural asset, however strict regulations need to be in place and be implemented with a strong arm to ensure no further sprawl at the expense of the delicate ecosystem. A lot of the camps are owned but not run by locals. I urge consumers to make informed decisions and pressure the businesses they choose to visit to always hire locally and adhere to conservation laws.

Let one thing be clear though, there is no overtourism in Amman. It is my city, and most tourists pass through only scratching the surface. There is so much depth in the city that very few tourists have explored

Although a few streets may be considered touristic areas, there is a wealth of history and culture to the city that is rare and unique around the world. Most tourists don't get to see that part.

In your view, what is the best way to measure the success of sustainability initiatives in travel and tourism?

In our work we did benchmark assessments, then quarterly reports, then bi-yearly assessments. We were looking at a number of people benefiting from the tourism sector through our work both directly and indirectly.

We were also assessing income levels and using local average as a benchmark, we look at age groups and gender of people working with us, and finally, we looked at milestones in their growth. For example, they left a government job and started working full time, or they put their kid through college, etc.

All said, the hardest thing to measure, which is the one thing that makes our method stand out, is the pride and dignity that our partners have in their work. If anyone figures out how we can translate that into metrics, please let us know!

Clearly, greenwashing is such a fad and now with the excitement over the SDGs I expect to be seeing a lot more of it. I think we stress the measurement of the final result, rather than the process in smaller milestones in a way that pressures complex big businesses to falsify data to ride the wave. The reality is that being a sustainable business is complex, hard, and takes time.

For example, a large-scale hotel going green means they first have to go through their existing stock of plastic shampoo dispensers, rather than throwing them out. It is a better decision for nature conservation if we are stating facts. But as consumers or industry specialists, we look judgmentally at the plastic dispenser, not realizing the complexity of such changes.

Taking that into account, I think we need to focus more on the content of how we go through this journey to become more sustainable as a sector, rather than fixating on the final destination. It's important to have goals, but we have to spend more time mapping out the path to get there.

Link to the interview: https://sustainability-leaders.com/muna-haddad-interview/

Nada Roudies on How Morocco Is Developing a More Sustainable Tourism

Morocco | Nada Roudies has played an active role in her 19-year career in the Moroccan public sector, holding various positions like the Vice Minister of the Ministry of Tourism (2012–2017) before devoting herself to consulting as an international expert in tourism and sustainable development. At the international level, she has headed various initiatives like the global partnership for sustainable tourism from 2013 to 2015, co-lead for the Sustainable Tourism Program of the 10-year framework of Programmes on Sustainable Consumption and Production of the United Nations (10YFP-STP) and the African Charter for sustainable tourism in partnership with the UNWTO. As a consultant, Nada provides strategy consulting in different sectors such as tourism policies, sustainable development, marketing and communication strategies for international organizations (such as UNWTO and GIZ) and governments.

Areas of expertise: sustainable development, sustainable tourism, Africa, UNSDG 12 (Resource Efficiency), UNSDG 8 (Equal & Fair Economic Opportunities).

The following interview with Nada was first published in October 2015 on Sustainability-Leaders.com.

F. Kaefer, *Sustainability Leadership in Tourism*, Future of Business and Finance,
https://doi.org/10.1007/978-3-031-05314-6_81

Nada, when did you discover your passion for sustainability?

I think sustainability is a passion that I developed gradually through my work towards sustainable tourism, both at national level with the establishment and implementation of Morocco's sustainable tourism strategy, and at international level when promoting sustainable tourism as a vector of development to the United Nations.

An important part of my motivation are the men and women (from tourism companies, NGOs, associations, etc.) I get to meet: Really passionate people who are fighting every day to create profitable tourism businesses while minimizing their environmental footprint and supporting local communities. It is rewarding to accompany them in their adventure and to provide them with the appropriate tools and institutional frameworks.

Your main insights from presiding the Global Partnership for Sustainable Tourism (GPST, 2013–2015)?

The global partnership was a very interesting network through which we shared experiences and started developing tools to advocate and promote sustainable tourism globally. The main achievements during this period were:

* The screening criteria for sustainable tourism projects, which we developed for financing institutions to adopt and use for their evaluations;
* Our guidelines for a life-cycle approach to sustainable tourism strategy: from design to implementation.

The Global Partnership for Sustainable Tourism also provided technical assistance to the Caribbean region, as well as countries in Africa and Asia, helping them design and evaluate their sustainable tourism strategies.

We participated in many events related to sustainable tourism and organised two international symposiums as part of our general assemblies (one in Agadir, Morocco and the other in Namibia).

We also contributed to the creation of the Sustainable Tourism Program under the 10YFP on Sustainable Production and Consumption framework.[1]

Your vision for sustainable tourism in Morocco for the next years?

Sustainability is a key part of our current ten-year tourism development strategy in Morocco. To spread the benefits of touristic activities, we designed our tourism strategy in a way that supports development across the country, not just in one or two popular destinations. We also took into account environmental constraints and pressure levels, to make sure our destinations remain livable for local communities.

To make tourism development in Morocco more sustainable, we have been developing regulations and incentives for our investors and tourism operators, both of which encourage a life-cycle approach, where sustainability is present in all stages of the tourism product.

---

[1] https://www.unep.org/explore-topics/resource-efficiency/what-we-do/one-planet-network/10yfp-10-year-framework-programmes

In addition, we are developing a set of indicators at national and regional level to measure and monitor sustainability in tourism, and to be able to demonstrate and communicate our commitment to sustainability.

Our main challenge is to establish sustainability as approach for tourism development in Morocco and to get all stakeholders on board: institutions, professionals, media, tourists and the Moroccan citizens.

Which achievements as Secretary General at the Moroccan Ministry of Tourism are you most proud of?

As General Secretary I had to manage and lead many projects which I am quite proud of. But the human epic that it represents is certainly the most rewarding.

As a Moroccan woman, although not necessarily excessively feminist, being part of those who can demonstrate that Moroccan women have their place next to the men in high governmental functions in our country is an achievement in itself.

But to remain in the subject of sustainability, I am really proud to be participating in the positioning of Morocco as a leading destination regarding sustainable tourism. I am particularly proud of the leading role I assumed within the GPST and now in the 10YFP-Sustainable Tourism as a co-lead.

How does the Ministry of Tourism help companies and destinations within Morocco to become more sustainable?

As Ministry of Tourism, and to ensure a minimum level of sustainability in the sector, we put in place a mandatory regulatory and normative framework (specifications for investors, hotel classification standards, etc.).

We are aware that this may present additional costs for operators and a need for capacity building. We thus organize with the regions building awareness and capacity events, we develop good practice guides for the various touristic activities, we develop incentives (technical or financial assistance) for SMEs who want to adopt a sustainability approach or to obtain certification, and we encourage initiatives through our Sustainable Tourism Trophies.[2]

We also try to introduce sustainable tourism in training courses in our hotel and tourism schools.

Where do you see the main challenges in Morocco and the Greater Maghreb region for sustainability in travel and tourism?

In my opinion, the main challenge for sustainable tourism in destinations at the moment is to move from the promotion of responsible tourism as a niche product for small and specific segments, to a destination-wide strategy which involves all the stakeholders.

Another challenge is to change the perception that large numbers of visitors and tourism growth automatically mean irresponsible tourism. Changing this image will require work on the tourism product itself, but also awareness-raising among consumers.

Ultimately, much of our success regarding a more sustainable tourism will depend on whether and how we train and educate the young generations.

---

[2] https://www.trophees-tourisme-durable.ma

When advising your government colleagues and the private sector on implementing sustainable tourism strategies, what main challenges have you encountered?

Many challenges of course, including:

- the additional costs, the belief that it can have a return on investment
- the need of technical assistance and building capacity
- the size of the market we can address

Link to the interview: https://sustainability-leaders.com/interview-nada-roudies/

Colombia | As a development and tourism advisor, Natalia has a wide range of experience in the industry. She is the Country Representative for the Canadian Executive Service Organization CESO-SACO in Colombia, a leader in COMUNITUR, a platform for information and exchange for local development and tourism initiatives, and a Global Sustainable Tourism Council official trainer. Natalia studied Government, Finance and International Relations at Universidad Externado de Colombia and holds a master's degree in Environmental Intervention from the Social Psychology Faculty of the University of Barcelona. With more than 15 years of experience in tourism, development and international relations, she has worked with communities, public and private sector, events, capacities strengthening

© The Author(s), under exclusive license to Springer Nature Switzerland AG 2022
F. Kaefer, *Sustainability Leadership in Tourism*, Future of Business and Finance,
https://doi.org/10.1007/978-3-031-05314-6_82

for adults and organizations, and as a lecturer. Natalia is co-founder of the recently created network Women and Sustainability, an ecofeminist and a constant learner.

Areas of expertise: sustainable tourism, tourism business, community-based tourism, Latin America, UNSDG 8 (Equal & Fair Economic Opportunities), UNSDG 12 (Resource Efficiency).

The following interview with Natalia was first published in April 2015 on Sustainability-Leaders.com.

Natalia, what was your view of sustainability and tourism when you first started your professional career?

When I first heard about sustainability, I saw a proposal to do things in the right way. I started to work in tourism in Machu Picchu, Peru, in a hotel that was offering guided tours to add value to their guests; the tours focused on nature and culture.

As a tour guide, I had the opportunity to experience the region, and learned about orchids, birds and the Inca culture. This is how I fell in love with tourism.

For me at that time, sustainable tourism was a practice to show others the marvelous nature and local culture; an instrument for learning about and protecting the places we visit.

Now at the beginning of 2015, what has changed?

My view on sustainable tourism has not changed, but my understanding of it has been growing as I learned about new concepts and ideas. What has changed is that since that time my work became my passion; and I keep learning every day to improve my skills and my knowledge about sustainable tourism.

One big change occurred when I started to work with local communities in rural Colombia. During that time, I learned that sustainability cannot happen without people and individuals; without them—without all of us—it is impossible to speak of sustainable destinations. Without the involvement of local communities and consideration of their socio-economic situations in particular, sustainability won't happen.

Your key insights as sustainable tourism advisor?

Every job, every community, every individual and experience are different. And when your job becomes your passion, the world becomes a small place, where great things can happen.

How do sustainable tourism approaches differ between Canada and Latin America (Colombia), your principal working areas?

Every place and culture is different; in Colombia and Latin America generally, we have still so much to learn regarding sustainability. We have to meet basic needs and solve local problems that pose huge challenges.

We want to strengthen our links with countries like Canada; places with higher living standards, and where people and regional leaders have traditionally been more respectful of nature and cultures.

I truly believe that tourism offers the opportunity to bring people and ideas together, to improve knowledge and mutual understanding.

Where do you see the main challenges for sustainable tourism in Colombia? And Canada?

In Colombia, we need stronger commitment of public institutions to provide basic infrastructure and improve living conditions for locals. And we need the tourism industry and professional associations to commit to sustainability as master strategy and a way of doing things right.

Canada leads the world in terms of favourite destinations to visit, according to a National Geographic environmental stewardship survey. Those destinations represent not only Canada's natural beauty and remarkable diversity; the country's commitment to sustainable tourism is also well supported by government resources, industry associations, partnerships and non-profit organizations.

In your view, what is the best way to measure sustainable tourism performance in hotels and destinations?

Well, I think the best way is to establish mechanisms that would help monitor destinations on a regular basis; a mechanism that would offer results to support improvements and corrections. So far, most countries and regions use their own standards and guidelines for hotels and destinations.

Very recently, the Global Sustainable Tourism Council has developed global criteria for sustainable tourism (hotels, tour operators and destinations), which are now considered the best international standards to follow and apply.

What advice would you give newcomers to sustainable tourism consulting?

Love what you do, learn every day, and don't give up on sustainability. We are on our way, and we are becoming an ever-stronger force.

Your favorite tourism, travel or sustainability book right now?

The UNWTO and the Klagenfurt University in Austria launched the *International Handbook on Tourism and Peace*, which features several experiences around the globe. This book was awarded at ITB in 2015 and includes my article on community-based tourism in the Uraba-Darien region of Colombia.

Your preferred news sources to stay up to date on sustainable tourism?

The GSTC and the UNWTO are the main global institutions, but there are also other sources with a strong focus on nature, the Rainforest Alliance for example. Other topics to look for are responsible tourism, social responsibility and human development.

The great thing about tourism (also challenging) is that it is very diverse in that it involves various economic sectors, a wide range of stakeholders, issues and topics. This makes it difficult to find one source for all, but you can use the GSTC and UNWTO to track other sources, institutions, individuals and databases.

Also, there is a growing network of initiatives dedicated to sustainability in tourism, including the Sustainability Leaders Project, Hopineo, the South American Sustainable Tourism Network, Green Destinations, and so on.

Why did you decide to become a university lecturer alongside your consulting with CaLatam?

I've been a tourism consultant for a long time and usually that implies workshops and training. The university is new for me, and I found this very challenging and motivating. I think it is very useful to stay up to date and this work makes me think about teaching methodologies, how to develop critical thinking and motivate students to be great professionals—being able to influence those things is just amazing.

CaLatam (Canada-Latin America) is a vehicle to bridge two regions engaged in sustainable tourism. I am interested in facilitating exchanges, expanding the network, and learning from each other. Through CaLatam we are creating a bridge to transfer intelligence and expertise—we have much to learn from each other.

How does CaLatam help tourism companies/destinations around the world?

We help build bridges between Canada and Latin America, specifically Colombia. We support communities through our CSR program, create networks through events and build capacities through training, while conducting strategic assessments for organizations, communities and destinations.

We are supporting the Uraba-Darien region of Colombia in their process to build a culture of peace through tourism, and support emerging networks and innovation together with students and volunteers. We spread the word about tourism, peace, sustainability, responsibility and ethics.

Why did you decide to represent the GSTC in Colombia?

My leitmotiv is Think Global Act Local; GSTC is the perfect place to share, meet and exchange ideas for global sustainability. I'm very grateful for this opportunity and support GSTC's mission in Latin America, specifically in Ecuador and Colombia.

Why did you co-found the South American Sustainable Tourism Network?

We have a great cultural and natural richness here and sustainable tourism is a huge opportunity for our destinations. The SAST Network was born in Brazil, after a GSTC meeting in Bonito, by people who share the same passion and want to strengthen sustainability in our countries. We want to build a network strong enough to support local initiatives, and are currently looking for allies and supporters to join us in making this a reality.

Finally, which are the leading academic institutions in the sustainable tourism field in Canada and Latin America?

In Colombia, Externado University is one of the main academic institutions leading the sustainable tourism field. In Canada, Ryerson University in Toronto, Algonquin College in Ontario and the University of Northern British Columbia.

Link to the interview: https://sustainability-leaders.com/interview-natalia-naranjo/

Switzerland | In 2016 Olivier Cheseaux, an UoAS architect passionate about heritage, founded Anakolodge with a concept based on the respect of the vernacular architectural heritage of the Alpine mountains. Our ancestors were building to address their needs when space was becoming scarce for human beings and livestock. This survival architecture was gradually lost when modernity, comfort and mechanisation took hold of human beings. In Val d'Hérens, the remnants of traditional agricultural buildings are still visible. Ruins of two barns, two granaries and two sheds were rescued and then restored in complete respect for local architectural tradition typical to the Valais region. These contemporary transformations—without any artifice, a fake sense of 'oldness' or imitation—showcase this heritage. Having celebrated its fifth anniversary in 2021, Anakolodge has hosted visitors from all over the world. With the Covid 19 pandemic, Swiss guests looking for nature and authenticity have come in large numbers to recharge their batteries.

Areas of expertise: sustainable business, green hotel, hospitality, mountain tourism, Switzerland, UNSDG 12 (Resource Efficiency).

The following interview with Olivier was first published in September 2018 on Sustainability-Leaders.com.

Olivier, as architect you specialize in rescuing old buildings and turning them into contemporary Swiss mayens—rustic mountain chalets. Why?

The duty to remember. . . our ancestors worked under difficult conditions to build these objects of 'survival'. Life in the mountains was complicated, cold, snow, famine. . . It was important to me to restore them and to give them a second life. It was first necessary to understand each object and its unique nature, before transforming them into a place to live, a tourist place.

You are the creator of anako architecture. Briefly, what does this concept stand for?

For me, the location is always the starting point of an architectural project. You don't build the same object twice because each plot is unique! Anako architecture is the starting point of this whole philosophy of respect for a place, for the heritage, for the people who will live there.

Anako is the name of an old orejone Indian and shaman. The people of Anako lived in perfect harmony with nature. Their homes were fitting perfectly in the native American forests.

My idea of architecture has an identical purpose: to create an accommodation that respects and preserves its location. For me, each place deserves its own architecture, just as each anakolodge mayen is unique.

Sharing platforms like Airbnb have become quite a headache for many popular city destinations, but also present massive opportunities for less touristic, remote places. Which has been your experience? Do you use them for the promotion of Anako Lodge?

Airbnb is a global player in innovative tourism. Anakolodge was inspired by the sharing philosophy behind Airbnb. Excesses encountered in urban destinations do not apply to a concept like ours. We offer our mayens to tourists to give those buildings new life.

Through Airbnb it was possible to advertise Anakolodge around the world without being affected by off-peak seasons. Holiday seasons are not the same in Switzerland, Australia, or China. This allows us to be open 365 days a year.

Which aspects of building Anako Lodge did you find the most challenging?

To make the builders and carpenters appreciate the heritage. I wanted to recover everything; old beams, boards. . . It is easier for them to use new materials and get rid of old stuff. I had to fight to get there. From the tourism perspective, no one except my wife believed in my vision to offer a return to basics, tranquility, NOTHINGNESS!

Getting financial support was difficult, because my project did not meet the expectations of bankers, used to calculating investment opportunities based on immediate profit. Anakolodge is a long-term project, expected to run over at least two generations.

In times where the economy has become more local, circular and based on usage rather than ownership, how can villages use this momentum for securing their economic viability and well-being of their residents?

Precisely, here in this valley the tourist economy was based on sales of holiday homes. New constructions are prohibited since March 2013. At Anakolodge we work with local people, the grocer, the restaurateur, the cheese monger. And that's totally new.

The added value created by Anakolodge is spread over several actors, including local people, and that is our philosophy. We work together, we increase our skills and we all experience benefits from tourism!

Social businesses like Swiss Youth Hostels have for a long time practiced a business model which goes beyond financial gain. Can you briefly describe the business model and philosophy of Anako Lodge?

I always say that Anakolodge is a mixture between a youth hostel and a time-share cottage. I don't want to produce something; I just want to connect different people and partners. If you want a service, you contact the grocer, the restaurant owner, the ski school. If you want to stay in the mayen, to cook on a wood stove, everything is planned for.

My primary purpose is not short-term profit but sustainability and above all the pleasure to please. I find it very satisfying to meet guests enjoying being here, disappointed having to leave the mayens ultimately, and looking forward to coming back.

What role do architects and real estate developers play in sustainable tourism planning and development at a destination level?

Currently, none! The developers build to sell without caring much about sustainability.

Anakolodge is a small drop in the tourism ocean, but at least it's a little answer... I hope that architects and investors will follow the same path in the coming years.

To your mind, is there a good synergy between destination managers, property developers and architects in Switzerland, in terms of sustainability strategies?

It's a difficult question... in my opinion no. At the moment, profit takes precedence over sustainability. A whole way of thinking needs to be reviewed and reinvented. In my opinion, there is unfortunately nothing currently in place to create this dynamic. Sadly.

Which trends do you observe in architecture right now, which might support or hinder a more sustainable tourism industry?

I see more and more young, private individuals or architects who want to make a difference. But often these projects do not see the light of day because investors don't support them.

Which 3 bits of advice could you share with other architects and real estate developers in terms of how to succeed with sustainability?

I don't like to give advice; I prefer to share my experiences. You have to believe in your project, find partners who want something sustainable, respectful and who agree to work for a living.

Sharing the fruit of your labour with local partners, your family, that is what is sustainable.

Anything else you'd like to mention?

People must wake up. Our planet is in survival mode and our politics and the economy only focus on short-term gains. But people can make a difference, even with small initiatives like Anakolodge.

Anako said, "Nature is lent to you by your great-grandchildren. . ."

Link to the interview: https://sustainability-leaders.com/olivier-cheseaux-interview/

The Netherlands | Paul Peeters is a Professor at the Centre for Sustainability, Tourism and Transport (CSTT) of Breda University of Applied Sciences, The Netherlands. His publications cover a wide range of topics, including climate scenarios, system dynamic approaches to tourism research and modelling, air transport, tourism transport mode choice, modal shift, policymaking, tourism climate mitigation and adaptation policies, as well as transport technological developments. He has written reports and advised several UN bodies like ICAO, UNEP, UNFCCC, the European Commission, the European Parliament, the Dutch government and parliament, and several NGOs.

Areas of expertise: climate change, responsible travel, tourism research, aviation, sustainable tourism, UNSDG 13 (Fight Climate Change)

The following interview with Paul was first published in March 2016 on Sustainability-Leaders.com.

Paul, do you remember the first time you heard about sustainability in a tourism context—and your initial thoughts?

As a child, I wanted to become an entomologist, but I ended up as an aircraft engineer at Fokker Aircraft. After four years at Fokker and worrying news about failing resources and acid rain destroying our forests, I started a career as a

507
F. Kaefer, *Sustainability Leadership in Tourism*, Future of Business and Finance,
https://doi.org/10.1007/978-3-031-05314-6_84

researcher, first in wind energy, followed by sustainable transport and, finally, 13 years ago as a professor of sustainable transport and tourism. I had become familiar with sustainable development before, but only then did I really learn the definitions of 'sustainable tourism'. And these definitions surprised me! So much seemed to be drenched in kerosene!

Has your view on sustainability and tourism changed since then?

When I started research in this area, I thought a whole range of environmental issues would be at stake. But I learned that energy and climate change were actually the most serious ones, with water and waste being problematic in some destinations.

I found it very difficult to get this message to the 'sustainable tourism research community' that was mainly engaged in nature-based, community-based and pro-poor tourism, off-setting carbon emissions and eco-labels for accommodations and destinations. These forms of sustainable tourism depended strongly on long-haul flights from the rich west to the poor south. Really difficult to understand!

Despite the many certifications, initiatives and events dedicated to sustainable tourism, we are finding it difficult to identify truly sustainable destinations. Why is it so difficult to put sustainable development theory into practice?

Most studies just look at international tourism, even though 80% of global tourism is domestic. Furthermore, transport issues are generally ignored. Probably because most tourism studies only take into consideration the destination point of view.

From a destination perspective, investing in a local airport is generally beneficial, because you will take visitors away from other destinations. However, from a global perspective, the conclusion will be different: globally, investing in airports will not generate more tourism trips, it just redistributes source markets and destinations and increases the distances tourists travel. Some destinations gain, others lose.

The focus on the international and destination perspective seriously hampers our understanding of sustainable tourism development, because both the main environmental problem, transport, and the main solution, domestic and short-haul tourism, are put outside the equation. So, it seems the theory fails from the beginning.

One major headache linked to the industry's sustainability is tourism transport and its implications for climate change. Do you see progress in this area?

On the one hand, some big tour operators embrace carbon disclosure. In the Netherlands, the branch organisation ANVR asked us to develop Carmacal, a tool to very precisely assess the carbon footprint of package tours. Not only flights but all forms of transport, half a million accommodations with different estimated footprints. This tool provides accurate data for a carbon label for tour packages and shows where and how to efficiently reduce emissions.

But on the other hand, there are 12,000 large jets currently on order and this fleet will fly at least until 2060, based on 2010s technology.

Based on a global tourism emissions model, I recently calculated that global tourism is set to emit some 300 GT CO_2 between 2015 and 2100, which is 30% of the global budget for sustainable development. Can it ever be sustainable to take so much of this budget, also needed for cooking, heating and lighting the homes of billions of poor people?

What are your main insights so far as director of the Centre for Sustainable Tourism and Transport at NHTV Breda University?

Tourism students and scholars all share one thing: love to travel. Most of our tourism students and teachers travel far more than average. One problem of this might be that we deliver generations of tourism managers with a long-distance travel mind-set, thus developing products that invite people to travel more than, strictly spoken, is necessary.

Paul, you are an engineer by training. Two questions: what is the main technical barrier stopping us from enjoying sustainable aviation, and what is the most promising solution?

The challenge of every aircraft designer is to create an aircraft that transports as much payload over as long as possible distances at the lowest possible cost. Every additional ton of fuel for a certain flight will cost you ten paying passengers. The best way to achieve this? Fuel efficiency! Therefore, the fuel efficiency of current aircraft is rather close to the aerodynamic, thermodynamic and mechanical laws of nature.

It may be that some 30-40% can be gained upon the newest generation of Airbus NEO's and Boeing Dreamliners, but that will be it. And, it will take about half a century to bring us there, a timespan that may show 3—4 times the current volume of air transport.

How different is this for trains, cars and ships, where near-zero energy and real zero emissions are technically feasible and which some railway systems, like those Switzerland and Sweden, have used for decades already?

Your thoughts on the outcome of the United Nations Climate Change Conference (COP 21) in Paris December 2015?

In general, COP21 was very successful, all countries in the world agreeing on setting ambitious goals and pathways to get there. However, for tourism, it was disappointing because aviation was exempted and left to the International Civil Aviation Organisation (ICAO). This organization is very much dominated by national industry interests. ICAO's new fuel efficiency standard will not save much fuel.

The other program, market-based measures, just involves offsetting the increases of emissions after 2020 for international flights. Therefore, net emissions stay constant after 2020. With such supplier behaviour, the genuine green traveller has no other option than to abandon flying altogether and discover the exciting world of train and bus travel.

Link to the interview: https://sustainability-leaders.com/interview-paul-peeters/

Australia | Dr. Paul Rogers is one of Asia-Pacific's most experienced tourism for development practitioners. Completed in 1997, his PhD centred on tourism, conservation and development issues in Nepal's Sagarmatha (Mt Everest) National Park. Paul is a long-term expert consultant with the United Nations World Tourism Organisation, the World Bank, ADB and numerous other international organisations. He has worked extensively in Nepal, Bhutan, Laos and the Greater Mekong Region, as well as Myanmar, North Korea, North-West Africa and Australia. Paul is co-founder of Planet Happiness—a tourism and big data project of the non-profit, the Happiness Alliance. Inspired by several assignments in Bhutan, Planet Happiness focuses the attention of all tourism stakeholders on well-being, to use tourism as a vehicle for development that demonstrably strengthens destination sustainability and the quality of life of host communities.

Areas of expertise: ecotourism, sustainable development, sustainable tourism, Asia, UNSDG 15 (Forests & Biodiversity), UNSDG 8 (Equal & Fair Economic Opportunities)

© The Author(s), under exclusive license to Springer Nature Switzerland AG 2022   511
F. Kaefer, *Sustainability Leadership in Tourism*, Future of Business and Finance,
https://doi.org/10.1007/978-3-031-05314-6_85

The following interview with Paul was first published in March 2016 on Sustainability-Leaders.com.

Paul, when did you discover your passion for sustainability?

I'd say my passion for sustainability and wanting to 'do the right thing' has been innate.

It really awakened as a backpacker in Asia and crystallized while on the Everest base camp and Kala Patar trek: watching the sun come up behind Everest at 5.30 am. Things started to take on more meaning while wandering back down those majestic alpine valleys (that I didn't want to leave behind).

What was your view of sustainable tourism when you first started your professional career?

With that intense experience—immersed in the enormity and scale of the Himalayan landscape and the spirit of the people who lived within it—I wanted to know how tourism could be developed to create win-wins for everyone. So, I started researching, reading all I could on tourism, development and protected areas, and ended up with a PhD on 'ecotourism' to the Mt Everest National Park and its environs.

To me, ecotourism should be intrinsically sustainable—or it ain't ecotourism!

But at the same time, I realized (and continue to realize) that sustainable / responsible tourism is incredibly complex. It's a process—a journey which we'll unlikely achieve or reach. It's a dynamic system in constant need of attention.

You are a senior tourism expert at LuxDev. What is your job about? And which achievements are you most proud of?

The Luxembourg Development Cooperation, via its bilateral cooperation agency LuxDev, is playing its part to achieve three SDG's [Sustainable Development Goals] and make the world a better place.

I've been contracted to develop a tourism human resource development strategy for Myanmar—and that's what I'm focused upon right now. Development of the strategy is a key objective of the Myanmar Tourism Master Plan. That's a piece of work I'm proud to have worked on, with the rest of the team and all who contributed to it, including the Minister, the government and other stakeholders that joined our many workshops.

I'm also really pleased with the Lampi Marine National Park Ecotourism Plan, which I worked on last year for Myanmar's only marine national park. But while Instituto Oikos are doing a great job of supporting the government, the Park and its local community including the Moken (sea gypsies), the complexities of implementing the plan are somewhat daunting and more technical and financial resources are needed to really secure a better future for the park!

The Lampi plan is the first ecotourism plan for any of Myanmar's protected areas, and this was produced shortly after completing the Myanmar Ecotourism Policy, which I also worked on. It centers upon the tourism & protected areas relationship.

You have a lot of consultancy experience linked to tourism policy and planning and have worked for UNWTO, SNV, International Labour Organisation, Asia Development Bank and UNDP, among others. What are the main challenges of consulting at such a high level? And which project/achievement are you particularly fond of?

I'm really lucky to have had these opportunities to work on such a fantastic range of assignments—it's been a serendipitous journey! I guess the challenges are about making the plans real & implementable, which is never easy.

Take the Myanmar Tourism Master Plan as an example—the union minister had already set in place a responsible tourism policy, which the master plan had to align with. So, our team drafted a plan that sets out what the ADB felt was needed to achieve responsible tourism. The plan recognizes that donor and development partner support is needed to fully implement the plan.

It's unfortunate that not enough donors and development partners are into the tourism sector, to give responsible tourism the support it deserves. Development needs are complex and there are so many competing priorities!

In previous interviews, we learned that the focus is shifting to destinations as key change agents for sustainability in tourism. Do you agree? Which are the main challenges at destination level?

Absolutely—totally agree!! It's the direction that the Myanmar HRD strategy is taking.

Key challenges vary from one destination to the next—it all comes down to the local context. The qualities of the destination, the character of the public, private, civil society, NGO/development partner stakeholders—and, last but not least, the appetite and enthusiasm of local champions, which are critical in terms of implementation.

Actually, maybe the last word here should be the ability to enforce rules and regulations. So many locations I've worked in lack the human and financial resources to implement laws, rules and regulations. This, and stakeholder coordination, are generally the two greatest challenges.

Paul, you recently said that we should move from Millennium Development Goals to the 'Wellbeing Economy Goals'. Can you explain it?

Yup. I cannot believe we've replaced the MDGs [Millennium Development Goals] with the SDGs! How dry is that? How are we going to inspire and motivate the masses to engage in the Sustainable Development Goals?

The majority of people I come in contact with have no idea what Sustainable Development Goals are about—or interest to be inspired by it. But if we look at Bhutan and the story of Gross National Happiness, people are curious and immediately drawn to it.

My time in Bhutan has taught me that 'Happiness' translates to 'Wellbeing'. And a little more contemplation leads us to conclude that the pursuit of wellbeing is essentially the same as sustainable development.

So, if you're looking to inspire and engage civil society in sustainability debates—are you going to succeed by talking about the sustainable development goals, or by sharing a story, an engaging narrative about Gross National Happiness and Bhutan?? Think about it!

It seems to me 'Wellbeing' is a term people are open and ready to relate to. So, if we talk about the 'Wellbeing Economy Goals'—the WEGs—it will be far easier to popularize interest in critical issues and debates to shape our all futures.

The last point on this, I was at the truly inspiring 'High-level Meeting on Happiness and Well-being' convened by Bhutan at the UN in NYC on 2nd April 2012. It was a fantastic occasion, and I continue to be drawn to big picture projects in support of this agenda. I'm working on one in particular that has great potential but needs 'the stars to align'.

It takes so much synergy to really make a difference. Conversations are ongoing with a number of change-makers, and I'll get back to you if that alignment shapes up!

Can you name two books linked to sustainability and tourism that you highly recommend?

One of the first passages that really enthused me when I began this journey was *Making the Alternative Sustainable: Lessons from Development for Tourism*, by Emanuel De Kadt in Smith & Eadington's *Tourism Alternatives: Potentials and Problems in the Development of Tourism*.

Not linked to tourism, but the key to the sustainability and well-being agenda is *The Spirit Level: why greater equality makes societies stronger* by Wilkinson & Pickett. That's a great read!

If you had to start your professional journey all over again, knowing what you know now about sustainability and tourism, what would you do differently?

Wow. Good question. I really don't know. By slowing down and talking to local people much more, I guess. By taking a step back, and being more patient and less hard-headed when things go awry. And by learning how to meditate and do some yoga. I still need to learn both!

Link to the interview: https://sustainability-leaders.com/interview-paul-rogers/

Peter Richards on Responsible Tourism and Community Development in Asia

Thailand | Peter Richards has 20 years of hands-on experience at the crossroads of responsible tourism and community development: grassroots to global levels. During this time, he has worked in the private and non-profit sectors, with governments, NGOs, academia, tourism associations, tour operators and guides, local communities, hotels and restaurants. His core skills include sustainable tourism project management, training and coaching, developing and marketing inspiring local experiences, facilitating win-win partnerships between diverse stakeholders in international tourism supply chains, and helping businesses and organisations to raise sustainability standards. He is passionate about motivating stakeholders across sectors and cultures to work together and achieve results.

F. Kaefer, *Sustainability Leadership in Tourism*, Future of Business and Finance,
https://doi.org/10.1007/978-3-031-05314-6_86

Areas of expertise: responsible tourism, tourism business, Asia, destination development, community-based tourism, UNSDG 8 (Equal & Fair Economic Opportunities).

The following interview with Peter was first published in September 2016 on Sustainability-Leaders.com.

Peter, you first came into contact with the concept of sustainable tourism through your work as Regional Responsible Tourism Coordinator at Intrepid Travel in Southeast Asia (2000-2002). What were your initial impressions of the concept?

Before working with Intrepid, I had worked for a year as an English teacher in a rural village about twenty kilometers outside Chiang Mai, Thailand. I lived on site, in the school. After school, most of the teachers went home. Initially, I felt quite isolated in the school campus. However, the local school staff: cooks, housekeepers and gardeners, began to notice that I was interested in village life. They took me under their wing and invited me to various events, such as temple fairs, house-warming parties, and other interesting events.

When I joined Intrepid as a tour leader, I was eager to share this simple, welcoming side of Thailand, far away from the red light and full moon clichés. Intrepid was one of the first tour operators to have a responsible travel policy. This was a key reason why I applied to join the company. I thought: "Yes, Intrepid wants to share the same Thailand that I do."

At that time, many international tour leaders worked in SE Asia. Most tour leaders loved the countries where we worked and treasured the local people we met.

Responsible tourism was something very immediate and practical to us. We wanted to solve concrete problems which we saw, such as school children without books or shoes, or 'hilltribe' villagers without blankets in the winter. We also took 'irresponsible behavior' by tourists quite personally!

Looking back, I still appreciate our raw passion for SE Asia and the efforts which we made. However, in hindsight, our initiatives were often based on partial-understandings and over-simplification. They were 'Band-Aid solutions' which did not help to solve problems at their root causes. Under the leadership of Jane Crouch, Intrepid Responsible Tourism (RT) Coordinators began doing more research, consulting with local and international organisations, and supporting initiatives which were being rolled out systematically. Intrepid's RT now has much greater impact.

My personal highlight as RT coordinator was developing ethnic hill tribe language sheets and training tour guides to help travelers communicate with villagers during trekking trips. Travelers cross the world hoping for authentic, local experiences. Villagers are often interested in meeting travelers. They are also sensitive about how tour guides present their cultures. However, it is not unusual for professional guides to interpret local life on behalf of community members, rather than actively facilitating cultural exchange.

These experiences ignited my interest in community-based tourism. Community-based tourism (CBT) prepares communities to welcome guests on their own terms and share their ways of life in their own words.

In your opinion, how has sustainable tourism changed or developed in the region and internationally in the past 15 years?

I think sustainable tourism has changed massively and for the better. The focus of sustainable tourism has shifted from essentially small-scale ecotourism and CBT projects, which explicitly defined themselves as an alternative to mass tourism, to a much broader focus on how to make all types and scales of tourism more sustainable and responsible.

Companies and supporting organisations are approaching sustainability more strategically and systematically, from the office to the field. Extensive resources and tools have been developed to help every size of tourism business raise their sustainability performance. We have better tools to manage the negative impacts of mass tourism (e.g., improving waste management or reducing water and energy consumption), and motivate big tour operators, hotels, restaurants, etc. to contribute towards the Responsible Tourism promise of 'better places for people to live in and better places for people to visit'. For example, through sourcing local products, offering decent work and welfare to local people, robust CSR, etc.

Meanwhile, many of us working on CBT and ecotourism have learned a lot from past successes and failures. CBT has attracted fair criticisms when projects have failed or underachieved. In some cases, projects initiated by NGOs were located too far from tourism hubs and flows to attract visitors. In other cases, project teams had not provided community members with business skills to manage their enterprises or built viable market linkages. Private sector initiatives also hit obstacles when staff lacked the expertise to mobilize or train community members, mediate conflict, manage tourism benefits transparently, etc.

NGO's have moved from activism to strategic engagement with industry and government. At the same time, tourism businesses and government have become more interested in sustainable tourism and local community benefits, and more open to working with NGO's.

Now, across Asia, we can see many examples of better-designed CBT and ecotourism, based on partnerships with responsible tour operators, and technical support by social workers. A good example is the partnership between Peak DMC and Action Aid in Myaing, Myanmar.

I believe that there is a very important role for long-term, committed organisations, which facilitate stakeholder cooperation through partnerships and pilot projects. In the UK, The Travel Foundation have done fantastic work, engaging massive tour operators to fund pilot projects, which benefit local people and the environment. CBT-I (Thailand), and INDECON (Indonesia) have worked for over 20 years to create opportunities for local people to participate in, benefit from, and influence the direction of tourism development.

At the regional level, Wild Asia has made an important contribution through their annual Responsible Tourism award, based on the Global Sustainable Tourism Criteria (GSTC), which celebrates its 10-year anniversary this year.

Compared to 15 years ago, we have much deeper understanding about who needs to work together, and what they have to do to raise sustainability performance.

Perhaps the most important challenge for the immediate future is how to add much more value when marketing sustainable tourism, in order to successfully incentivize and reward sustainable suppliers?

After over a decade working in sustainable tourism development in Thailand, you began work in May 2015 as Consultant on Cultural Tourism Development and Market Access for the International Trade Center in Myanmar. What drew you to this project, and what is it all about?

Kayah is Myanmar's smallest state, situated in the hilly, eastern part of the country. It is a fascinating area, inhabited by nine ethnic groups, which weave a tapestry of livelihoods, languages, beliefs, customs, costumes and cuisines across the state. Kayah had been closed to visitors for over 50 years, due to fierce conflict with the military government and between ethnic armed groups. During the past decade, these groups have gradually moved from conflict towards greater trust, cooperation and peace. The state capital, Loikaw, is now served almost every day with flights from Yangon by Air KBZ and Myanmar National Airlines; or travelers can take a scenic boat ride from Nyaungshwe / Inle Lake to Pekong.

There are high hopes for tourism as an engine for socio-economic development and as a multi-stakeholder process with the potential to foster further trust, partnerships and peace. The Netherland's Trust Fund (NTF) Inclusive Tourism project in Myanmar is a partnership between The Netherlands Government, The Ministries of Hotels and Tourism and Commerce of Myanmar, and The International Trade Center (ITC). NTF III aims to 'enhance trade competitiveness of the tourism sector in Kayah state, by integrating local producers and service providers into tourism value chains, helping inbound tour operators in Myanmar to develop inclusive, cultural tours, and promoting Kayah in international markets'.

The project is really exciting, because it is an integrated suite of strategic actions at many different points along tourism supply chains: from local communities to high-value markets. Activities include training local, ethnic minority communities and small food and crafts producers to develop, manage and operate fun, interactive, cultural tourism activities; linking these experiences and products with local operators and guides in the state capital of Loikaw; marketing the new tours to tour operators in Yangon; and building the capacity of these Yangon tour operators to successfully and responsibly market their offer to EU tour operators.

The project is also supporting a national and state level branding campaign and training public sector partners and Myanmar's tourism associations. It's very ambitious!

Potjana Suansri (former director of Thailand CBT-I) and I are working with a local ITC team to mobilize and train community members, develop community-based cultural tours, and support tour operators in Loikaw and Yangon to market them. We were drawn to this project as an opportunity to share two decades of experiences in Thailand with our neighbors in Myanmar to help build on best practices and avoid past failures.

It is also a privilege to be able to work with Kayan (long neck Karen) people, to develop respectful cultural exchange programs, which we hope will be a stark contrast to the sad, 'human zoo' situation which Kayan refugees had to face in Thailand for many years.

Most inspiring is training over 20 young, Kayah professionals and students, across ethnic divides, to work as a team and become CBT trainers, so they can replicate lessons learned, after the project ends.

Myanmar has only recently opened its door to tourism, but it is quickly catching up with popular neighboring destinations like Thailand and Laos. What are private and public stakeholders doing to promote sustainability in the country's tourism development?

Overall, it's an inspiring time to work in Myanmar. Many tourism stakeholders are sincerely committed to making a contribution to national development. People are eager to share their lives and cultures with the world. However, there is also concern about how rapid tourism growth may impact traditional cultures and ways of life, and the environment.

The Myanmar government has done a lot of things right. As a result of strong teamwork between national and international organisations, responsible tourism and community benefits are now embedded in Myanmar's tourism policies. Initial reform in the tourism sector was initiated in 2011. A partnership with Hans Seidel Foundation (Germany) lead to the Nay Pyi Taw Responsible Tourism Statement (2012). Subsequently, an ambitious process of workshops and consultations informed the Myanmar Responsible Tourism Policy (2012). Other, early milestones were the Policy on Community Involvement in Tourism, and an excellent brochure on Do's and Don'ts for tourists developed by Tourism Transparency.

Since then, there have been an increasing number of sustainable tourism initiatives in Myanmar. Private sector associations, including The Union of Myanmar Travel Association (UMTA) and Myanmar Tourism Federation (MTF) are partnering with international projects. They are encouraging members to attend sustainable tourism trainings, workshops, etc., organised by international development partners including CBI, ITC, GIZ, and UK AID.

SST Tours, one of Myanmar's first ecotourism pioneers, and a nascent Myanmar CBT Network also organize regular discussions to network and share sustainable tourism knowledge. Many individual tour operators are also active, joining FAM trips and organizing their own surveys.

Initially, a cautious approach was taken to opening local communities to tourism. Six pilot 'community involvement in tourism' (CIT) projects were initiated, of which the ITC project is one. In 2015, the NLD won a landslide victory. The current Minister of Hotels and Tourism (MOHT), U Ohn Muang is highly supportive of sustainable and community-based tourism. CBT is now included as a key strategy under the MOHT's 100 Day Plan.

Personally, I think the decision to begin with only 6 destinations was smart, because stakeholders can now learn and apply lessons from six different models, according to what best fits their context.

For those interested to learn more, there are very interesting resources online at Tourism Transparency and The Business Innovation Faculty (BIF).

Another exciting innovation is the recent establishment of the Myanmar Responsible Tourism Institute (MRTI). And, if you would like to follow the step-by-step development of cultural community-based tourism in Kayah, you are most welcome to join the Kayah Inclusive Tourism group on Facebook.

In your experience, what is the biggest challenge faced when developing community-based cultural tourism products?

A big overall challenge is that successful CBT development requires a balance of community development and tourism skills. Because community development and tourism are such different education and career paths, teams developing CBT often lack one or the other. As a result, there can be big 'blind spots,' which cause projects to fail or under-perform.

Team members with tourism backgrounds will focus on products and markets. They will notice if a community is well-located, nearby or on route between tourism hubs. They will have enough experience to notice whether a community is honestly exceptional enough to compete, or only 'quite a pleasant place'. They will see opportunities for innovative product development and have the skills and contacts to get CBT products to market.

On the other hand, the social dimension to CBT makes it different from most other products. CBT is usually managed and operated by a village tourism club, sometimes working in partnership with a tour operator. Roles may include coordinators, community tour guides, cooks, transport providers, expert artisans or musicians. Work is often shared between active community members through queues and rotation systems. A percentage of income may be contributed towards a community fund, which is used to fund various local community development and conservation initiatives, ensuring broader spread of benefits.

For this to succeed, people need to work together as a team.

Crucial success factors for CBT, which are not immediately obvious, include how well people work together? How much do people trust each other? Even if a village is exceptionally beautiful, CBT will fail if enough community members seriously distrust their leaders or each other. On the other hand, local people may be too busy or even disinterested to develop tourism. How will you know this?

Your social workers must invest enough time and energy to earn trust. They need the skills to conduct a community study which is sharp enough to identify hidden potentials for cultural experiences and hidden weaknesses among people. They need to help community members to understand tourism and why tourists want to visit their village. They need to help find a balance between aspects of local life which community members feel proud and comfortable to share with guests and real activities and services which tour operators and tourists will actually be interested to buy. They must identify who is interested, available and prepared to take responsibility when bookings are made and tourists begin to arrive.

If you could give three tips to someone wanting to start a community-based cultural tourism tour operation, what would they be?

The person who taught me the most about CBT is Potjana Suansri, founder of Thailand Community Based Tourism Institute. For Potjana, CBT is principally an adult education process. Local people can develop a wide range of transferable skills, earn additional income and share their ways of life. Based on observing her, and my own experiences, I would say:

First is respect. Even if your initiative is explicitly aiming for 'poverty alleviation,' never give community members the impression that you consider them to be poor. CBT succeeds when local people take ownership and responsibility for programs which they feel proud, comfortable and confident to offer visitors. Successful working relationships, where all parties pull their weight and take responsibility, are based on mutual respect. People know when they are respected. So, take time to get to know people. Show sincere interest in their lives, and confidence in their ability. Support them with the necessary skills to get the job done (e.g., interpretation, cooking, taking and confirming bookings, simple accounting).

Second, do not underestimate the amount of time it can take to build community members' skills to welcome tourists and deliver a great experience. Short trainings are not sufficient. Effective ways to build skills include study tours to existing CBT destinations, trainings (e.g., hospitality, hygiene, guiding, accounting), follow-up coaching, and helping community members to reflect on experience and improve. For example, encouraging community members to reflect on the experiences of welcoming families, as distinct from students or senior travelers. How are the visitors different? How should levels of physical challenge or complexity of interpretation be adapted to different ages, fitness levels, etc.?

Last but not least, there is no tourism without tourists. If you are a community worker (NGO, consultant, academic, etc.), then approach professional tour operators and tour guides and invite them to join CBT development activities from the beginning. This will give tour operators the chance to meet community members, build trust, start to build relationships and provide some feedback on product development. A key step is training professional tour guides to work as a team with local community guides. Professional tour guides are used to being a 'one-stop service' for tourists' every need. It can feel counter-intuitive to share the spotlight with local community guides; however, for CBT to succeed, community members need space to learn from experience, including making mistakes.

Getting tour operators, government agencies and local communities to come together to develop and promote sustainable tourism is no small feat, yet you have done it successfully on a number of occasions. What were the major challenges you faced in fostering this multi-stakeholder cooperation? How did you overcome them?

Most projects I have worked on have aimed to develop and market new, local, sustainable tourism experiences. Multi-stakeholder cooperation along international tourism supply chains are a team effort. It is crucial to thoroughly map stakeholders, along the chain, from end to end; and try to ensure that you have enough people and resources to engage with each group. Having enough resources to work from the field to the market is never easy.

A key success factor is to facilitate a space where stakeholders are open to listen, and try new things. This is not easy, especially if you are dealing with confident, senior people, with decades of experience. It is also very difficult in formal situations (like big meetings), where people feel exposed.

People will often listen to each other more willingly if they have shared positive experiences together. Organisation a study tour or a FAM trip can be a great way to draw people away from a formal environment, and get them talking and thawing out.

Holding workshops, small group activities which mix participants across stakeholder groups, and requiring discussion and cooperation are very useful in getting people to listen to each other. It's also good to get people working together on something practical, quickly, rather than waste time disagreeing on points of principle! In 2010, when CBT-I (social workers) worked in partnership with the Thailand Ecotourism and Adventure Travel Association (tour operators) we started the project with many opposing views. Burning the midnight oil until 11 pm for days at a time fostered mutual respect, and finally we were a great team.

In the villages, differences in power and social status mean that outside stakeholders often seriously underestimate community members' potential to be 'partners' rather than simply 'products.' This situation can be improved by building community members' confidence and skills to communicate, negotiate and cooperate with outside stakeholders. Our team have developed a simple guide training system, Safety, Story, Service. We also conduct joint training with community members and tour operators and guides, so that these stakeholders have lots of opportunities to work together, and develop teamwork.

As a facilitator, I often find myself helping to negotiate a workable compromise between the demands of tour operators and communities. I try to help stakeholders empathize with each other's roles and limitations. Communities need to understand: Who are tourists? Where are they from? Why do they travel? Why are they interested to visit a village? What are the positive and negative impacts of tourism? What kind of commitment is CBT?

However, I also keep in mind that, although we hope that tourism will be an opportunity for community members, we cannot assume that it will be. CBT requires time and effort. Tour operators also need to be flexible to the reality of life in rural villages. If necessary, the project has to slow down to the speed of the village. Then, my job is to explain clearly to tour operators or donors why this has to happen, and hope that they listen!

Link to the interview: https://sustainability-leaders.com/interview-peter-richards/

Rachel Dodds on Sustainable Tourism in Canada

Canada | Rachel has worked in the tourism industry for over 25 years. She has helped stakeholders including NGOs, governments, destinations and the industry to move forward on sustainable tourism issues. Rachel has published over 60 journal articles and three books including a recent book on overtourism. Rachel is passionate about helping the tourism industry become more sustainable. She is the Director of Sustaining Tourism, a boutique consultancy, and a Professor at Ryerson University in Canada. Rachel has travelled to over 80 countries and lived on 4 continents. She has worked on sustainable tourism projects around the world with governments, businesses, non-profits and community groups. Her key areas of focus are overtourism, tourism management and research into consumer behaviour.

Areas of expertise: island destinations, tourism research, tourism business, North America, UNSDG 11 (Sustainable Human Settlements), UNSDG 8 (Equal & Fair Economic Opportunities).

The following interview with Rachel was first published in May 2015 on Sustainability-Leaders.com.

Rachel, when did you discover your passion for sustainable tourism?

© The Author(s), under exclusive license to Springer Nature Switzerland AG 2022   523
F. Kaefer, *Sustainability Leadership in Tourism*, Future of Business and Finance,
https://doi.org/10.1007/978-3-031-05314-6_87

I have always loved to travel, but I think my passion really started when I was visiting Mexico at the age of 15 and saw raw sewage run down the beach into the ocean. I thought, how can these places do this?!

Years later, I lived in Australia on the Gold Coast and the highrises left the beach in shade by 3 pm every day, and each year the beach had to be replenished from dredged sand because all the natural protection barriers for erosion had been removed and the beach was washing away.

This was when I really decided that the only career I wanted to do was to work with businesses and destinations to make tourism more sustainable.

How has your view of sustainable tourism changed over time?

My view of sustainable tourism has certainly expanded. I used to think policy was the way to achieve change, but now realize that change can happen top down or bottom up—it is about leadership as well as passionate people who understand the risk of doing nothing. I used to also get discouraged by the lack of action, but now I have no shortage of amazing examples of businesses and people who are inspiring.

Why did you decide to set up Sustaining Tourism and work as a consultant alongside your lecturing career?

I was a consultant first. I have been consulting since 1998. Funnily, I never intended to be an academic and got my PhD so I could work for the development banks. I have since done consulting work for development banks and realized that I prefer to work as a consultant and didn't want to be an employee, so when I got an offer to teach and research it seemed like a great fit.

I love doing both—consulting keeps me grounded to the industry and what is going on, and teaching keeps me real and inspired as I always have some fabulous students. I love to research and find answers to questions or try to prove things, so I think I have the best of all worlds.

Your main insights from researching and consulting on sustainable tourism?

I believe my main personal insight is that there is always more to do, therefore my work will never be done. The more I learn, the more I realize what I don't know. I also think we can always improve on processes and there are always advancements in technology or changing political interests that we must take into consideration in order to make change happen and tourism more sustainable.

I find that in research, many people write or research about different things, but they are all talking about sustainability—just with a different label. I believe sustainable tourism cross-sects with multiple other disciplines and interest areas—tourism is dependent on resources so climate change, water issues and sustainable livelihood issues all need to be considered as the forefront.

Tourism and hospitality cannot be insular in its approach and disciplines, such as geography, political science, engineering, marketing, entrepreneurship and others all intersect. I work hard to try to publish in different disciplinary journals so that sustainable tourism becomes more widely understood and discussed, as often the focus is on the more visible industries, like extractives, but it doesn't mean it is any more important.

As a consultant, I realize that selling the business case is the most important. Each destination or company or policy maker has different goals and visions so I have learned that although I may see sustainable tourism as an opportunity, I must not assume everyone else does or even wants to.

As a professor I try to get my students to understand that we can't keep measuring tourism success based on visitor numbers—increasing yield is much more important than increasing numbers. I also think elaborating on the difference between growth and development needs a lot of attention as we often assume growth is development.

In 2010 you published a book on Sustainable Tourism in Island Destinations, together with Sonya Graci. In a nutshell, what is the book about?

The book offers examples of sustainable tourism in islands, both good and bad.

We worked hard to try to discuss issues in layman, rather than colloquial academic writing, and outlined the key issues for tourism and the need for change.

A number of island destinations are portrayed as case studies, including from the Caribbean, South East Asia, North America, Europe and Africa.

Your favorite sustainable tourism examples/success stories?

Oh, I have so many! Well, I love what Fogo Island Inn is doing in Newfoundland and lots of Caribbean islands are really improving their supply chain issues when it comes to local food production in particular. I think the Azores is doing some great things too.

Other than just island examples, I was really impressed by the Hacienda Tres Rios resort in Mexico, near Playa del Carmen. They hire all local people and pay fair wages, and are really working to be ahead of the curve when it comes to protecting the cenotes and diverting waste from landfill, which is a huge issue there.

Goodness, I could go on forever, I usually try to post good practices on my website so people can see a myriad of different good examples to be inspired by.

In your view, what is the best way to measure the success of sustainability initiatives in travel and tourism?

That is a question I couldn't answer in a paragraph! I think accountability is key and not just on financial, environmental and social indicators, but of all stakeholders too.

Inclusion of all stakeholders is key, but accountability and engagement are where change happens.

I also think it is about transferability—there are some really fabulous five-room lodges in remote parts of the world, but if they don't really affect many people, or other destinations don't learn from them, then they don't really have that much impact.

Your advice for newcomers to sustainable tourism research?

My advice is to do what you are passionate about... don't get discouraged by rejections and remember Rome wasn't built in a day (oh my goodness—I sound like my father!)

Your favorite tourism, travel or sustainability book?

The *Routledge Handbook of Tourism and Sustainability* is quite comprehensive, but I also love *The Final Call* by Leo Hickman, published in 2007—it was a really plain great read that showcased how awful tourism can be, but is motivating you to change at the same time.

How can we better connect sustainable tourism researchers and professionals?

I think applied research is a start—industry needs practical solutions and so ensuring research is widely distributed—not just in academic journals but through industry reports, online publications, seminars and conferences.

I try to keep my website up to date too and your website is a great tool to connect both!

Link to the interview: https://sustainability-leaders.com/interview-rachel-dodds/

New Zealand I Rachel is a passionate environmental professional. Her past work has combined sustainability with a community commitment to achieve 17 years of environmental improvement for the Kaikoura Community. More recent work has included natural disaster recovery to improve social capital and environmental health. Rachel is passionate about preserving the environment through the protection of special places and ecosystems. Rachel lives off the grid and grows her own food. She believes that no matter where you live, one can reduce their footprint by being mindful of our environment, our consumption and considerate to everything on the planet.

Areas of expertise: destination sustainability, New Zealand, urban sustainability, UNSDG 12 (Resource Efficiency), UNSDG 15 (Forests & Biodiversity).

The following interview with Rachel was first published in July 2018 on Sustainability-Leaders.com.

Rachel, following your studies of Environmental Science, you have over the last 12 years spent much of your time and energy on promoting sustainability at the Kaikoura District Council in New Zealand. Do you remember what first got you interested in the natural environment?

© The Author(s), under exclusive license to Springer Nature Switzerland AG 2022   527
F. Kaefer, *Sustainability Leadership in Tourism*, Future of Business and Finance,
https://doi.org/10.1007/978-3-031-05314-6_88

I was passionate about the environment already as a child. I had a childhood filled with exploring the natural environment and became aware of the issue of endangered species early. My goal was to save blue whales, and I became aware of environmental degradation affecting habitat loss. I can't tell you when, but I can tell you it has been part of my thinking since childhood. *Silent Spring* was discussed in my house when I was a child.

When was the first time you heard about "sustainability" linked to tourism?

I visited Cape Tribulation in Australia when I was in my late teens. We had picked up an Aboriginal hitchhiker and hearing him speak about the changes to make way for resorts made me think of environmental impacts and tourism. Maybe sustainability wasn't part of the conversation then, but the impact on cultural identity and changes to the environment was.

Has your view on sustainable tourism changed since then, through your experience of working in Kaikoura?

Working in Kaikoura has consolidated my ideas of sustainability. It has also changed my view toward tourism, as a lot is to be gained through responsibility, education and experiences. People feel ownership and want to act more sustainably if they have been part of something bigger.

The sustainability (or lack thereof) of destinations is receiving more and more attention, as visitor numbers grow. In your experience, which hurdles are the most difficult to overcome when trying to enhance the sustainability performance of a destination like Kaikoura?

Politics is the biggest issue. Politics affects spending money, conservation and sustainable policy. Politicians are often reluctant to make radical changes or investment that may have a social or environmental benefit, even if the changes are indirect.

The other issue is embedding sustainability into decision-making. This is why businesses are so important in destinations. Business owners can influence public decision-makers and lead by example.

Kaikoura has become well-known around the world as an urban sustainability champion, and an early adopter of the EarthCheck tourism sustainability programme. How did you manage to get this far?

Kaikoura has worked with EarthCheck over the years to improve performance. The work started with educating the Community. The Community had to be on board and embrace the journey. Business leaders in the Community were also instrumental, as they lead by example. Many businesses use sustainable practice in everyday behaviour, from laundry to recycling. Every little action has a positive effect.

Using local government planning documents, sustainability was embedded in the culture of the Community. Decisions were made based on the environmental impact, whilst environmental restoration programs were championed. Our long-term commitment to sustainability means that we have been able to improve environmental performance for the Community over 17 years and counting.

More resilience in the face of natural disasters, such as stormwater inundations or earthquakes, is one of the benefits often associated with sustainable destination management. Having experienced both in Kaikoura, would you agree?

Sustainability definitely made people more resilient. I think there is more that could be done to improve both Community and people's resilience—rainwater tanks for every house for water storage, solar panels for energy resilience. At a Community level, land set aside for stormwater retention areas to reduce the impact of flooding and increase biodiversity.

In Kaikoura, there will be a time when the Community can reflect on what helped them and what they could do better. That time has not yet come, but I hope that the opportunity will be taken up when it does.

Which trends do you observe in destination management and development at the moment, linked to sustainability?

Sadly, sustainability is seen as an add-on for Community recovery. The rebuild is trying to incorporate sustainability into building design and development. However, it is up to the individual and there are some lost opportunities.

Is sustainability performance becoming a key factor for regional and destination competitiveness and reputation? Where are the links?

I think sustainability is the one thing that destinations cannot afford to overlook. People are more educated and critical now when it comes to sustainability. To remain competitive and retain reputations, destinations have to continually improve their performance. This means investment and keeping abreast of the latest technology. Status quo is no longer acceptable.

In your view, is New Zealand on the right track, as a destination? Which are the key challenges, or "hot" topics at the moment?

I think New Zealand may be on the wrong track, as the nation is very dependent on oil and economic growth. New Zealand has also become very dependent on primary industry exports and tourism. Both of these are very dependent on oil prices. When oil shortages arise, this will directly affect New Zealand's earning, as a country. I don't think we are ready of this.

In addition, primary industries are being targeted for climate change initiatives. I think it is time to speak to the urban populations about sustainability and the contribution we all make to climate change. New Zealand needs to look inward. We all have the ability to make small changes to improve our footprints and protect our environment.

One of the projects you were involved with in Kaikoura is the Trees for Travellers initiative. In a nutshell, what was it about?

Trees for Travellers is a carbon offset program, where a visitor can purchase a tree to go toward offsetting the carbon from their trip. However, the program is so much more than that: it is an intergenerational commitment to the environment. An indigenous, locally sourced tree seedling is purchased. This is planted on public land in Kaikoura and goes towards re-forestation and re-creation of indigenous biodiversity. The tree is tagged so the visitor can come back and visit, see the tree grow and contribute to restoring a piece of Kaikoura's natural environment.

Your advice for destination managers keen to make their place more sustainable?

Trust your heart, there is no one size fits all. You have to find what your Community is passionate about, as it is the Community that leads the commitment. Do it because living responsibly and looking after our planet is the right thing to do. Don't do it for commercial gain, but remember there are some people you have to sell the concept to, and that commercial gain might be the only thing they understand.

If you could turn back time and start all over again, knowing what you know now about public administration, sustainability and the travel business, would you do anything differently?

Every organization needs an environmental champion. If they are able, that champion needs to take a strategic approach to sustainability. However, in some organizations it is better that it happens piecemeal. When people see the success of environmental projects, they buy into it more.

If I could do things over, I would make the sustainability issue more public, so people can openly debate the costs and merits.

Link to the interview: https://sustainability-leaders.com/rachel-vaughan-interview/

Nepal I Raj is currently working on tourism recovery in Nepal and Norway. He is doing his part to help overcome the pandemic, working with a citizen collaborative called Covid Alliance for Nepal, connecting and synergising skillsets amongst the Nepalese, from ground level work to international lobbying for vaccine equity. Raj is supporting crisis communications in Nepal and around the world as a consultant, and working in the transformational travel space as an ambassador of the Transformational Tourism Council in Nepal and India. He is also currently the Chapter Leader of Travel Massive in Kathmandu, helping start-ups and budding businesses retain their interest in travel.

Areas of expertise: responsible tourism, tourism business, crisis management, Asia, UNSDG 8 (Equal & Fair Economic Opportunities)

The following interview with Raj was first published in December 2017 on Sustainability-Leaders.com.

Raj, as a passionate adventure traveller, do you remember the first time you heard about sustainable tourism? When did you decide to engage professionally in tourism and use it as a tool to support the sustainable development of your home country, Nepal?

I had not heard about sustainable tourism until after I started my company, socialtours, back in 2002. My vision was to see and support tourism development in different ways, and to do this by following the principle of linked prosperity. In practice, I focused on environmental sensitiveness, cultural sensitiveness and contribution to the local economy. I now know that those are also the main principles of Responsible Tourism.

I believe the first time I came into contact with sustainable tourism would have been one year later, when I got engaged with development agencies, which were using the term quite a lot.

Has your view on tourism and sustainability changed since the establishment of your company, socialtours?

Nothing much has changed in my views and opinions. If anything, I am going deeper and deeper into the concept of business responsibility, and sustainable practices are part of that. I was, and always will be, a practitioner, so will find ways to learn more about sustainability by running the company, or through my consultancy work.

Which have been the main entrepreneurial or business challenges that socialtours has had to overcome?

Visibility is probably the biggest challenge. The tourism marketplace is extremely busy. Getting visibility is difficult. We have overcome this challenge to a large extent by creating a very strong brand and image for the company, which reflects what we do, but it is a continual challenge.

As a tour operator, are there any barriers that you think prevent local communities in Nepal to participate more significantly in the travel industry?

There are actually no real barriers for Nepalese local communities to participate in the travel industry. The only possible barrier is lack of accurate information to help communities realise where the demand is, and to match demand with strong products developed using an informed mechanism.

Communities tend to experiment. Some of those experiments fail and others succeed. With more accurate information, more communities would succeed.

How do you measure your success in terms of sustainability performance?

We do so much work in sustainability that measuring it all has become a chore. So, we decided to just build some stories around what we do, and to continue doing what we are doing. Getting lengthy reports done does not seem to justify the time and money it takes to maintain those. In many ways, we think that we have gone beyond that.

Which sustainability challenges do you think will be the most crucial for socialtours and Destination Nepal to address in 2018?

While new tourism products develop in Nepal all the time, making sure that those follow the principles of sustainability by minimising negative impacts is probably the biggest challenge in Nepal.

As Nepal moves ahead, there will be more and more travellers coming in. Delivering sustainable products to those without degrading the natural environment or the local cultures is going to be an ongoing challenge, especially ensuring that there is adequate compensation for the local economy.

At socialtours, the challenge is to keep the process dynamic and to continue our work towards being as sustainable as tour operator as we can. Committing the necessary effort and focus is a constant challenge. Sustainability also needs to be fun and connected to the ethos of the company—another challenge.

A key part of your Responsible Tourism Policy at socialtours is 'educating the traveller' and 'spreading the word'. How do you communicate sustainability internally (b2b, e.g., suppliers), and externally (b2c, potential clients/travellers)?

We have built quite a reputation for the sustainable practices that we do in the company. We invite suppliers to come and learn from us, and most of them enjoy these experiences. This helps in spreading the word among suppliers.

Externally, with clients, we raise awareness via trip briefings in our office, email communications and feedback mechanisms that are entrenched in how we work.

As co-founder of the community initiative NepalNOW, can you tell us about the campaign's achievements so far, in terms of using tourism as a tool for crisis recovery and alleviation, following the 2015 earthquake?

NepalNOW was a major step in the right direction for Nepal after the Earthquake. When images of destruction and stories of damage covered the internet, the NepalNOW campaign aimed to offer a different narrative through positive stories and images. The initiative brought all the players together—having one website and campaign helped to do that.

NepalNOW started as a travel community initiative and was eventually taken up by the government. This was another big success because it then had adequate and continued funding to make it work.

Reflecting on your experience with NepalNOW, which lessons or advice could you share with destinations around the world facing the consequences of natural disasters?

As soon as disaster hits, a one stop information source that provides accurate information and starts positive campaigns is a must.

The I AM IN NEPAL NOW campaign with travellers that we started in Nepal has since been replicated in Ecuador and currently in Bali. Putting positive images online and letting willing travellers contribute towards restoring tourism really helps. And getting the community to co-share the campaign helps to bring everyone together behind the sole focus of bringing back the travellers. This should be practiced everywhere where tourism takes a hit.

Which travel trends do you observe internationally that might support or hinder a better, more responsible tourism in Nepal?

I think we have only just started moving in the right direction. More and more travellers are looking for experiential experiences and this is a very promising sign. New generations of travellers want experiences and not just attractions, which is really good, too.

If the travel industry follows up on these trends and makes more and more responsible products that are environmentally & culturally sensitive, while ensuring contributions to the local economy, the travellers will follow. This will create a good win-win situation for responsible tourism.

Finally, Raj, your advice to tour operators who want to engage with responsible/sustainable tourism business practices but don't know where to start?

FOCUS. Do not think about too many things at the same time. Focus on some principles, kickstart practice on these principles and try to go deeper and deeper into it. Get it internalised in your company, your products and the delivery of the products. That passion for focus will never fail you and will automatically create a reputation for the business that you can then ride on.

Link to the interview: https://sustainability-leaders.com/interview-raj-gyawali/

South Korea | Randy Durband is CEO of the Global Sustainable Tourism Council (GSTC), the UN-created NGO that manages global standards for sustainable tourism. He brings a unique blend of skills and experience based on two career paths he has taken in travel and tourism, spanning 40 years. He has been in senior leadership positions with major outbound tour operators as the President of Travcoa, Clipper Cruise Lines, Executive VP of Tauck and now a second career in sustainable tourism. He is a frequent advisor and speaker to governments and businesses on sustainable tourism policies and implementation. Randy has served on many tourism boards and committees in Asia, North America, Europe, Africa and as a judge for leading sustainable tourism awards.

Areas of expertise: sustainable tourism, destination sustainability, certifications, UNSDG 11 (Sustainable Human Settlements), UNSDG 12 (Resource Efficiency)

The following interview with Randy was first published in January 2015 on Sustainability-Leaders.com.

© The Author(s), under exclusive license to Springer Nature Switzerland AG 2022
F. Kaefer, *Sustainability Leadership in Tourism*, Future of Business and Finance,
https://doi.org/10.1007/978-3-031-05314-6_90

Randy, what was your view of sustainability and tourism when you started your professional career?

Working for a tour operator offering packaged tours, I was proud of the fact that group tours on a motorcoach are an energy-efficient mode of transport. That was before there was much discussion at the industry level of sustainability practices and policies, so my views were not nearly as broad and deep on sustainability as they are today.

Now at the beginning of 2015, what has changed?

It has changed completely because now there is well-established literature and debate regarding sustainability in travel and tourism. We've made a lot of progress in identifying needs and opportunities, and many subsectors are trying to do the right things, but we have not made strong enough progress by enough of the largest enterprises in travel.

As a leader in the sustainability field, which have been your main insights?

One is the lack of full recognition that sustainability is complex and needs to be thought of as a journey, with continuous improvement and adjustments to ever-changing conditions.

Another relates to the fact that travel and tourism is a highly fragmented industry with much variance in attitudes and knowledge about sustainability. That results in much confusion about terminology, goals, impacts, costs. . . really all facets of that challenging journey to sustainability.

Another insight, or should I say strongly held opinion, is that I see effective destination management as essential to take sustainability in tourism to the next level. Aviation may be somewhat separated from this, but all other subsectors of the travel & tourism industry can and should be impacted by management approaches at the destination level. We need a combination of awareness-raising and reasonable regulation—in that order—to require sustainable practices and behaviours by travellers and travel providers. Global and national standards should be applied within the local context.

Do you share the view that sustainability has become mainstream?

Perhaps talking about sustainability has become mainstream. But true sustainability in policy and practice is in my view still in its infancy. Most of the discussion and best behaviours are limited to a too-small percentage of this very large industry.

Where do you see the main challenges for sustainable travel and tourism?

To my mind, the top challenge for both sustainability in general and travel & tourism, in particular, is that the demand for the sustainable product must shift from soft to hard.

Ask most consumers if they care about doing the right thing for planet and people and they'll say yes. But their purchase behaviours are based mostly on price and convenience to them. They need to factor in their good intentions at the point of purchase more than they do now.

And the main opportunities?

The top opportunity is for travel providers to get better at fusing traditional sales messaging with a new language that inspires the traveller to buy the products that are better for the planet and people. I hope and expect that those marketers of travel who get that messaging right will enjoy a competitive advantage.

In which region(s) of the world do you see most interest and momentum for sustainability?

Of the 3 interest groups—travellers, the private sector selling travel, and the public sector managing destinations—all seem to have the highest levels in the entire northern half of Europe. The second tier might include the rest of Europe and Australia. Most of the rest of the world, especially the two global giants of the USA and China, have a very long way to go. There has been some good ecotourism and adventure travel movement in the Americas, Africa, and elsewhere.

Which has been your favourite sustainability book in 2014?

I still keep 2013s *Overbooked* by Elizabeth Becker top of mind as the best book in the recent past. Nothing I'm aware of came close in 2014.

Why did you decide to lead the Global Sustainable Tourism Council?

I believe in the mission of GSTC to develop and disseminate baseline standards for sustainability. Referring back to my earlier comments about my strong view of the need for focus on sustainable destination management as a key driver of improved sustainability in travel and tourism, the November 2013 release of GSTC's Destination Criteria provides the global framework to move in that direction. That excites me.

Early in 2014, GSTC needed a leader, and we agreed that it was a great fit for me to step in. Though I lack expertise in the technical issues of standards, accreditation, certification, etc., I do know something about the business of travel on all seven continents, am motivated, and have good organizational skills, so I thought I could make a contribution.

How does the GSTC support tourism businesses and destinations?

By providing a common framework for understanding what should be done to be able to claim they are working towards sustainability. By providing tools for learning, assessing, measuring, and managing for sustainability.

Your experience with the many different rating and accreditation organizations?

There are more 150 certification schemes for hotels throughout the world. They range in quality from excellent to very poor. That difference in quality is a key reason why the GSTC was originally created, to assess the quality of standards used by those certification schemes.

GSTC Recognition simply means that the standard applies each of the GSTC Criteria; while GSTC Approval has a more qualitative side to it, verifying that the policies and processes of the certification scheme are sound.

Link to the interview: https://sustainability-leaders.com/interview-randy-durband/

UK | Rebecca Hawkins has worked in the travel industry for the last 25 years. She started her career in hotels and landscape management before writing a PhD and joining the World Travel & Tourism Council. She has worked with the largest travel companies in the world helping them create responses to the sustainability challenges raised (initially) by the Rio de Janeiro Earth Summit. She continues to work regularly for large companies as well as international conservation organisations, destinations, travel trade associations and charities, advising on a wide range of issues from reducing plastic pollution, conserving water, enhancing the local contribution of tourism and destination management. Recently, she has been on projects that bring creative and tourism professionals together to develop experiences that reflect the essence of the place. Rebecca delivers much of this work within her own business (Responsible Hospitality Partnership Ltd) and she also teaches at Oxford Brookes University.

Areas of expertise: hospitality, tourism business, UNSDG 12 (Resource Efficiency), UNSDG 8 (Equal & Fair Economic Opportunities)

The following interview with Rebecca was first published in September 2016 on Sustainability-Leaders.com.

Rebecca, when did you first become aware of sustainable tourism?

F. Kaefer, *Sustainability Leadership in Tourism*, Future of Business and Finance,
https://doi.org/10.1007/978-3-031-05314-6_91

The foundations were probably laid by my Gran who, like so many of her generation, had a waste not want not ethos. She was a school cook and could make a meal out of almost anything. In her view, there was no excuse for wasting food (or anything else come to that). All that was needed was a bit of creativity and willing kitchen helpers (often her grandchildren) to cook up a storm.

College was the point at which I first came across 'sustainability' as something that people wrote books about (although most called it environmental management back then). I started college on a history course, but quickly discovered it wasn't for me. I found myself in the environment section of the library one day. By chance, I picked up *Silent Spring* (Rachel Carson) and went from there to James Lovelock's *Gaia* all in the same afternoon. And that was my light bulb moment. I realized that 'sustainability' wasn't just a quirky lifestyle choice, but an academic discipline in its own right, underpinned by serious theoretical principles. I changed course and became absolutely passionate about 'sustainability' as it applied to pretty much everything.

Post college, my early jobs were in tourism and hospitality (I have always loved travel). I ended up landing a contract as Research Assistant at Bournemouth University and this involved writing a PhD. This gave me the freedom to really examine the environmental and social impacts of tourism just as sustainable tourism was starting to emerge on the policy agenda. It also gave me a great grounding in implementing a research proposal (I learned as much from the failures as successes). In fact, I often claim that writing my PhD was the single most important factor in forming my career.

What made you decide to pursue a career in sustainable tourism? How did you get started in the industry?

A combination of having a PhD and luck! After my PhD, I worked in various tourism roles and then saw a job advertised working for the World Travel & Tourism Council, running their environmental research center. I have never wanted a job as desperately as I wanted that one. I applied and about a week later had been offered the job. I couldn't believe that I had landed my dream job with two of the leading visionaries in the sector as my bosses.

I spent the next five years working with 70 of the world's biggest travel and tourism companies understanding their responses to the environmental issues, writing policy documents (including the interpretation of Agenda 21 for the travel and tourism industry) and testing solutions in companies and destinations. I loved every minute, and I still get a huge thrill from my work today.

You have been working as a Research and Consultancy Fellow at Oxford Brookes University for almost two decades. Can you tell us a little about your work there?

I have run my own business for 20 years, but also really enjoy the academic environment and so have always kept one foot in that door. I am a Research and Consultancy Fellow for Brookes, and I mainly supervise PhD students. I love the enthusiasm that my students come with and the sense that they can really make a difference through their research. I find that my consultancy complements my academic role perfectly: many RHP clients become case studies for PhD students and the academic environment provides a safe space in which commercial companies can explore some of the 'harder to tackle' aspects of sustainability.

I love the point in the PhD journey at which students become the 'expert' in their chosen field. I have had students working on a wide range of topics, including understanding the mechanics of staff behaviour change, assessing how the imagery we use conveys messages about sustainability to consumers, analyzing the significance of boundary setting in environmental reports and reviewing the role of corporate governance procedures and practices. Many of my ex-students are now teaching in universities across the globe, ensuring sustainable tourism is firmly on the academic agenda.

In 2010, you founded the Responsible Hospitality Partnership (RHP); what led you to start the organization, what is its mission and how do you set about achieving it?

RHP is my third business venture and has its origins in my two earlier companies. Our mission is to help tourism organisations reduce costs, improve staff morale and build trust in their brands by operating in an environmentally and socially responsible way.

I firmly believe that change is best delivered from within, and so we work with companies to help them adapt their DNA to embrace sustainability principles. Our client base includes micro-enterprises through to multinational businesses, trade associations and governmental agencies. We deliver a wide range of support including providing energy and water management (we have developed our own software platform), food waste prevention support, the development of guidelines and voluntary agreements and the delivery of staff motivational training.

Over the course of our work, we have reduced carbon emissions within our client businesses significantly, implemented measures to engage staff in food waste prevention and used our stakeholder facilitation process to develop voluntary agreements that go way beyond our wildest expectations in terms of human rights as well as environmental protection.

I am really proud of what we have achieved, but I feel there is still so much more to do with the difficult to reach areas of the sustainability agenda. These include things like human rights protection (the tourism sector has a large number of minimum wage and zero-hour contract employees), embedding sustainability throughout supply chains to ensure that destinations maximize the value from tourism, ensuring that we measure what matters and ensuring that sustainability commitments permeate throughout all parts of an organisation rather than those that are owned or managed alone. It is great that we have organisations like ITP and ABTA who are also very engaged in the dialogue on these issues.

While recycling, water and resource management are well-known topics within the tourism field, you have given lectures and spearheaded several projects regarding food waste—a topic that gets significantly less attention when it comes to sustainability in the hospitality industry. In your experience, what can hospitality businesses do to manage food waste, and what role does that play in their overall sustainability efforts?

Food waste is tricky and quite unlike energy and water management, which are relatively easy to tackle (for the most part you can install a device, make sure meters are read and you get the savings—cost and environmental). Food waste is a tough nut to crack for a number of reasons, including the fact that nobody likes to admit that they waste food; we work in a sector that is predicated on plenty and so there can be resistance to discussing things like portion size and plate waste; effective waste prevention requires behaviour change throughout an organization (from head office to each individual unit), and there is uncertainty among businesses about how to talk to clients about these issues.

To prevent food waste, CSR teams need to be able to reach the hard to get at areas of the business (including procurement teams and menu planners), to motivate staff, to ensure changes in practice don't impact on product quality or on food waste up the supply chain, to introduce effective ways of measuring change (not cheap or easy) and to communicate success. On our side is the fact that staff are often passionate about and value food.

Food donation schemes and the fact that they help those in need are also a point of great pride. But food donation is just one piece in the waste prevention jigsaw and needs to be paired with other initiatives to produce cost and environmental savings. When we really engage teams, we can get significant savings (up to 30% in some settings) and ensure that people in need benefit. But it needs real commitment throughout a business, and there are a number of hard-to-reach areas around food waste and the food cycle that we have as yet barely touched on.

Combine the complexity of this issue with the fact that most businesses judge the significance of food waste as an issue by the cost of disposal, and you realize why it is not the key priority for many businesses. In reality, the cost of food waste disposal accounts for less than 4% of the real costs of food waste—most of the costs of food waste lie in food procurement and staff time processing food that is subsequently thrown away. When judged by disposal cost alone, food waste accounts for a significantly smaller proportion of turnover in most hospitality businesses than energy and water.

Food waste prevention, however, is about a lot more than cost savings. It is a strategic issue that has vital moral and environmental dimensions.

From a moral standpoint, the amount of food we throw away globally is sufficient to provide food for every man, woman and child who is chronically undernourished worldwide (although of course, the food that is wasted is often not in the areas in which there is the greatest hunger). As such, food waste prevention can play a key role in improving national and global food security.

Environmentally, the carbon emissions of meat production alone (14–18% of the global total) dwarf those of any single element of the tourism sector (aviation accounts for around 3% of total emissions).

The Out-of-Home (restaurant, pub, café, institutional feeding etc.) market accounts for almost half of spending on food in the UK and a similar proportion in other developed countries. So, businesses that are really committed to reducing their environmental footprint (let alone improving sustainability) need to fully engage in the food waste agenda and find ways to make a genuine difference to these up-stream moral and environmental issues.

In your view, where do most businesses go wrong when working across the sustainable tourism agenda? What advice do you give them?

Responsible business programs should support long term business success and build trust in a brand. To deliver these outcomes, however, programs need to be more than skin deep. We do help businesses that just want to save a bit of money on energy, water or waste to achieve their ambitions and will promote the environmental improvement that results from these programs. But:

A bit of resource efficiency does not equate to a fully-fledged commitment to sustainability or responsible business. Businesses that want to really leverage from the trust and quality benefits of sustainability will place these issues at the center of all business decisions (alongside HR, finance, quality control, etc.) and will embrace not only the environmental but also the social elements of this agenda.

In fact, we think that social justice will be the next big issue for many businesses. Those who engage in the dialogue around these issues will be those who are genuinely committed to long-term sustainability (some companies are already dealing with these issues).

What three pieces of advice would you give to someone wanting to start a career in the responsible tourism industry?

That's difficult to answer without resorting to clichés.

- Be passionate about what you do (but don't become a censorious eco-bore)
- Learn the skills of listening, engaging and motivating … most staff teams are more than able to create solutions to sustainability dilemmas. They just need a bit of facilitation and support along the way
- Be prepared to start in a mainstream tourism or hospitality role and to convince your employer that the sustainability path is one that will pay dividends (financially, reputation-wise, morally)

Link to the interview: https://sustainability-leaders.com/interview-rebecca-hawkins/

Switzerland | When it comes to sustainability in the Flims LAAX region, you cannot avoid Reto Fry. Grown up in Flims, Reto is a passionate snowboarder and fly fisher, studied tourism management and is the proud father of Louie and Sofia Fry. He has been employed at Weisse Arena Group (WAG) since 2010 to reduce the company's environmental footprint. As the Head of Greenstyle, he develops environmental visions and strategies, apart from implementing measures to run a sustainable alpine destination. Alongside his job as the environmental representative at WAG, he is also the founder and managing director at the Greenstyle Foundation, a non-profit organisation that works for the maintenance and protection of the environment in the local area. One of Reto's greatest dreams is to cover all the energy

© The Author(s), under exclusive license to Springer Nature Switzerland AG 2022
F. Kaefer, *Sustainability Leadership in Tourism*, Future of Business and Finance,
https://doi.org/10.1007/978-3-031-05314-6_92

needs of Flims, Laax, Falera, Trin and Sagogn with 100% local renewable energy, transforming the destination from an energy consumer to a producer.

Areas of expertise: destination sustainability, mountain tourism, UNSDG 12 (Resource Efficiency), UNSDG 13 (Fight Climate Change)

The following interview with Reto was first published in October 2018 on Sustainability-Leaders.com.

Reto, you have been involved with tourism and sustainability for several years now. Do you remember what first got you interested in the topic?

Fossil fuels are a finite resource which we unfortunately heavily depend on, although local, renewable alternatives are available. Not to mention oil spills, causing catastrophic damage to humans and nature, harmful emissions contributing to climate change, and all the money we send to warmongering countries, etc. For me, it was always a mission to reduce oil consumption as much as possible and to find energy solutions which are based on renewable, infinite and more environmentally friendly resources.

Secondly, I grew up in Flims/Laax, where most jobs (including mine) depend on the tourism sector. Our biggest capital is our unique landscape and nature. We make mountains and lakes accessible and enjoyable for our guests. But we also need to find a way to protect them, to make sure tourism is sustainable.

How has your understanding of sustainability changed since taking on the position of environmental officer at Weisse Arena Gruppe (WAG)?

In the beginning, my plan was to change many things mainly from an ecological point of view. However, I had to learn that sustainability is not just a welfare program. Ecology, economy and sociology are strongly connected.

Of course, I knew that already from my university studies, but I needed some time to understand how our need to create jobs, our alpine tourism, and ecological needs can work together. That's why our vision is not just, 'Laax wants to become the world's first self-sufficient alpine destination'; it is also 'through financially viable initiatives'. It has to be good for the environment and the local economy.

WAG describes its environmental vision as 'greenstyle'. In a nutshell, what is this about? And why the focus on sustainability and environmental best practice?

Reto Gurtner, CEO and President of Weisse Arena Gruppe, is the visionary leader of this forward-thinking company. One day he called me and said he needs an environmental officer and if I would be interested. That was in 2010 when we came up with the first version of the Greenstyle concept. It consisted of topics relevant for WAG, like Energy, Waste, Water, Food & Purchasing, Transportation, Biodiversity and Communication.

Then we met Ross Harding, an Australian sustainability warrior. Together, we defined our vision: LAAX wants to become the world's first self-sufficient alpine destination through financially viable initiatives.

Based on this vision we developed a strategy which follows the principle to reduce, reuse, and recycle. We have set demanding, but doable, goals until 2023, including:

- Reduce CO_2 emissions by at least 30%
- Reduce electricity consumption by 0.4 GWh
- Reduce residual waste by half

Why the focus on sustainability in the first place?

To us, success means maximising our positive social impact as we minimize our environmental impact.

We want to create a place where people can travel to without feeling guilty, and who are willing to pay the costs for this benefit.

We know that about 90% of our guests are saying that sustainability is very important, and really appreciate our effort. Sustainability is now a part of guest satisfaction.

Human-caused climate change is real, and thus mandatory for us to fight it.

On the other hand, reducing energy consumption is not just good for the environment, but also for saving money. So why shouldn't we do it?

Which destinations around the world serve you as examples?

Aspen Snowmass and Vail resorts are two inspiring companies in winter sports destinations. From Aspen Snowmass, I got the idea of founding an environmental foundation: the Greenstyle Foundation is dedicated to preserving and developing the environment across the destination Flims/Laax.

We also have a yearly climate kick boot camp in LAAX, where innovative start-ups receive coaching and can present their ideas.

Climate change and seasonality are two challenges frequently mentioned in connection with destination sustainability, especially mountain tourism. How do you approach those at WAG, and across the destination?

Right now, winter tourism in Laax is generating—within only 4 months—about 90% of our revenue.

For us, moderate climate change (a rise of global temperatures by no more than two degrees) has two dimensions. First, some tourism experts predict that more people from cities will want to enjoy cooler temperatures in mountain destinations, which could be a positive change. This is actually something we already witnessed this summer.

On the other hand, the absence of snow in lower areas will make resorts vanish. Even though a big part of our slopes is on snow-secure levels, at above 2000 m altitude, we already experience the consequences of the changing climate: winter seasons are becoming shorter.

What role can marketing play as a tool for (more) sustainable destinations and responsible travel?

Marketing and communication are probably the most important facilitators of sustainability. It is not about being perfect, but about doing something and being transparent. Only that way people will support you, for example by buying the right goods and giving you constructive feedback on how to improve.

All too often tourism businesses and destinations develop well-meaning sustainability strategies but ultimately fail at implementation. To your mind, what does it take to succeed? Which are the key pitfalls to avoid?

This is quite clear to me now, but I needed about 7 years to find out—and I am still learning:

- Have a forward-thinking leader
- Create a demanding vision
- Work out a strategy on how you want to achieve your vision
- Set realistic goals
- Communicate: make sustainability an always-present topic within the company, and across the destination
- Don't take your work too seriously, but also never give up
- Don't forget to celebrate even the smallest goals you have achieved
- Don't waste time trying to convince people who are 'blind'
- Find partners who are open to new ideas and changes
- Take your time. Sustainability needs to be a company principle

Which aspect of making Laax as a destination more sustainable do you find the most challenging?

Snow grooming. Right now, we do not have any renewable alternatives to eliminate CO_2 emissions. We are trying to use less fuel. All snow groomers in LAAX are equipped with an engine data system that indicates the optimum speed for minimal diesel consumption.

But also, the height and depth of the entire skiing and snowboarding region were measured using 3D technology. Because of this, a 3D snow depth measurement system in the snow groomers shows where there is a lot of snow and thus facilitates slope preparation with maximum efficiency.

We hope to have new solutions, like hydrogen/electric snow grooming vehicles, for the next strategy goals 2023–2030. If that problem is solved, then a carbon-neutral operation of a winter destination based on renewable energy is possible.

Looking ahead, which trends do you think will likely impact the sustainability of winter tourism in mountain regions?

In my view, human-caused climate change will have the biggest impact on the sustainability of winter tourism, and the whole planet.

However, society is becoming smarter, and will soon be ready for a transformation from fossil fuels to a future based on renewable, infinite energy resources. People are going to invest more in sustainability and sustainable products or experiences. Forward-thinking businesses will profit even more from a sustainable business approach.

Link to the interview: https://sustainability leaders.com/reto-fry-interview/

UK | Richard Butler is an Emeritus Professor at the University of Strathclyde Business School. He was trained as a geographer at the Universities of Nottingham (B.A.) and Glasgow (PhD) and taught at universities in Canada (University of Western Ontario), England (Surrey), Italy (CISET, Venice), Holland (Breda NH University) and Scotland (Strathclyde). His teaching experience has been in tourism, recreation and resources management and his principal research interests have been in destination development, tourism in islands and remote areas, indigenous peoples and tourism and the many aspects of sustainable development in a tourism context. He has published over 20 books on tourism, and over 200 articles, papers and chapters on various issues in tourism. He is a past president of the International Academy for the Study of Tourism and the Canadian Association for Leisure

© The Author(s), under exclusive license to Springer Nature Switzerland AG 2022 549
F. Kaefer, *Sustainability Leadership in Tourism*, Future of Business and Finance,
https://doi.org/10.1007/978-3-031-05314-6_93

Studies, and in 2016 was awarded the UNWTO Ulysses Prize for Excellence in the Creation and Dissemination of Knowledge.

Areas of expertise: sustainable tourism, tourism research, sustainable development, UNSDG 11 (Sustainable Human Settlements), UNSDG 12 (Resource Efficiency).

The following interview with Richard was first published in October 2016 on Sustainability-Leaders.com.

Richard, your academic background is in geography. How did your studies in this field lead you to eventually devote your professional life to tourism and sustainability?

Geography is about the world, how humans relate to and affect the natural environment and its response. We are concerned with spatial patterns and distributions, 'why what is where' is a good summary. It is very logical, therefore, for tourism geographers to be interested in how tourism develops, what it means for people and places, and what the patterns may be in the future. We are thus interested in how transport systems develop, how people become more mobile and what mobility means in the context of tourism for example, why do people go to certain places, what physical features are attractive to people, and what needs to be done to protect them.

In your opinion, how has the concept of sustainable tourism changed over the years?

It has become even more meaningless than when it was first mooted in 1987. It now encompasses such a wide list of topics, interests, methods and ideologies as to be virtually useless in any real sense. It has become a substitute for action, an all-encompassing term for anything even the slightest shade of green, regardless of whether it really is sustainable or not. It has given rise to far too many places, facilities, and services claiming to be sustainable when they are far from it.

Nowhere that requires most of its visitors to fly long haul or travel long distances generally should claim to be sustainable. We have cheapened the term by trying to get all the benefits without paying any of the costs that would be involved in anything truly sustainable. Like 'ecotourism', it is a term thrown at anything and everything.

You have published 17 books on tourism and hundreds of articles in professional journals. Which themes, issues, or trends do you find most interesting and relevant to the current landscape of sustainability in the tourism industry?

- First and above all, changes in patterns, in impacts, in tastes, in numbers.
- Second, the relationships between the natural environment and human development and use of areas and methods of mitigating the undesirable effects of tourism and on tourism.
- Third, the links between power, politics, peace, war, and tourism; so much tourism is related to the heritage of conflicts—over land, religion, resources, and beliefs—but we pay little attention to these links.

- Fourth, remote areas and tourism, particularly in the context of islands, and the issues associated with development—not just of tourism—on people and their environments.

One of your areas of expertise is destination management; in your opinion, what are the main challenges destinations face in implementing a successful sustainable tourism strategy?

The unsolvable problem of ever-increasing numbers and the impacts of such growth along with the refusal of the industry and governments to acknowledge that numbers are a problem and have to be faced and, in some cases, limited.

Recent protests in places like Venice show clearly that residents of such places are well aware of the problems from excessive numbers, but no agency is willing to cut or even halt an increase in numbers.

Related to that is the lack of joined-up thinking and planning, thus, for example, a new air service may be started to a location before any planning for tourism has taken place in that location, with the result that tourism rises and development takes place without any planning or controls.

What have you learned teaching the next generation of tourism professionals?

Unfortunately, most of them have a limited background in terms of deep knowledge in any traditional discipline. I regret that it is possible to go through university programs to the PhD level, taking essentially only tourism modules and programs and thus getting little in-depth conceptual knowledge about other subjects.

I do not see tourism as a discipline, I see it as a fascinating topic to study, one that deserves a high level of intellectual study and research of the highest integrity but not something to gain an undergraduate degree in.

Do you think that the information and insights gained within the world of academia are reaching the individuals involved in the operational side of the tourism industry? How can we better bridge the gap between sustainable tourism research and its practical application?

Not really, but many of us teaching in tourism have had no experience of the operational side of the industry. Tourism is much more than simply the business of tourism, and nowhere is this more evident than in sustainable aspects of tourism.

However, I doubt very much that the tourism industry (or any other industry) is or can be very serious or committed about sustainable development as long as they are growth-oriented, profit-driven, and short-sighted in terms of planning and development. This is not meant to be unfair criticism, the industry has to focus on its goals and needs, including for almost all industries making a return on investment and keeping owners/shareholders content. But, running an operation that is self-sustaining in terms of operation and perhaps construction, labour, and supplies does not make that operation sustainable if people travel thousands of miles to get there.

People fail to grasp the difference between sustainable development and operations, often brilliantly done, and unsustainable tourists coming to the operation thus making it unsustainable in the overall sense.

What changes do you foresee for the tourism industry in the next decade regarding sustainability?

Not much, more noise and publicity and awards, and very little in terms of making tourism as a whole any more sustainable.

Quite frankly, sustainable tourism is impossible, and we should focus on making operations more sustainable and acknowledge that tourism is an industry that deep down is impossible to make sustainable.

We would have to stop long-haul travel, which would be a disaster to many developing countries and regions and also drastically limit choices for tourists, both of which are undesirable for many reasons, particularly poverty alleviation. Not everything can be sustainable.

Anything else you'd like to mention?

Despite all the comments above, I strongly support the concept of sustainability and believe it can be applied in some areas of tourism, not many but some. We do the concept no good by pretending it is applicable in every situation in every place for every form of tourism. Ironically, we could move closer to sustainability by reducing many of the niche forms of tourism which have major carbon emission problems and boosting mass tourism which, per capita, is often more sustainable than ecotourism.

Link to the interview: https://sustainability-leaders.com/interview-richard-butler/

Richard Hammond on How to Communicate Responsible Tourism

UK | Richard runs Green Traveller, a UK-based media production agency that promotes sustainable transport and tourism through words, photos and film. Richard manages a team of freelance writers, photographers and videographers to provide digital content for a range of businesses, NGOs and organisations in the travel and tourism industry to promote sustainable tourism in destinations. Clients have included the World Travel and Tourism Council, The Travel Foundation, The Long Run, Brittany Ferries, Eurostar and several national tourist boards, such as France, Greece, Spain, Switzerland and Visit Britain/Visit England. Richard also runs the Green Holidays[1] website he founded in 2006 for UK travellers looking for a more sustainable holiday. It includes guides to flight-free travel from the UK, as well as car-free guides and holidays that contribute to biodiversity conservation and local economic empowerment. The website also includes Green Traveller's Guides to over 50 destinations worldwide.

---

[1] https://www.greentraveller.co.uk

© The Author(s), under exclusive license to Springer Nature Switzerland AG 2022   553
F. Kaefer, *Sustainability Leadership in Tourism*, Future of Business and Finance,
https://doi.org/10.1007/978-3-031-05314-6_94

Areas of expertise: ecotourism, responsible tourism, marketing, sustainability communication, UNSDG 12 (Resource Efficiency).

The following interview with Richard was first published in October 2016 on Sustainability-Leaders.com.

Richard, when did you first learn about the concept of sustainable tourism?

In the summer of 1996, during a conservation volunteer expedition to Mauritius and Rodrigues, organised by Raleigh International. Part of the 3-month, hands-on trip included three weeks working on an ecotourism scoping project to manage visitors to a bird-breeding area in Rodrigues. Each day, en route to an area of land we were helping to conserve, we'd pass a Tropicbird chick nesting at the foot of a tree.

One morning, however, as we passed the nest, all we could see were bits of fluff. It looked as if the chick had been taken by a predator. It made us all feel a bit low, but later that day, we heard a commotion overhead and looked up to see two magnificent White-tailed Tropicbirds flying across the clear blue sky, with a third smaller bird trailing behind, flying somewhat awkwardly. It was the chick on its maiden flight. It was such a simple but evocative experience... it was my 'light bulb moment'.

Tell us about the professional journey that led you to create Greentraveller Limited in 2006.

After gaining a master's in Publishing in 1994, I worked for several years in publishing, first as a researcher for Harper Collins Broadcasting and then as an editor in natural history and travel publishing for Dorling Kindersley, New Holland and A&C Black.

In 1999, I helped establish the online news portal TravelMole (including launching a sustainable tourism newswire), which gave me an early and valuable insight into the huge influence of internet publishing and e-newsletters.

In 2003, I was approached by the Guardian newspaper to write a regular feature on 'green holidays'. It began as a monthly feature, and then became weekly. I was inundated with emails about the column, so in 2006, I set up the website Greentraveller.co.uk, which was initially a blog and online forum for readers to debate the issues.

The site quickly attracted a lot of interest, and the potential for it to become a commercial enterprise soon became obvious. So in 2009, after I'd completed writing a book on green travel (*Clean Breaks—500 New Ways to See the World*, published by Rough Guides, co-written with Jeremy Smith, the editor of Travindy), I raised capital from investors to launch Greentraveller as a limited company, using the website as a focal point for businesses to promote green places to stay as well as holidays in Europe organised by tour operators to destinations that could be reached overland from the UK, by train, foot, or passenger ferry.

This month (October 2016), we will be celebrating our 10th anniversary by widening the portfolio to include holidays further afield that benefit biodiversity conservation and the economies of local communities. It's an exciting time in the development of the company, and we are really looking forward to promoting holidays that genuinely make a difference, from community-run ethical walking holidays in India to safari camps in Africa and conservation holidays in South

America. This development has come about because of demand from our existing customers.

There's a lot of confusion in this area, and we feel we are well-placed to provide clarity for holidaymakers to help them find operators that are offering products that are genuinely contributing to biodiversity conservation and the economies of local communities.

Greentraveller consists of two main parts: the sustainable tourism consultancy/ media agency and the consumer website www.greentraveller.co.uk. Can you tell us a little about each service and why you chose to combine them both under the Greentraveller umbrella?

Greentraveller Media is our agency and produces marketing campaigns, visitor guides and social media trips for both the public and private sector. Our latest campaign has been for the Swiss Tourist Board to promote the Grand Train Tour of Switzerland.

We also produce high quality online videos (we've produced over 100 in the last 4 years) for hoteliers, B&B owners, tour operators, councils, transport companies and tourist boards. We've filmed with drones and a variety of specialist kit to gain technically challenging footage.

Video is a brilliant medium to promote sustainable tourism as, when done well— showing rather than merely telling—it can truly inspire more sustainable choices.

If a picture can paint a thousand words, how many can a video?

Our consumer website (Greentraveller.co.uk) features green places to stay, single-day activities (such as canoeing, conservation projects, arts and crafts) and longer multi-day trips organised by tour operators that promote low-impact travel contributing to biodiversity and the economies of local communities. The website has been described in the national press as the first place to look for green holidays, yet we were most proud when it was shortlisted as the 'Best Consumer Website' in the British Travel Press Awards alongside the travel pages of The Telegraph, Daily Mail, CNN, and Wanderlust Magazine. This demonstrated that our site is no longer considered niche, but is able to compete with the big players. The site also includes our award-winning blog, which we were thrilled to see included in last year's Expedia's Top 30 Blogs for its 'digital influence'.

In essence, then, they are two sides of the same coin: one aimed specifically at the consumer looking to plan and book a holiday themselves and the other aimed at businesses and organisations looking to promote those kinds of holidays.

Communication and public relations are where many travel companies and destinations struggle, especially when it comes to communicating their sustainability efforts. What challenges have you helped companies overcome over the years and how?

I think one of the greatest challenges is to put aside the worthiness of sustainability and focus instead on communicating the benefits of going green.

In our experience, holidaymakers don't want to be lectured on how to save the planet, but respond more enthusiastically to messages that convey how a greener holiday can be a better holiday—both for the destination and the traveler.

We've helped companies reposition their messages to reflect this through imagery and storytelling. For instance, posturing on about how procuring local food reduces food miles isn't going to pull in punters, whereas using imagery and video to show holidaymakers harbour-side watching fisherman bring in their catch for the day, cooking the catch on an open fire, sprinkling herbs and spices on it, then eating it fresh... that's a far more compelling way to promote a sustainable holiday choice.

What criteria do you use when deciding which providers to feature on your consumer website? How do you ensure that the companies live up to your requirements?

We screen accommodation owners on their efforts to deal with 5 main areas:

• Energy use
• Water conservation
• Waste management
• Procurement of local, seasonal food
• How they encourage guests to use more sustainable transport

For tour operators in Europe, we work with them to fillet out only those trips that can be reached overland by train and/or ferry, and for those trips further afield, we ask them to provide information on how they as a company and/or their trips contribute to biodiversity conservation and the economies of local communities. Most of these trips are also bookable as ground-only trips—many visitors to our website, of course, are not UK-based.

We're not a certification or grading scheme (we do draw upon the grading of accredited schemes to assess businesses), but we do use our network of travel writers to review featured accommodation and trips. While form-filling is useful, there's nothing quite like actually visiting a property or going on a trip to see first-hand what it's like.

Transportation is an issue that plagues the sustainable travel industry. You have written several articles on the topic of sustainable transportation and assisted in the marketing efforts of the Local Sustainable Transport Fund (LSTF). First, can you tell us about the LSTF?

The Local Sustainable Transport Fund is a UK Government initiative that aims to stimulate changes to local transport to cut carbon emissions and create local growth (see the 2011 white paper 'Creating growth, cutting carbon'). The funding included both capital funding and revenue funding. In total, the Department for Transport awarded funding to 96 sustainable transport packages from 77 local authorities between 2011 and 2015.

In 2013, we held a workshop at the Nottingham Conference Centre (incidentally, one of the best conference venues I've ever been to!) on how to use new media and marketing to deliver behavioral change. In his welcome address, the then Parliamentary Under-Secretary of State, Department for Transport, Norman Baker MP spoke of the need for 'strong, clear marketing' to increase the benefits of new and improved infrastructure and services provided by LSTF.

In essence, it's about getting people off their dependence on petrol cars and putting more bums on buses and bikes.

In your opinion, how can the sustainable tourism industry tackle the issue of green transportation and its integration into overall sustainability planning both for individual tourism operators and at a destination level?

Most forms of motorized transport involve some form of carbon emissions, so transport has to be at the heart of destination sustainability simply because it is often the single most influential environmental factor. I've been on the judging panels of various awards and have felt encouraged seeing the recent developments made in sustainable transport.

As judge of the Eurostar Ashden Awards for Sustainable Travel in 2014, I was impressed with the work Ecotricity has done to help kick-start the uptake of electric cars in the UK, creating a new 'Electric Highway' of extensive charging points across the UK motorway system, which has dramatically extended the geographical reach of electric vehicles.

As one of the judges of the WTTC's Tourism for Tomorrow Awards, I was delighted to learn about the carbon management tool 'Carmacal Carbon Calculator' produced by ANVR, The Netherlands Travel Trade Association, which aims to measure the carbon footprint of all aspects of a tour package. Innovations like these are game changers.

Any other advice you would like to give?

Given the amount of greenwashing out there, I think it's really important to be armed with factual information. So often I read applications for awards that don't include statistics which demonstrate measurement and performance over time—it's all too easy to be a generalist.

A particular bugbear of mine is when people use the word 'local'... it's virtually meaningless without context (e.g., is it local to a 3-mile radius, a 20-mile radius, a region, or a country). I much prefer it when I see descriptions of local business names to provide a meaningful context to the notion of 'local'.

Link to the interview: https://sustainability-leaders.com/interview-richard-hammond/

Shannon Guihan on Tourism Sustainability and Destination Overcrowding

Canada | A tourism development expert, Shannon has worked as a guide, a destination development consultant and now leads the direction on sustainability for the more than 40 travel brands that make up The Travel Corporation. Shannon holds an MSc in Tourism and Environmental Management, where she focused on conduct practices for marine-based adventure tourism. Currently, Shannon leads the TreadRight[1] team, which oversees The Travel Corporation's efforts to MAKE TRAVEL MATTER® through the group's sustainability strategy, How We Tread Right. Shannon also leads the TreadRight Foundation, the group's not-for-profit dedicated to supporting community-based projects. With more than 15 years of international experience working at multiple levels of development and delivery, Shannon's primary focus is on the impact that travel providers place on destinations globally, and the way that businesses can develop, implement and measure tactical approaches to mitigate that impact.

---

[1] https://www.treadright.org

© The Author(s), under exclusive license to Springer Nature Switzerland AG 2022
F. Kaefer, *Sustainability Leadership in Tourism*, Future of Business and Finance,
https://doi.org/10.1007/978-3-031-05314-6_95

Areas of expertise: overtourism, sustainable tourism, tourism business, UNSDG 11 (Sustainable Human Settlements)

The following interview with Shannon was first published in May 2017 on Sustainability-Leaders.com.

Shannon, your professional career has included work for companies, non-profits and, since 2011, consulting with Bannikin. Do you remember what got you interested in tourism in the first place?

I absolutely do. I'm from Newfoundland, Canada, a province on the East Coast. My first job, at 16, was as a sea kayak guide. I guided all throughout my undergraduate degree and became rather involved with tourism on the island, which I quickly understood to be 'happening'. The cod fisheries on the island were closed in 1992, and the growth of the tourism industry was, in part, a reaction, so it was happening quickly and, from my perspective on the water, without check.

This clear shift from a heavy to a soft industry was not exclusive to Canada, and is the reason why I pursued an MSc in Tourism & the Environment. To understand the checks.

How has your view on tourism and sustainability changed over time? Your key insights?

In short—it has to be market-driven and consumer demand-proven. I think we're seeing a shift towards the former these days, but I'm afraid I remain disappointed with the latter.

Among other things, you are currently Program Director at the TreadRight foundation in Geneva, Switzerland. In a nutshell, what's this foundation all about?

The TreadRight Foundation is a not-for-profit organization funded solely by the companies that make up The Travel Corporation (TTC). The foundation seeks to encourage responsible tourism through its travel brand partners, while supporting local organisations protect wildlife and communities.

TreadRight has taught me much about scale and a market driven approach. TTC's leadership is genuinely committed to moving the needle on sustainability, and do so while remaining successful. The power they have to implement change is impressive, and I've learned that scale can be a positive thing.

Overtourism has become a critical sustainability issue for many popular destinations now, especially cities but also national parks. In your view, what can destination managers do to address this?

I maintain that overtourism is the elephant in the room for tourism professionals and destinations, and the answer will be a genuine challenge, or certainly what I believe the answer to be: regulatory requirements to guide the market. I believe strongly in a user pays model, and in limiting visitor numbers in a significant way.

The market, both consumers and providers, might disagree with me, but I believe that tourism is a privilege, not a right. And for the sake of many parts of the world, that needs to not be an elitist approach, but a responsible one.

The UN have declared 2017 the International Year of Sustainable Tourism for Development. In your view, which are the main destination challenges right now regarding the sustainability of travel and tourism?

I think this is twofold:

Destination marketers hold the keys to the kingdom these days, and they value visitation growth as their core KPI. On the other hand, the world's top destinations are struggling with too many travellers. The coming to terms of those two camps, in my opinion, is our biggest challenge.

Culinary travel was the focus of your work with Michelin Food & Travel from 2009-2011. To your mind, which are three essential ingredients for an authentic— perhaps even transformative—travel experience involving food or drinks?

I would suggest that the same elements are required to ensure the authenticity of any experience.

- Research—know where you are going, truly. Buy a guidebook, actually read it. It's something overlooked far too often, I think, not understanding the place you are surrounded by. That's on the traveller.
- Don't rush. It's ok if you don't see everything you want to on this trip—whether you make it back to a destination or not—convince yourself that you will, you might actually see a place. This goes for product development experts as well.
- Go off season. You'll be surprised with how much more you see.

Imagine you could turn back time and start all over again. Knowing what you know now about business, travel and sustainability, what would you do differently?

This one has made me think—and truly I'm not sure. I actually think that the conversation happening amongst industry executives at the moment is promising. From the WTTC to the UNWTO—I feel we're finally starting to see a genuine interest in the rooms that make decisions. Perhaps had I my time back, I would have used the understanding of how decisions actually get made a little more wisely.

Your thoughts on the current state of tourism sustainability in North America, Canada especially?

I don't think that it's too far off from what we see globally, however Canada is in a somewhat unique situation. As it recovers from a decline in US arrivals, and benefits from all that's happening in the States and Europe to further drive visitors north—the growth will continue.

2017 is Canada's 150th, and the marketers have taken hold of this brilliantly. So, we have growth caused by multiple factors, to areas many of which do not yet see overcrowding as a factor.

And we are still not addressing these challenges well. Parks Canada is thinking about it—they are actively trying to manage overcrowding through proactive communications planning. I don't see signs of addressing this challenge elsewhere at the moment.

Your advice to destination developers/managers eager to embrace sustainability and to turn it into a reputational and competitive advantage?

Solitude—it sells. Look at Iceland—that's what they've been selling since the mid 2000s. The marketers recognised it—the problem is that the managers failed to protect it. Use that advantage to sell an experience that's priced to maintain quality, and available in quantities that are manageable, and that protect the product—the destination.

Work with your regulators—the disconnect between what a destination has—such as Tulum—a quiet, relaxed corner of Mexico not yet overrun with resorts—needs to be appreciated by those regulators. They need to understand what value they have today, and how not to degrade that value proposition. Learn to speak their language.

Link to the interview: https://sustainability-leaders.com/interview-shannon-guihan/

India | Shivya Nath has been writing at the intersection of responsible travel, conscious living and self-discovery for over a decade. She is the founder of the award-winning travel blog—The Shooting Star,[1] a bestselling author and a leading advocate of sustainability. She has appeared on the cover of National Geographic Traveller India and been featured by The Washington Post among travellers changing the way we think about the world. Her work has appeared on BBC Travel, Conde Nast Traveller and many leading national and international publications. During the

---

[1] https://the-shooting-star.com

© The Author(s), under exclusive license to Springer Nature Switzerland AG 2022
F. Kaefer, *Sustainability Leadership in Tourism*, Future of Business and Finance,
https://doi.org/10.1007/978-3-031-05314-6_96

pandemic, she co-founded Voices of Rural India[2]—a passion project that aims to build digital storytelling skills and alternate livelihoods among rural communities across India.

Areas of expertise: responsible travel, sustainability communication, sustainable tourism, Asia, UNSDG 11 (Sustainable Human Settlements), UNSDG 12 (Resource Efficiency).

The following interview with Shivya was first published in November 2020 on Sustainability-Leaders.com.

Shivya, you have been travelling around the world for over eight years now. How has the pandemic affected your work as a travel blogger?

Back in 2013, I gave up my rented apartment and sold most of my belongings. In the years since I've felt at home in many places around the world but not put down roots anywhere.

Having no possessions and no commitment to a single place felt liberating on many levels. . . until I found myself in lockdown! In a way, this unprecedented crisis has challenged everything about my life philosophy in the past seven years.

I don't own a house or a car, and until a few months ago, I didn't even own any cooking equipment. I've long believed in the sharing economy to find homes and rides around the world. COVID came as a total shock to my existence.

I've also dedicated almost a decade of my life to my travel blog, The Shooting Star. It's my primary source of living, yes. But it's also my passion, my bridge to the world and one of the few constants in my nomadic life. And what is travel blogging at a time when we can't travel, right?

The travel industry (including travel blogging) has taken a big hit in the past 7 months. Personally, my blog traffic has dwindled, on-going projects have been put on hold and potential assignments postponed indefinitely.

Yet I remain cautiously optimistic.

I believe that sometime in the distant future, we will travel again. Borders will re-open, businesses that survive will emerge stronger and we'll get our passports stamped. And when that happens, travel blogging—especially the kind that's rooted in sustainability—will become more important than ever.

To be honest, in the middle of a pandemic that has shattered many travel and life plans, I feel really grateful about the choices I've made. I'm glad I didn't put off my dream of slow travelling the world on my own terms. I'm grateful I didn't build a bucket list to tick off only once I retired.

Do you remember what first got you interested in tourism and sustainability?

My perspective on travel began to change as I travelled to under-the-radar destinations and spent time with local communities. It made me realise that travelling is not just about sightseeing and photos, but a deeper change in your world view.

The world of travel has changed immensely since I set out on my first solo trip and began my journey as a travel writer/blogger back in 2011. Travelling has

---

[2]https://www.voicesofruralindia.org

become more accessible, flights are cheaper than ever before, and Instagram has changed the way we see the world.

I've gone back to places I'd fallen in love with, only to find them ruined by mass tourism, the plastic menace and other pitfalls of tourism. That gradually led me down the path of sustainability—first re-evaluating my own travel footprint, then writing about responsible travel choices as a more immersive way to experience the world.

In the age of overtourism and in the midst of a climate crisis, responsible tourism is not just a pressing need to protect the incredible natural and cultural heritage of our world. It is also the only way we can still find authentic experiences, engage meaningfully with locals and savour the pristine beauty (or what remains of it) on our planet.

What motivated you to take up 'passion projects' to work towards the betterment of the environment and communities? Was there a specific moment or epiphany?

I think it was a gradual evolution of the connection I made with places and local communities on my travels.

Back in 2011, I set out on my first solo trip in India. The destination was Spiti Valley, a barren, high altitude mountain desert in the Trans-Himalayas. My mission was to volunteer with Spiti Ecosphere,[3] a tourism-driven social enterprise in the region, and immerse in other ways of life. That trip changed my life.

One week after that trip, I resigned from my full-time desk-based job in Singapore and started charting out a different path in life.

Six years later, when I finally went back, I nearly cried when the shared taxi deposited me in Kaza, the administrative capital of Spiti. The town that I remembered with only a couple of shops and guesthouses, a handful of travellers, and nothing but the barren mountains all around, had changed beyond recognition. Taken over by chaotic concrete construction, shops, tourists and mounds of plastic waste.

So, in collaboration with Spiti Ecosphere—the organisation I had first volunteered with—and their volunteers, I decided to focus on one big issue: plastic bottles.

Thus began a passion project, I Love Spiti, committed to reducing plastic waste in Spiti. We built a life-size art installation with plastic bottles to create awareness, began the conversation with local hotels, restaurants and shops, installed public water refill points in 4 villages with the support of LifeStraw,[4] and are currently working on a revenue stream to be able to compile and send plastic waste from the valley down to the plains for recycling at the end of each tourist season.

Voices of Rural India is one of your most recent initiatives. Why the focus on rural areas? And how is this initiative benefiting the communities you write about?

---

[3] http://www.spitiecosphere.com

[4] https://lifestraw.com

The conversation about Voices of Rural India (VoRI) began sometime in May this year when the scale and intensity of COVID-19 were finally sinking in—and the impact of the pandemic-induced lockdown on tourism was becoming evident.

It gradually became obvious that tourism is unlikely to recover in the foreseeable future (in advance of a vaccine) and that despite the loss of livelihoods, rural communities in India are choosing to remain closed to the outside world. As urban dwellers with easy access to the digital world, we can continue to work, study and scout new opportunities online. In rural parts of India however, the lack of digital skills and tailored opportunities continue to be a challenge.

This context sparked the idea of Voices of Rural India.

VoRI is essentially an effort to turn this unprecedented crisis into an opportunity to create alternate livelihoods by upgrading digital skills in rural India, while also preserving grassroots knowledge that is slowly disappearing.

In the short-term, Voices of Rural India is creating a revenue stream for affected communities through digital journalism. The storytellers—typically guides, home-stay hosts, people involved in tourism, and youth and women from the community—are paid a fee directly in their bank accounts for every story accepted for publishing.

In the long run, it aims to develop digital storytelling skills at the grassroots level, along with becoming a repository of local culture and knowledge, documented in local voices.

Even though India is theoretically open for domestic travel now, many people are choosing not to, and many communities have decided to remain closed to tourism. The next best thing is to travel virtually, through the words, photos and videos of the very people we travel to meet.

Vegan travel is a comparatively new market niche and a topic you also like to write about. What potential, do you think, does this form of travel have in the context of promoting responsible travel and more sustainable tourism?

I transitioned into a vegan lifestyle over four years ago, after learning about the animal cruelty involved in using animals for food and other lifestyle products like leather and silk.

As someone who has faced sexism and racism first hand on my travels, I was shocked to realise that I was blissfully unaware of my own speciesism—the idea that we wish freedom and love for some animals (birds, dogs, etc.) and think it's okay to brutally use other animals (cows, goats, etc.) to satisfy our taste buds.

Along the way, I also learned that animal agriculture accounts for 14.5% of global carbon emissions and is extremely water and land use intensive. According to many scientific reports, our food choice can be one of the most powerful ways to fight climate change as an individual.

So, for me, veganism and sustainable travel go hand in hand. The vegan movement has taken hold in most cities around the world, and will only become stronger with time. As conscious travel grows, so will the demand for travel options that are not just environmentally-friendly but also rooted in compassion for all living beings.

Indian consumers haven't embraced sustainable tourism yet. How can we encourage tourism industry players and travellers within the country to embrace sustainable travel?

It's an uphill journey, but with growing awareness, especially among the younger generations, I hope we'll slowly get there.

India has some commendable sustainable tourism initiatives that leverage tourism not just as a means to create sustainable livelihoods but also as a way to invest in clean energy, wildlife conservation, waste management, etc. But they're the exceptions, not the norm.

First, we need a radical shift of mindset among travellers, to view travel not as a way to create an enviable Instagram feed, but as a journey to discover other ways of being, both physically and within. When we start to see travelling as an extension of life rather than an escape from it, we'll build more meaningful connections with nature and the people we meet along the way—and feel compelled to take measures to protect both.

Second, we need big policy changes within the tourism sector in India. Instead of building destinations for tourists, we need to create better places to live—with a focus on sustainable infrastructure, green spaces, heritage preservation, conservation of biodiversity and ownership of natural resources. With such a shift in policy, we'll automatically create better destinations to visit.

And third, we need the industry to evolve and take ownership of the natural and cultural heritage of the country, becoming its custodians rather than mere beneficiaries.

What can sustainable tourism brands and 'green' destinations do to share their efforts and experiences more effectively with travellers?

I think the biggest communication challenge for sustainable tourism brands is to not use the word 'sustainable'. Except for a niche set of travellers, many still think of sustainable/responsible/green travel as boring. Brands need to leverage all available channels—from their website and social media to influencer marketing and any form of advertising—to position sustainable travel as more immersive, edgy and desirable.

We need to bring everyone who influences travel decisions (celebrities, bloggers, Instagrammers etc.) into the conversation, and hold them accountable for the kind of travel they promote. We need a movement to convert 'responsible travel' into simply, travel.

What do you think will be the role of niche bloggers and social media influencers in travel and tourism, following the coronavirus pandemic?

When we emerge from this crisis, we need to ensure we don't walk into another one. That means the future of travel is all about becoming more conscious of how we promote destinations, reduce our individual carbon footprint and use our tourism money to meaningfully support local communities and businesses.

Bloggers and social media influencers must create awareness and lead by their own choices, or risk becoming obsolete.

In your experience, which regions or destinations are the most pro-active in how they practice and promote sustainable tourism?

Bhutan is a glowing example of a destination that has prioritised quality over quantity and forests over 'development'. It truly uses tourism as a means for sustainable development.

One can argue that Bhutan's daily fee is a deterrent for many travellers who simply cannot afford it. However, that fee not only minimises the negative consequences of tourism in Bhutan but also allows the sizeable revenue generated through tourism to be reinvested in the country with a focus on environmental and social well-being.

On the other hand, travellers who can afford to explore Bhutan are offered rewarding 'living natural classroom' experiences different from any of its neighbouring Himalayan destinations.

Unfortunately, many destinations in India that started as exemplary examples of community-oriented and environmentally-conscious tourism have fallen prey to mass tourism—with price-sensitive travellers choosing cheaper travel options that create little to no benefits for the local people and ecology.

Link to the interview: https://sustainability-leaders.com/shivya-nath-interview/

Madagascar | Sonja Gottlebe has 30 years of experience in tourism in Madagascar, a famous biodiversity hotspot. She has built three ecolodges and manages adventure trips all over the island. Operating in a poor and fragile country like Madagascar means sustainability in tourism is essential and not a market trend. Sonja and her husband Patrice have launched a Malagasy non-profit organisation that aims to improve the social, health and environmental conditions of the communities living near the lodges. Like many places in the world, Madagascar got hit hard by the economic consequences of the pandemic. Sonja decided to study again to further her knowledge in tourism and recently got a professional certificate in sustainable tourism from GSTC. She also attended a training program in Community-Based Tourism Trainer delivered by the International Trade Centre.

Areas of expertise: sustainable business, ecotourism, hospitality, responsible tourism, tourism business, UNSDG 15 (Forests & Biodiversity), UNSDG 8 (Equal & Fair Economic Opportunities)

The following interview with Sonja was first published in September 2020 on Sustainability-Leaders.com.

Sonja, what led you to dedicate your life to the development of sustainable tourism experiences in Madagascar?

I grew up in Madagascar with my parents. As my father is a botanist, I spent my childhood in nature and was taught that sustainability in life is essential. Ever since I started as a young hiking guide with the German Alpinist Association (DAV),[1] our responsibility to interact with Malagasy people and fragile biodiversity was evident.

You started Boogie Pilgrim in 1989. How has your understanding of sustainability changed since you first got involved?

Over 30 years ago, those concepts didn't exist in tourism. Later on, ecotourism was born, and many other definitions came up: green tourism, responsible tourism, pro-poor tourism, equitable tourism, etc.

Madagascar, which has a strong French colonial heritage, found its way to sustainable tourism development quite late. Once ecotourism started to be fashionable, operators jumped on the band wagon, often practicing greenwashing. It took another decade for the realization to sink in that sustainability wasn't just a new marketing concept. Back in those days, I felt pretty lonely, and was sometimes called the 'German eco-fanatic'.

But then things started to change, even on our island! At Boogie Pilgrim, I had to involve local communities from the beginning, especially with our three lodges in very remote regions. Local procurement and employment were the only way to make our activities sustainable in all ways.

Apart from the economic gains of locals who are involved with Boogie Pilgrim, what are some of the social benefits to these communities?

Our lodges create jobs for the neighbourhood. We did lots of training to get them to where they are now—language skills, food and beverage training, hospitality management, guide training, boat skippers, maintenance, etc. In each location, we raised funds to build schools for their community and support medicare. A Malagasy family is about ten people on average, including three generations living under the same roof. Our lodges have made a direct impact on about 300 locals from the Bush House and Tsara Camp.

Our tour operator business isn't as directly involved with the locals as the lodges, but still—we are Fair Trade Tourism[2] approved. Most of our itineraries support projects by visiting and staying in community-based accommodation by small Malagasy suppliers wherever possible.

The current coronavirus pandemic has put travel upside down. Apart from the significant challenges it poses, do you think it also presents us with opportunities, for instance regarding tourism sustainability in Madagascar?

The sustainability of tourism in Madagascar is endangered by the coronavirus crisis. As our sector is not a government priority, we do not get any substantial support to survive. Many companies will close before the end of the year. The public sector is using definitions like 'sustainable tourism' all over, often without knowing

---

[1] https://www.club-arc-alpin.eu/en/about-us/member-associations/dav/
[2] http://fairtradetourism.org

what it means! But they noticed that international donors and NGOs like to hear it, to give money away.

But still, the only way to develop tourism in a place like Madagascar is by respecting fair trade rules, equity, sustainable principles, and environmental protection.

There is a strong interest now in enjoying open spaces, following months of isolation at home during this pandemic. What trends in travel are we going to see in the coming months and years?

Madagascar has the ideal profile, naturally. Our destination is known for huge empty spaces, great national parks, and biodiversity. Due to our location and destination characteristics, we have been protected from overtourism since the start.

Madagascar is a big playground for adventure tourism and nature lovers. Before the crisis, we hardly reached 300,000 tourists in 2019—a small amount considering that we are the fourth biggest island in the world!

Reducing poverty through conservation and tourism: What are the common challenges locals face in terms of benefiting from revenue generated through tourism, in Madagascar?

Tourism is the most fragile economic sector in a country like Madagascar. I had to deal with other local crises before COVID: political instability, sanitary crisis (cholera, plague, dengue). Tourism is the first to get hit every time because of media coverage leading to the foreign community stopping to travel almost immediately.

Local communities in Madagascar which depend too much on tourism are experiencing existential difficulties right now. Tourism employees cannot expect any help from the government, which means that a crisis like the coronavirus pandemic can have a severe impact on their lives.

Those who are independent workers, like tour guides, have to make their living during a fairly short season, facing a crisis of some sorts almost every five years.

Which lessons or key insights from Boogie Pilgrim might serve other tourism businesses or destinations around the world?

After 30 years of experience in tourism in Madagascar, we decided to launch Boogie's Solidarity.[3] It's a Malagasy nonprofit organization created by my husband, Patrice Raoull and myself as CEO and project manager at Boogie Pilgrim Group— both of us long-time advocates of sustainable tourism and experienced with humanitarian work.

Through the association, we aim to improve the precarious social, health, and environmental situations of the communities, primarily in the vicinity of our lodges. We noticed that developing sustainable tourism for real needs a separate entity from a company. It was too expensive and complicated to create a foundation, so this non-profit association will do the job.

---

[3] http://boogiesolidarity-madagascar.org/en/

The idea is also to be independent of the company, in case tourism faces any difficulties. Boogie's Solidarity can go its own way, find donors for projects, and be operational even at times with low income through tourism.

Link to the interview: https://sustainability-leaders.com/sonja-gottlebe-interview/

Peru | President and CEO of Society for Sustainable Tourism & Dev. Inc., Susan Santos de Cárdenas specialises in sustainable tourism development and stewardship initiatives with Community Social Responsibility for the grassroots. A savvy tourism professional and hotelier with more than 25 years of experience managing sales, marketing, operations, events—of MICE and human resources for small and luxury hotel resorts, tour operators, travel agencies, lifestyle events and publications, in the Philippines, Singapore, Peru, Japan and Indonesia. She is a keynote speaker and resource person for learning and capacity-building workshops in ecotourism, community-based tourism and agritourism. Susan was the adviser for Local Government Units (LGUs) in the Philippines and currently guides national, regional and community-based organisations in Southeast Asia. A founding Board Member of the Asian Ecotourism Network, Green Destinations Southeast Asia Partner and Representative, and UNWTO Multi Advisory Council—MAC Partner.

Areas of expertise: climate change, tourism development, sustainability, island destinations, UNSDG 11 (Sustainable Human Settlements), UNSDG 8 (Equal & Fair Economic Opportunities)

© The Author(s), under exclusive license to Springer Nature Switzerland AG 2022   573
F. Kaefer, *Sustainability Leadership in Tourism*, Future of Business and Finance,
https://doi.org/10.1007/978-3-031-05314-6_98

The following interview with Susan was first published in May 2016 on Sustainability-Leaders.com.

Susan, do you remember what first attracted you to sustainable tourism?

I was first attracted to sustainable tourism when I started to work with Inkaterra,[1] Peru's eco-pioneer and conservation leader since 1975, long before ecotourism became trendy. I was inspired by Inkaterra founder & CEO Jose Koechlin, who is a supporter and promoter of scientific research for biodiversity as the baseline of profitable conservation, education and economic growth of local communities.

I approached Inkaterra, Rainforest Alliance and Sustainable Travel International, who are also trailblazers in promoting Sustainable Tourism, for support to start my ground work in the Philippines.

With motivation and enhanced knowledge gained through participation in relevant workshops, I was able to build my personal brand to work on green solutions to global issues—ultimately, from policies to practices, for grassroots growth underlining cultural, community and ecological legacy.

Having successfully engaged in numerous Sustainable Development projects in many countries, with projects ranging from Eco lodges to travel agencies, what advice would you offer to newcomers who want to follow your example?

As the saying goes, leadership is not a position, but action and by example, so even if new graduates are starting out in their careers, they should look up and emulate role models and advocate sustainability ethos.

My advice is to commit to making tourism sustainable—it is our responsibility as industry players to achieve sustainability and resilience, in times of global warming and climate change. All those in tourism and hospitality must work to embrace sustainable tourism and best green practices.

Among your many successful projects, can you tell us more about 'The Coron Initiative'?

The Coron Initiative (TCI) was conceived during my first visit to Coron, Palawan, the Philippines in June 2010. During that trip, I was introduced to Al Linsangan, Executive Director of the Calamianes Conservation & Cultural Networks Inc. (CCCNI). Together, we established the formal agreement to adopt and implement TCI, which was later co-organized by our Society for Sustainable Tourism & Development.[2]

As the project name suggests, it is just a start of a movement, an evolution of culture for sustainability not only in Coron and Calamianes, but also the rest of the Philippines' 7107 islands, where most of the tourism resources rely on its rich and stunning coastal and marine ecology.

The purpose of TCI is to provide a replicable model for future tourism development in a country where island biodiversity is the main attraction and whose development requires careful planning that adheres to sustainable tourism guidelines.

---

[1] https://www.inkaterra.com

[2] https://sstdi.org

I believe that we need to educate and espouse tourism growth that does not destroy our rich natural heritage which attracts tourists to visit in the first place. We must ensure we don't only protect the environment for future generations, we also improve governance and stewardship.

TCI is still very much a work in progress with a lot of room to grow in education, capacity building, implementation of best practices, mentoring, and championing community leaders.

A sustainable tourism development in destinations can only be achieved through participation of all local residents. There is a need for a willingness and ability for all stakeholders to work with this kind of bottom-up approach.

In this context, both environment and social NGOs—including our own—have an important role to play, putting pressure on the industry and facilitating legislation, agreements and local participation for community development.

What motivated you to introduce conservation as sustainable tourism framework for Coron & Calamianes Islands?

With its natural beauty and rich indigenous culture, Coron is the fastest developing prime tourist destination in the Philippines. Palawan was proclaimed as a UNESCO Man and Biosphere Reserve in 1990. Coron was submitted in 2006 for consideration to be designated as a World Heritage Site to protect this jewel. Growing national and international tourist numbers and economic activity have led to more infrastructure projects, inter-island transport, trash dumping, commercial fishing, wastewater, and deforestation.

Learning the lessons from Boracay Island, where I worked as a pioneer in the early 90s for 10 years, I foresaw that Coron as a destination should not go the way of Boracay in unsustainable development, where there is no regard for preserving and protecting its spectacular beaches (though vulnerable), which attracted tourists in the first place.

Your thoughts on the current state of tourism in the Philippines, regarding sustainability?

The Philippines' Department of Tourism (DOT) has a history of established guidelines in Ecotourism Development since 1999, and its Tourism Act of 2010. This act recognizes that sustainable tourism development is integral to the country's social and economic development. Although this act is good in theory, unfortunately, it is not practised by all. Many tourism destinations in the Philippines lack participation from the host communities and local stakeholders. To date, only a handful of destinations actually follow those Ecotourism principles, if at all.

Throughout my 20+ years of experience in the Philippine tourism industry, new resorts, hotels, event venues have opened and grown with no concern for sustainability. Issues such as climate change add to the country's vulnerability, considering that most tourism activities occur on islands and in coastal areas.

The DOT should start to engage in promoting sustainability as a hallmark for both public and private stakeholders in destinations and host communities. It should promote tourism investment that strives to adopt environmentally sound technologies and sustainability measures, such as renewable and efficient energy, water use efficiency, ecologically solid waste management, and organic farming to provide a livelihood to local communities.

The Philippine tourism industry should promote projects which are compatible with the cultural identity of the local population's way of life. Furthermore, the tourism sector should always make sure it acts in accordance with the cultural heritage, and respect the cultural integrity of tourism destinations. This might be accomplished by defining codes of conduct for the industry and hence providing investors with a checklist for sustainable tourism projects.

The DOT should help fundraising for local NGOs to enable them to engage in a dialogue on tourism and provide industry education programs, which encourage responsible consumption, natural resource use, environmental protection and local culture conservation.

On the other hand, local governments should harmonize laws on sustainable tourism, including regulations, fee standards, licensing, etc. so that they will be more focused on social and environmental efforts to sustain tourism in the host community or province.

From your experience, what are the main challenges when trying to educate public and private stakeholders about environmental conservation and sustainable tourism development?

While both public and private stakeholders may be willing to 'go green' and practice sustainability principles, there is a conflict between the pursuit of economic gains and social and environmental responsibility. Stakeholders lack information and education on the requirements of sustainable tourism and how to integrate economic potentials with environmental and social requirements.

In the context of sustainable development of tourism, local governments play an important role as those in charge of conserving and managing resources. The challenge is to ensure that both local governments and private investors are willing to support sustainable tourism development guidelines.

LGUs [local government units] have the mandate to craft their own tourism plans, which sets out the priorities over the medium to longer-term and intend to contribute to community well–being. But because LGUs are subject to political elections, their focus tends to be more short-term. Change of regime can affect the adoption of the sustainable tourism guidelines for legislation.

For instance, the LGUs of Coron & Calamianes have yet to include sustainable tourism measures in their Municipal Tourism Code, and Environmental Conservation laws are yet to be integrated.

In my talks and presentations for the private sector (hotels, F&B, MICE, travel agents, tour operators) on the Triple Bottom Line principle of sustainability— People, Planet, Profit—they all agree on the values, but when it comes to application, they falter, as these principles are in conflict with their focus on rapid return on investment.

The hurdle is to mediate between fast profit vs. responsible practices. Here we are, educating on 'limited carrying capacity' and longer stay 'quality tourism' as opposed to quick turnover, short stays with massive and maximized tourism arrivals, which are the priority of most, if not all.

It is critical that both public and private sector players commit to and embrace sustainable tourism.

As a Board Member of the Asian Ecotourism Network, which changes do you hope to see in Asia regarding sustainability and tourism?

I hope to see all the members of the AEN formulate a bespoke Sustainable Ecotourism framework for Asian destinations, in accordance with the UN World Tourism Organization and the Global Sustainable Tourism Council. With my experience from Inkaterra, I can contribute to the formulation of these guidelines, based on ecological research and inventories, conservation and sustainable tourism best practices.

AEN should also adopt a streamlined version of sustainable ecotourism guidelines set by UNWTO and GSTC. My contribution to helping this streamlining process is culled from my active participation in conferences and workshops, such as the Extractive Industries Transparency Initiative (EITI) where we see sustainable tourism as an alternative to destructive and artisan mining in protected areas, and the recent UNESCO World Congress for Biosphere Reserves.

Link to the interview: https://sustainability-leaders.com/interview-susan-santos-de-cardenas/

USA | Tensie Whelan is a Clinical Professor of Business and Society at NYU Stern School of Business, where she launched the Center for Sustainable Business, whose mission is to prepare individuals and organisations with the knowledge, skills and tools needed to embed social and environmental sustainability into core business strategy. Her previous work includes serving as the President of the Rainforest Alliance, Executive Director of the New York League of Conservation Voters, Vice President of the National Audubon Society, Managing Editor of Ambio, a journal of the Swedish Academy of Sciences and a journalist in Latin America. She was most recently appointed to the board of InvestIndustrial SPAC and is an Advisor to the Future Economy Project for Harvard Business Review. Tensie holds a B.A. from New York University, an M.A. from American University and is a graduate of the Harvard Business School Owner President Management (OPM) Program.

Areas of expertise: sustainable tourism, sustainable business, tourism research, UNSDG 11 (Sustainable Human Settlements)

The following interview with Tensie was first published in January 2015 on Sustainability-Leaders.com.

F. Kaefer, *Sustainability Leadership in Tourism*, Future of Business and Finance,
https://doi.org/10.1007/978-3-031-05314-6_99

Tensie, in 1991 you wrote your first book on sustainable tourism—what was your view of sustainability and tourism back then?

I saw both the negative impacts of mass tourism (on water quality, fragile ecosystems, cultures) and the opportunity to do it right, as I was living in Costa Rica at the time, which was pioneering ecotourism.

Now at the beginning of 2015, what has changed?

Well, Costa Rica, unfortunately, has begun to turn to mass tourism on its coasts, which demonstrates the pressures that the continued increase in travel brings. Tourism is becoming a greater part of many countries' GDP, so environmental and social consideration often take a backseat.

On the other hand, there has been high level focus on improving the sustainability of tourism in Mexico, Ecuador, Kenya and other countries. The industry is also beginning to take it on—Travelocity has a green travel site, many hotel chains are incorporating sustainable practices, and ecotourism has grown substantially.

Since Costa Rica has been a pioneer in ecotourism, can you explain more about the situation there?

Fortunately, there are a lot of very positive developments there as well. More than 320 hotels (small, chain, boutique, eco) and tour operators have been certified by the Costa Rican Certification for Sustainable Tourism, using a standard led by the government and recognized by the Global Sustainable Tourism Council (GSTC).

Many of these businesses also comply with the Rainforest Alliance's certification standard for sustainable tourism.

Additionally, one hundred and eight beaches in the country obtained the Blue Flag certification; these beaches have organized committees with private and public representation that work in conservation, education and health issues.

Rural community-based tourism is also very well-organized, helping to provide local jobs and income.

Currently, the Rainforest Alliance is working with the Costa Rican System of Conservation Areas, providing training to communities and nature guides around 10 key conservation areas.

We are also working with the National Chamber of Ecotourism and Sustainable Tourism, the Network of Private Reserves and the National Chamber of Rural Community-Based Tourism and Citi Foundation on a project aimed at training hundreds of employees of tourism SME on best management practices across the country.

As an award-winning leader in the sustainability field, which are the main lessons for you personally, and as President of the Rainforest Alliance?

Finding a sustainable path is not easy when all the incentives are built in the other direction. Personally, for example, I want to find inexpensive clothes. But I know that often means they are produced in substandard working conditions. Unsustainable products should be taxed, and sustainable ones should get tax breaks to make it easier for people to buy them.

Real progress is being made by civil society and the private sector on some key sustainability issues—but not fast enough. And the government is mostly missing in action.

Do you share the view that sustainability has become mainstream?

Sustainability is not yet mainstream, but it is getting there. And yes, that is very good. As long as sustainability is niche and a 'nice to have' rather than a 'must-have' we will not move far enough or fast enough.

Why should hotels and other tourism businesses opt for the RA verified seal and accreditation over all the other programs now available?

Any system that complies with the Global Sustainable Tourism Criteria of the Global Sustainable Tourism Council (as we do) is a good bet. Working with us, however, means that you are working with an international NGO with a reputation for passion and pragmatism, for credibility and impact. And we are a great partner who will help you improve!

Why did you decide to lead RA, rather than continue to work as journalist, researcher or management consultant?

I like working to solve issues over a sustained period with a team of colleagues and friends. Journalism is a license to learn, but you can't fix things. And consulting is fascinating, but ultimately, you can't ensure that anything gets implemented. I chose to join RA because we are achieving change at scale—working in 60 plus countries, with 2.3 million producers and 6000 companies, in more than 180 million acres of land.

Link to the interview: https://sustainability-leaders.com/interview-tensie-whelan/

Spain | Tila Braddock has been in the tourism industry since 1998 with a goal to create a sustainable business to make a positive ecological impact. Along with his wife and business partner Michelle, Tila established Finca de Arrieta[1] with a philosophy to reinvest into the environment and use eco-friendly methods wherever possible. For 12 years, Finca de Arrieta has been powered by solar energy powering 17 holiday homes, 2 swimming pools and a desalination plant, making freshwater for 70 people daily. Finca de Arrieta is the inspiration that created the Lanzarote Retreats portfolio of beachside holiday homes. Tila's latest project is in the famous surfing town of Newquay, England with a UK-based company MiTie Limited in a 3-year project converting and upcycling a beautiful church into a sustainable, eco-chic holiday home.

Areas of expertise: ecotourism, green hotel, hospitality, UNSDG 12 (Resource Efficiency).

The following interview with Tila was first published in January 2020 on Sustainability-Leaders.com.

Tila, Lanzarote Retreats' Eco Village Finca de Arrieta has become the place to stay when exploring the volcanic island. Do you remember what first got you interested in the tourism business, and sustainability?

Yes, it was when I first stayed in Tarifa (in 2003), windsurfing with my brother. I stayed at a beach-side chic resort and that was the moment I was inspired to create something similar in Lanzarote, but with a sustainable angle: an eco-retreat with lower impact structures, using minimal materials to construct!

Why did you establish Lanzarote Retreats, together with your wife Michelle?

After buying the 30 000 meters of land in front of Arrieta beach, we converted a huge water tank into what is now called the Eco Luxury Villa. It was an amazing family home for us, but feeling the recession in 2007, we decided it was too big a project to have just for the family. Hence, we thought it would be worth trying to rent out a sustainable, Mongolian Yurt!

The strong interest from many visitors, looking for a different way to experience Lanzarote and its magical North—away from the hotel resorts and mass tourism—motivated us to expand.

How do you 'live' sustainability at the eco village—what makes it special?

Finca de Arrieta runs off-grid, exclusively on the many hours of Lanzarote sunshine, using tracking solar panels (that rotate based on the sun's position) and some small wind turbines. It's a powerful twin 48v system running seventeen properties (all with fridges, lights), two swimming pools with pumps, the all-important salt chlorinators and water desalination plant, that makes 3 tonnes of fresh water daily. The system has never gone wrong in 12 years!

Some of the day-to-day activities at the Eco Village that are aimed at making the space more environmentally friendly are reducing the number of plastic water bottles needing recycling (by providing a refill station for guests in the honesty shop).

Rain water is scarce in Lanzarote, so guests use eco-products which allows us to re-use waste water for the trees/cacti in the Finca. We reuse the salty bi-product from

---

[1] https://www.lanzaroteretreats.com

the desalination plant (to top up our pools) and of course the use of electric/hybrid cars by staff and guests.

Also, the kitchen food waste gets fed to the Finca animal residents... ducks, donkeys and chickens (in return you can pick up your fresh eggs for breakfast). There are many other examples and we keep looking for new ways.

What kind of traveller are you targeting?

The concept has attracted numerous amounts of different travellers from around the world, with or without knowledge on sustainability.

We would be happy to keep welcoming more open-minded professionals or retired couples, those looking for a quiet place to work while away (e.g., writers), the many families who are keen to educate their children on all the advantages and joys of off-grid living and want to enjoy a more intimate family time. And of course, we love to have the active sport-loving travellers come and stay.

What they all end up having in common, is a need to be in a tranquil place, in touch with nature, to be able to switch off from everyday life and stresses... away from the busy tourist resorts in the south!

What are the economic benefits of environmental sustainability for a hotel?

There are strong economic benefits, as well in that we are continually looking on where to consume less, recycle & re-use and save in a way beneficial to all.

The most evident one is the amount we save on the electricity and water bills! For example, the electricity bill for 17 properties would probably be about 100€ each a month (not including on grid amenities). So that's a monthly saving of about 1700€ for the Finca.

We make our own water, saving three truckloads (ca. 13,000 L per truck) a week at 110€ each = 330€ = a month 1320€! ... and the quality of water is better now than before!

The Guests in the Finca have the option to use Toyota Prius hybrid/electric cars (which probably is the most economical car for a family holiday).

Some of the staff, cleaners & maintenance use electric cars to get tasks done in different properties, throughout the island, saving approx. 140€ a week.

We also save by re-using the salty bi-product from our desalination plant to top up our pools and by using waste water for the plants. So just in energy, water and petrol there's a saving of approximately 3500€ a month. So that's over 40000€ a year!

Which are the main challenges you've encountered in developing and managing Finca de Arrieta eco village?

The authorities weren't as enthusiastic at the start and tried to create bureaucratic obstacles. Any project of this kind presents its hurdles, one of which was trying to finance it. I'm lucky that I've managed to work closely with my suppliers, who have been very patient as well as enthusiastic over this project!

Having spent more than 30 years on Lanzarote, would you consider the island a good destination for sustainability-conscious travellers?

Lanzarote is the ideal island for sustainable projects, with its unspoilt, as-mother-nature-intended landscape, its hours of great sun, wind, waves and even geothermal activity!

The island was among the first in the world to become a UNESCO Protected Biosphere Reserve, in 1993, and signed up for the Biosphere Responsible Tourism program in 2015.

The impressive collaboration between nature and man is continuously demonstrated and lived, especially through the legacy of the artist/architect/ecological pioneer, Cesar Manrique, who helped shape and mould Lanzarote (setting it aside from other Canary Islands, which is very visible in the striking unchanged landscape and lack of over-development).

Which are the main topics and concerns linked to tourism sustainability at the moment, in the Canary Islands?

The other Canary Islands haven't benefited from the likes of activists such as Cesar Manrique, to help maintain the environment by developing a tourism product that is based on sustainability. Hence, they have, to fit in with the rapidly growing tourism to the islands, ended up over-developing their coastlines and have rapidly changed their landscape.

Each business in the Canaries (not only in the tourism sector), needs to be more aware of the negative impact on the Environment and cultural heritage of the Archipelago. Awareness/responsibility needs to be more widespread amongst all (businesses, staff, locals, as well as visitor to the islands), assisting to minimise the impact caused by waste, emissions and consumption. This of course without affecting the quality of service or the experience offered to the sun seekers to this stunning, most southern part of Europe.

Over the years you've hosted many celebrities and influencers at the eco village. If you had to choose just one encounter, which one would it be?

We have attracted a lot of journalist/press attention with our project over the years, welcoming everything from Europe's most popular magazines/newspapers, to even the modern social media influencers!

Who really stood out for me: the McEwan sisters, Caitlin (8) and Ella (10), the 'Plastic free youth pioneers', that came back for another visit to the Eco Village after a successful petition on Change.org, to get Burger King UK (and almost McDonalds), to stop giving away plastic toys with children's meals.

The sisters decided to start the petition after learning what a serious harm plastic waste does and even won the first ever Youth Pioneer Plastic free award! These young environmentalists have given us hope and are an example to all of us, especially in moments when we feel that by ourselves, we cannot make a difference!

How do you market and communicate your sustainability features to potential visitors—and how do you engage them once they're at the eco village?

All our guests are invited to go to see the Eco Village and learn more about off grid living, even if they stay in our holiday homes outside of the Eco Village.

The environment at the finca encourages our guests of all ages, to connect and communicate more between each other, to be social outdoors in the communal spaces.

30% of our new guests come through word-of-mouth recommendations. At Lanzarote Retreats we try to communicate our brand and ideas through social media too. We do a monthly newsletter to our client base, which has proved to be both informative and effective. As these newsletters also get shared, more and more people are aware of our existence.

We believe that, if we keep giving the best service while at the same time not jeopardising on the sustainability aspect, we will have the guest revisiting and sharing their positive experiences of sustainable tourism!

Now over 40% of our business comes from returning clients. That's why we have never felt the need to pay to advertise, if it can still somewhat work in the old-fashioned way.

You take pride in creating a warm and inspiring culture at Lanzarote Retreats as a workplace. What role does social sustainability play nowadays in the context of sustainable tourism?

There are many ways that we try to share the concept of sustainability, with staff, guests and even local University of Tourism students, that come to learn about the benefits of sustainable tourism.

We rely on our guest feedback forms, that we get from every visit, and we try to act on feedback where possible. We work on the same principle with all our team/ staff. We listen to them and hopefully empower them in their role. We believe we are as successful as the team we build around us. That is why it's important we show them gratitude and invest in them!

Anything else you'd like to mention?

We are lucky to operate on a small, tourism-dependent island that has miraculously managed to avoid the fate of so many other holiday spots. Lanzarote is what it is today, because they managed to create an amazing destination for tourists but maintained the traditional way...which is a blend of man and the stunning nature here!

The diverse people that live here, have a great quality of life. They have opportunities and live the tranquil Lanzarote outdoor lifestyle, in one way or another connecting with nature (for example by catching some waves after work).

The Island offers the best climate/landscape to train and compete throughout the year, hence it has become Europe's number 1 sporting destination. Countless international sporting events and competitions are held here throughout the year, meaning we have a lot of visitors that come out here to be active, so like the locals, making the most of the relaxing environment, climate and the landscape that Lanzarote has to offer!

There are many ways we can all share what we know about sustainability with our community or work. The Sustainability Leaders Project helps us to highlight the possibilities, shows where others are on the scale of sustainability and how we can all live this way if we choose to.

Link to the interview: https://sustainability-leaders.com/lanzarote-retreats-inter view-tila-braddock/

Australia | Tony brings unique skills and experience gained through senior executive roles in protected area management and tourism development and management. He was the founding director of Ecotourism Australia[1] and served continuously on its board from 1991 to 2016. He is Vice-Chair of the Global Ecotourism Network and is on the board of the Asian Ecotourism Network.[2] An environmental scientist and urban and regional planner by profession, he established Tony Charters and Associates in 2004 and provides strategic policy advice and master planning to leading industry operators and government, particularly in project planning, development and investment; destination development; and sustainable management. Tony has undertaken national ecotourism master planning exercises for the Philippines, Sri Lanka and Saudi Arabia. He is UN-FAO's Senior Ecotourism Expert. In 2014, Tony was awarded an Order of Australia (AM) by the Governor-

[1] https://www.ecotourism.org.au

[2] https://www.asianecotourism.org

F. Kaefer, *Sustainability Leadership in Tourism*, Future of Business and Finance, https://doi.org/10.1007/978-3-031-05314-6_101

General for his services to the ecotourism industry, nature-based recreation and environmental education.

Areas of expertise: ecotourism, destination development, tourism business, Australia, UNSDG 11 (Sustainable Human Settlements), UNSDG 12 (Resource Efficiency)

The following interview with Tony was first published in June 2016 on Sustainability-Leaders.com.

Tony, you began your professional career as an environmental scientist and urban planner. How do these areas of expertise support your work today in sustainable tourism and planning?

I was in the second year's intake at the newly established Griffith University in the late 70s. At that time, 'environmental science' was seen by many as a 'pseudo-science'. And yet, time and time again over the years, the multi-disciplinary approach adopted in that undergraduate degree course proved to be extremely valuable and relevant.

To this day, this approach is so important—bringing specialists together, under-standing their language and applying environmental science to tourism planning. Sustainable tourism and ecotourism rely on a light touch to the environment. Environmental science and protected area planning are a natural fit for this work.

Some tourism companies promote sustainable practices for publicity to attract customers rather than engage in real social and environmental work. In your experience, is such 'greenwashing' still on the rise, for example in Australia?

I think there are still elements of greenwashing, but this is not just an Australian thing—it is a global trait. The Eco Certification scheme in Australia, now 20 years in operation, was developed in part to enable consumers and industry distribution agencies to select operators who were taking their eco credentials seriously.

Fortunately, the consumer is becoming more knowledgeable about sustainability, but we still have a long way to go. Social media are accelerating the adoption of good practices. Transparent feedback through public reviews exposes bad (and good) practices very rapidly.

Your work has played a pivotal role in the development and implementation of various eco-certification programs. How important is it for a tourism business to obtain certifications, such as Eco Certified in Australia?

Eco Certification was designed to be very instructional and transparent. It was also designed to be iterative and to recognize and reward the innovative practices of ecotourism operators. It was made to be both practical and affordable.

There have been many very pure and technically competent certification schemes that have come and gone, because they were not attuned to the practical and economic realities of running an ecotourism operation.

Competent certification schemes equip operators to develop more financially viable and environmentally sustainable businesses. We are seeing the benefits of Eco Certification throughout the industry in Australia. It is now common for 25–40% of the Australian Tourism Awards to go to Eco Certified operations across a myriad of award categories.

As a consultant, you provide innovative solutions for sustainable outcomes that address climate change. Which are the most critical aspects linked to sustainable tourism development that popular tourist destinations, such as Australia, need to tackle?

Without question in my mind, education and information about climate change are the most important things the tourism industry can contribute. The scale of the climate change problem is beyond the tourism industry's capacity to significantly influence in a direct way—although it clearly has a role to play. However, in the field of education and interpretation, it can play a major role.

Many of the world's most valued tourism destinations are being or will be damaged by the impacts of climate change. Whether this is from more frequent typhoons and cyclones, wildfire, coral bleaching, inundation, reduced snowfall, or other impacts, they affect a high proportion of our most valued tourism destinations.

The tourism industry can therefore play an important role in bringing the climate change mitigation story to travellers, encouraging them to live in a more sustainable way so as to give these amazing destinations a chance at survival. There is no more powerful way of getting these messages through than at the source.

Do you have any success stories you'd like to share that demonstrate how tourism can act as a positive force for change?

I am fortunate to be a finalist judge for two international tourism awards, the Tourism for Tomorrow Awards (WTTC) and the World Legacy Awards (National Geographic and ITB). Having been in that role for over 10 years has given me a unique insight into hundreds of outstanding examples of ecotourism as a force of good for communities, economic development and, importantly, the personal growth of travellers.

I recommend going to the websites of these two award programs to read the case studies of the finalists and winners—they are truly inspirational.

In reviewing the hundreds of applicants for awards over the years, it is very pleasing to note that inspiration and innovation come from operators of all scales: from micro enterprises to global brands and across all nations.

Because ecotourism focuses on what is different and unique about the nature and culture of a destination, it has the capacity to nurture diversity and tolerance to a level perhaps beyond any other form of tourism.

Having successfully engaged in sustainable tourism projects in Australia and internationally, what career advice would you offer newcomers to the industry?

Understand that tourism is a business and that for it to succeed, it must be profitable. And try to get experience across both the private and public sectors. Having that broader experience has been invaluable to me over the years, as it helped me to see the perspectives and motivators of both sectors.

As a Board Member of the new Asian Ecotourism Network (AEN), where do you see the Network's greatest opportunity and benefits? And what are the potential risks?

AEN represents a very significant proportion of the world's ecotourism industry through the nations represented on its Board and throughout its members and followers. Asia, including Asian Pacific nations such as Australia and

New Zealand, has huge potential for ecotourism. The diversity of natural and cultural heritage in the region is extraordinary. The thirst for knowledge about building sustainable ecotourism ventures in this region is insatiable. AEN can help to spread an understanding of ecotourism to communities, start-up ventures and existing operators. Sharing knowledge and building networks is an invaluable role for AEN.

The major risk involved in setting up any new organisation is building viability. With the support of the board members and key supporters such as DASTA in Thailand, the viability of the organisation is very sound.

Link to the interview: https://sustainability-leaders.com/interview-tony-charters-ecotourism-australia/

UK | Tricia Barnett, co-director of Equality in Tourism,[1] works to ensure women's equality throughout the tourism sector. The organisation won The To-Do Award for Human Rights in Tourism in 2021. Tricia has initiated and developed a unique pilot project in Tanzania that trains impoverished farming women in Kilimanjaro to transform their livelihoods by supplying hotels with quality horticultural produce. As director of Tourism Concern, she established it as a global organisation for advocacy and change. Accolades include being highly commended for the Greatest Contribution by a Person for Responsible Tourism, a Lifetime Achievement Award from the British Guild of Travel Writers, runner up for the Observer's Green

---

[1] https://www.equalityintourism.org

© The Author(s), under exclusive license to Springer Nature Switzerland AG 2022
F. Kaefer, *Sustainability Leadership in Tourism*, Future of Business and Finance,
https://doi.org/10.1007/978-3-031-05314-6_102

Campaigner of the Year; identified by Travel Trade Gazette as one of the 50 most influential women in the UK industry. Tricia received an honorary doctorate from The University of Brighton.

Areas of expertise: sustainable development, sustainable tourism, UNSDG 5 (Empowering Women), UNSDG 8 (Equal & Fair Economic Opportunities)

The following interview with Tricia was first published in August 2017 on Sustainability-Leaders.com.

Tricia, you have been involved in responsible tourism and human rights advocacy for many years. Do you remember what brought you to the topic?

How can I forget? It was provoked by a long holiday hitching on my own in Cuba where I had the most brilliant, open, hospitable experience. This was followed the next year by a miserable, fraught, hostile holiday in Jamaica where even taking a bus could be problematic.

Soon afterwards, I signed up as a very mature student to study anthropology. I took the opportunity to research into tourism to Jamaica. Why were the two experiences so very different? I learned that tourism had come to Jamaica following the abolition of slavery in the late 19th century and replicated the social patterns of the plantation economy.

I got deeper into the subject and began to understand tourism's impacts on people's human rights. I followed up by signing up to the first-ever Masters' degree in the Anthropology and Sociology of Travel and Tourism. Tourism Concern then provided the vehicle to campaign for a sustainable, participatory and just tourism.

How has your view on tourism and its human rights 'performance' changed?

Critical to our work has been partnering with people and groups in the South who approached us for support and advocacy. Often, they live under repressive regimes and they don't have the freedom to campaign openly. Their stories haven't changed.

I feel sad to see the same patterns of abuse repeated over and over again. Governments and industry still fail to recognise that human rights are a fundamental element of any sustainable approach to development.

At Tourism Concern we produced two reports on tourism and human rights, the first prompted by the 50th anniversary of the UN's Universal Declaration of Human Rights. Little had improved by the time we produced our second report. As with all our reports we listed recommendations and calls for action to ensure people's protection.

Examples of human rights abuses include people's right to water. There are huge problems of appropriation, depletion and pollution of water by hotel and resort developments because of unregulated tourism. This is true, particularly for island communities. People's environments, living standards, livelihoods and development opportunities are undermined.

Working conditions and labour rights are also commonly abused by every sector of the tourism industry globally.

We also produced very well researched reports on the business case for the tourism industry to understand why a business needs to take a human rights approach. They covered risk management, competitive advantage, social sustainability, business leadership and ethics.

What led you to co-found Equality in Tourism: Creating change for women, and what is it about?

It took me far too long to recognise that sustainable tourism would never be achieved without gender equality.

We set up Equality in Tourism to drive change in the development and practice of global tourism by empowering women through the involvement of local communities, and by engaging stakeholders at every level.

I know that tourism can create positive change in communities if all members enjoy equal access to the industry and its benefits. However, the way that the tourism industry operates magnifies the disadvantages and constraints that women face in their lives.

Gender inequality significantly undermines the potential of the tourism industry.

We established our non-profit organisation because we want to work with all sectors of the industry and with local communities to help ensure women enjoy an equal share in the global tourism industry.

We use gender analysis to understand the norms that structure relations and processes within institutions and society to take the industry forward so that it can be genuinely sustainable.

You ran Tourism Concern for an impressive 20 years, which are your key insights and lessons from that period?

- That only regulation, nationally and globally, will push the industry forward into the 21st century so that it respects people's human rights. When the industry is doing alright it sees no reason to change.
- That at a local level, a rights-based approach to tourism development is far more sustainable than a top-down approach.
- That operators are so powerful and competitive they bargain the rack rate at destination hotels down to extraordinarily low levels: ensuring that those at the bottom end of the pyramid never earn a living wage, that the environment doesn't benefit, and communities lose out.
- That local people really can benefit from tourism, and everyone can share its benefits when there's a real commitment to supporting local entrepreneurship and fair working conditions.
- That it's so hard to get the holidaymaking public on board—especially when you're operating on a shoestring! However committed someone is to the environment or human rights, when they're on holiday most people cut off from these concerns and only want to have a good time.

The UN has declared 2017 the International Year of Sustainable Tourism for Development. In your view, which are the main challenges right now regarding the sustainability of travel and tourism? And how are organizations like the UN helping to overcome those challenges?

The UNWTO was responsible for declaring this International Year and relating it to the UN's Sustainable Development Goals (SDGs). In brief, we are particularly concerned with Sustainable Development Goal (SDG) 5—achieve gender equality and empower all women and girls. I shall extract just one section of our response. Their paper asks:

What specific changes in tourism policy, business practices and consumer behaviour are needed to address these issues and contribute to sustainable development?

Our response:

First, it should be noted that this report fails to grasp the importance of SDG5 for achieving all other development goals, as outlined consistently across UN policy discourse. Without tackling gender inequality in a meaningful and substantive way, tourism's potential to contribute to the other SDGs will be substantively reduced.

Second, the paper only discusses a relatively narrow aspect of gender equality issues in tourism—employment and decent work. This misses a large range of how tourism interacts with gender inequalities. In particular, it depicts a limited framing of gender issues as related to economic empowerment, leaving out discussions of political empowerment and broader questions on structural inequalities.

Third, the paper does not demonstrate an understanding of gender analysis or gender inequality and—as such—it is not clear how tourism is meant to contribute to achieving SDG5, nor the role of institutions in working toward this. Moreover, mixing 'women' and 'youth' is an unhelpful conflation, and further serves to marginalise the importance of gender issues to both the SDGs and the tourism sector.

Fourth, it is well documented that gender equality cannot be achieved without dedicated resources. In order to make progress in this area, a budget is required in order to advance tourism's contribution to SDG5.

Our recommendations:

Building on these points, we (Equality in Tourism) recommend the following actions by UNWTO in order to address the gaps in its work on gender equality and to increase the potential of tourism to contribute to SDG5:

• Establish a dedicated budget for gender equality and SDG5 within UNWTO to ensure that necessary actions and programmes can be carried out. This should be done in collaboration with specialists in gender and tourism in order to ensure that UNWTO work matches international norms and standards on gender equality.
• Update the Global Report on Women in Tourism 2010. This was originally intended to be a triennial report. As such, it was expected that this report would have been updated in 2013 and 2016.
• Review the literature and best practice on gender equality in tourism and update the Discussion Paper accordingly, drawing on the available expertise and knowledge in the field

Which of the many projects and campaigns during your work with Tourism Concern have you personally found the most rewarding?

Although we operated on a minimal budget, we achieved a very great deal, both as campaigners and publishers.

Our teachers pack and video on The Gambia was requested by almost every school in the country.

Approaches to us by porters in Kilimanjaro, Machu Picchu and the Himalayas about their inhumane working conditions resulted in us working with them and the trekking industry to produce The Trekking Porters' Code on Working Conditions, which is still in use today and even led to a change in the law in Peru and strict regulation on Kilimanjaro. Porters no longer take up loads of over 40 kilos, or wear flip flops and sleep in caves—all on one meal a day and leftovers.

Aung San Suu Kyi asked us to try and prevent tourists from coming to Burma until democracy was restored. Tourism Concern joined forces with Burma Campaign UK to successfully persuade many UK tour operators to pull out of the country because of tourism's links with mass human rights abuses. The UK Government also called for tour operators to withdraw.

Displacement of people from their homes and livelihoods was a key campaign. We were approached for help by residents too afraid to speak openly. We managed to stop a mega-development on Zanzibar that would have displaced 20,000 farmers and fishermen from their homes.

Imagine you could turn back time and start all over again. Knowing what you know now about the tourism business, what would you do differently?

Not a lot! We worked on a pittance. We could have done so much more if we had been well funded. But I'm still no wiser about how you raise funding to challenge tourism development that has negative impacts. I was and am on a constant learning curve.

Your 3 bits of advice to hotel or tour managers eager to empower women as part of their ambition to become socially responsible, sustainable businesses?

You can start by looking at your own workplace and carry out your own gender audit. Survey everything from your board (if you have one), your executives, your managers—all the way down the hierarchy. Is there a fair balance between men and women? Does your company have policies on gender equality and equal opportunities, on training and mentoring, on child care provision, on the flexibility of working hours and place? Are these policies genuinely operational and do male and female employees receive equal pay for equal work?

Then, if you're up to it, it's good to check out whether there are such policies in place for those you contract in as well as throughout your supply chain.

After all this, if you'd like the help of Equality in Tourism—we'd be delighted to take you further.

Link to the interview: https://sustainability-leaders.com/interview-tricia-barnett/

Vassilis Katsoupas on Running a Sustainable Restaurant in Greece

Greece | Vassilis operates a thematic restaurant in the region of Zagori in Epirus Greece, specialising in mushrooms. Almost all the food served is locally sourced and produced along with some gathered in the wild, such as wild mushrooms, herbs and vegetables. He also grows mushrooms, seasonal vegetables and herbs on his small farm. Vassilis has been in the restaurant business from an early age following in the footsteps of his father. He has a B.A. in political science and a master's in environmental studies at York University in Toronto and worked in environmental advocacy for several years before starting his own business.

Areas of expertise: hospitality, sustainable business, Greece, UNSDG 12 (Resource Efficiency).

The following interview with Vassilis was first published in July 2019 on Sustainability-Leaders.com.

F. Kaefer, *Sustainability Leadership in Tourism*, Future of Business and Finance, https://doi.org/10.1007/978-3-031-05314-6_103

Vassilis, before starting your culinary adventure with your restaurant Kanela & Garyfallo in Greece, you lived in Canada and worked in the media business. What triggered this change of career path—and location?

Well, as the old adage goes, life is what happens to you while you are busy making other plans... My decision to move here was triggered by a divorce. I always had this romantic dream to move back to my family's place of origin in the mountains of Epirus, in northern Greece, where I spent many memorable summer holidays in my youth.

Epirus is a very green place with wonderful nature and landscapes, and it always felt to me as paradise on earth, beckoning for my return. This feeling became even stronger after the changes in my family life.

Kanela & Garyfallo is unique in that you specialize in mushrooms: wild ones and those cultivated on your own farm. Where does the passion for mushrooms come from?

I would often spot mushrooms while trekking and loved eating them, but my interest really developed while taking a course in bio-remediation and finding out about the potential uses of different plants and fungi for restoring polluted soils.

Later, I became even more fascinated through my contact with the renowned mushroom researcher and author Paul Stamets. Few people realize how important fungi are for our biosphere and how useful they can be to our communities as sources of food, medicine or even environmental restoration technologies.

On the other hand, I had inherited a culinary 'gene', so to speak, as I literally grew up in our family-owned restaurant and have worked forever in this business. Combining my education with my family tradition, I came up with the idea of teaching people about mushrooms—one bite at a time.

Which would you consider the main challenges for a small restaurant, in terms of operating sustainably?

The main challenge is deciding where to start. There are so many issues to be addressed in terms of sustainability!

The obvious choice would be to focus on sourcing sustainably grown food, but minimizing energy and water use, as well as food waste, are equally daunting issues.

Striking the right balance is extremely important. One needs to focus on achieving maximum impact while balancing operational costs and to continue building up on steps taken in the right direction.

Sustainability values and criteria must be integrated in all operational modalities, but it is equally important to focus on the continuous need for teaching and communicating effectively with customers, suppliers and staff about the value of doing so!

Food waste is a topic which we discussed at length in our interview with Benjamin Lephilibert.[1] To your mind, how can small restaurants with limited financial means be smart about production and disposal of food waste?

---

[1] https://sustainability-leaders.com/benjamin-lephilibert-interview/

Being smart starts with basic grandma wisdom! You know, older generations were very skilled at minimizing waste and maximizing the productive use of every bit of food on hand.

At Kanela & Garyfallo, we follow several well-tested methods: Fresh, quickly spoiling foods are bought in modest quantities to minimize spoilage, and supplies are managed carefully to ensure new arrivals get in the back of the line.

We buy in season in bulk to ensure best quality and price, and fill up our pantry with various preserves and condiments. In the kitchen, we make sure to utilize fully all usable food scraps by making our own vegetable and meat stocks and improvising various condiments.

Leftovers are donated to friends with chicken and geese, in return for an occasional egg basket. Table leftovers are gathered for the dogs guarding our farm, and the patio cats in charge of endless customer entertainment.

In the front of the house, we follow a sensible eating strategy—our aim is to please, not to increase sales at the expense of our customers and the environment. We offer most items on the menu in half portions and warn against excessive orders, clearly stating our aim to reduce food waste. As an incentive, complimentary desserts are promised to those who leave enough room.

In our experience, the single most effective strategy to minimize food waste at the table is to suggest clients to eat in typical Greek family style, sharing all food, opting for a variety of small portions and repeat orders. This way, customers enjoy their dining experience more, as they get to taste a variety of dishes without overeating. And generally, the bill tends to be lower compared to what it would have been otherwise.

Food provenance—where food is sourced from—is another topic receiving considerable attention now. Which rules or guidelines do you follow in this regard, at Kanela & Garyfallo?

Our main rules are to stay local, seasonal and as far away from industrially produced food as possible. We are also not interested in exotic, out of season food or produce that has to travel long distances, even if it is labelled 'organic'.

We produce shiitake and oyster mushrooms, as well as some specialty vegetables and herbs in our farm 8 months of the year, skipping our low season, the cold winter months, to save on energy costs.

Our associate collectors supply us throughout the year with a variety of seasonal wild vegetables, fruits, flowers, herbs and more than 30 different kinds of wild mushrooms. We buy in bulk and preserve the excess by freezing, drying and pickling, to have on hand throughout the year.

Several local producers supply us with fresh vegetables, pulses, grains, eggs and dairy products, as well as locally grown, free range beef, lamb, goat and chicken.

Overall, about 80% of our food supplies is grown or collected within a 20 km radius. Buying local makes environmental as well as economic sense and helps us to be socially responsible. By relying on local producers and collectors we have strengthened our local community and provided venues for the development of sustainable local economies.

Which other tricks and tips would you share with us, in terms of how restaurants can become more sustainable?

In terms of the bigger picture, I think those in the gastronomy business have a responsibility to educate ourselves as well as our staff and customers about healthier, more sustainable eating. This does not necessarily mean lecturing people. I prefer giving our customers more options to discover how they can make a difference on a daily basis through something as basic as eating, whether at home or in a restaurant.

We do not need to re-invent the wheel. All we need is to study the tradition of the cucina povera in the European south and most other rural places around the world. Since the Stone Age, people have been very inventive at making the most out of the little they had on hand. It is only in the last century of quick paced industrialization, urbanization and rampant consumerism that we have been weaned off sustainable eating traditions.

One of the most pressing issues we need to address—especially in the developed, affluent world—is unsustainable meat consumption. Restaurants must offer more meatless options and educate customers to eat meat smarter.

At Kanela & Garyfallo we offer a menu that is 75% vegetarian and vegan. All meat used is local and comes from free range animals, not factory farms. We follow a nose to tail policy, using all parts of the animal, not just the premium cuts.

Little meat, lots of vegetables, grains, legumes and fruits: the essence of the Mediterranean diet is the roadmap to sustainability.

Which trends do you witness at the moment, which might support or hinder a more sustainable restaurant scene and culinary experience?

I must admit of having a love-hate relationship with the whole gourmet glamorization of the restaurant scene, the rush to develop ever more intricate, complicated and beautiful abstractions served as exclusive culinary experiences.

I am fascinated by the art and finesse of many celebrated gourmet chefs, but I think there is a sharp contradiction between that scene, the way most people eat and the way our society relates to nature's bounty.

To me food is about love, family, sharing and belonging. Eating around a table with people you care about, sharing food and thoughts, is a cardinal humanizing experience, but it is also a critical moment of interaction with our natural world and the way we view ourselves in relation to the Earth.

There is tremendous waste, over-processing of food and drive to impress with extravagant materials from the other side of the planet or the production of super complicated, altered states of food through endless, tortuous, technological interventions.

To support a more sustainable restaurant scene and culinary experience, we need more simplicity, more spirituality, more respect for nature's bounty and less catwalk.

If you had to pick one, which dish would you want to be served yourself—for example on your special day?

That would be a grilled mushroom stuffed with 'Kayanas', a quick and simple Greek peasant dish of scrambled eggs cooked with a reduced confit of tomato, pepper and courgette, sprinkled with some thyme, hot chili and a tiny bit of crumbled feta cheese! A whole lot of earthy umami...

Which restaurants or chefs have served you as inspiration, in how they approach sustainability?

I am inspired by all those who put local ingredients, heirloom varieties, local cuisine first. Think Global, Cook Local!

If given the chance to turn back time—is there anything you'd do differently?

Time runs in one direction, and I am more concerned with the future. We do as we can, always making mistakes and hopefully learning from them.

If given the chance to go back in time, I would probably try to do everything differently, but chances are I would still end up making similar mistakes and loving similar things in life.

Link to the interview: https://sustainability-leaders.com/vassily-katsoupas-interview/

Mexico | Vicente is the founder and CEO of Sustentur, a social enterprise focused on sustainable tourism that works to increase the competitiveness of companies and destinations while preserving natural resources, improving the quality of life in local communities and promoting wellness. Vicente has experience in launching and leading sustainable tourism projects; has coordinated projects for World Bank, Inter-American Development Bank, and international development agencies, in addition to government, private companies and rural communities. He is an international lecturer and advisor on sustainability and tourism issues. Vicente is also the founder of the Sustainable and Social Tourism Summit, the most important sustainable tourism event in Latin America, held annually in Mexico. He has participated twice in TEDx events as a speaker on travel and sustainable tourism topics. In 2020, the Global Shakers Initiative recognised Vicente as one of the 40 leaders in sustainable tourism in the world.

F. Kaefer, *Sustainability Leadership in Tourism*, Future of Business and Finance, https://doi.org/10.1007/978-3-031-05314-6_104

Areas of expertise: sustainable tourism, tourism business, sustainable development, sustainability communication, Latin America, UNSDG 12 (Resource Efficiency)

The following interview with Vicente was first published in June 2015 on Sustainability-Leaders.com.

Vicente, do you remember the first time you heard of the concept of sustainability? What was your initial reaction?

Yes, it must have been in 1996 or 1997 when I began my bachelor's degree in Tourism in Mexico City. In the beginning, like most of the terms you hear when you are studying, 'sustainability' seemed a little vague. Even after studying it in more detail and hearing about it in various forums and conferences, up to a few years ago, it was difficult to find practical examples of the application of sustainability, especially in the tourism sector.

Now, in mid-2015, what has changed?

Many things. Now there is a more robust framework for sustainability, a commitment from the private sector and public institutions. Many organizations work on the issue, and proactive communities show that new models of development are possible.

However, when you review the conclusions of international summits, such as Rio + 20, we realize that there is still a long road ahead to achieving a development that is viable from an environmental, social and economic standpoint.

Before becoming CEO of SustenTur, you worked as Sustainable Tourism Coordinator at the Mesoamerican Reef with WWF (2009–2012). What is your best memory from this time?

That was an interesting period because it was the moment that my eyes were opened to the conservation of biodiversity. Even though I had worked with issues relating to ecotourism and development planning, working with the world leader in conservation allowed me to involve myself to a larger extent in the issue. This allowed me to better understand the need to link different economic sectors in order to achieve sustainable development.

I can't remember a specific moment, but what I enjoyed most were the field trips with colleagues from the four countries of the Mesoamerican Reef. Together we realized the pressures that exist for this important ecosystem. I was positively surprised how many people working in conservation are committed to preserving this treasure, representing public, private and social sectors.

What is the greatest challenge in working for an international NGO like WWF?

I think the greatest challenge is integrating a tourism agenda into the objectives and guidelines of the organization. With a presence in more than 100 countries, global challenges like climate change, unsustainable fisheries, resource consumption, illegal hunting, deforestation, among others, get the attention of organizations like WWF.

And tourism, being an important economic sector, is not really integrated as a priority into the policy of the organization. Documenting the importance of working in this activity is vital in achieving support from the organization, and to advance the search for sustainability in tourism.

What achievements related to your time at Amigos de Sian Ka'an are you most proud of?

At Amigos de Sian Ka'an, my work focused more on the development of projects with local communities. There are many achievements that I am proud of, especially those that had to do with observing the community making giant steps towards the consolidation of their projects.

But definitely, the day that I am most proud of during that position was the 8th of May, 2014, when Tianguis Turístico de México launched Maya Ka'an, the new ecotourism destination in Quintana Roo, and certified 7 tourism companies as sustainable businesses.

Why did you start SustenTur, and what is its main function?

I decided to start SustenTur precisely because of the need to communicate the initiatives in sustainable tourism.

In Latin America, there is no mature consciousness of what the communication of sustainable tourism means—not just to improve the image of the businesses, 'boast' their achievements, or serve niche markets—but also as a way to educate, raise awareness and give examples that both large and small companies and destinations, like small private or community projects, can benefit from the strive towards sustainability.

At SustenTur, we advise companies, destinations, governments and communities to communicate their sustainable tourism projects, and in the case that they don't have any or don't know how to implement them, we advise them on how to achieve their objectives.

What are the most important aspects of communicating sustainability?

First, communicate genuine achievements. Second, really get to know the target audience, because each of them requires specialized communications: tourists, investors, association colleagues, NGOs, governments, etc.

What communication approach does SustenTur suggest?

An approach that aims to educate their target audience about sustainability through differentiated strategies and works with tourism destinations, private companies, community projects and civil organizations.

In your opinion, how will the sustainable tourism agenda evolve in coming years? What trends are you observing?

It will continue to advance in terms of international interest in the issue, the increasing regulation local governments put on development, and the market trends.

For 2030, I envision a tourism sector that:

- Commits further to generating experiences instead of just activities, trying to attract the new 'tribe' of travellers that Amadeus lists in its report—Future Travel Tribes 2030.
- Views sustainability at a destination level more than individual, generating better results in environmental and social issues.

- Commits to increase the number of international certifications for businesses in the sector.
- Generates better communication with the tourist over the impact that their vacations have and how to minimize them.
- Generates more Public-Private alliances and pushes the creation of social businesses for the benefit of the sites in which they develop.

What is the best way to measure the success of initiatives that promote sustainable tourism?

With impact indicators: environmental, social and economic. Of course, we should create ways to measure progress in activities and specific actions, but it is best to have indicators or indices that measure if the tourism is or isn't, in reality, contributing something to the area in which it operates.

The Travel and Tourism Competitiveness Report by the World Economic Forum is a great tool for this, but we can also use methodologies like Social Return on Investment or other functions, to measure impact.

Link to the interview: https://sustainability-leaders.com/interview-vicente-ferreyra-acosta/

UK | Vicky has worked in tourism since the mid-1990s and in sustainable tourism since the mid-2000s, managing destinations, tours, marketing and eCommerce for tour operators, travel agents, media and NGOs. A love of nature, combined with a fascination with internet technologies and mainstream jobs, led to witnessing the industry's negative impacts, and opportunities for sustainable development for communities and conservation. Vicky has worked in community and conservation volunteer tourism and charity challenges. She has a master's degree in Responsible Tourism, qualifying as a ranger in Africa and freelancing including sustainable tourism accreditation. Vicky brought her knowledge and skills together to launch Earth Changers,[1] a curated collection of some of the best positive impact, sustainable tourism, which won Travelmole UK Best Responsible Travel & Tourism Website, 2019. Vicky also works on select consultancy, develops the European Ecotourism Network and is a Board Trustee of a sustainable development charity in Madagascar.

---

[1] https://www.earth-changers.com

© The Author(s), under exclusive license to Springer Nature Switzerland AG 2022
F. Kaefer, *Sustainability Leadership in Tourism*, Future of Business and Finance,
https://doi.org/10.1007/978-3-031-05314-6_105

Areas of expertise: sustainable tourism, tourism business, entrepreneurship, Africa, UNSDG 15 (Forests & Biodiversity), UNSDG 12 (Resource Efficiency)

The following interview with Vicky was first published in September 2019 on Sustainability-Leaders.com.

Vicky, there seems hardly an organization or initiative devoted to sustainable tourism which you haven't been involved with over the years. Do you remember what first brought you to the topic?

Oh, there's many! And many vivid penny-drop moments, along a constant gradual evolution. I worked in travel web development and marketing, in London, from the late 90s. The Internet was still new with few jobs, only really for big brands in the mass market. Around 2001, migrating analogue TV mass market media Teletext Holidays to digital, I got a work deal to Kenya, excited to go on safari.

I didn't expect the horror of piles of plastic waste along roads and a British clientele only interested in tanning, eating and drinking as much as they could, with no desire for interaction outside the all-inclusive walls with people, places, culture or nature. Many I asked had booked last minute through my company. It wasn't why I worked in tourism, and I felt responsible.

Retrospectively, I realised the damage I had witnessed as a mountain lover in the mid-90s, working in ski and summer hiking holidays in the French Alps as a resort manager for a UK tour operator, seeing first signs of permafrost melting and dealing with drunken clients and impacts on mountain communities.

How have your views on responsible tourism changed since then, and through your many years of experience in strategy and at the coal face of tourism?

After my Kenya trip in 2001 I became aware of the huge disparity between the profit-focused industrial model of tourism and the attitude of responsible tourism I naturally had, defined in 2002 in The Cape Town Declaration by Harold Goodwin, later my professor.

I moved to oversee ecommerce marketing for Virgin Holidays, being one of the few tour operators getting involved in sustainable tourism, but I had a calling to return to Africa. I did that in 2006, volunteering in lion conservation and community development and travelling through Southern Africa. I saw where tourism could help but didn't.

In Botswana, I met a lady my exact age with multiple children, married to an abusive guy with multiple wives, all HIV positive. She couldn't finance or access support, yet here she was next door to luxury tourism. I went on to work in positive impact charity challenges and volunteer tourism, for hosts and guests, saw what worked and what didn't.

I combined on-the-ground, strategic and academic insights for my MSc thesis in 2012 on product, digital marketing and greenwashing (supervisor Xavier Font) that was published and since became a top 10 download of all time in the Journal of Sustainable Tourism.

I believe I was also the first student at Leeds Beckett ever to be threatened with litigation over the publication of a thesis! I'm proud of that, I put data to some big company exploitation that was damaging people and places. We had to jump through some legal hoops to publish (by anonymizing the companies) but it put onus on the principles and got a lot of international media attention.

There have been some big changes in the industry since, such as in many discontinuing orphanage tourism, not least because consumer awareness has grown and changed demand, so if I have helped contribute to that change, great.

Responsible tourism was a new concept when I started, and I've developed with it as we learn more how terrible tourism can be and how we can make it better. My sense of in/justice, responsibility, duty and resolve to change the sector has only strengthened.

Reflecting on your many jobs and responsibilities to date, which have you found the most rewarding? And which the most challenging?

Managing charity challenges has been incredibly rewarding but also tough, in all ways. Many guests have experienced a big loss and channeled all the energy of grief into raising money for charities at home which may have supported loved ones. They're still vulnerable, often completely out of their comfort zones having never experienced a developing country, culture or nature, and they're concerned about the physical challenge, such as hiking up Kili. But they come to realise the emotional challenge is greater. Having poured everything into this trip, they have a sudden realization of 'what now?'. Most break down at some point. But that's the break-through, to the rest of their lives, which brings them immense healing, strength and pride.

It's a privilege to witness and support, and to see them connect with locals in destinations who enable and support that process and perspective. It's truly life-changing transformative tourism all-round and amazing to work with. The role and responsibility of it is vastly undervalued.

Staff management of tour operator teams in ski resorts was probably the most challenging as a 23 year old manager! Of course staff mean well, just get into a LOT of personal scrapes while you try to keep them on track in customer-facing jobs. I'm proud of a zero staff turnover in all my seasons coupled with great customer reviews and profit centre reports.

In tourism, I've also worked with commercial organisations who didn't care, and NGOs who didn't get market requirements. I'm a do-er, with a start-up and creative mentality and drive, so bureaucratic approaches challenge me when I just want to get on with the job. I'm not interested in power egos, manipulation and Machiavellian politics.

As a female, hitting the glass ceiling is also a challenge I've experienced many times and is still prevalent in this sector, with majority male boards and female-strong lower echelons—I look forward to equality.

EarthChangers is your new initiative: What is it all about?

Earth-Changers.com is focused on positive impact tourism, promoting life-changing places with world-changing people for extraordinary experiences with purpose.

I've worked on it for around three years now, and it's been live over two. We work with specialists in the sector who deliver incredible support to local communities and conservation through tourism, often feeding locally into the

Sustainable Development Goals. It can be a beautiful and luxurious lodge in a private protected area, a really raw bush camp, ocean sailing conservation expedition, or a tour of local social enterprise, appealing to different budgets and demographics. The commonality is psychographic, the absolute commitment to and evidence of sustainability for local purpose, people and place.

Start-up is hard work, let alone in competitive tourism, especially when consumers haven't hugely embraced sustainable tourism yet. I'm doing a lot of work to raise awareness and educate on sustainable tourism as a whole, to consumers and as a model to media and trade organisations who follow Earth Changers too. I felt it was missing in the market and it's the right thing to do.

My appointment as a Defra Ambassador [UK Department of Environment, Food & Rural Affairs] was wonderful, unexpected recognition. I'm making in-roads, good for the sector as a whole, and aiming to scale-up Earth Changers with the right support.

It's not always easy to align one's ideals with one's needs—such as earning enough to be able to pay a mortgage. Which values guide your work (and life)?

I follow my passion, work hard to get to where I want to be and develop expertise. Money follows. In 2006 I knew I wanted to create a sustainable tourism start-up, but it wasn't the right time in the market, and I needed to be freer financially and better able to discern sustainability.

For a decade I trained as a ranger in Africa, did my MSc while working full-time in travel ecommerce, freelanced for NGOs and conservation... When knee injury and reconstruction incapacitated me in 2016, I felt it was the right time, and I was ready, probably like no one else in the world. Passion-driven doesn't mean no consideration or knowledge of what I was getting into. Being able and open to side-project consultancy helps.

It's not easy to follow passion, get out of comfort zone, go back to basics, not earn money whilst endlessly grafting and making enormous sacrifices, challenging your own mentality, getting back up from knock-backs. But status, material possessions, corporate ladder, power... ultimately don't matter to me. What does is difficult to distill into just a few 'values' words for me and Earth Changers' but...

Adventure: With ups, downs and unknowns, life's one big adventure trip in life-long learning to master your own responses. I'm an optimistic realist and a queen of finding silver linings.

Connection—to my purpose, places, people and spirituality. Emotional connection can only be felt, not replicated inauthentically.

Integrity—honest and strong principles. You can't fake it, it greenwashes out. I believe in an equal, respectful, fair and just world. Whether you're the Queen or the bin man, everyone has the same right to be here and take up the same space in the Earth's ecosystem. We belong to it, not it to us.

I'm intuitive so can be spontaneous—life's too short for me. But risks are carefully calculated with security of years of grafting and savvy financial management.

Perhaps typical for sustainability, I plan for the long haul and work on delayed gratification with a relatively simple life based on need not want.

On your website you describe how difficult it was to find a job in 'sustainable tourism'. Looking at the industry now, has this changed? What advice would you give to graduates of sustainable tourism programs?

I finished my MSc in 2012 and even in that short time, more so since starting research on Earth Changers just 3 years ago, I see many more opportunities. There are some awesome jobs now—if only I wasn't committed already!

The market has grown, which brings many more opportunities, but also a downside: A decade ago responsible tourism consisted of less people more united by shared dreams in reaction to mass market negative impacts. Some may have called it cliquey. Now, it's amazing to see sustainable tourism growing in scope and mainstreaming, so it's bigger and more powerful collectively, but I feel it has inevitably diluted the connection of shared vision. It's important for us to keep that.

Interest has possibly also grown disproportionately to consumer action, meaning there are more qualified people than number of actual jobs, so it's more competitive and brings a greater need to stand out.

I would advise graduates get three things:

- Experience from the bottom up, coal face managing tourists in destinations, as there's no better way to understand who wants what, why, when and how—key to product and marketing.
- Critical thinking in an objective, structured, academic approach—key to sustainability.
- Networking—and not just online! Talk to real people—relationships are key.

I would say, work hard—there are no short-cuts—and pay it forward. I have made huge sacrifices in my life by voluntarily working nights, weekends and holidays for years on end—whether organising meet-ups or tweet-ups, writing blogs, managing charity challenges, or pro bono helping non-profits with marketing. But this has all been valuable learning and experience gained. When the jobs come, you're more rounded on your CV and capable.

Qualified peers appreciate the skills and knowledge, and, with the growth of responsible tourism and its value, so will clients.

Online tools and social media profiles have long become a key part of tourism marketing. To your mind, which are the dos and don'ts in using those—which crucial factors should tourism businesses and destinations keep an eye on?

Do—have clear key values at the core of all communications. Get everyone in the organization to help define those, bottom up. As with all responsible tourism, involve all stakeholders—take them with you.

Do—have a clear strategy. Social media can be great, but if you're not careful it's just reactive and sucks time.

Do—raise awareness; engage conversation, and nurture relationships.

Don't—expect direct returns. It's not that it can't or won't sell, but it's more indirect, nurturing through time, trust and consistent messaging.

Don't—go purely for follower numbers.

Do—look at reach and engagement KPIs. It's a more holistic approach like offline PR than online ROI. It takes time, but is often the first place people find out about you or look for you.

Don't—expect someone with a lot of followers, not experienced and/or qualified in sustainable tourism, to talk with depth, compassion, honesty and diplomacy about challenging sustainable tourism issues.

Do—go for aligned values. You'll get better engagement, which algorithms work to, and ultimately more sales. Quality over quantity, as we say in our Earth Changers Manifesto.

Having been involved with Africa for many years, which are the main opportunities there right now, with the potential to help businesses or destinations become more sustainable?

Yes, I think many in Africa should be teaching the rest of the world that, not vice versa!

Clearly, it's a big continent and every country and region has its unique situation.

Threats are clear: Climate change, poverty, water shortages, food accessibility, malnutrition, health, education, human rights, sustainable communities, political injustice, corruption, conservation of species, poaching. . . . the list goes on. All the sustainable development goals and more.

Like anywhere in the world.

But those threats mean opportunities for creative solutions, and African people have amazing energy and are generally incredibly entrepreneurial, tenacious, creative, resourceful, and thus sustainable, to come up with those.

The Long Run,[2] for whom I worked, is full of incredible pioneers and initiatives in private protected areas in Africa (and elsewhere) which act as models worldwide, including to me as inspiration and motivation for Earth Changers.

I'm also a Trustee for a sustainable development charity in the south east of Madagascar. It's probably the rawest, least developed place for tourism I've been. Faced with utter battering by climate change and poverty, 5 days of relatively inaccessible overland from the capital, the population is so unsupported, yet so open. They have embraced sustainable lobster fishing and female empowerment embroidery projects which have revolutionized local livelihoods, family income, food, health and education. It's great to witness.

How does your home country, United Kingdom, fare in terms of tourism sustainability?

From an Earth Changers standpoint, we have less supply of positive impact tourism. That's not because, as people may expect, we're a wealthy country without need in fact nearly 33% of children and 22% of the public are in poverty (58% of those in 'persistent poverty').

As an island with bad reputation for weather and rail network, there's much demand to go abroad.

---

[2] https://www.thelongrun.org

But we have amazing traditional rural B&Bs and farm stays that I spent childhood holidays in, and pioneering home-grown organisations like Unseen Tours, with whom I've seen different areas of London, even where I've lived for years, through the eyes of the homeless, bringing me a whole new perspective and understanding of our political and social systems.

We have Green Tourism (for whom I have also worked!), one of the world's original and largest green awards certification programmes, established in 1997 and now with a couple of thousand certified properties and venues.

And now we are seeing wonderful new concept hotels growing, like Good Hotel who have indeed been good enough to host our WTM Responsible Tourism Networking for the last 2 years, and hopefully will again this year. It's got great aligned values and is perfectly located next door to Excel.

But with awareness of climate change growing exponentially here, there are various things afoot.

Our Government Department for Environment, Food and Rural Affairs (Defra) has a 25 Year Environment Plan, plus this year launched the #YearOfGreenAction—for which I was invited to be an ambassador and am delighted to ensure sustainable tourism is represented.

In fact, tourism in the UK falls under the Government Department for Digital, Culture, Media and Sport which has a Green GB Week the same week as WTM this year, a good opportunity to bring aims together.

Now we have a Parliamentary Inquiry just launched into the impacts of tourism—both in the UK and of UK tourists abroad. I expect this will raise awareness of sustainable tourism in the general public.

You are very active in the sustainable tourism community—on conferences, with awards and managing online communities (#RTTC). How has this scene changed in recent years—where are we headed?

A lot more people are aware and paying attention: reading, listening, watching. Though like anything online, the level of passive following dwarfs active participation. But people are learning.

As sustainability consciousness grows, so sadly but almost inevitably does greenwashing. And that undermines those who are sustainable, because it encourages scepticism and judgement of greenwashing when you're not, and this in turn promotes greenhushing in fear of sticking heads above the parapet and getting shot down.

But it also helps people to be confident asking questions to better discern. Earth Changers helps that—it's influencing people by raising awareness, teaching them what to look for, and other suppliers how to improve, as well as promoting our pioneering partners.

Awards-wise, I've recently been a judge on the ITTAs for WTM and for TTG. There are some amazing initiatives, achieving great impacts and it's very tough to judge between top contenders.

If there's one thing, I'd recommend awards entrants do, it's read and address in answers the actual questions, not just pitch what you want in hope!

New initiatives are emerging all the time. Currently, I'm also helping (re)develop the European Ecotourism Network[3] and connecting it with the Global Ecotourism Network hub. It's early days for EEN and we're keen to grow a knowledge sharing network.

Tourism professionals sometimes avoid engaging with 'sustainability', since it is not something usually part of their KPIs. Reflecting on your own experience, which advice can you share in terms of how to deal with 'selling' sustainability to supply chain, staff and clients?

I'd say people are people, talk to them in laymen's terms, engage with them on a human basis. Start with the why—explain what the purpose is. People buy into a people, passion, a common vision and mission. Passion combined with vision is powerful. Allow and enable stakeholders to contribute to the process, bottom-up. Provide the tools, structure, information and education for people to empower themselves.

Of course, sometimes as an experienced and/or qualified person you must step in to provide expertise and steer, but you don't have to dictate from the front. On my charity challenges, I always walk at the back and let the local guides be with the Alpha guests at the front. That way, I come upon everything I need to, without missing important things otherwise happening behind me.

Sustainability is not a competition, it's not about beating others, it's about achieving the goal as a team and enjoying the journey.

Link to the interview: https://sustainability-leaders.com/vicky-smith-interview/

---

[3] http://www.ecotourism-network.eu/en-welcome

The Bahamas | Vik Nair is President and Professor (Sustainable Tourism), at DISTED College, Penang, Malaysia, having previously served as Dean of Graduate Studies and Research and Professor in Sustainable Tourism at the University of The Bahamas (UB). A programme Leader for the Responsible Rural Tourism Network, he is also a consultant with many national and international projects in South-East Asia and the Caribbean. A well sought-after keynote speaker, his exceptional research achievements with more than 350 publications to his credit has earned him many international and national accolades. Vik successfully led a team to write the country report for The Bahamas for UNDP's Sustainable Development Goals in 2017. He is currently working on numerous consultancy projects in The Bahamas with Tourism Development Corporation and the Department of Environmental Planning and Protection in developing community-based tourism and ecotourism. His research specialisation is in sustainable and responsible tourism, rural tourism, ecotourism management, environmental management, community-based tourism and green tourism.

Areas of expertise: community-based tourism, destination development, responsible tourism, Caribbean, Asia, UNSDG 12 (Resource Efficiency)

The following interview with Vik was first published in June 2019 on Sustainability-Leaders.com.

Vik, having been involved in research on ecotourism and community-based tourism for many years now, do you remember what first got you interested in the topic?

It was by chance that I got involved in this field. With a first degree in science horticulture, I worked for a few years in oil palm plantations and golf course landscaping. My exposure to this career helped me understand how the unsustainable management of our environment will result in negative long-term consequences to mankind. Hence, I continued my postgraduate studies in environmental impact assessment and my doctoral research in ecotourism management.

With the knowledge that I gained, I continued to work with local communities and various stakeholders to educate them to be responsible to their surroundings—environmentally, socio-culturally, and economically. Also, I published extensively to advocate my passion and worked on various award-winning projects.

How has your view on the potential of tourism to support rural communities changed over the years?

Over the years, I realised that sustainable tourism is not just about finding the equilibrium as far as the three pillars of sustainability are concerned. The 3Ps—profit (economy), people (socio-cultural), and planet (environment)—are reliant on the fourth 'P', which is politics.

The potential of tourism to support rural communities is dependent on the politics of the country.

Politics here also means the governance, accountability, policies, and ethics. With no political will, the 3Ps will fail miserably. Thus, winning the support of whoever is the government of the day is critical for rural communities to reap the potential that tourism can bring.

Tourism is a good alternative income for many rural communities—if it is executed correctly with support from the government.

Empowering Bahamians through entrepreneurship was the topic of a recent seminar you offered. In a nutshell, what do you propose?

Although tourism is the biggest GDP contributor in The Bahamas, the benefits that trickle down to the local community are minimal. The Minister of Education of The Bahamas in 2019 indicated that almost 70% of high school students in The Bahamas do not end up in universities or colleges. They go directly into the workforce with poor skills and no entrepreneurship acumen.

Thus, in my seminar I propose a toolkit to take them step-by-step in preparing, developing, and sustaining community-based tourism initiatives in their neighbourhood, village, township, or even in some of the outer islands in The Bahamas.

The Bahamas is an archipelagic nation that has a rich culture and heritage unique to each island. This toolkit will empower Bahamians, especially the youth, to be innovative and succeed as entrepreneurs in developing an authentic community-based tourism product that is currently lacking in the country. These young and innovative entrepreneurs can transform their existing natural and man-made assets within their community into a tourism product that is marketable.

Community-based tourism is also the focus of a new partnership between the Bahamas Tourism Development Corporation and the University of The Bahamas. Briefly, how did the partnership come about, and what are its objectives?

The partnership between the Bahamas Tourism Development Corporation and the University of The Bahamas is timely, as both parties have a similar aim in bringing enlightenment in the community and showing a new and refreshed face to tourism that can benefit them directly. Through this partnership, a series of workshops will be organized for all relevant stakeholders and members of local communities to better manage their products and attract the bulk of cruise tourists that The Bahamas receives, to visit these villages and experience real Bahamian culture and heritage.

Having worked for many years in Malaysia, and now The Bahamas—how do both countries approach tourism sustainability—are there any differences?

As a fairly rich and politically stable nation with no natural disasters, Malaysia has sufficient funding and expertise (with more than 100 universities and 400 colleges) to drive the tourism industry to be more sustainable. The Malaysia homestay experience programme, a product of community-based tourism, won the UNWTO Ulysses Award for Innovation in Public Policy and Governance in 2012.

In The Bahamas, the lack of funding, poor capacity, and hazards such as hurricanes have resulted in making tourism sustainability more challenging. Compared to many other countries in the region, The Bahamas is politically stable. Despite this, the continuous politicization of every single issue in the country and the unstable employer-union-government environment has also made the country slow in its development. Tourism sustainability in The Bahamas should not just be about the economy (which is important) but also about the long-term impact on communities and the natural environment.

In an archipelagic nation like The Bahamas, it is more challenging to manage and replicate fundamental resources in each island. The country's economic and environmental vulnerabilities amplify The Bahamas' sensitivity to harmful external shocks. These vulnerabilities emanate from a limited resource base; high dependence on export earnings; a small domestic market; and high susceptibility to climate change and natural and environmental disasters.

So, the approach to tourism sustainability is certainly different.

Which part of developing sustainable CBT is the most difficult, in your experience?

On average, the gestation period for community-based tourism to succeed takes up to 5 years. This can be a challenge, as most communities engaged in CBT projects want to see immediate results. Only those who can persevere can succeed in this form of tourism. Hence, preparing the community to stay focused and not lose hope in their community-based tourism endeavour can be difficult, if tourism is the main income and is not regarded as an alternative source of income for the community.

In many successful community-based tourism models in Southeast Asia, the women and the youth are involved in the business, whereas the men are typically the primary breadwinners as they are involved in other mainstream careers like agriculture, fishing, local industries, etc. Thus, the local community is able to have a sustained income all through the low and high tourism seasons. Nonetheless, reaching this sustained state can be one of the most difficult tasks.

Which are the main challenges in Malaysia and The Bahamas as destinations, currently hindering more sustainable tourism?

Finding the right balance between overtourism and undertourism is a challenge that is not only facing Malaysia and The Bahamas as destinations. It has become a major issue in many destinations across the globe, hindering a more sustainable and responsible tourism.

Finding the right capacity for a destination without damaging its very existence can be a challenge if tourism is a major contributor to the GDP, as can be seen in both Malaysia and The Bahamas. Politicising the actual economic impact of tourism in both countries results in over-exploitation of tourism resources, hence rendering them unsustainable.

Joao Ministro of the Algarve in Portugal in his interview[1] ***highlighted how tourism can be a great tool for preserving and strengthening cultural traditions. But it can also destroy them. How can communities—especially in rural areas— seize the potential of tourism and at the same time avoid its pitfalls?***

He is spot on. In my community-based tourism toolkit that I am introducing to The Bahamas, the critical stage of developing community-based tourism, especially in rural areas, is the first stage which is preparing the community.

Assessing community needs and readiness for tourism is critical. Why should the community be involved in tourism? This is the key question that needs to be asked and answered before a community is ready to embrace tourism.

The community needs to be educated on how tourism can be a great tool for preserving and strengthening cultural traditions, but if they are ignorant of the potential pitfalls, it can destroy the authenticity of the community's culture. Only after the community agrees to proceed to work on tourism should they be educated and prepared for tourism.

Once a community decides to embrace tourism, educating, and preparing the community is crucial.

Would you consider tourism in its current state in The Bahamas sustainable? Which are the main issues?

Tourism in its current state in The Bahamas may not be sustainable. There are many challenges that need major interventions and collaboration between the government and the industry.

---

[1] https://sustainability-leaders.com/joao-ministro-interview/

The Bahamas cannot move away from the low yield cruise tourist, there must be a strategy in place to increase the receipt for these day-trippers. Despite the high arrivals, the average expenditure of cruise tourism in the country is USD$70–132. Hence, it is critical to focus on product development that will encourage these day-trippers to explore further and contribute more to the local economy.

The environment is the heart of tourism in The Bahamas. Hence, all efforts and legislation must be put in place—or tightened—to protect the environment and ensure the country is safe for tourists to explore.

Politics should not cloud the actual long-term consequences of tourism development on the pristine archipelagic small island destination state.

To your mind, what role does—or could—destination branding and marketing play, in facilitating sustainable rural and community-based tourism in The Bahamas?

Destination branding and marketing play a critical role in facilitating sustainable rural and community-based tourism in The Bahamas.

The 21st century tourist expects more from a destination. The experiential part of their visit is critical. They want to be—or be seen to be—responsible when they visit a destination (both in rural and urban settings).

Hence, social media and travel advisory sites (e.g., TripAdvisor, Expedia) play a key role in facilitating these rural tourism sites. No tourist is going to risk coming to an off-the-beaten rural destination if the word-of-mouth indicates a negative image. So, branding and marketing are key.

Which broader trends and developments do you observe, likely to positively—or negatively—impact the work of community tourism organizations and responsible tourism advocates in the Caribbean?

On 13 December 2012, the one-billionth tourist had arrived. Today the number is exceeding 1.2 billion. From a survey carried out with these tourists to understand better what the 21st-century tourist expects, the following top five actions were indicated:

- Buy local
- Respect local culture
- Protect heritage
- Save energy
- Use public transport

Destinations that can fulfill these expectations will be successful in attracting tourists. Hence, responsible tourism is the way forward for all tourism offerings in the Caribbean. Community-based tourism fits perfectly this expectation of the 21st-century tourist.

As a professor and dean of graduate studies at the University of The Bahamas, do you experience a growing interest in the topic of sustainability among students or colleagues?

Indeed, there is a growing interest in the topic of sustainability and responsibility in the university community at large. In October 2018, we officially opened the GTR

Campbell Small Island Sustainability Research Centre (SIS-RC) at the University of The Bahamas to address the needs of the country.

Small Island Sustainability is a signature programme at the University of The Bahamas with various degree options that will provide students the opportunity to interact with each other and create inter-disciplinary synergies for dynamic research. The importance of this new research centre and the role it will play in the future growth and development of the country—as far as sustainability is concerned— cannot be denied.

Which gaps in research would you like to be addressed—which urgent topics need to be investigated further?

The dire consequences of climate change that will increasingly affect tourism seasonality, as well as tourism product design, are two important lines of study that need to be conducted.

Despite being an important agenda item that has also been highlighted in the UN Sustainable Development Goals, there seems to be a lack of seriousness among the policymakers in accepting these research findings.

What experiences have you made with community-based tourism in Southeast Asia—which communities have applied it particularly successfully, in your view?

I have had many good experiences during the two decades I have worked with local communities in Southeast Asia. Every country has a different approach in managing successful community-based tourism.

Although one size does not fit all, there is a commonality in the development process and life cycle of community-based tourism projects. For example, the local community in Miso Walai, Kinabatangan in Sabah, Malaysia[2] successfully introduced a community cooperative to run its community-based tourism (CBT), which eventually expanded its economic benefits to all members of the community who participated in the programme. The cooperative allowed the community to move away from relying on government handouts or subsidies. Shareholders in the cooperative enabled higher community involvement in the decision-making processes, increased tourism income distribution, solidified the sense of ownership, strengthened social cohesion, and inevitably increased community support. This model should be benchmarked.

What lessons have you learned along the way of developing and promoting the concept?

It is critical in community-based tourism that the community understands what tourism is all about. Tourism is not a panacea for all their problems. Hence, asking the right questions to the community is crucial.

For instance, why tourism? What is their current lifestyle? What are the expectations of the community? What is their current source of livelihood and socio-economic condition, and what is the long-term prospect of their current source of livelihood?

---

[2] https://www.mescot.org/village_homestay_home.htm

The answers to these questions are fundamental for the community to comprehend before they decide on tourism as an alternative source of income, a tool to justify conservation/preservation efforts, and/or a training ground for future participants in other economic sectors.

Only then can a situational analysis be carried out to understand the community's values, attitudes, concerns, aspirations and expectations of community-based tourism.

Link to the interview: https://sustainability-leaders.com/vikneswaran-nair-interview/

Thailand | Passion for travel has been in Willem's blood since a young age. His professional career in the travel industry started at a student travel organisation in the 1980s. He moved to Thailand in 1987 and founded Khiri Travel in 1993. The DMC now operates in eight countries and has a dedicated charity, Khiri Reach. Willem founded YAANA Ventures in 2016 to create and grow a portfolio of sustainable hospitality brands in Asia. In its portfolio today, besides Khiri Travel, are Anurak Lodge in Thailand, Cardamom Tented Camp in Cambodia; GROUND Asia, specialising in service-learning projects for schools and universities; Naruna Retreats, providing personal leadership development; and HMP Master, a new cloud-based solution for boutique hotel management. Luxury tented camp development company Visama Lodges and The Orient & Occident Company are focused on quality, regenerative hospitality.

Areas of expertise: conservation, ecotourism, responsible tourism, sustainable development, tourism business, UNSDG 11 (Sustainable Human Settlements), UNSDG 15 (Forests & Biodiversity)

The following interview with Willem was first published in June 2020 on Sustainability-Leaders.com.

Willem, having dedicated many years to developing destinations and tourism businesses in Southeast Asia—what got you interested in the region? Why did you decide to dedicate your career to tourism?

My love for travel developed during my early teens when I used to go on solo bicycle trips. These trips started in my birth country, The Netherlands, followed by trips through Germany, Belgium, Luxembourg, and the UK.

When I was 17, I bought my first flight ticket to Indonesia and I fell in love with Asia when I travelled there. I soon returned to other top destinations when I landed a job at a student travel organization in The Netherlands.

Later, I got a job offer at a small DMC in Thailand in 1987, and even though I had planned for maybe a year of gaining experience, I ended up staying and set up Khiri Travel.

Your Cardamom Tented Camp in Cambodia was a Changemaker finalist at World Travel and Tourism Council's Tourism for Tomorrow Awards 2019. What does such recognition mean to you personally, and how does it benefit your organisation's efforts to conserve wildlife and forests?

Cardamom Tented Camp is the result of a unique partnership between the Minor Hotel Group's Golden Triangle Asian Elephant Foundation,[1] the Cambodia-based nature conservation organisation Wildlife Alliance[2] and our YAANA Ventures.[3]

It is a not-for-profit project, with proceeds going to the Wildlife Alliance rangers who help to protect the 18,000 hectares of rainforest that the camp is located in.

We're convinced that tourism is a force for good. To be a finalist at the WTTC's Tourism for Tomorrow Awards, after only one year of opening is a recognition that the concept has proven correct. The international attention that the event gave the project made a big positive impact.

Khiri Travel was the first DMC in Southeast Asia to be Travelife-certified. How has this certification helped your business to achieve the triple bottom line of sustainability: people, planet, and profit?

We were, together with one other, the first to get certified in some of our larger destinations a few years back, and we are now the first to get certified in all seven destinations which we operate in.

The triple bottom line implies that the impact on all three essential parts of the business is measured. Travelife certification helps in setting up a framework that keeps challenging the business to operate in a measurable and more sustainable way.

---

[1] https://www.helpingelephants.org

[2] https://www.wildlifealliance.org

[3] https://yaana-ventures.com

It also shows that the certification is not the end result—it's just a step in a continuing journey of doing things better.

Wildlife conservation is under severe stress right now due to the absence of ecotourism because of coronavirus restrictions. How is this affecting your own operations and ecotourism experiences offered through your company?

The total ban on travel, going on for months now, continues to have a devastating impact on every aspect of society, including the efforts on nature conservation. First in line are projects that get almost exclusively funded by tourism. A prime example of that is the elephant camps in Thailand. The feeding and medical care of these captive elephants is very expensive.

Our friends at the Golden Triangle Asian Elephant Foundation have already taken in many of the elephants of the camps that could no longer afford the upkeep—but they themselves are the foundation of a business largely dependent on tourism, so funds start to run dry there too.

Our own Cardamom Tented Camp has started a Fundraiser Campaign to ensure the rangers keep getting the support over the next crucial months, hoping travel will pick up after that.

Efficient management of crowds through technology is one of your suggestions for a sustainable tourism recovery after the COVID-19 crisis. With many countries, especially developing nations, waiting to recover lost ground with regards to revenue through tourism, do you think there will actually be an emphasis on countering overtourism?

Indeed, technology can be deployed to tackle overtourism, in particular at specific sites like Angkor, the Grand Palace, or National Parks. For example, to make entrance tickets only available online, with a limited number of visitors per time slot. This is already done successfully for years in museums in the US and Europe.

Better spread of tourists during the day, and perhaps even dynamic pricing, will result in better yield at the same time. As some effort is needed to access some locations, tourists are encouraged to visit other sites instead, spreading the tourism dollar.

The private sector and NGOs need to play a leading role in encouraging governments to start implementing such technology.

Visitor dispersion is seen as the panacea for overtourism; but how can destinations ensure that the locals of the 'new' destinations are prepared to receive the influx of visitors when they are already aware of the negative implications of mass tourism?

New destinations can start with the implementation of best practices for visitor management. There is an enormous number of resources available. On a local level, the decision to develop a tourist site needs, of course, to be in agreement with the local community, who have to be aware of the positive effects, such as job opportunities and income diversification, as well as ways to mitigate any possible negative effects.

Which destination in Asia do you think has done a commendable job in tackling overcrowding through tourism? What can other destinations learn from them?

Well, there is always Singapore as an example, but that nation-state may not always be relevant compared to its much larger neighbours. Some private museums, such as the Jim Thompson's House (in Bangkok) can be looked at as examples of good practice.

Bhutan is often cited as an example with their limitations on visitors, but again, this is a very specific example that is not going to work everywhere and thus difficult to recommend to most other countries. By and large, as destinations, all Asian nations need to pioneer this on their own.

How have your views on tourism sustainability changed over the years, and through your experience of working with destinations?

In the early years of Khiri Travel, we felt it was just logical to give back to the community. There were much less tourists, 20–25 years ago. The communities we visited on our tours often clearly needed some help. Travelers also had another mindset, knowingly visiting very poor countries.

As development came along, for the casual visitor, it's sometimes harder to understand why support is still needed. We have set up our charity Khiri Reach to highlight some of the causes that continue to need support.

As for operational sustainability, there have been massive developments. Public knowledge about climate change, pollution, and human rights has given a tremendous boost to the cause of sustainability. Certification programs organized under GSTC have fine-tuned the efforts.

Your three bits of advice for sustainable tourism entrepreneurs or travel business owners in Asia, how to support wildlife conservation, and at the same time ensure their own economic sustainability?

Make sure your business has a higher purpose than just financial profits, while at the same time making sure it is profitable. Without money in the bank, you cannot sustain a business.

From early on, try to build your business around sustainability criteria, outlined by GSTC-aligned certification bodies or even the UN Sustainable Development Goals (SDGs). It makes your certification process so much easier later on.

When your business is not directly involved in wildlife conservation, which is likely, you can still help by raising awareness amongst your staff and clients, and by supporting conservation efforts.

Link to the interview: https://sustainability-leaders.com/willem-niemeijer-interview/

Germany | Dr. Willy Legrand is a Professor of Hospitality Management at the IU International University of Applied Sciences in Germany. Over the past two decades, he has established over two dozen undergraduate and graduate courses on sustainable development in tourism and hospitality management in Europe, South and North America, the Middle East and Asia. He is the lead author of *Sustainability in the Hospitality Industry: Principles of Sustainable Operations* 4th edition. Legrand is also the lead author of *Social Entrepreneurship in the Hospitality Industry: Strategies for Change* which is a timely addition to tackling the many challenges facing the industry in the coming decade. He further co-chairs the HospitalityNet World Panel on Sustainability in Hospitality. The panel tackles all pressing sustainability issues and is supported by 100+ industry professionals, consultants and scientific experts. The aim is to get sustainability anchored more than ever into the tourism and hospitality conversation and decision-making.

Areas of expertise: green hotel, hospitality, sustainable tourism, tourism business, UNSDG 12 (Resource Efficiency), UNSDG 13 (Fight Climate Change)

The following interview with Willy was first published in October 2019 on Sustainability-Leaders.com.

© The Author(s), under exclusive license to Springer Nature Switzerland AG 2022   629
F. Kaefer, *Sustainability Leadership in Tourism*, Future of Business and Finance,
https://doi.org/10.1007/978-3-031-05314-6_108

Willy, having been involved in research and teaching on hospitality and sustainability for many years, do you remember what first got you interested in the topic?

I grew up on a farm. The careful handling of nature and use of resources such as water and soil and the constant monitoring of weather conditions during planting or harvesting season was part of my everyday life.

I chose to study geography and continued with an MBA with a specialization in Corporate Environmental Management before moving on with my PhD. Ultimately, my desire was to combine my interest and knowledge of the functioning of our environment with the hospitality industry.

Following years of experience working in the hospitality industry, here too I quickly noticed the imperative to carefully handle natural resources; our prime input into all our operations in form of food, energy, water, construction material etc.

How has your view on the hotel industry changed over the years, especially in terms of its sustainability?

The hotel industry has implemented myriads of measures pertaining to mitigating environmental impacts over the past two decades. Many of those are certainly good when assessed in isolation and have provided hotel owners and managers with good return on their investments.

Still, the sum of those measures is by far not enough to reach any climate goals. I am still missing the long-term commitment towards a decarbonized industry. Operational budgeting in hotels is a yearly affair and most hotels have a strategic planning horizon of no more than 5 years. However, those hotel companies (alongside with key partners such as investors, owners, architects) are involved in developing and operating hotel properties built with an operational value spreading over 20, 30, 40 or more years.

The argument here is that of those hotels opening today (and there are globally anywhere between 8 and 10 of them per day) which are not carbon neutral, none will reach carbon neutrality in the next 20, 30 or 40 years. At least not before new investments in major renovations are taking place in its core components of heating, cooling, ventilating and insulating.

The industry cannot wait to start in 2030 to shape the hotel sector of 2050. But our current goals are, at best, set to 2025. The 90% reduction in emissions by 2050 as mandated by the International Tourism Partnership[1] needs actions today.

You just launched the World Panel on Sustainability in Hospitality. What is this all about?

The idea behind the HospitalityNet World Panel on Sustainability is to gather leading experts and thought leaders from around the world in the field of hospitality management to respond to controversial ideas or questions facing the hospitality industry.

Faced with increased environmental and societal challenges, the hospitality industry has a responsibility to act and experts, including industry professionals,

---

[1] https://sustainablehospitalityalliance.org

consultants, governmental and non-governmental experts, researchers and academics, all explore ideas, solutions and strategies required by the industry.

The goal is to get sustainability anchored more than ever into the hospitality conversation and decision-making.

Over the last years, you have gained considerable expertise in climate change mitigation and adaptation strategies in the hospitality sector. Which would be your advice for hoteliers, on how to approach this?

In order to keep pace with international efforts to limit global mean temperature (below 2 °C and to avoid a 1.5 °C) as set by the Paris Agreement, our industry must consider partial decarbonisation by 2030 and a close-to-complete decarbonisation by 2050.

Since a large portion of our carbon emissions are linked to the way we design, develop, construct, refurbish, retrofit as well as heat, cool and ventilate our buildings; the larger picture involves our current supply of hotel properties (see below: existing properties) as well as the construction pipeline (see below: new properties).

The topic is relatively complicated in the hotel world due to the many parties involved in creating the 'hotel' experience. This includes investors, owners, architects, designers, developers, brands, operators that may all be different entities.

To keep it simple, here are a few practical points to consider:

For existing hotel properties (independent owner-operator):

- **Gathering data & creating a baseline:** The owner and operator of an existing hotel property should have a clear understanding of the property in terms of energy usage. This is about assessing current performance which includes conducting an energy audit with some key performance indicators such as kilowatt-hours per square meter per year ($kWh/m^2/year$), kWh per guest-night and kilograms (or tons) of Carbon Dioxide equivalent emissions per year (tCO_2e). For smaller properties, online auditing tools are available to get a start. This can then be the baseline from which to measure progress.
- **Benchmarking:** Hotel owners and operators must continue monitoring performance monthly throughout the year to understand how energy data behave and fluctuate. Benchmarking between properties or within a similar segment or geographical location can facilitate decision making on where or what to improve. Here too, online benchmarking tools can help.
- **Assessment, operationalization & evaluation:** Once a clear snapshot of the property's energy performance has been made, decisions must be made about improvement potential along with return-on-investment calculations. Since three-quarters of energy consumed in hotels is used towards heating, cooling, ventilating, lighting and providing hot water, those are the areas to tackle. From low-hanging fruits with quick payback period such as installing lighting control systems or replacing lightbulbs to heat recovery systems, isolation improvement and boiler improvement, calculations on the investment required alongside the savings per year in terms of (1) Megawatt-hour, (2) tCO_2e and (3) financial should be undertaken and priorities set for the next 5 to 10 years.

- **Consider Nearly Zero Energy Hotel (neZEH):** The EU's Energy Performance of Buildings Directive requires all new buildings to be nearly zero-energy (NZEB) by the end of 2020 and existing buildings to transition towards NZEB by 2050. There are many initiatives helping current hotels towards a transition to neZEH. This is the inevitable future, which needs planning today. In setting priorities and drafting a plan, the hotel owners and managers should consider both investments in energy efficiency and renewable energy towards reaching a positive energy balance.

For new hotel properties (hotel brands):

- Since much of the construction pipeline (globally more than 3000 hotels opened their doors last year) involves the large hotel brands, when looking at new property development one critical aspect is the commitment of all partners in creating a carbon-neutral hotel industry. The operators of hotels are not necessarily well placed in influencing any decisions being made during the hotel development stages. As such, the key issue here is to foster better coordination between investors, developers, brands and operators so that sustainability has a chance to be enacted from the very first planning stages such as the initial feasibility. The Hotel for Owners of Tomorrow (HOT) is an example of how such coordination can take place.
- Once the parties are at the table, then a carbon-neutral hotel or positive energy balance properties can be planned—they already exist and have made the case for a sound return on investment.

Which hotel brands would you consider leaders in sustainability right now?
Considering that sustainability covers the environmental, social and economic pillars, it is difficult to settle for one example. Rather there are many players in the hotel industry, which are pushing the boundaries of what and how a sustainable hotel can be.

Also, one aspect that I keep stating: we are part of a global industry, and thus, sustainable hospitality does not translate into 'one company trying to do its very best in a given market'. Rather, it is an entire industry that stands up to face the environmental and societal challenges by exploring ideas, solutions and strategies of how to develop future hotels and how to manage operations in a sustainable way.

With this in mind, here is a very short list of interesting and notable companies for specific areas of operations:

- The Explorer Hotel Group in southern Germany and Austria shows the world that it is possible to go passive housing with hotel operations, achieve carbon neutrality and provide a decent return to allow for expansion from one property opened in 2010 to a planned 40 hotels by 2027.
- The Boutiquehotel Stadthalle in Vienna, Austria is also an interesting case of a hotel achieving carbon neutrality in an urban setting. Similar to Explorer Hotel

Group, it uses passive housing and has excellent examples of upcycling within its interior design.

- Scandic Hotels is showing the other hotel chains what can be done in terms of accessibility of its properties. Working together with guest with physical disabilities and people with special needs, the chain devised a 159-point standard following the journey of a guest in a hotel.
- Inkaterra in Peru is a pioneer in combining hospitality services, nature conservation, education and local community involvement. The seven properties all contribute to sponsoring biodiversity inventories in natural areas where they are located.
- Soneva is an interesting brand to watch as it seeks to bridge sustainability and luxury. The company has invested in carbon mitigation projects for more than a decade now, funding windmill installation, tree planting and supplying energy efficient cookstoves. The company has ditched the branded plastic water bottle for reusable glass bottle using their own filtered water also a decade ago, long before the public outcry around plastic waste.
- Otherwise, innovation in hotel concepts is always exciting. For example, the modular accommodation Flying Nest by Accor, which uses marine containers as mobile rooms to be assembled where needed—an original way to reduce the construction costs and related emissions of fixed properties.
- Treehouse hotels are also interesting to watch. The Resort Baumgeflüster in northern Germany caters to all of us who have a so-called 'nature deficiency syndrome' offering guest a great reconnection to nature.

Which part of making a hotel business (more) sustainable is the most difficult, in your experience?

The greatest hindrance in enacting sustainability is the inherent complexities surrounding ownership, brands and operators in the hotel industry. While ownership and management are often under one roof in the private hotel industry, this is not the case for most hotel chains.

The parties involved in the hotel investment and development are not the same operating the property, which in turn may be running under a brand. Achieving substantial changes, including a clear plan towards carbon neutrality equates to significant investments required in existing or new hotel infrastructure. This is often the responsibility of the investors and/or owner. This may translate in directing capital away from shareholders' short-term gains and growth plans and is thus not particularly popular.

In terms of risk assessment, investing in carbon neutrality is the best option but there is a lot of reluctance due to short-term economics.

The combination of the hotel industry's structural business model and the emphasis on fast economic returns work as the greatest barrier to rapid and necessary changes in light of the climate emergency.

Which are the main trends likely to impact hospitality professionals in the years ahead, linked to sustainability?

I think we are seeing three push factors fostering change in our industry and that will continue to create waves in the years to come.

- Consumer-led campaigns have already shown what they can achieve last year by pushing many hospitality companies to consider alternatives or simply ban single-use plastics. It is more likely that we will continue seeing those campaigns across the travel industry calling for a change on any issue from food waste to travel miles.
- Governments, via taxation or legislation, will continue to increase the pressure. The many carbon pricing initiatives in place or planned in more than 45 countries or the ban on single-use plastics in the EU coming into effect as of 2021 are two recent examples.

  We will see this trend continuing and with the EU's Energy Performance of Buildings Directive which requires all new buildings to be nearly zero-energy (NZEB) by the end of 2020 and existing buildings to transition towards NZEB by 2050, investors, developers, owners, architects, brands and hotel operators may finally sit around one table when planning the future of this sector.
- The hospitality industry's self-regulation and voluntary codes of conduct will continue to be popular approaches in dealing with sustainability challenges, but with increased scrutiny from civil society.

Benjamin Lephilibert in his interview[2] emphasized the need to be more proactive regarding the reduction of food waste in hotels. Is this a topic which hoteliers care about, in your experience?

I hope they do! Many hoteliers (and guests) have realized the foolishness surrounding wasting food. Some have yet to understand the implications. Food waste has implications at many levels. From a business point of view, it has direct financial impacts, as waste must be removed from hotels.

Additionally, food wasted has originally been ordered, stored, transferred, prepared, cooked and possibly served. Anyone in the hotel industry would tell you that the sum of all those steps accounts for a large amount, most notably in terms of labour-hours and energy costs.

Of course, the bigger picture includes agricultural inputs, food transformation, cool storage and transportation chains all with extremely large footprints.

Food waste may be due to inaccurate forecasting, bad storage, wrong preparation techniques or inadequate portion size leading to guest plate waste. There is room for improvement in the hotel industry on the topic of food waste.

To your mind, what role does—or could—destination branding and marketing play, in facilitating more sustainable tourism in Europe and elsewhere?

Destination branding plays a role in the extent to which it can use some of its marketing funds to invest in its destination, rather than external marketing events, focusing on bringing more visitors.

---

[2] https://sustainability-leaders.com/benjamin-lephilibert-interview/

In a very connected, instagramable world, pristine destinations will be marketed, whether they want to or not. Additional funds for marketing are then best used in sustainability initiatives at the destination. Cleaning beaches, improving cycling paths, facilitating food networks within the destination, facilitating sustainable trainings for businesses are just a few initiatives which have direct impacts in making the destination more sustainable. This is actually nothing new, it's already happening.

Over the last 15 years, you have dedicated much energy and time to transform the hospitality management curriculum. Why do we need a change, and what progress have you seen so far?

The world in which we live demands hoteliers to not only possess knowledge in the traditional fields of operations, finance, marketing, consumer relationship, technology and communication but also increasingly in stakeholder relationship, environmental management as well as ethical and social responsibility. The hospitality management curriculum has a duty to be at the forefront of all those topics to provide future graduates with all the tools necessary to adequately deal with the changes taking place in the internal and external environment.

Over the past 15 years, my contribution has been toward the new generation of hoteliers. I have installed sustainability programmes in universities in the US, South America, the Middle East and here in Europe.

From the 4 to 5 students who would choose sustainability in hospitality elective course 10 years ago; I now fill classes. The level of knowledge is high—students can tell you that an average EU-based hotel consumes anywhere between 200 and 400 kWh/m^2/year and that greywater recycling is not rocket science.

My textbook *Sustainability in the Hospitality Industry: Principles of Sustainable Operations* is now in its third edition and a useful tool used in class and a great read for the industry.

The coming generation of hoteliers is acutely aware of the environment in which they lead the industry—and this makes me hopeful.

Anything else you'd like to mention?

The millennial youth come with an understanding that the world cannot continue being used the way we use it—and that matters to the future hoteliers as well (or any business for that matter)—so whether it is about future recruiting of staff or engaging with the future customers, choosing sustainability is to stay relevant to both groups.

Link to the interview: https://sustainability-leaders.com/willy-legrand-interview/

UK | Dr. Xavier Font is a Professor of Sustainability Marketing at the University of Surrey and Professor II at the University of the Arctic in Norway. He researches and develops methods of sustainable tourism production and consumption. He has published widely about sustainable tourism certification and consulted on sustainable product development, marketing and communication for several United Nations agencies, the International Finance Corporation, the European Commission, Visit England, Fáilte Ireland, Visit Wales, VisitScotland and WWF among others. He has supervised 17 PhDs to completion. Xavier co-edits the Journal of Sustainable Tourism and has conducted over 150 courses for more than 3000 businesses on how to market and communicate sustainability. He is currently the Principal Investigator for the University of Surrey for the €23M Interreg project Experience to develop low season sustainable tourism visitor experiences in the UK and France.

Areas of expertise: sustainability communication, tourism research, sustainable tourism, UNSDG 11 (Sustainable Human Settlements).

The following interview with Xavier was first published in February 2015 on Sustainability-Leaders.com.

Xavier, what was your view of sustainability and tourism when you first started your professional career?

© The Author(s), under exclusive license to Springer Nature Switzerland AG 2022　　637
F. Kaefer, *Sustainability Leadership in Tourism*, Future of Business and Finance,
https://doi.org/10.1007/978-3-031-05314-6_109

Brought up in the Costa Brava, north of Barcelona, I remember thinking 'there must be another way' better than mass tourism. I thought of sustainability as an alternative form of tourism, very different from what I could see on a daily basis impacting negatively on our landscape.

Now at the beginning of 2015, what has changed?

I have come to realise that every type of tourism product must be more sustainable, and that all companies must take responsibility for it. But I no longer think sustainability is a niche, a highly different product, but I see it as a journey. Focusing on the niche aspect allows the big players to get away with murder.

Your main insights from working as sustainable tourism researcher and advisor?

I believe most of us are driven by routines and easy choices—we don't intend to be unsustainable as consumers or suppliers of tourism services, just as we don't intend to get fat from eating the wrong things. Purposefully asking individuals to change their behaviour will only affect a small percentage of the population.

An alternative approach is needed to help most people consume more sustainable products, based on making their daily choices easy. This also means that suppliers of tourism services need to learn to redesign their products thinking of sustainability as part of quality management.

Do you share the view that sustainability has become mainstream?

Not enough, in my view. We are going in that direction, slowly. We have an increasing share of sustainable consumption—buying products that happen to be more sustainable, because the sustainable supply chain management of international firms are going that way. This has to be good, but quite often customers are not aware that their tea, chocolate, timber or clothing are made with sustainable products. So that's not enough.

What is missing is the next step, sustainable consumerism—the old idea that customers will choose products because they are sustainable. I don't see that segment growing that fast.

The compromise is getting customers to be aware that a product is better for them, that it is of higher quality, because it is in some way sustainable. The purists would say this is selling your soul, though. I think it is living in a market economy.

Where do you see the main challenges for sustainability?

We need to move away from subsidies and demonstration projects that are often not financially viable, and incorporate sustainability in the processes of production and consumption—UNEP has been working on that line for a few years now, for example.

But then, the environmental engineers need to work with the marketing and consumer behaviour teams to design products that customers want to buy, not just what is technically feasible. And the marketing people need to learn how not to greenwash. I despair every time I read an environmental policy because most are rubbish, frankly.

Where are the main opportunities?

In the marketing of sustainability as a normal product, implicitly sold as more quality.

In which regions of the world do you see most interest and momentum for sustainability and sustainable tourism?

I see pockets in different places in the world. I know Europe better, of course, where I would like to think that both large and small companies are more aware of their responsibilities, and some of them are doing a great job.

Which has been your favourite tourism, travel or sustainability book in 2014? And which books are on your reading list for 2015?

I enjoyed *Sustainability Marketing: A Global Perspective* by Frank-Martin Belz and Ken Peattie. It is no longer new, but it is still the best for this topic.

I am now reading Daniel Dennett's *Intuition Pumps And Other Tools for Thinking* because I need to read widely to find my inspiration.

Why did you decide to set up respondeco[1] and work as a consultant alongside your lecturing career?

Because I saw a gap in the market when analysing the poor communication practices, we saw. In close to 100 interviews of sustainability awarded companies, we found hardly any made conscious use of their sustainability effort to market their products, and when they did so, it was too blunt or too obvious. There was no understanding of persuasive communication applied to sustainability.

Then VisitEngland contracted me to write the manual *Keep It Real*, and this has now grown into a business.

We have now run over 60 courses in several countries, and analysed hundreds of websites, and we find companies really need to learn how to communicate the fun and quality behind sustainability.

In a recent study, we found the more sustainable companies only communicate 30% of what they do—we call this greenhushing, that is purposefully keeping quiet about what you do because you think customers will not care.

In your view, where is sustainable tourism research and academia headed?

Good question. My concern is that research effort is not optimised, because it is driven by personal agendas rather than societal needs. So, all too often, I fear that sustainability research is really going nowhere. At least nowhere that has an impact on society. Compare it with medical research, would you really start a project there that didn't have a very specific, needs-based focus?

We need fewer publications but with more substantial evidence and societal impact.

Which academic journals do you think are the most important for emerging sustainable tourism scholars and students?

The Journal of Sustainable Tourism[2] of course, but then a wide range of tourism and hospitality journals cover this field: Tourism Management, International Journal of Hospitality Management, Annals of Tourism Research, Tourism Geographies, for example.

Sustainable vs. responsible tourism—where to draw the line?

---

[1] https://www.respondeco.org

[2] https://www.tandfonline.com/toc/rsus20/current

Simple: responsibility is the process and attitude; sustainability is the goal. Nobody is sustainable, but I don't want to do business with someone who is irresponsible.

Which are the most important aspects when communicating sustainability in tourism?

Put the customer at the centre of the sustainability experience, and tell them how they will have more fun. But do it honestly and without exaggeration.

What should tourism businesses and destinations never do in terms of communication?

Be boring, speak to the environmental police. Your customers are not members of Amnesty International and Greenpeace that will go out of their way to buy green, they are normal people who have busy lives and want to get good value for money for their holidays.

What communication approach does respondeco suggest?

Incorporate sustainability into your usual marketing practices, by making sustainability look normal. Remember that communicating you are sustainable isn't the objective, but to use it as a tool to make customers act in a certain way, be it book with you, change their behaviour, or simply feel more satisfied with the choices they were already going to make.

Use every channel possible, and remember you communicate all the time, not just to first time customers before booking.

Any sustainability communication success stories you'd like to share?

There's this business I visited in Wales that had said on their bedroom browser that due to past customer complaints of being cold, now they had insulated their roof. I was worried about already. I asked at reception what they had insulated with, and they said sheep's wool.

I asked if the sheep were from Wales, and they pointed out of the window and said, "our own flock, out there". So, I rewrote the message for them "Our 300 sheep you see out of the window gave away their winter woolly coats to insulate our roof and keep you warm at night. Sleep tight". They tell me customers now ask if they can buy wool for their own lofts.

Link to the interview: https://sustainability-leaders.com/interview-xavier-font/